Balancing Greenhouse Gas Budgets

Balancing Greenhouse Gas Budgets
Accounting for Natural and Anthropogenic Flows of CO_2 and other Trace Gases

Edited by

Benjamin Poulter

Josep G. Canadell

Daniel J. Hayes

Rona L. Thompson

ELSEVIER

Elsevier
Radarweg 29, PO Box 211, 1000 AE Amsterdam, Netherlands
The Boulevard, Langford Lane, Kidlington, Oxford OX5 1GB, United Kingdom
50 Hampshire Street, 5th Floor, Cambridge, MA 02139, United States

Copyright © 2022 United States Government as represented by the Administrator of the National Aeronautics and Space Administration. Published by Elsevier Inc. All Other Rights Reserved.

No part of this publication may be reproduced or transmitted in any form or by any means, electronic or mechanical, including photocopying, recording, or any information storage and retrieval system, without permission in writing from the publisher. Details on how to seek permission, further information about the Publisher's permissions policies and our arrangements with organizations such as the Copyright Clearance Center and the Copyright Licensing Agency, can be found at our website: www.elsevier.com/permissions.

This book and the individual contributions contained in it are protected under copyright by the Publisher (other than as may be noted herein).

Notices
Knowledge and best practice in this field are constantly changing. As new research and experience broaden our understanding, changes in research methods, professional practices, or medical treatment may become necessary.

Practitioners and researchers must always rely on their own experience and knowledge in evaluating and using any information, methods, compounds, or experiments described herein. In using such information or methods they should be mindful of their own safety and the safety of others, including parties for whom they have a professional responsibility.

To the fullest extent of the law, neither the Publisher nor the authors, contributors, or editors, assume any liability for any injury and/or damage to persons or property as a matter of products liability, negligence or otherwise, or from any use or operation of any methods, products, instructions, or ideas contained in the material herein.

ISBN: 978-0-12-814952-2

For information on all Elsevier publications
visit our website at https://www.elsevier.com/books-and-journals

Publisher: Candice Janco
Acquisitions Editor: Peter J. Llewellyn
Editorial Project Manager: Timothy Bennett
Production Project Manager: Surya Narayanan Jayachandran
Cover Designer: Matthew Limbert

Typeset by STRAIVE, India

Dedication

The ideas and concepts presented in this book come from extensive years of research and collaboration carried out by colleagues around the world. We dedicate this book to two of these close colleagues, Dr. Vanessa Haverd and Dr. Bob Scholes. Vanessa was an exceptionally talented and engaging scientist at CSIRO, Australia, and a world leader in investigating the role of the biosphere in the carbon cycle and developing land-surface models. Bob was a dynamic and inspirational mentor to many scientists in Africa, leading South Africa's first greenhouse gas inventory in 1995, and many global assessments on climate change and biodiversity. Vanessa and Bob both played key roles in the Global Carbon Project and their influence is found across many of the chapters of this book.

Contents

Contributors	xv
Foreword	xxi
Preface	xxiii
Acknowledgments	xxv

Section A
Background

1. **Balancing greenhouse gas sources and sinks: Inventories, budgets, and climate policy**
Josep G. Canadell, Benjamin Poulter, Ana Bastos, Philippe Ciais, Daniel J. Hayes, Rona L. Thompson, and Yohanna Villalobos

1	The human perturbation of the carbon cycle and other biogeochemical cycles	3
2	Inventories of anthropogenic GHG: The foundation of the Kyoto protocol and the Paris agreement	4
3	GHG budgets: Constraining GHG sources and sinks	9
4	Supporting the global stocktake and the net-zero emissions policy goals	12
5	A new generation of technologies and observations to constrain global and regional GHG budgets	14
6	Extending the carbon budget and accounting frameworks to meet broader policy information needs	18
	Acknowledgment	22
	References	22

vii

viii Contents

Section B
Methods

2. CO$_2$ emissions from energy systems and industrial processes: Inventories from data- and proxy-driven approaches

Dustin Roten, Gregg Marland, Rostyslav Bun, Monica Crippa, Dennis Gilfillan, Matthew W. Jones, Greet Janssens-Maenhout, Eric Marland, and Robbie Andrew

1 Introduction		31
2 Overview of inventory approaches		34
2.1	Emission estimation from energy statistics	34
2.2	Activity data	37
2.3	Emission factors	39
2.4	Spatial and temporal emission disaggregation	39
2.5	Other approaches	42
3 Uncertainty		43
3.1	Emission estimation	43
3.2	Spatial and temporal modeling	46
3.3	Other sources of uncertainty	47
4 Examples of emission estimates and products		47
4.1	IEA	47
4.2	bp	48
4.3	CDIAC-FF	48
4.4	EDGAR	49
4.5	Global carbon project	50
4.6	GCP gridded fossil emissions dataset	50
5 Summary		52
References		52
Further reading		56

3. Bottom-up approaches for estimating terrestrial GHG budgets: Bookkeeping, process-based modeling, and data-driven methods

Benjamin Poulter, Ana Bastos, Josep G. Canadell, Philippe Ciais, Deborah Huntzinger, Richard A. Houghton, Werner Kurz, A.M. Roxana Petrescu, Julia Pongratz, Stephen Sitch, and Sebastiaan Luyssaert

1 Introduction to bottom-up (BU) approaches		59
1.1	Definitions and discrepancies	63
2 Bottom-up methodologies		64
2.1	Stock-change versus flux-based accounting	64
2.2	Bookkeeping methodology	66
2.3	Process-based methodology	70
2.4	Data-driven methodologies	74
3 Relevance to Stock-Change and flux-based accounting		76
3.1	Uncertainties	77

Contents **ix**

	3.2 Comparisons between approaches	78
4	Conclusions	78
	References	78

4. Top-down approaches

Rona L. Thompson, Frédéric Chevallier,
Shamil Maksyutov, Prabir K. Patra, and Kevin Bowman

1	**Introduction**	87
2	**Measurements of greenhouse gases in the atmosphere**	89
	2.1 Ground-based measurements	89
	2.2 Satellite measurements	91
3	**Atmospheric modeling**	95
	3.1 Modeling atmospheric transport and chemistry	95
	3.2 Types of atmospheric transport models	97
	3.3 Relating surface fluxes to atmospheric mixing ratios	98
4	**Inversion concepts**	100
	4.1 Bayes' theorem and its application to optimizing fluxes	100
	4.2 Introduction to different optimization methods	103
	4.3 Ensembles for estimating the posterior uncertainty	110
	4.4 Estimating prior flux and observation uncertainties	111
	4.5 Boundary conditions	115
5	**Application to land biosphere CO_2 fluxes (NEE)**	117
6	**Application to fossil fuel emissions of CO_2**	120
7	**Application to CH_4 fluxes**	126
8	**Application to other GHG fluxes**	130
9	**Sources of error**	132
	9.1 Transport errors	132
	9.2 Aggregation errors	135
10	**Validation of flux estimates from inversions**	137
11	**Summary and conclusions**	139
	Acknowledgments	140
	References	140

Section C
Case Studies

5. Current knowledge and uncertainties associated with the Arctic greenhouse gas budget

Eugénie S. Euskirchen, Lori M. Bruhwiler, Róisín
Commane, Frans-Jan W. Parmentier, Christina Schädel,
Edward A.G. Schuur, and Jennifer Watts

1	**Introduction and background: Arctic ecosystems**	159
2	**Methodologies**	162
	2.1 Components of the greenhouse gas budget of terrestrial arctic ecosystems	162

x Contents

	2.2 Methodologies for flux estimation in the Arctic	163
	2.3 Top-down and bottom-up methods for estimating carbon fluxes in the Arctic	172
	2.4 Terrestrial ecosystem and land surface models in the Arctic	173
	2.5 Review of Arctic GHG estimates by sector and associated key uncertainties	176
3	**Uncertainty and reducing uncertainty**	183
4	**Perspective and future opportunities**	184
	4.1 The current status of the GHG budget of the arctic terrestrial and marine environments	184
	4.2 Future perspectives: Improving the Arctic GHG budget	185
	Acknowledgments	186
	References	186

6. Boreal forests

Daniel J. Hayes, David E. Butman, Grant M. Domke, Joshua B. Fisher, Christopher S.R. Neigh, and Lisa R. Welp

1	**Carbon in boreal forests**	203
	1.1 The major components of the boreal forest carbon budget	205
2	**Estimating carbon stocks and fluxes in boreal forests**	206
	2.1 Sampling boreal forest carbon stocks	208
	2.2 Sampling boreal ecosystem carbon fluxes	210
	2.3 Carbon emissions from wildfire	210
	2.4 Carbon in the aquatic system	211
3	**Carbon accounting in boreal forests**	212
	3.1 National forest inventories	213
	3.2 Carbon in harvested wood products	214
	3.3 Managed vs unmanaged forest lands	215
	3.4 The role of remote sensing in boreal forest inventories	216
4	**Regional-scale modeling**	217
5	**Synthesis**	218
	Acknowledgments	221
	References	221

7. State of science in carbon budget assessments for temperate forests and grasslands

Masayuki Kondo, Richard Birdsey, Thomas A.M. Pugh, Ronny Lauerwald, Peter A. Raymond, Shuli Niu, and Kim Naudts

1	**Introduction and background**	237
2	**Methodologies for flux estimations in temperate regions**	239
	2.1 Net carbon flux estimations	239
	2.2 Components of the carbon budget in temperate regions	243

Contents **xi**

3 Review of the carbon budget of temperate forests and
grasslands 252
  3.1 Adjustments for the carbon budget 253
  3.2 Carbon budget assessment 255
4 Uncertainties in carbon fluxes 257
  4.1 Reliability and uncertainty in observational methods 258
  4.2 Uncertainty in components of the temperate carbon
budget 261
5 Perspective and future opportunities for policy
decision-making 262
  5.1 Progress over past decades 262
  5.2 Future perspective 263
  5.3 Toward policy-driven carbon budgets 263
  References 264

8. Tropical ecosystem greenhouse gas accounting

*Jean Pierre Ometto, Felipe S. Pacheco, Mariana Almeida,
Luana Basso, Francisco Gilney Bezerra, Manoel Cardoso,
Marcela Miranda, Eráclito Souza Neto, Celso von
Randow, Luiz Felipe Rezende, Kelly Ribeiro, and Gisleine
Cunha-Zeri*

1 Introduction and background: Tropical ecosystems 271
  1.1 General description 271
  1.2 Understanding changes in carbon cycling and storage 272
2 GHG budget in the tropics 273
  2.1 Components of the greenhouse gas budget tropical
ecosystems 273
  2.2 Methodologies for flux estimation in the tropics 275
  2.3 Top-down and bottom-up methods for estimating
carbon fluxes in the tropics (modeling) 283
  2.4 Terrestrial ecosystem and land surface processes
in the tropics 285
  2.5 GHG emissions from tropical forest deforestation
and degradation 288
  2.6 Review of tropical GHG estimates by sector 289
3 Uncertainty and reducing uncertainty 293
4 Perspective and future opportunities 296
  Acknowledgments 298
  References 298

9. Semiarid ecosystems

*Ana Bastos, Victoria Naipal, Anders Ahlström,
Natasha MacBean, William Kolby Smith,
and Benjamin Poulter*

1 Introduction and background: Global drylands and
semiarid ecosystems 311

xii Contents

1.1	Ecology	312
1.2	Threats	313
2	**Methodologies**	**314**
2.1	Components of the greenhouse gas budget of semiarid ecosystems	314
2.2	In situ based methodologies for flux estimation in semiarid ecosystems	316
2.3	Atmospheric inversion monitoring of semiarid ecosystems	318
2.4	Remote sensing	319
2.5	Land surface modeling of semiarid ecosystems	322
2.6	Soil erosion	323
3	**Future perspectives**	**326**
	Acknowledgment	**327**
	References	**327**

10. Urban environments and trans-boundary linkages

Kangkang Tong and Anu Ramaswami

1	**From science to policy for urban carbon accounting**	337
2	**Four carbon accounting approaches for individual cities**	339
2.1	Purely territorial carbon accounting approaches	340
2.2	Community-wide trans-boundary infrastructure supply-chain carbon footprinting approaches	348
2.3	Consumption-based carbon footprinting approaches	358
2.4	Total community-wide carbon footprinting	362
3	**Accounting biogenic carbon from land use and land-use change in individual cities**	363
4	**From individual cities to initiatives for all urban areas' carbon accounting**	366
	References	370

11. Agricultural systems

Stephen M. Ogle, Pete Smith, Francesco N. Tubiello,
Shawn Archibeque, Miguel Taboada,
Donovan Campbell, and Cynthia Nevison

1	**Introduction**	375
2	**Carbon stocks, flows, and emissions in agricultural systems**	379
2.1	Cropping systems	379
2.2	Livestock systems	381
2.3	Lateral transport and supply chains	383
3	**Methodologies**	383
3.1	Inventories of agricultural soil C stock changes, N_2O and CH_4 emissions	384
3.2	Inventories of emissions from livestock systems	387

Contents **xiii**

3.3	Lateral transport and supply chains	387
3.4	Top-down inversion methods	389
4	Improving regional GHG inventories for agriculture	390
5	Conclusions	392
	Acknowledgments	393
	References	393

12. Greenhouse gas balances in coastal ecosystems: Current challenges in "blue carbon" estimation and significance to national greenhouse gas inventories

*Lisamarie Windham-Myers, James R. Holmquist,
Kevin D. Kroeger, and Tiffany G. Troxler*

1	Background	403
2	What limits traditional AFOLU estimation approaches in coastal ecosystems?	404
3	IPCC guidelines for national-scale estimation of coastal wetland carbon	407
4	Improving application of the IPCC NGGI guidelines in the United States	412
5	Implications for the scale of GHG estimation	419
6	Implications for carbon cycle science on coastlines	421
7	Final thoughts	421
	Acknowledgments	422
	References	422

13. Ocean systems

Peter Landschützer, Lydia Keppler, and Tatiana Ilyina

1	Summary	427
2	The ocean as a sink/source of GHGs to the atmosphere	428
3	Preindustrial (or natural) carbon budget based on inverse estimates	431
4	Anthropogenic perturbations and the contemporary global carbon sink	434
5	Regional marine carbon sink	438
6	Storage of anthropogenic carbon	441
7	Variability of the ocean GHG uptake	442
8	Future outlook	445
	References	447

xiv Contents

Section D
Forward Looking

14. Applications of top-down methods to anthropogenic GHG emission estimation

Shamil Maksyutov, Dominik Brunner, Alexander J. Turner, Daniel Zavala-Araiza, Rajesh Janardanan, Rostyslav Bun, Tomohiro Oda, and Prabir K. Patra

1	Introduction	456
2	Using inverse estimates of non-CO_2 GHG emissions in national reporting	456
3	Methane emissions detection at facility and basin scale	458
4	Large point source emission monitoring using satellite observations	463
5	Precision and sampling requirements for future satellite observations	465
6	Developing global high-resolution transport modeling capability for analysis of the satellite and ground-based observations of anthropogenic greenhouse gas emission	467
7	Developing high-resolution emission inventories for inverse modeling	471
8	Summary	474
	References	475

15. Earth system perspective

Lesley Ott and Abhishek Chatterjee

1	Introduction and background: What is an earth system model?		483
2	Carbon cycle modeling in the context of earth system models		485
3	Data assimilation in earth system models		488
	3.1	Data assimilation related to numerical weather prediction	489
	3.2	Data assimilation related to land and ocean carbon cycle	489
	3.3	Data assimilation related to atmospheric carbon observations	490
4	Future direction for carbon cycle science, earth system modeling, and DA applications		491
	References		492

Index 497

Contributors

Numbers in parentheses indicate the pages on which the authors' contributions begin.

Anders Ahlström (311), Department of Physical Geography and Ecosystem Science, Lund University, Lund, Sweden

Mariana Almeida (271), Earth System Science Centre/National Institute for Space Research, Sao Jose dos Campos, São Paulo, Brazil

Robbie Andrew (31), CICERO Center for International Climate Research, Oslo, Norway

Shawn Archibeque (375), Department of Animal Sciences, Colorado State University, Fort Collins, CO, United States

Luana Basso (271), Earth System Science Centre/National Institute for Space Research, Sao Jose dos Campos, São Paulo, Brazil

Ana Bastos (3, 59, 311), Department of Biogeochemical Integration, Max Planck Institute for Biogeochemistry, Jena, Germany

Francisco Gilney Bezerra (271), Earth System Science Centre/National Institute for Space Research, Sao Jose dos Campos, São Paulo, Brazil

Richard Birdsey (237), Woodwell Climate Research Center, Falmouth, MA, United States

Kevin Bowman (87), Jet Propulsion Laboratory, California Institute of Technology, CA, United States

Lori M. Bruhwiler (159), National Oceanic and Atmospheric Administration, Boulder, CO, United States

Dominik Brunner (455), Empa, Swiss Federal Laboratories for Materials Science and Technology, Dübendorf, Switzerland

Rostyslav Bun (31, 455), Department of Applied Mathematics, Lviv Polytechnic National University, Lviv, Ukraine; Department of Transport, WSB University, Dąbrowa Górnicza, Poland

David E. Butman (203), Department of Civil and Environmental Engineering, University of Washington, Seattle, WA, United States

Donovan Campbell (375), The University of the West Indies, Jamaica, West Indies

Josep G. Canadell (3, 59), Global Carbon Project, Climate Science Centre, CSIRO Oceans and Atmosphere, Canberra, ACT, Australia

Manoel Cardoso (271), Earth System Science Centre/National Institute for Space Research, Sao Jose dos Campos, São Paulo, Brazil

xv

xvi Contributors

Abhishek Chatterjee (483), NASA Goddard Space Flight Center, Greenbelt; Universities Space Research Association, Columbia, MD, United States

Frédéric Chevallier (87), Laboratoire des Sciences du Climat et de l'Environnement, Gif sur Yvette, France

Philippe Ciais (3, 59), Laboratoire des Sciences du Climat et de l'Environnement LSCE CEA CNRS UVSQ, Gif sur Yvette Cedex, France

Róisín Commane (159), Department of Earth & Environmental Sciences, Lamont-Doherty Earth Observatory, Columbia University, Palisades, NY, United States

Monica Crippa (31), European Commission, Joint Research Centre (JRC), Ispra, Italy

Gisleine Cunha-Zeri (271), Earth System Science Centre/National Institute for Space Research, Sao Jose dos Campos, São Paulo, Brazil

Grant M. Domke (203), US Department of Agriculture, Forest Service Northern Research Station, St. Paul, MN, United States

Eugénie S. Euskirchen (159), Institute of Arctic Biology, University of Alaska Fairbanks, Fairbanks, AK, United States

Joshua B. Fisher (203), Joint Institute for Regional Earth System Science and Engineering, University of California, Los Angeles, CA, United States

Dennis Gilfillan (31), North Carolina School of Science and Mathematics, Durham, NC, United States

Daniel J. Hayes (3, 203), School of Forest Resources, University of Maine, Orono, ME, United States

James R. Holmquist (403), Smithsonian Environmental Research Center, Edgewater, MD, United States

Richard A. Houghton (59), Woodwell Climate Research Center, Falmouth, MA, United States

Deborah Huntzinger (59), School of Earth and Sustainability, Northern Arizona University, Flagstaff, AZ, United States

Tatiana Ilyina (427), Max Planck Institute for Meteorology, Hamburg, Germany

Rajesh Janardanan (455), National Institute for Environmental Studies, Tsukuba, Japan

Greet Janssens-Maenhout (31), European Commission, Joint Research Centre (JRC), Ispra, Italy

Matthew W. Jones (31), Tyndall Centre for Climate Change Research, School of Environmental Sciences, University of East Anglia, Norwich, United Kingdom

Lydia Keppler (427), Max Planck Institute for Meteorology; International Max Planck Research School on Earth System Modelling (IMPRS-ESM), Hamburg, Germany; Scripps Institution of Oceanography, University of California San Diego, La Jolla, CA, United States

Masayuki Kondo (237), Institute for Space-Earth Environmental Research, Nagoya University, Nagoya; Center for Global Environmental Research, National Institute for Environmental Studies, Tsukuba, Japan

Contributors **xvii**

Kevin D. Kroeger (403), Department of Earth and Environment/Institute of Environment, Florida International University, Miami, FL, United States

Werner Kurz (59), Natural Resources Canada, Canadian Forest Service, Victoria, British Columbia, Canada

Peter Landschützer (427), Max Planck Institute for Meteorology, Hamburg, Germany

Ronny Lauerwald (237), UMR Ecosys, Université Paris-Saclay, Paris, France

Sebastiaan Luyssaert (59), Department of Ecological Sciences, Vrije Universiteit Amsterdam, Amsterdam, The Netherlands

Natasha MacBean (311), Department of Geography, Indiana University, Bloomington, IN, United States

Shamil Maksyutov (87, 455), National Institute for Environmental Studies, Tsukuba, Japan

Eric Marland (31), Department of Mathematical Sciences, Appalachian State University, Boone, NC, United States

Gregg Marland (31), Research Institute for Environment, Energy, and Economics (RIEEE), Appalachian State University, Boone, NC, United States

Marcela Miranda (271), Earth System Science Centre/National Institute for Space Research, Sao Jose dos Campos, São Paulo, Brazil

Victoria Naipal (311), Departement of Geosciences, École Normale Supérieure, Paris; Laboratory for Climate and Environmental Sciences (LSCE), Gif-sur-Yvette, France

Kim Naudts (237), Department of Earth Sciences, Vrije Universiteit Amsterdam, Amsterdam, The Netherlands

Christopher S.R. Neigh (203), Code 618, Biospheric Sciences Laboratory, NASA Goddard Space Flight Center, Greenbelt, MD, United States

Eráclito Souza Neto (271), Earth System Science Centre/National Institute for Space Research, Sao Jose dos Campos, São Paulo, Brazil

Cynthia Nevison (375), Institute for Arctic and Alpine Research, University of Colorado, Boulder, CO, United States

Shuli Niu (237), Key Laboratory of Ecosystem Network Observation and Modeling, Institute of Geographic Sciences and Natural Resources Research, Chinese Academy of Sciences, Beijing, People's Republic of China

Tomohiro Oda (455), The Earth From Space Institute, Universities Space Research Association, Columbia; Department of Atmospheric and Oceanic Science, University of Maryland, College Park, MD, United States; Graduate School of Engineering, Osaka University, Suita, Osaka, Japan

Stephen M. Ogle (375), Natural Resource Ecology Laboratory and Department of Ecosystem Science and Sustainability, Colorado State University, Fort Collins, CO, United States

Jean Pierre Ometto (271), Earth System Science Centre/National Institute for Space Research, Sao Jose dos Campos, São Paulo, Brazil

Lesley Ott (483), NASA Goddard Space Flight Center, Greenbelt, MD, United States

xviii Contributors

Felipe S. Pacheco (271), Earth System Science Centre/National Institute for Space Research, Sao Jose dos Campos, São Paulo, Brazil

Frans-Jan W. Parmentier (159), Department of Physical Geography and Ecosystem Science, Lund University, Lund, Sweden; Centre for Biogeochemistry in the Anthropocene, Department of Geosciences, University of Oslo, Oslo, Norway

Prabir K. Patra (87, 455), Center for Environmental Remote Sensing, Chiba University, Chiba; Research Institute for Global Change, JAMSTEC, Yokohama, Japan

A.M. Roxana Petrescu (59), Department of Ecological Sciences, Vrije Universiteit Amsterdam, Amsterdam, The Netherlands

Julia Pongratz (59), Ludwig-Maximilians-Universität Munich, München; Max Planck Institute for Meteorology, Hamburg, Germany

Benjamin Poulter (3, 59, 311), Biospheric Sciences Laboratory, NASA Goddard Space Flight Center, Greenbelt, MD, United States

Thomas A.M. Pugh (237), Department of Physical Geography and Ecosystem Science, Lund University, Lund, Sweden; School of Geography Earth & Environmental Sciences and Birmingham Institute of Forest Research, University of Birmingham, Birmingham, United Kingdom

Anu Ramaswami (337), China-UK Low Carbon College, Shanghai Jiao Tong University, Pudong New District, Shanghai, China; High Meadows Environmental Institute; M.S. Chadha Center for Global India, Princeton University, Princeton, NJ, United States

Peter A. Raymond (237), Yale School of the Environment, Yale University, New Haven, CT, United States

Luiz Felipe Rezende (271), Earth System Science Centre/National Institute for Space Research, Sao Jose dos Campos, São Paulo, Brazil

Kelly Ribeiro (271), Earth System Science Centre/National Institute for Space Research, Sao Jose dos Campos, São Paulo, Brazil

Dustin Roten (31), Department of Atmospheric Sciences, University of Utah, Salt Lake City, UT, United States

Christina Schädel (159), Center for Ecosystem Science and Society, Northern Arizona University, Flagstaff, AZ, United States

Edward A.G. Schuur (159), Center for Ecosystem Science and Society, Northern Arizona University, Flagstaff, AZ, United States

Stephen Sitch (59), University of Exeter, Exeter, United Kingdom

Pete Smith (375), Institute of Biological & Environmental Sciences, University of Aberdeen, Aberdeen, United Kingdom

William Kolby Smith (311), School of Natural Resources and the Environment, University of Arizona, Tucson, AZ, United States

Miguel Taboada (375), National Agricultural Technology Institute (INTA), Natural Resources Research Center (CIRN), Institute of Soils, Ciudad Autónoma de Buenos Aires, Argentina

Rona L. Thompson (3, 87), NILU—Norsk Institutt for Luftforskning, Kjeller, Norway

Kangkang Tong (337), China-UK Low Carbon College, Shanghai Jiao Tong University, Pudong New District, Shanghai, China

Tiffany G. Troxler (403), Department of Earth and Environment/Institute of Environment, Florida International University, Miami, FL, United States

Francesco N. Tubiello (375), Statistics Division, Food and Agriculture Organization of the United Nations, Rome, Italy

Alexander J. Turner (455), Department of Atmospheric Sciences, University of Washington, Seattle, WA, United States

Yohanna Villalobos (3), Global Carbon Project, Climate Science Centre, CSIRO Oceans and Atmosphere, Canberra, ACT, Australia

Celso von Randow (271), Earth System Science Centre/National Institute for Space Research, Sao Jose dos Campos, São Paulo, Brazil

Jennifer Watts (159), Woodwell Climate Research Center, Falmouth, MA, United States

Lisa R. Welp (203), Earth, Atmospheric, and Planetary Sciences, Purdue University, West Lafayette, IN, United States

Lisamarie Windham-Myers (403), US Geological Survey Water Mission Area, Menlo Park, CA, United States

Daniel Zavala-Araiza (455), Environmental Defense Fund, Amsterdam; Institute for Marine and Atmospheric Research Utrecht, Utrecht University, Utrecht, The Netherlands

Foreword

This book could not be more timely. The authors provide for the first time in one book the scientific foundations behind accounting for greenhouse gases. Political leaders around the world are now moving on from establishing their nation's commitments on tackling climate change to designing policies for delivering actions on the ground. The design of climate policies needs to be based on clear scientific insights using the latest observations and understanding. After working with policymakers over the past decade, I realize the genuine desire to understand how their policies will work and how they will be monitored in the context of real-world implementation. Sound and detailed knowledge is more important than ever before because methodologies are being developed to deliver the global stocktake that forms the first opportunity to check the progress of the international Paris Climate Agreement. I am glad to see that this book takes on the challenge of providing the background needed to put human actions in perspective with the environment, which will benefit scientists and policy advisors alike.

Greenhouse gas accounting has so far been based on self-reporting of emissions at the country level. This book puts the self-reporting of emissions into the context of balancing greenhouse gases, by looking at how emissions propagate in the environment and how they interact and interfere with the natural world. It shows how observations can be used to provide independent constraints on greenhouse gas emissions and the effectiveness of climate policy, and where the limits are. The book helps put human interventions in the context of the real world so that the richness and complexities of the natural environment can be adequately considered as the world transitions away from carbon-intensive activities.

Attempts to balance greenhouse gases arose initially from the need to explain the rise in CO_2 levels in the atmosphere, first measured directly at Mauna Loa Observatory in Hawaii. Emissions of CO_2 from human activities minus their absorption in the land and ocean natural environment (the carbon 'sinks') account for the growth of CO_2 in the atmosphere, constituting the global carbon balance (or budget). Global Carbon Budgets were reported in all six assessment reports of the Intergovernmental Panel on Climate Change (IPCC), with the first published in 1990. In 2004, the Global Carbon Project (GCP) operationalized the annual update of global carbon budgets. Alongside several other authors of this book, I was part of the pioneering international group of GCP

xxi

xxii Foreword

scientists who took on this task and led the annual budget update for 13 years. Annual carbon budgets started as an attempt to increase the support of the carbon research community for the climate policy process. It quickly turned into a platform for scientific exchanges and innovation as well as a keystone event in the annual climate calendar. GCP has expanded its budget analysis to other CH_4 and N_2O to provide a more comprehensive view of the impact of humans on climate. The budget balance approach is now well recognized as one of the most powerful scientific constraints, and the approach has been adapted to other aspects of the climate system, namely, sea level and heat. Balancing budgets now sets the boundary for cutting-edge research in climate change.

Today we have a relatively clear view of the global emissions of greenhouse gases and their partitioning in the environment, even though uncertainties remain. The real challenge now is to break this down at the regional level. Regional breakdowns would both provide more relevant information for decision-makers locally and help reduce the remaining uncertainties globally. The GCP fostered an initial effort in the early 2010s to assess the evidence at the time, under the first REgional Carbon Cycle Assessment and Processes (RECCAP) umbrella published in the journal *Biogeosciences*. Since then, new methods have been developed based on advances in computing including machine learning, new data including from satellite-based CO_2 sensors, and a further understanding of how the carbon cycle operates including in the built environment.

The authors of this book are at the forefront of our research field. Their book provides a comprehensive overview of the issues related to balancing greenhouse gases and how they can be resolved. There is a big need for the book. It brings together what we know about balancing greenhouse gases in one single place, and provides the background and support for the new generation of carbon cycle scientists and policy advisors who will implement the necessary actions to tackle climate change.

Corinne Le Quéré

Royal Society Professor of Climate Change Science, University of East Anglia, Norwich, United Kingdom

Preface

Since the adoption of the Paris Agreement in 2015, the science on climate change is clear; that CO_2 emissions must reach net zero by 2050 or earlier, and that CH_4, N_2O, and halogenated greenhouse gas emissions must be reduced significantly to avoid 1.5°C or 2°C warming over preindustrial levels. Our book *Balancing Greenhouse Gas Budgets: Accounting for Natural and Anthropogenic Flows of CO_2 and Other Trace Gases* is written for students, practitioners, and experts who are interested in gaining a deeper understanding of the history, methodologies, and applications used for greenhouse gas accounting and its increasing relevance in informing climate policy.

The contributors to this book are scientists who work at universities, private organizations, and federal agencies around the world and have spent their careers developing methods to track and quantify greenhouse gas emissions and removals as well as the underlying processes that regulate changes in greenhouse gas concentrations over time. This global perspective has helped shape the breadth of individual chapters and the teams writing them, as well as the topics that they cover. The book's scope encompasses the methods, regional cases, and the future outlook of greenhouse gas accounting. We address an increasing demand to better understand the synergies between policy-driven greenhouse gas inventories that aim to quantify "anthropogenic" influences on emissions and removals with science-based approaches that quantify fluxes from both natural and managed systems.

A variety of topics that are related to greenhouse gas accounting methodologies and their adaptation in different parts of the world are covered in the book. The diverse perspectives from the individual authors add to the comprehensiveness of the chapters, with authors representing each continent and bringing a range of experiences in local to national studies on climate change, including contributions to the United Nations Framework Convention on Climate Change and the Intergovernmental Panel on Climate Change.

Chapter 1 provides the context for greenhouse gas accounting, including a historical perspective on policy-driven versus scientific methodologies. Chapters 2, 3, and 4 provide a description of each of the main methodologies for greenhouse gas accounting, covering inventories, "bottom-up" process modeling, and "top-down" atmospheric inversion approaches. Chapters 5–11 provide examples of how these methodologies are applied to land regions with separate chapters for arctic, boreal, temperate, tropical, semiarid, urban, and agricultural

xxiii

xxiv Preface

systems. Chapters 12 and 13 address greenhouse gas accounting for aquatic systems, including the open ocean and nearshore coastal ecosystems. Chapters 14 and 15 cover "forward-looking" topics in greenhouse gas emissions estimation, including advances in atmospheric inversions and data assimilation.

We intend for this book to provide a foundation for greenhouse gas accounting that can be used in classes and coursework, and as a guide to informing local to national to global scale accounting frameworks, and as a reference for understanding the integration of policy and science-driven approaches. We also hope that the contributions within the book will help to advance the science needed to inform climate policies that require emission sources and sinks to be balanced in order to stabilize the Earth's climate.

Acknowledgments

The editors acknowledge support from their home institutions: NASA Goddard Space Flight Center, Earth Sciences Division, Maryland, United States; Climate Science Centre, CSIRO Oceans and Atmosphere, Canberra, ACT, Australia, and the Australian National Environmental Science Program—Climate Systems Hub; School of Forest Resources at the University of Maine, United States; and NILU—Norsk Institutt for Luftforskning, Kjeller, Norway.

Section A

Background

Chapter 1

Balancing greenhouse gas sources and sinks: Inventories, budgets, and climate policy

Josep G. Canadell[a], Benjamin Poulter[b], Ana Bastos[c], Philippe Ciais[d], Daniel J. Hayes[e], Rona L. Thompson[f], and Yohanna Villalobos[a]

[a]Global Carbon Project, Climate Science Centre, CSIRO Oceans and Atmosphere, Canberra, ACT, Australia, [b]Biospheric Sciences Laboratory, NASA Goddard Space Flight Center, Greenbelt, MD, United States, [c]Department of Biogeochemical Integration, Max Planck Institute for Biogeochemistry, Jena, Germany, [d]Laboratoire des Sciences du Climat et de l'Environnement LSCE CEA CNRS UVSQ, Gif sur Yvette Cedex, France, [e]School of Forest Resources, University of Maine, Orono, ME, United States, [f]NILU—Norsk Institutt for Luftforskning, Kjeller, Norway

1 The human perturbation of the carbon cycle and other biogeochemical cycles

Human activities have increased emissions and atmospheric concentrations of carbon dioxide (CO_2), methane (CH_4), nitrous oxide (N_2O), and halogenated gases, all of which are heat-trapping greenhouse gases (GHGs). Before the industrial revolution, the biospheric emissions of CO_2, CH_4, and N_2O from the Earth's land, ocean, and inland waters were roughly in a dynamic equilibrium with natural sinks. However, the combustion of fossil fuels, land clearing along with other agricultural and industrial activities have driven an imbalance in the global sources and sinks, with atmospheric CO_2, CH_4, and N_2O concentrations now 47%, 156%, and 23% higher than that in 1750, respectively (Canadell et al., 2021). This imbalance is unprecedented in the Earth system in two ways. First, the atmospheric concentrations of the three main GHGs were higher in 2019 than at any time in the past 800,000 years, at 409.9 ppm for CO_2, 1866.3 ppb for CH_4, and 332.1 ppb for N_2O, and current CO_2 concentrations are also unprecedented in the last 2 million years. Second, the rate at which the CO_2 concentration has been accumulating in the atmosphere during the Industrial Era is at least 10 times faster than any other 100-year period over the last 800,000 years, and 4–6 times faster than any other 1000-year period in the last 56 million years (Canadell et al., 2021). All these changes in GHG concentrations have led to the rapid warming of the planet, with impacts on almost all

Balancing Greenhouse Gas Budgets. https://doi.org/10.1016/B978-0-12-814952-2.00024-1
Copyright © 2022 Elsevier Inc. All rights reserved.

4 Section | A Background

aspects of the Earth system (IPCC, 2021), including the rapid transformation of terrestrial and marine ecosystems (Canadell & Jackson, 2021), with direct consequences for human health, food security, and regional economies.

Unlike the halogenated gases, which are mostly synthetically produced by the chemical industry (and have thus appeared relatively recently in the atmosphere), CO_2, CH_4, and N_2O have, in addition to anthropogenic emissions, large natural emissions and sinks from biogeochemical and chemical processes on land, in the ocean and in the atmosphere (Fig. 1). This makes the study of these three GHGs particularly complex as it requires the capability to separate anthropogenic emissions from natural sources and sinks. In addition, the natural sources and sinks are not stable but respond to human-driven environmental changes, including climate change as well as natural climate variability.

2 Inventories of anthropogenic GHG: The foundation of the Kyoto protocol and the Paris agreement

The United Nations Framework Convention on Climate Change (UNFCCC) was established in 1992 to combat "dangerous human interference with the climate system." At the first Conference of Parties (COP), the UNFCCC established a range of initiatives, including that all countries provide national-level inventories for all greenhouse gases, at annual intervals for developed countries (i.e., Annex 1) and every 2 years for non-Annex 1 countries (i.e., emerging economies and less developed countries). The least developed countries would choose their reporting years at their own discretion. The objective was to encourage the presentation of information and a national inventory of anthropogenic emissions by sources and removals by sinks of all GHG not already dealt with by the Montreal Protocol. The national communications are to be done in a consistent, transparent, and comparable manner, taking into account specific national circumstances. Fig. 2 shows an illustration of the attribution of the global anthropogenic GHG emission to activities and end-uses.

The inventories account for anthropogenic emissions of GHGs only, including those from agriculture and land-use change, using 1990 as the first year, or baseline. In 1997, the Kyoto Protocol (2005–20) provided a legally binding mandate for emission reductions, with the inventories providing the basis for determining whether countries met their targets to reduce emissions over two commitment periods. The IPCC provides the scientific basis for the methodologies used in the national inventories, first described in the 1994 IPCC Guidelines for National Greenhouse Gas Inventories, and later revised and refined in 1996, 2006, and 2019, with supplements for land use in 2003 and wetlands in 2013. The methodologies combine activity data and emission factors using either stock-change or gain/loss approaches to estimate emissions and removals, and are designed to use detailed information where available and more generalized information in cases where little or no data exist. A Tier system based on different levels of analytical complexity and data richness was

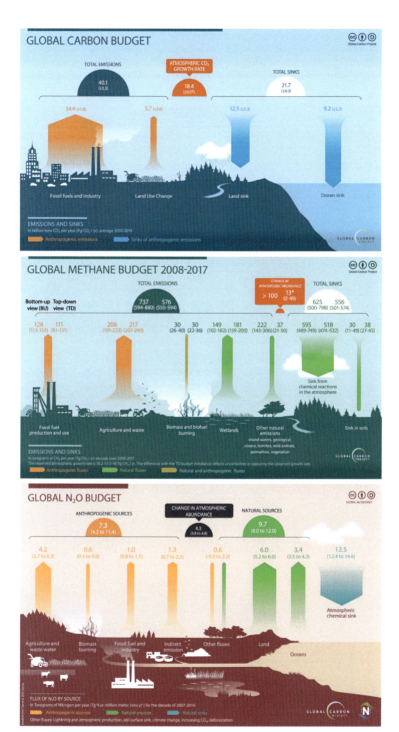

FIG. 1 Global biogeochemical cycles with their natural and human perturbation fluxes. Top panel: mean annual global CO_2 budget, 2011–19 (Global Carbon Atlas, 2021) based on (Friedlingstein et al., 2020). Middle panel: mean annual global CH_4 budget, 2008–17 (Saunois et al., 2020). Bottom panel: mean annual global N_2O budget, 2007–16 (Tian et al., 2020).

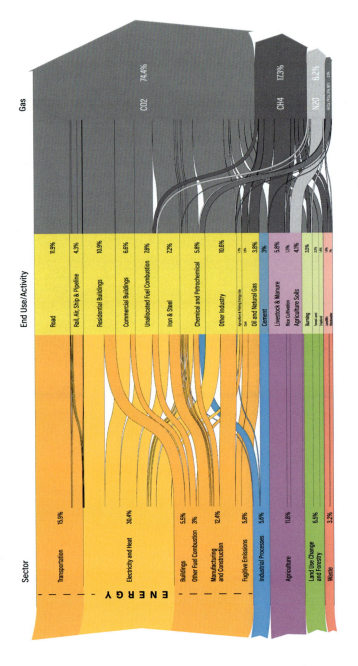

FIG. 2 Global greenhouse gas emissions are attributed to the main end uses and activities for 2016. (*Source: World Resources Institute.*)

developed, which is associated with a level of uncertainty. Tier 1 methods have higher uncertainty and use global average emission factors, Tier 2 methods have medium uncertainty, using regional emission factors, and Tier 3 methods have the lowest uncertainties and use local data and more sophisticated modeling approaches. The type of approach used affects the estimates and their uncertainties, with higher uncertainties for data-poor countries and lower uncertainties for data-rich countries.

While the inventories aim to provide estimates of direct anthropogenic emissions and removals, for some GHGs and sectors they inadvertently include indirect and natural processes. For example, inventories of CO_2 emissions from the land use, land-use change, and forestry (LULUCF) sector include the effects of climate change and CO_2 on ecosystems, and may also implicitly include (or not) natural disturbances (Canadell et al., 2007; Grassi et al., 2018). In addition, different countries use the proxy of "managed lands" for the emissions that they are responsible to cover with their inventories. The use of different definitions for what is designated as managed land has a large effect on what processes are included in the inventories and on the extent of land upon which the reporting is based (Grassi et al., 2018). For instance, the definition of managed lands includes almost the entire land area in the United States and European inventories, whereas, for Canada, Russia, and Brazil, large areas of land are considered unmanaged and excluded from the GHG inventory reporting. In tropical countries, the distinction between managed and unmanaged land is often unclear, and illegal logging along with forest degradation may occur on what is reported as unmanaged land and thus omitted from inventories, thereby contributing to uncertainty in the reporting.

These differences in definitions, methodology, and reporting make the inventories less comparable among countries and also lead to uncertainties on how genuine, additive, and effective GHG national mitigation targets are. For instance, errors occur in estimating GHG emissions from the agriculture and LULUCF sectors for not correcting indirect and natural processes, and because land ecosystems and management are unique to each country so too are the errors (Cui et al., 2021). This is in contrast to, for example, the calculation of CO_2 emissions from the combustion of fossil fuels, which depend on the type and quantity of fuel combusted with a narrower spread of potential uncertainties, albeit also requiring a regional/national focus (Liu et al., 2015).

The national GHG inventories for UNFCCC reporting are prepared by government bodies or other institutions in each country. For instance, in the United States, the Environmental Protection Agency (EPA) develops the national GHG inventory, with the development cycle beginning a year before submission to the UNFCCC. EPA scientists and experts contribute to the inventory, which then goes through an external review process. Atmospheric measurements are used to help refine emission factors and act as independent benchmarks for inventory uncertainty assessment. In Australia, the government updates the GHG inventories four times a year, with a 25-m resolution satellite-based

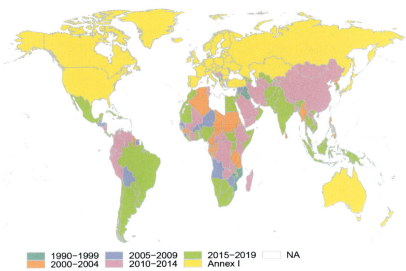

FIG. 3 The last period for which GHG emissions inventories submitted to the UNFCCC are available. The figure includes all types of inventories, from the very detailed and annually updated GHG inventories from Annex I countries (developed countries), to biennial reporting for some non-Annex I countries with a wide range in the quality of reporting and detail provided. Least developed countries submit inventories at frequencies chosen at their own discretion and often provide very limited detail. *(Source Minx, J., Lamb, W., Andrew, R., Canadell, J., Crippa, M., Döbbeling, N., et al. (2021). A comprehensive dataset for global, regional and national greenhouse gas emissions by sector 1970–2019. Earth System Science Data Discussions, 1–63. https://doi.org/10.5194/essd-2021-228.)*

system that is used to estimate emissions from land clearing and revegetation due to human activities. However, the frequency and quality of reporting are much reduced in developing and less developed countries, limiting the capacity of global analyses to track progress toward the objectives of the Paris Agreement (Fig. 3). All signatory countries to the Paris Agreement will be required to report on a regular basis from 2024 onwards.

The limitations of GHG inventories in many regions of the world limit the value of global estimates from the sum of national inventories and their use to constrain studies of the anthropogenic perturbation of the carbon cycle. Moreover, major disparities in flux estimates have been identified when comparing inventory data with other estimates using independent or partially independent approaches. For instance, the global anthropogenic net land-use CO_2 emissions reported by global biospheric models, as presented in the annual Global Carbon Project-Global Carbon Budget (Friedlingstein et al., 2020) and the IPCC Assessment reports (Canadell et al., 2021; Ciais et al., 2013), have a discrepancy of about 4 Gigatons of CO_2 per year when compared to the aggregate national estimates reported to the UNFCCC (Grassi et al., 2018). Inventory-based emissions estimates show that CH_4 emissions from coal mining in China declined between 2012 and 2016 (Gao, Guan, & Zhang, 2020; Sheng, Song, Zhang, Prinn, & Janssens-Maenhout, 2019), whereas atmospheric inversion estimates showed an increase in CH_4 emissions albeit at a slower rate for 2015 (Miller et al., 2019) and 2016 (Chandra et al., 2021). Deng et al. (2021) also suggest that GHG inventories grossly underestimate CH_4 emissions from oil and gas extracting countries in the Gulf region and Central Asia (Deng et al., 2021). Thompson et al. (2019) reported a faster rate of global N_2O emissions using atmospheric inversions than that estimated based on IPCC emission factors utilized in national reporting.

3   GHG budgets: Constraining GHG sources and sinks

GHG budgets are the compilation of estimates of all GHG sources and sinks constrained by mass balance, i.e., globally the change in atmospheric abundance is balanced by the sum of all sources and sinks. Regionally, there is also a mass balance constraint, but there are larger uncertainties due to the exchange of mass between the regional domain and the rest of the world. The mass balance constraint provides a means to help assess the magnitude of the net emissions reported to the UNFCCC and how uncertain they might be. In addition, research-driven GHG budgeting efforts, which are separate from the UNFCCC national greenhouse gas inventory reporting, can help assess inconsistencies and gaps in current emissions reporting, and support model development and benchmarking to assess the future evolution of GHG sources and sinks.

The scientific budget framework relies on the use of multiple flux estimates, fully or partially independent that are derived from the use of bottom-up and top-down approaches. The incorporation of multiple constraints from these different estimates also enables some level of redundancy to assess uncertainties in the budget calculations (Fig. 4). Bottom-up estimates are typically based on some combination of global and national GHG inventory-based fluxes, process-based biogeochemical models for land and ocean, bookkeeping models, upscaled flux products from observations, and remote sensing-based flux estimates including carbon stock changes over time. Top-down flux methods are based on inverse methods that use atmospheric GHG concentrations and transport models to optimize land-atmosphere and ocean-atmosphere flux estimates. In some cases, top-down methods can attribute net fluxes to major source categories, such as biogenic and fossil sources, based on prior knowledge of the spatial distribution of sources or by the classification of unique isotopic signatures (Chandra et al., 2021; Saunois et al., 2020; Zhang et al., 2021). With such an integrative framework, the construction of GHG budgets is the ultimate test of our knowledge on the magnitude and uncertainties of GHG sources and sinks.

The global carbon budget has been assessed by the IPCC Assessment Reports, while the Global Carbon Project has taken a leading role in the construction and further development of the science underpinning the global and

10 Section | A Background

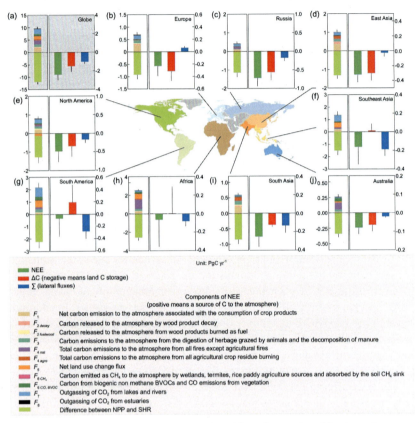

FIG. 4 Net carbon fluxes, storage changes, and lateral fluxes from trade and riverine carbon export to the ocean for the decade 2000s, based on the REgional Carbon Cycle Assessment and Processes-1 (Ciais, Bastos, et al., 2020; Ciais, Yao, et al., 2020). *NEE*, net ecosystem exchange; *ΔX*, carbon stock change; *PgC*, petagrams of carbon.

regional assessments, and extending the budgeting work to cover the three most important GHGs: CO_2, annually from 2009 (Friedlingstein et al., 2020); CH_4 (2007, 2016, 2020; Saunois et al., 2020); and N_2O 2020 (Tian et al., 2020). The latter efforts have led to the full incorporation of the three budgets in the last two assessment cycles of the IPCC (Canadell et al., 2021; Ciais, Sabine, et al., 2013).

The GCP-REgional Carbon Cycle and Processes study (RECCAP; Canadell, Ciais, Sabine, & Joos, 2012; Ciais, Yao, et al., 2020; Stavert et al., 2022) was established to support the development of budgets for regions and large nations, building from the initial efforts of the EU-project, CarboEurope, in the early 2000s. CarboEurope developed the first comprehensive GHG regional budgets for Europe (Schulze et al., 2009, 2010). Similar efforts were followed with the USA-State of the Carbon Cycle Report (USCCSP, 2007) and others in Australia (Haverd et al., 2013), China (Piao et al., 2009), among others.

Budgeting activities can vary widely in their approach, and different studies have developed their own methodology and workflow, using a range of datasets to estimate emissions and sources. The flexibility in these approaches means that data derived from multiple sources such as direct observations of emissions from flux towers and soil chambers, aircraft campaigns, satellites, process-based models, atmospheric inversions, and GHG inventories—can be combined to provide comprehensive assessments that contribute to the next-level research users and policymakers. A limitation of the current GHG budget approaches is the spatial resolution at which budgets can be developed, which is determined by data availability and model resolution, particularly for atmospheric inversions. This limits their current use in directly supporting the development of national GHG inventories, except for larger countries such as Australia, Brazil, China, Russia, and the United States. Likewise, atmospheric components of the budget have limits on their ability to partition anthropogenic from natural fluxes. However, new data platforms and model development will fundamentally change this premise in the years ahead making GHG budgets more central to national GHG assessments (see the section on New generation of technologies and observations).

RECCAP1 was the first global effort to develop a set of regional (10) and ocean basins (5) carbon budgets covering the entire globe over the decades of 1990–2009. The regionalization allows for the incorporation of regional knowledge on unique processes such as a detailed land-use change in South America and Southeast Asia, dust transport in Australia, and permafrost thawing in Russia while using regionally specific datasets and modeling capabilities. This effort led to the first global carbon budget constructed from the bottom-up (Ciais, Yao, et al., 2020), and provided new insights into the dynamics of the carbon cycle, including an estimate of the global soil heterotrophic respiration smaller than all previous estimates. This work highlighted the large effects of lateral carbon fluxes from crop and wood trade along with riverine-carbon export to the ocean. Building from the results in RECCAP1, a new carbon accounting framework has been developed (Fig. 5; Ciais, Bastos, et al., 2020), which is guiding the second phase of RECCAP. In addition to the CO_2 flux estimates in the previous effort, RECCAP2 also includes estimates of CH_4 and N_2O fluxes along with stock change estimates based on new satellite products and observation-based data, and extends the assessment to 2010–19.

The process of reconciling top-down and bottom-up estimates is key to constraining uncertainties in the overall budget by way of identifying inconsistencies and methodological problems, missing fluxes, and/or overlapping estimates (double counting). Wider use of atmospheric constraints has shown to be of great value in detecting major gaps between reported emissions and what is being observed in the atmosphere (Shen et al., 2021), identifying hotspot regions (Zhang et al., 2020), and supporting the development of improved

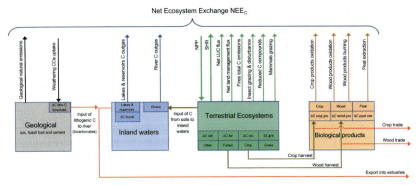

FIG. 5 Carbon budget framework and individual flux components as developed for the REgional Carbon Cycle Assessment and Processes-2 (Ciais, Bastos, et al., 2020).

emission factors (Thompson et al., 2019). Together, the benefits of the scientific budgeting approach underpin the set of best practices in monitoring, reporting, and verification (MRV), which are central to compliance under the UNFCCC.

4 Supporting the global stocktake and the net-zero emissions policy goals

In 2015, the Paris Agreement was adopted at the 21st COP of the United Nations Framework Convention on Climate Change (UNFCCC). The goal of the Agreement is to limit global warming to well below 2.0°C and pursue efforts to limit the temperature increase to 1.5°C compared to preindustrial temperatures. Article 4 stipulates that in order to achieve the long-term temperature goals, "parties aim to reach global peaking of greenhouse gas emissions as soon as possible,... and to undertake rapid reductions thereafter in accordance with the best available science, so as to achieve a balance between anthropogenic emissions by sources and removals by sinks of greenhouse gases in the second half of this century."

The Paris Agreement is different from the Kyoto Protocol because it is the first agreement to establish a specific global mean temperature goal at which global warming should be limited. Importantly, the Paris Agreement, via IPCC, also established a tangible link to the underlying biogeochemical requirements to stabilize the climate system. These requirements can be directly linked to emissions targets that nations, regions, and their global aggregate need to achieve: i.e., a balance between GHG sources and removals, otherwise known as net-zero emissions. This was achieved by better understanding the linear relationship between cumulative GHG emissions and global temperature change. Although the Agreement specifies "anthropogenic" fluxes, from an Earth perspective, the climate will stop warming only when at least all CO_2 emissions (regardless of whether they are from human activities or natural sources) reach net-zero (emissions equal to sources), and other GHGs emissions decrease very significantly (IPCC, 2021). Reaching net-zero GHG emissions, that is for all non-CO_2 GHG in addition to CO_2 emissions, would stop further global warming and lead to the slow decline in global mean temperature (IPCC, 2021).

This GHG balance also requires tracking any emerging biogeochemical-climate feedbacks, which might lead to the need to adjust the size and speed of mitigation efforts. Global net-zero CO_2 emissions need to be reached by 2050 for the 1.5°C goals and by halfway through the second half of this century for the well below 2°C goals (Arias et al., 2021).

To track progress toward these objectives of the Paris Agreement, the global stocktake was established to assess progress toward the multiple objectives on mitigation, adaptation, and finance flows every 5 years, with the first assessment due in 2023. Of particular relevance to the work on GHG budgets is the monitoring and tracking of the emissions reduction efforts encapsulated in the Nationally Determined Contributions (NDCs, or short-term mitigation goals) and the progress toward net-zero emissions (the long-term goal). Scientific budget assessments can support the quantification of mitigation activities both regionally and globally, but except for a small number of large countries such as Australia, China, the United States, and the European Union as a region, they are not currently conducted at the national level which is what would be required to support meeting the NDCs and net-zero emission targets. The work on GHG budgets includes assessing the combined effectiveness of the NDCs, improving national inventories, and contributing to the setting of more stringent NDCs over time. Globally, budgets over time will be able to detect changes in the strength of the natural sinks, currently removing 56% of all anthropogenic emissions, and emerging biogeochemical-climate feedback (Canadell et al., 2021; Friedlingstein et al., 2020).

Nature-based solutions to achieve mitigation targets and net-zero emissions goals (Griscom et al., 2017; Smith et al., 2016) are becoming ever more important, particularly those activities removing CO_2 from the atmosphere (CDR, carbon dioxide removal). Their growing importance comes from the large requirements for negative emissions in decarbonization pathways consistent with the Paris agreement (Fuss et al., 2014; IPCC, 2018). A thorough assessment of the potential for nature-based solutions is required as well as tracking the effects of rapidly changing climate and atmospheric CO_2 on the sink capacity of CDR activities and the natural biospheric sink are required (Canadell & Schulze, 2014; Roe et al., 2021; Walker et al., 2020).

A comprehensive and transparent approach to all sources and sinks of GHGs is also central to the development of carbon markets, certification schemes, green bonds, standards of carbon credits, and voluntary markets. They are all growing quickly as nations increase their mitigation commitments, pledges to net-zero emissions, and a broader spectrum of corporate and civil actors become involved.

5 A new generation of technologies and observations to constrain global and regional GHG budgets

Remote sensing is playing a growing role in supporting the monitoring, reporting, and verification of GHGs sources and sinks. Satellite systems are critical for filling in the gaps among in situ measurements (in both time and space), especially over large, remote, and inaccessible regions with otherwise sparse monitoring networks on the ground. The public opening of the Landsat archive (Wulder, Masek, Cohen, Loveland, & Woodcock, 2012) and its availability for analysis on cloud platforms (Gorelick et al., 2017) is allowing for unprecedented access to high resolution and up-to-date information on land cover and land-use change everywhere on Earth (Kennedy et al., 2018).

For regions and nations with vast land to cover or without a comprehensive inventory program, repeated satellite analysis of land-use change and above-ground biomass can be combined in a simple book-keeping model to track carbon stocks and fluxes for the purposes of GHG budgets (Tang et al., 2020). For instance, when Australia began to estimate emissions from land-clearing in the early 1990s, it relied on a number of disparate observations and reports from different states with no common methodologies, including analyzing the sales of Tordon, a chemical used to kill trees, to indirectly estimate the annual amount of land clearing (personal communication Prof. Graham Farquar, head of the National Greenhouse Gas Inventory in early 1990s). Since then, Australia has developed a tier-3 modeling approach to estimate GHG emissions and removals from land use, land-use change, and forestry with continent-wide information, and land-use transitions at 25 m resolution based on the NASA-USGS Landsat retrievals (Lehmann, Wallace, Caccetta, Furby, & Zdunic, 2013). Many other countries, particularly in more advanced and emerging economies have developed equally sophisticated systems to track the anthropogenic GHG sources, including the combustion of fossil fuels and land use including agriculture, land-use change (Dutra et al., 2012), and forestry.

Compared to the use of optical imagery to detect changes in land cover, active remote-sensing systems such as synthetic aperture radar are used to more directly measure above-ground C stocks from airborne and spaceborne platforms (Berninger, Lohberger, Stängel, & Siegert, 2018). Airborne laser scanning, or LiDAR, is capable of high-resolution biomass carbon mapping (Montesano et al., 2014) and is increasingly being used operationally in national forest inventories (Naesset, 2007; White et al., 2013). New products related to solar-induced fluorescence (SIF) and vegetation optical depth (VOD) are providing unique insights into productivity and biomass that previous multispectral and coarse-spatial resolution radar missions were unable to provide (Fan et al., 2019; Liu et al., 2015; Qin et al., 2021).

Advances in ground-based networks, space-based monitoring, modeling, and data assimilation techniques can contribute substantially toward reducing uncertainties and providing lower-latency information on emissions and removals. For example, the 2019 Refinement to the 2006 IPCC Guidelines for National Greenhouse Gas Inventories (Gitarskiy, 2019) now includes guidelines on how atmospheric inversions can be used in emissions estimation. Specifically, atmospheric GHG inversions can be compared with national GHG inventories to provide quality assurance of the inventory by verifying total emissions for particular categories and gases, helping countries to target areas of uncertainty.

Ground-based networks have expanded and become more coordinated with one another over the past decade. These include observing networks such as NEON, GLEON, SAEON, Fluxnet, Euroflux, Ameriflux, Ozflux, and Asiaflux. Trace gas measurements provided by these networks are the basis of training datasets for machine learning or artificial intelligence algorithms used to develop data-driven, gridded time series of greenhouse gas emissions. Gaps in measurements still exist, particularly in the tropics and high latitudes, where logistics are challenged by remote access and difficult weather conditions. Measurements of methane are sparse globally, making uncertainty for wetland emissions relatively high.

Over the past few years, a new satellite constellation of space-based remote-sensing platforms has emerged to provide insight into plant function (Stavros et al., 2017) and other important measurements such as forest canopy height, canopy vertical structure, and GHGs column retrievals. National space agencies (e.g., NASA, ESA, ASI, JAXA) and private companies (Maxar, Planet, DLR) have launched a range of optical, lidar, and radar instruments as free-flyers or as hosted payloads using the International Space Station. These missions include NASA's Ice, Cloud, and land elevation satellite (ICESat-2) (Markus et al., 2017), the Global Ecosystem Dynamics Investigation (GEDI) (Dubayah et al., 2020), the Orbiting Carbon Observatory-2 OCO-2 and OCO-3 (Crisp et al., 2017; Eldering, Taylor, O'Dell, & Pavlick, 2019), ESA's Sentinel 1, 2, and 5, the TROPOspheric Measuring Instrument (TROPOMI) (Veefkind et al., 2012), and JAXA's GOSAT (Greenhouse gases Observing SATellite) series (Kuze et al., 2016). Commercial spacecraft include MAXAR's Worldview high-resolution instruments, which can be used for mapping individual shrubs and trees in open-canopy forests, and Planet's 'Doves' providing daily 3-m resolution observations of land cover and land cover change. On the horizons are the joint NASA-ESA Biomass mission and NASA's NISAR mission, which combined with GEDI and ICEsat-2, will provide full global coverage of biomass and $\sim 30 m$ resolution, and ESA's FLEX mission, which will provide improved estimates of photosynthesis activity and plant health and stress conditions.

The new technologies for satellite measurements of GHGs such as CO_2 and CH_4 provide new opportunities to study the global carbon cycle, particularly in regions that have been historically poorly covered by ground-based carbon observations. For instance, the Greenhouse gases Observing SATellite (GOSAT) (CO_2, CH_4) (Kuze et al., 2016), the Orbiting Carbon Observatory-2

(OCO-2) (CO_2) (Crisp et al., 2017), and the global carbon dioxide monitoring satellite (TanSat) (CO_2) (Yang et al., 2021) now provide column GHG retrievals across the world that can be used to quantify carbon flux exchanges (Fig. 6A). The NASA OCO-3 launched to space in 2019 (Eldering et al., 2019; Taylor et al., 2020) offers more dense information than OCO-2 at northern and southern mid-latitudes, while the upcoming NASA Geostationary Carbon Cycle Observatory (GeoCarb) (Moore et al., 2018) will focus on the Americas' carbon cycle with the total concentration of CO_2 and carbon monoxide at a 5–10-km horizontal resolution. The CO2MVS constellation will track CH_4 and CO_2 and is under development by ESA and the European Organization for the Exploitation of Meteorological Satellites (EUMETSAT).

Although satellite-based column GHG estimates are no replacement for precise and accurate atmospheric observations (Masarie, Peters, Jacobson, & Tans, 2014), their continuous improvement and broad coverage can provide new insights on regions such as the tropics and the Southern Hemisphere where atmospheric observations are too sparse (Peylin et al., 2013).

The value of the new GHG satellite data to develop regional carbon budgets is evident, as an example, for the Australian continent (Fig. 6). Atmospheric inversions rely on one GHG monitoring station (Cape Grim), located at a coastal site with onshore winds intended to provide global baseline measurements of atmospheric GHGs and pollutants (Pearman, Fraser, & Garratt, 2017). It is, therefore, free, by design, of continental influences, limiting its capability to constrain the fluxes across the continent in global atmospheric inversions (Haverd et al., 2013). Observations from the OCO-2 satellite retrievals have enabled to perform a first-of-its-kind regional atmospheric inversion (Fig. 6B, C; Villalobos, Rayner, Thomas, & Silver, 2020) informed (as a prior) by the land-atmosphere fluxes from a highly parameterized biospheric model for Australian conditions (CABLE; Haverd et al., 2018) (Fig. 6); the use of both bottom-up and top-down constraints has highlighted the importance of savanna ecosystems of the north and semiarid regions of the southeast as strong carbons sinks.

As a much denser and higher resolution sampling of GHG concentrations, biomass, vegetation structural characteristics, and flux measurements are becoming more widely available; the comparison between national inventories and comprehensive scientific GHG budgets can be done on a regular basis. More spatially and temporally explicit modeling is an important extension from current GHG inventories to address the needs for mitigation and adaptation.

Modeling approaches have advanced to take advantage of increased computational capacity, available at lower cost, and open-science mandates that provide faster access, transparency, and data equity. Collaborative tools such as Slack, Github, and others now provide instantaneous sharing of ideas and code, leading to less delay in the availability of data. Governmental coordination, such as the European Copernicus Program, the European Space Agency's

GHG inventories, budgets, and climate policy **Chapter | 1** 17

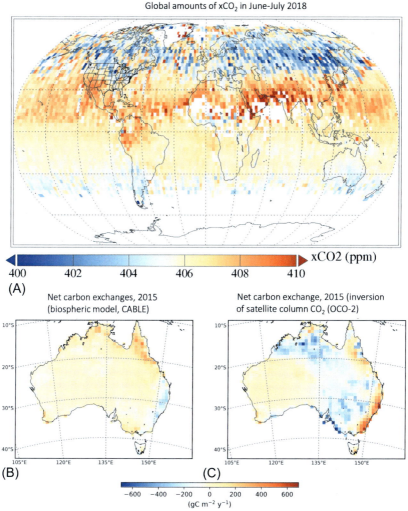

FIG. 6 (A) CO_2 column-averaged dry air mole fraction (ppm) June–July 2018, OCO-2 Lite (version 9). (B) Net carbon exchange for Australia in 2015 estimated by the Community Atmosphere Biosphere Land Exchange (CABLE) model, used as a prior flux estimate for the inversion in (C). It includes fire emissions but excludes emissions from the combustion of fossil fuels and land-use change. (C) Net carbon exchange for Australia in 2015 based on an inversion of column CO_2 from the Orbiting Carbon Observatory-2, including all CO_2 fluxes except fossil fuel emissions. *((A) Source: OCO-2 Science Team, Michael Gunson, A. E. (2018). OCO-2 level 2 bias-corrected XCO2 and other select fields from the full-physics retrieval aggregated as daily files, retrospective processing V9r. Greenbelt, MD, USA, Goddard Earth Sciences Data and Information Services Center (GES DISC). (C) Based on Villalobos, Y., Rayner, P., Silver, J., Thomas, S., Haverd, V., Knauer, J., et al. (2021). Was Australia a sink or source of CO2 in 2015? Data assimilation using OCO-2 satellite measurements. Atmospheric Chemistry and Physics, 21, 17453–17494. https://doi.org/10.5194/acp-21-17453-2021.)*

18 Section | A Background

Climate Change Initiative, and NASA's Carbon Monitoring System has enabled greater collaboration and data access for advancing greenhouse gas science.

6 Extending the carbon budget and accounting frameworks to meet broader policy information needs

The history of greenhouse gas accounting is explained by the information needs of a wide range of stakeholders. The stakeholders include scientists, policy-makers, and regulators at the city, state, national, and global scales. Corporations and nonprofit organizations have also been interested in determining their own carbon footprints and budgets, while the GHG budget framework has been important in higher education. Accounting approaches have evolved to inform policy, provide monitoring and enforcement, and develop a better understanding of ecosystems and the Earth system while contributing to theory and benchmarking used in model development and applications.

In addition to *global and regional carbon and GHG budgets* described in the preceding sections (Global: Friedlingstein et al., 2020; Saunois et al., 2020; Tian et al., 2020; Regional: Canadell et al., 2012; Petrescu et al., 2021; Petrescu et al., 2021; Saunois et al., 2020), the carbon budget and accounting frameworks have been extended to fulfill other information needs leading to the development of *cumulative carbon budgets* (Friedlingstein et al., 2020; Quéré et al., 2016); Fig. 7, and *remaining carbon budgets* (Matthews et al., 2020; Rogelj et al., 2016); Fig. 8.

Cumulative carbon budgets refer to the total amount of carbon emitted over a period and their partition between the atmosphere and the CO_2 sinks on land and in the ocean (Fig. 7). The cumulative carbon budget provides information on the historical role or responsibility of the various components of the budget leading to or avoiding climate change. The cumulative component of emissions has also been used to assign the responsibility to individual countries for their historical cumulative emissions (and contribution to global warming), as opposed to assessments based on their current emissions. This approach was initially discussed in the FCCC and is known as the Brazilian proposal, which did not succeed in being adopted. The inclusion of both cumulative CO_2 emissions and sinks has also been used to determine the net effect of individual regions and countries on the observed anthropogenic radiative forcing and climate change to date (Ciais et al., 2013; Fu et al., 2021).

The *remaining carbon budget* (Allen et al., 2009; Matthews et al., 2020; Rogelj et al., 2016) is the total amount of net CO_2 emissions that human activities can still release into the atmosphere while keeping the global mean surface temperature to a specific level, for instance, to 1.5°C or well below 2°C relative to preindustrial temperatures as pursued by the Paris Agreement (Fig. 8). The "net" term is important because the gross flux emissions can be larger than the permissible budget for a given temperature objective, if an equal amount of

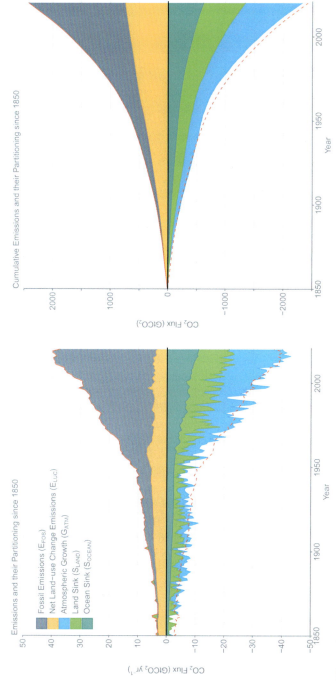

FIG. 7 Global cumulative carbon budget for all major CO_2 sources and sinks for the period 1850–2020 (Friedlingstein et al., 2021).

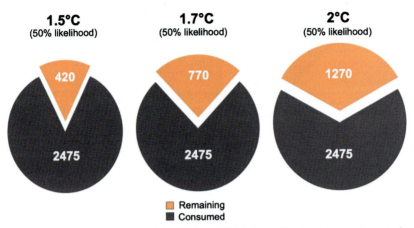

FIG. 8 Total carbon budget to 1.5°C, 1.7°C, and 2°C global surface mean temperature starting from 1850 to 2021, with the partition of the historical carbon budget and the remaining carbon budget. The remaining carbon budget since January 2022. Total CO_2 emissions include emissions from the combustion of fossil fuels, land-use change, and cement production. Budgets based on IPCC methods (Canadell et al., 2021) and updated data from Friedlingstein et al. (2021). *(Source: Global Carbon Project.)*

emissions are removed from the atmosphere at the same time, e.g., through carbon dioxide removal strategies.

The estimate of these remaining carbon budgets is possible because of the quasi-linear relationship between cumulative anthropogenic CO_2 emissions and global mean temperature, which implicitly accounts for the role of CO_2 sinks and assumes the rest of GHGs will also decline in line with preestablished choices under socioeconomic pathways (IPCC, 2021). This quasi-linear relationship implies that in order to stabilize the climate system at any given level, global CO_2 emissions must be reduced to net-zero at a time in the future depending on the temperature objective.

Remaining carbon budgets, often termed carbon budgets for short, have also been used to drive climate action at the level of nations (Committee on Climate Change, 2017), economic sectors (Steininger, Meyer, Nabernegg, & Kirchengast, 2020), and in the corporate and finance world (Strauch, Dordi, & Carter, 2020). Globally, the remaining carbon budget is underpinned by geophysical observations and modeling, however, the share of that global budget to countries, corporations, or economic sectors is based on socioeconomic and ethical arguments. Some quantitative approaches have been developed to encapsulate some of those aspects (Raupach et al., 2014). Regardless of how far the application of this framework is extended, its fundamental value has been to bring attention to the fact that emissions are a finite resource in a future carbon-constrained world, and that how much and how quickly budgets are spent is critical to design the transitions to a net-zero carbon economy.

On emissions accounting, several extensions exist beyond the *territorial-based emissions* accounting (i.e., carbon emissions originating in the territory considered), which underpins the national GHG emission inventories reported to the UNFCCC and budgets in the preceding section. One of such extensions gives the *consumption-based emissions* defined as the emissions associated with the consumption of products and services in a country regardless of where the emissions from the production of those products and services occurred (Andrew, Davis, & Peters, 2013; Barrett et al., 2013; Davis & Caldeira, 2010; Peters, Davis, & Andrew, 2012). Consumption-based emissions accounting helps to identify potential outsourcing of emissions to other countries as a strategy to decarbonize national economies or as an unintended outcome from trade imbalances between countries. It also highlights countries whose territorial emissions are enhanced by the production of products and services consumed (exported) outside of the country. For instance, Switzerland has three times higher consumption emissions than territorial emissions, while China, India, and Vietnam all have higher territorial emissions than emissions from goods and services consumed in the country (Global Carbon Atlas, 2021).

Extraction-based emissions assign the emissions from the combustion of fossil fuels to the country where the fuels were extracted, not the country where they are combusted (Lafleur, 2018). It is also known as the embedded carbon in extracted fossil fuels. This accounting approach highlights the responsibility of those countries extracting and trading fossil fuels to the rest of the world, and so fueling anthropogenic climate change. It also highlights that while more than 200 countries are consumers of fossil fuels, there are far fewer that extract and export them, leading to the suggestion that a climate treaty targeting fossil fuel-producing countries would enhance the effectiveness of the Paris agreement that requires 192 countries to agree and comply (Asheim et al., 2019; Erickson, Lazarus, & Piggot, 2018). Extraction-based emissions accounting has also allowed to assess the gap between the amount of carbon in fossil fuels planned for extraction by producer countries and the permissible emissions for a temperature target. Currently, the amount of carbon embedded in fossil fuels planned to be extracted by governments by 2030 is more than twice the amount of permissible carbon emissions to limit global warming to 1.5°C (SEI, IISD, ODI, E3G, 2021).

Building on all this acknowledge, additional carbon accounting frameworks have been developed to support a diverse set of jurisdictions and actors responsible for actionable mitigation and adaptation activities, notably cities. Over the past two decades, a system of scope levels has helped the cities to define and calculate their territorial emissions (scope 1), adding emissions produced elsewhere for energy consumed in the city (scope 2), and further incorporating a broader consumption-based carbon footprint including various levels of the supply chain for goods and services consumed in the city (scope 3). Emerging new accounting frameworks support multiscale action that incorporates sustainable development goals in addition to emission reduction targets (Ramaswami et al., 2021).

All these frameworks and accounting approaches at different scales and levels of comprehensiveness need to link to each other and add up to the global totals of emissions, to the observed accumulation in the atmosphere, and to the mitigation goals. Both anthropogenic and natural GHG fluxes need to be tracked and understood as we pursue the stabilization of the climate system by reaching net-zero carbon emissions first, along with a very significant reduction or net-zero emissions for the rest of GHGs.

Acknowledgment

JGC acknowledges the support by the Australian National Environmental Science Program—Climate Systems hub. We thank Peter Briggs for preparing the figure files.

References

Allen, M. R., Frame, D. J., Huntingford, C., Jones, C. D., Lowe, J. A., Meinshausen, M., et al. (2009). Warming caused by cumulative carbon emissions towards the trillionth tonne. *Nature, 458* (7242), 1163–1166. https://doi.org/10.1038/nature08019.

Andrew, R. M., Davis, S. J., & Peters, G. P. (2013). Climate policy and dependence on traded carbon. *Environmental Research Letters, 8*(3). https://doi.org/10.1088/1748-9326/8/3/034011.

Arias, P. A., Bellouin, N., Coppola, E., Jones, R. G., Krinner, G., Marotzke, J., et al. (2021). Technical summary. In V. Masson-Delmotte, P. Zhai, A. Pirani, S. L. Connors, C. Péan, S. Berger, … B. Zhou (Eds.), *Climate change 2021: The physical science basis. Contribution of working group I to the sixth assessment report of the intergovernmental panel on climate change* Cambridge University Press.

Asheim, G. B., Fæhn, T., Nyborg, K., Greaker, M., Hagem, C., Harstad, B., et al. (2019). The case for a supply-side climate treaty. *Science, 365*(6451), 325–327. https://doi.org/10.1126/science. aax5011.

Barrett, J., Peters, G., Wiedmann, T., Scott, K., Lenzen, M., Roelich, K., et al. (2013). Consumption-based GHG emission accounting: A UK case study. *Climate Policy, 13*(4), 451–470. https://doi. org/10.1080/14693062.2013.788858.

Berninger, A., Lohberger, S., Stängel, M., & Siegert, F. (2018). SAR-based estimation of aboveground biomass and its changes in tropical forests of Kalimantan using L- and C-band. *Remote Sensing, 10*(6). https://doi.org/10.3390/rs10060831.

Canadell, J., Ciais, P., Sabine, C., & Joos, F. (2012). *Regional carbon cycle Assessement and processes (RECCAP)*. https://bg.copernicus.org/articles/special_issue107.html.

Canadell, J. G., & Jackson, R. B. (2021). Ecosystem collapse and climate change. In J. G. Canadell, & R. B. Jackson (Eds.), *Ecological* Springer. https://doi.org/10.1007/978-3-030-71330-0_1.

Canadell, J. G., Kirschbaum, M. U. F., Kurz, W. A., Sanz, M. J., Schlamadinger, B., & Yamagata, Y. (2007). Factoring out natural and indirect human effects on terrestrial carbon sources and sinks. *Environmental Science and Policy, 10*(4), 370–384. https://doi.org/10.1016/j. envsci.2007.01.009.

Canadell, J. G., Monteiro, P. M. S., Costa, M. H., Cotrim da Cunha, L., Cox, P. M., Eliseev, A. V., et al. (2021). Global carbon and other biogeochemical cycles and feedbacks. In V. Masson-Delmotte, P. Zhai, A. Pirani, S. L. Connors, C. Péan, S. Berger, … B. Zhou (Eds.), *Climate change 2021: The physical science basis. Contribution of working group I to the sixth assessment report of the intergovernmental panel on climate change* Cambridge University Press.

Canadell, J. G., & Schulze, E. D. (2014). Global potential of biospheric carbon management for climate mitigation. *Nature Communications, 5,* 1–12. https://doi.org/10.1038/ncomms6282.

Chandra, N., Patra, P. K., Bisht, J. S. H., Ito, A., Umezawa, T., Saigusa, N., et al. (2021). Emissions from the oil and gas sectors, coal mining and ruminant farming drive methane growth over the past three decades. *Journal of the Meteorological Society of Japan. Ser. II, 99*(2). https://doi.org/10.2151/jmsj.2021-015.

Ciais, P., Bastos, A., Chevallier, F., Lauerwald, R., Poulter, B., Canadell, P., et al. (2020). *Definitions and methods to estimate regional land carbon fluxes for the second phase of the REgional carbon cycle assessment and processes project (RECCAP-2). Geoscientific model development discussions, September* (pp. 1–46). https://doi.org/10.5194/gmd-2020-259.

Ciais, P., Gasser, T., Paris, J. D., Caldeira, K., Raupach, M. R., Canadell, J. G., et al. (2013). Attributing the increase in atmospheric CO_2 to emitters and absorbers. *Nature Climate Change, 3*(10), 926–930. https://doi.org/10.1038/nclimate1942.

Ciais, P., Sabine, C., Bala, G., Bopp, L., Brovkin, V., Canadell, J., et al. (2013). Carbon and other biogeochemical cycles. In T. F. Stocker, D. Qin, G.-K. Plattner, M. Tignor, S. K. Allen, J. Boschung, … P. M. Midgley (Eds.), *R. L4673 (Trans.), Climate change 2013: The physical science basis. Contribution of working group I to the fifth assessment report of the intergovernmental panel on climate change* (pp. 465–570). Cambridge University Press. https://doi.org/10.1017/CBO9781107415324.015.

Ciais, P., Yao, Y., Gasser, T., Baccini, A., Wang, Y., Lauerwald, R., et al. (2020). Empirical estimates of regional carbon budgets imply reduced global soil heterotrophic respiration. *National Science Review, June.* https://doi.org/10.1093/nsr/nwaa145.

Committee on Climate Change. (2017). *Meeting carbon budgets: Closing the policy gap. June* (pp. 1–203). www.theccc.org.uk/publications%0Ahttp://www.sciencedirect.com/science/article/pii/B9780123838322000815.

Crisp, D., Pollock, H., Rosenberg, R., Chapsky, L., Lee, R., Oyafuso, F., et al. (2017). The on-orbit performance of the orbiting carbon observatory-2 (OCO-2) instrument and its radiometrically calibrated products. *Atmospheric Measurement Techniques, 10*(1), 59–81. https://doi.org/10.5194/amt-10-59-2017.

Cui, X., Zhou, F., Ciais, P., Davidson, E. A., Tubiello, F. N., Niu, X., et al. (2021). Global mapping of crop-specific emission factors highlights hotspots of nitrous oxide mitigation. *Nature Food, 2* (11), 886–893. https://doi.org/10.1038/s43016-021-00384-9.

Davis, S. J., & Caldeira, K. (2010). Consumption-based accounting of CO_2 emissions. *Proceedings of the National Academy of Sciences of the United States of America, 107*(12), 5687–5692. https://doi.org/10.1073/pnas.0906974107.

Deng, Z., Ciais, P., Tzompa-sosa, Z. A., Saunois, M., Qiu, C., Tan, C., et al. (2021). *Inventories against atmospheric inversions, august* (pp. 1–59).

Dubayah, R., Blair, J. B., Goetz, S., Fatoyinbo, L., Hansen, M., Healey, S., et al. (2020). The global ecosystem dynamics investigation: High-resolution laser ranging of the Earth's forests and topography. *Science of Remote Sensing, 1*(September 2019), 100002. https://doi.org/10.1016/j.srs.2020.100002.

Dutra, A. P., Ometto, J. P, Nobre, C., Montenegro, D., Almeida, C., Celia, I., et al. (2012). Modeling the spatial and temporal heterogeneity of deforestation-driven carbon emissions: The INPE-EM framework applied to the Brazilian Amazon. *Global Change Biology, 18,* 3346–3366. https://doi.org/10.1111/j.1365-2486.2012.02782.x.

Eldering, A., Taylor, T. E., O'Dell, C. W., & Pavlick, R. (2019). The OCO-3 mission: Measurement objectives and expected performance based on 1 year of simulated data. *Atmospheric Measurement Techniques, 12*(4), 2341–2370. https://doi.org/10.5194/amt-12-2341-2019.

Erickson, P., Lazarus, M., & Piggot, G. (2018). Limiting fossil fuel production as the next big step in climate policy. *Nature Climate Change, 8*(12), 1037–1043. https://doi.org/10.1038/s41558-018-0337-0.

Fan, L., Wigneron, J.-P., Ciais, P., Chave, J., Brandt, M., Fensholt, R., et al. (2019). Satellite-observed pantropical carbon dynamics. *Nature Plants.* https://doi.org/10.1038/s41477-019-0478-9.

Friedlingstein, P., Jones, M. W., O'Sullivan, M., Andrew, R. M., Dorothee, C. E. B., Hauck, J., et al. (2021, November). *Globla carbon budget 2021.* ESSDD. https://doi.org/10.5194/essd-2021-386.

Friedlingstein, P., Sullivan, M. O., Jones, M. W., Andrew, R. M., Hauck, J., & Meeresforschung, A. H. P. (2020). *Global carbon budget 2020.* ESSD.

Fu, B., Li, B., Gasser, T., Tao, S., Ciais, P., Piao, S., et al. (2021). The contributions of individual countries and regions to the global radiative forcing. *Proceedings of the National Academy of Sciences of the United States of America, 118*(15), 1–6. https://doi.org/10.1073/pnas.2018211118.

Fuss, S., Canadell, J. G., Peters, G. P., Tavoni, M., Andrew, R. M., Ciais, P., et al. (2014). Betting on negative emissions. *Nature Climate Change, 4*(10), 850–853. https://doi.org/10.1038/nclimate2392.

Gao, J., Guan, C. H., & Zhang, B. (2020). China's CH4 emissions from coal mining: A review of current bottom-up inventories. *Science of the Total Environment, 725.* https://doi.org/10.1016/j.scitotenv.2020.138295, 138295.

Gitarskiy, M. L. (2019). The refinement to the 2006 IPCC guidelines for national greenhouse gas inventories. *Fundamental and Applied Climatology, 2,* 5–13. https://doi.org/10.21513/0207-2564-2019-2-05-13.

Global Carbon Atlas. (2021).

Gorelick, N., Hancher, M., Dixon, M., Ilyushchenko, S., Thau, D., & Moore, R. (2017). Google earth engine: Planetary-scale geospatial analysis for everyone. *Remote Sensing of Environment, 202,* 18–27. https://doi.org/10.1016/j.rse.2017.06.031.

Grassi, G., House, J., Kurz, W. A., Cescatti, A., Houghton, R. A., Peters, G. P., et al. (2018). Reconciling global-model estimates and country reporting of anthropogenic forest CO_2 sinks. *Nature Climate Change, 8*(10), 914–920. https://doi.org/10.1038/s41558-018-0283-x.

Griscom, B. W., Adams, J., Ellis, P. W., Houghton, R. A., Lomax, G., Miteva, D. A., et al. (2017). Natural climate solutions. *Proceedings of the National Academy of Sciences of the United States of America, 114*(44), 11645–11650. https://doi.org/10.1073/pnas.1710465114.

Haverd, V., Raupach, M. R., Briggs, P. R., Canadell, J. G., Davis, S. J., Law, R. M., et al. (2013). The Australian terrestrial carbon budget. *Biogeosciences, 10*(2), 851–869. https://doi.org/10.5194/bg-10-851-2013.

Haverd, V., Smith, B., Nieradzik, L., Briggs, P. R., Woodgate, W., Trudinger, C. M., et al. (2018). A new version of the CABLE land surface model (subversion revision r4601) incorporating land use and land cover change, woody vegetation demography, and a novel optimisation-based approach to plant coordination of photosynthesis. *Geoscientific Model Development, 11*(7), 2995–3026. https://doi.org/10.5194/gmd-11-2995-2018.

IPCC. (2018). In V. Masson-Delmotte, P. Zhai, H.-O. Pörtner, D. Roberts, J. Skea, P. R. Shukla, ... T. Waterfield (Eds.), *Global warming of 1.5°C. an IPCC special report on the impacts of global warming of 1.5°C above pre-industrial levels and related global greenhouse gas emission*

GHG inventories, budgets, and climate policy Chapter | 1   **25**

pathways, in the context of strengthening the global response to the threat of climate change (p. 616).

IPCC. (2021). Summary for policymakers. In V. Masson-Delmotte, P. Zhai, A. Pirani, S. L. Connors, C. Péan, S. Berger, ... B. Zhou (Eds.), *Climate change 2021: The physical science basis. Contribution of working group I to the sixth assessment report of the intergovernmental panel on climate change.*

Kennedy, R. E., Yang, Z., Gorelick, N., Braaten, J., Cavalcante, L., Cohen, W. B., et al. (2018). Implementation of the LandTrendr algorithm on google earth engine. *Remote Sensing, 10* (5), 1–10. https://doi.org/10.3390/rs10050691.

Kuze, A., Suto, H., Shiomi, K., Kawakami, S., Tanaka, M., Ueda, Y., et al. (2016). Update on GOSAT TANSO-FTS performance, operations, and data products after more than 6 years in space. *Atmospheric Measurement Techniques, 9*(6), 2445–2461. https://doi.org/10.5194/amt-9-2445-2016.

Lafleur, D. (2018). *Aspects of Australia's fugitive and overseas emissions from fossil fuel exports.* http://hdl.handle.net/11343/214153.

Lehmann, E. A., Wallace, J. F., Caccetta, P. A., Furby, S. L., & Zdunic, K. (2013). Forest cover trends from time series Landsat data for the Australian continent. *International Journal of Applied Earth Observation and Geoinformation, 21*(1), 453–462. https://doi.org/10.1016/j.jag.2012.06.005.

Liu, Z., Guan, D., Wei, W., Davis, S. J., Ciais, P., Bai, J., et al. (2015). Reduced carbon emission estimates from fossil fuel combustion and cement production in China. *Nature, 524*(7565), 335–338. https://doi.org/10.1038/nature14677.

Liu, Y. Y., Van Dijk, A. I. J. M., De Jeu, R. A. M., Canadell, J. G., McCabe, M. F., Evans, J. P., et al. (2015). Recent reversal in loss of global terrestrial biomass. *Nature Climate Change, 5*(5), 470–474. https://doi.org/10.1038/nclimate2581.

Markus, T., Neumann, T., Martino, A., Abdalati, W., Brunt, K., Csatho, B., et al. (2017). The ice, cloud, and land elevation satellite-2 (ICESat-2): Science requirements, concept, and implementation. *Remote Sensing of Environment, 190,* 260–273. https://doi.org/10.1016/j.rse.2016.12.029.

Masarie, K. A., Peters, W., Jacobson, A. R., & Tans, P. P. (2014). ObsPack: A framework for the preparation, delivery, and attribution of atmospheric greenhouse gas measurements. *Earth System Science Data, 6*(2), 375–384. https://doi.org/10.5194/essd-6-375-2014.

Matthews, H. D., Tokarska, K. B., Nicholls, Z. R. J., Rogelj, J., Canadell, J. G., Friedlingstein, P., et al. (2020). Opportunities and challenges in using remaining carbon budgets to guide climate policy. *Nature Geoscience, 1–13.* https://doi.org/10.1038/s41561-020-00663-3.

Miller, S. M., Michalak, A. M., Detmers, R. G., Hasekamp, O. P., Bruhwiler, L. M. P., & Schwietzke, S. (2019). China's coal mine methane regulations have not curbed growing emissions. *Nature Communications, 10.* https://doi.org/10.1038/s41467-018-07891-7, 303.

Montesano, P. M., Nelson, R. F., Dubayah, R. O., Sun, G, Cook, B. D., Ranson, K. J. R., et al. (2014). The uncertainty of biomass estimates from LiDAR and SAR across a boreal forest structure gradient. *Remote Sensing of Environment, 154,* 398–407. https://doi.org/10.1016/j.rse.2014.01.027.

Moore, B., Crowell, S. M. R., Rayner, P. J., Kumer, J., O'Dell, C. W., O'Brien, D., et al. (2018). The potential of the geostationary carbon cycle observatory (GeoCarb) to provide multi-scale constraints on the carbon cycle in the Americas. *Frontiers in Environmental Science, 6*(OCT), 1–13. https://doi.org/10.3389/fenvs.2018.00109.

Naesset, E. (2007). Airborne laser scanning as a method in operational forest inventory: Status of accuracy assessments accomplished in Scandinavia. *Scandinavian Journal of Forest Research, 22*(5), 433–442. https://doi.org/10.1080/02827580701672147.

Pearman, G. I., Fraser, P. J., & Garratt, J. R. (2017). CSIRO high-precision measurement of atmospheric CO_2 concentration in Australia. Part 2: Cape grim, surface CO_2 measurements and carbon cycle modelling. *Historical Records of Australian Science*, *28*(2), 126–139. https://doi.org/10.1071/HR17015.

Peters, G. P., Davis, S. J., & Andrew, R. (2012). A synthesis of carbon in international trade. *Biogeosciences*, *9*(8), 3247–3276. https://doi.org/10.5194/bg-9-3247-2012.

Petrescu, A. M. R., McGrath, M. J., Andrew, R. M., Peylin, P., Peters, G. P., Ciais, P., et al. (2021). The consolidated European synthesis of CO_2 emissions and removals for the European Union and United Kingdom: 1990–2018. *Earth System Science Data*, *13*(5), 2363–2406. https://doi.org/10.5194/essd-13-2363-2021.

Petrescu, A. M. R., Qiu, C., Ciais, P., Thompson, R. L., Peylin, P., McGrath, M. J., et al. (2021). The consolidated European synthesis of CH_4 and N_2O emissions for the European Union and United Kingdom: 1990–2017. *Earth System Science Data*, *13*, 2307–2362.

Peylin, P., Law, R. M., Gurney, K. R., Chevallier, F., Jacobson, A. R., Maki, T., et al. (2013). Global atmospheric carbon budget: Results from an ensemble of atmospheric CO_2 inversions. *Biogeosciences*, *10*(10), 6699–6720. https://doi.org/10.5194/bg-10-6699-2013.

Piao, S., Fang, J., Ciais, P., Peylin, P., Huang, Y., Sitch, S., et al. (2009). The carbon balance of terrestrial ecosystems in China. *Nature*, *458*(7241), 1009–1013. https://doi.org/10.1038/nature07944.

Qin, Y., Xiao, X., Wigneron, J.-P., Ciais, P., Canadell, J. G., Brandt, M., et al. (2021). Annual maps of forests in Australia from analyses of microwave and optical images with FAO forest definition. *Journal of Remote Sensing*, *2021*, 1–11. https://doi.org/10.34133/2021/9784657.

Quéré, C. L., Andrew, R. M., Canadell, J. G., Sitch, S., Korsbakken, J. I., Peters, G. P., et al. (2016). Global carbon budget 2016. *Earth System Science Data*, *8*(1), 1–45.

Ramaswami, A., Tong, K., Canadell, J. G., Jackson, R. B., Stokes, E., & (Kellie), Dhakal, S., Finch, M., Jittrapirom, P., Singh, N., Yamagata, Y., Yewdall, E., Yona, L., & Seto, K. C. (2021). Carbon analytics for net-zero emissions sustainable cities. *Nature Sustainability*, *4*(6), 460–463. https://doi.org/10.1038/s41893-021-00715-5.

Raupach, M. R., Davis, S. J., Peters, G. P., Andrew, R. M., Canadell, J. G., Ciais, P., et al. (2014). Sharing a quota on cumulative carbon emissions. *Nature Climate Change*, *4*(10), 873–879. https://doi.org/10.1038/nclimate2384.

Roe, S., Streck, C., Beach, R., Busch, J., Chapman, M., Daioglou, V., et al. (2021). Land-based measures to mitigate climate change: Potential and feasibility by country. *Global Change Biology*, *27*(23), 6025–6058. https://doi.org/10.1111/gcb.15873.

Rogelj, J., Schaeffer, M., Friedlingstein, P., Gillett, N. P., Van Vuuren, D. P., Riahi, K., et al. (2016). Differences between carbon budget estimates unravelled. *Nature Climate Change*, *6*(3), 245–252. https://doi.org/10.1038/nclimate2868.

Saunois, M., Stavert, A. R., Poulter, B., Bousquet, P., Canadell, J. G., Jackson, R. B., et al. (2020). The global methane budget 2000–2017. *Earth System Science Data*, *12*(3), 1561–1623. https://doi.org/10.5194/essd-12-1561-2020.

Schulze, E. D., Ciais, P., Luyssaert, S., Schrumpf, M., Janssens, I. A., Thiruchittampalam, B., et al. (2010). The European carbon balance. Part 4: Integration of carbon and other trace-gas fluxes. *Global Change Biology*, *16*(5), 1451–1469. https://doi.org/10.1111/j.1365-2486.2010.02215.x.

Schulze, E. D., Luyssaert, S., Ciais, P., Freibauer, A., Janssens, I. A., Soussana, J. F., et al. (2009). Importance of methane and nitrous oxide for Europe's terrestrial greenhouse-gas balance. *Nature Geoscience*, *2*(12), 842–850. https://doi.org/10.1038/ngeo686.

SEI, IISD, ODI, E3G, and U. (2021). *The production gap report*. https://doi.org/10.1049/ep.1978.0323.

GHG inventories, budgets, and climate policy **Chapter | 1** **27**

Shen, L., Zavala-Araiza, D., Gautam, R., Omara, M., Scarpelli, T., Sheng, J., et al. (2021). Unravelling a large methane emission discrepancy in Mexico using satellite observations. *Remote Sensing of Environment*, *260*. https://doi.org/10.1016/j.rse.2021.112461, 112461.

Sheng, J., Song, S., Zhang, Y., Prinn, R. G., & Janssens-Maenhout, G. (2019). Bottom-up estimates of coal mine methane emissions in China: A gridded inventory, emission factors, and trends. *Environmental Science and Technology Letters*, *6*(8), 473–478. https://doi.org/10.1021/acs.estlett.9b00294.

Smith, P., Davis, S. J., Creutzig, F., Fuss, S., Minx, J., Gabrielle, B., et al. (2016). Biophysical and economic limits to negative CO_2 emissions. *Nature Climate Change*, *6*(1), 42–50. https://doi.org/10.1038/nclimate2870.

Stavert, A. R., Saunois, M., Canadell, J. G., Poulter, B., Jackson, R. B., Regnier, P., et al. (2022). Regional trends and drivers of the global methane budget. *Global Change Biology*, *2021*(28), 182–200. https://doi.org/10.1111/gcb.15901.

Stavros, E. N., Schimel, D., Pavlick, R., Serbin, S., Swann, A., Duncanson, L., et al. (2017). ISS observations offer insights into plant function. *Nature Ecology and Evolution*, *1*(7). https://doi.org/10.1038/s41559-017-0194.

Steininger, K. W., Meyer, L., Nabernegg, S., & Kirchengast, G. (2020). Sectoral carbon budgets as an evaluation framework for the built environment. *Buildings and Cities*, *1*(1), 337–360. https://doi.org/10.5334/bc.32.

Strauch, Y., Dordi, T., & Carter, A. (2020). Constraining fossil fuels based on 2 °C carbon budgets: The rapid adoption of a transformative concept in politics and finance. *Climatic Change*, *160*(2), 181–201. https://doi.org/10.1007/s10584-020-02695-5.

Tang, X., Hutyra, L. R., Arévalo, P., Baccini, A., Woodcock, C. E., & Olofsson, P. (2020). Spatio-temporal tracking of carbon emissions and uptake using time series analysis of Landsat data: A spatially explicit carbon bookkeeping model. *Science of the Total Environment*, *720*. https://doi.org/10.1016/j.scitotenv.2020.137409, 137409.

Taylor, T. E., Eldering, A., Merrelli, A., Kiel, M., Somkuti, P., Cheng, C., et al. (2020). OCO-3 early mission operations and initial (vEarly) XCO2 and SIF retrievals. *Remote Sensing of Environment*, *251*(September 2020), 112032. https://doi.org/10.1016/j.rse.2020.112032.

Thompson, R. L., Lassaletta, L., Patra, P. K., Wilson, C., Wells, K. C., Gressent, A., et al. (2019). Acceleration of global N_2O emissions seen from two decades of atmospheric inversion. *Nature Climate Change*, *9*(12), 993–998. https://doi.org/10.1038/s41558-019-0613-7.

Tian, H., Xu, R., Canadell, J. G., Thompson, R. L., Winiwarter, W., Suntharalingam, P., et al. (2020). A comprehensive quantification of global nitrous oxide sources and sinks. *Nature*, *586*(7828), 248–256. https://doi.org/10.1038/s41586-020-2780-0.

USCCSP. (2007). *The north American carbon budget and implications for the global carbon cycle.* SOCCR (Issue November). papers3://publication/uuid/D24EB98E-A06A-489A-9F7A-F1ABBDC42108).

Veefkind, J. P., Aben, I., McMullan, K., Förster, H., de Vries, J., Otter, G., et al. (2012). TROPOMI on the ESA Sentinel-5 precursor: A GMES mission for global observations of the atmospheric composition for climate, air quality and ozone layer applications. *Remote Sensing of Environment*, *120*(2012), 70–83. https://doi.org/10.1016/j.rse.2011.09.027.

Villalobos, Y., Rayner, P., Thomas, S., & Silver, J. (2020). The potential of orbiting carbon Observatory-2 data to reduce the uncertainties in CO_2 surface fluxes over Australia using a variational assimilation scheme. *Atmospheric Chemistry and Physics*, *20*(14), 8473–8500. https://doi.org/10.5194/acp-20-8473-2020.

Walker, A. P., De Kauwe, M. G., Bastos, A., Belmecheri, S., Georgiou, K., Keeling, R. F., et al. (2020). Integrating the evidence for a terrestrial carbon sink caused by increasing atmospheric CO_2. *New Phytologist*. https://doi.org/10.1111/nph.16866.

28 Section | A Background

White, J. C., Wulder, M. A., Varhola, A., Vastaranta, M., Coops, N. C., Cook, B. D., et al. (2013). A best practices guide for generating forest inventory attributes from airborne laser scanning data using an area-based approach. *Forestry Chronicle, 89*(6), 722–723. https://doi.org/10.5558/tfc2013-132.

Wulder, M. A., Masek, J. G., Cohen, W. B., Loveland, T. R., & Woodcock, C. E. (2012). Opening the archive: How free data has enabled the science and monitoring promise of Landsat. *Remote Sensing of Environment, 122*, 2–10. https://doi.org/10.1016/j.rse.2012.01.010.

Yang, D., Liu, Y., Feng, L., Wang, J., Yao, L., Cai, Z., et al. (2021). The first global carbon dioxide flux map derived from TanSat measurements. *Advances in Atmospheric Sciences, 38*(9), 1433–1443. https://doi.org/10.1007/s00376-021-1179-7.

Zhang, Y., Gautam, R., Pandey, S., Omara, M., Maasakkers, J. D., Sadavarte, P., et al. (2020). Quantifying methane emissions from the largest oil-producing basin in the United States from space. *Science Advances, 6*(17), 1–10. https://doi.org/10.1126/sciadv.aaz5120.

Zhang, Z., Poulter, B., Knox, S., Stavert, A., McNicol, G., Fluet-Chouinard, E., et al. (2021). Anthropogenic emissions are the main contribution to the rise of atmospheric methane (1993–2017). *National Science Review.* https://doi.org/10.1093/nsr/nwab200.

Section B

Methods

Chapter 2

CO$_2$ emissions from energy systems and industrial processes: Inventories from data- and proxy-driven approaches

Dustin Roten[a], Gregg Marland[b], Rostyslav Bun[c], Monica Crippa[d], Dennis Gilfillan[e], Matthew W. Jones[f], Greet Janssens-Maenhout[d], Eric Marland[g], and Robbie Andrew[h]

[a]Department of Atmospheric Sciences, University of Utah, Salt Lake City, UT, United States, [b]Research Institute for Environment, Energy, and Economics (RIEEE), Appalachian State University, Boone, NC, United States, [c]Department of Applied Mathematics, Lviv Polytechnic National University, Lviv, Ukraine, [d]European Commission, Joint Research Centre (JRC), Ispra, Italy, [e]North Carolina School of Science and Mathematics, Durham, NC, United States, [f]Tyndall Centre for Climate Change Research, School of Environmental Sciences, University of East Anglia, Norwich, United Kingdom, [g]Department of Mathematical Sciences, Appalachian State University, Boone, NC, United States, [h]CICERO Center for International Climate Research, Oslo, Norway

Key messages

- Country emission estimates can be obtained using fuel statistics.
- Emissions data products are available for atmospheric modeling applications, and the gridded emissions are obtained via spatial modeling, such as emission disaggregation/downscaling.
- Ideally, the inventories need to be assessed using an independent evaluation.

1 Introduction

In the latest embodiment of international accord on limiting greenhouse gas (GHG) emissions, the Paris Agreement of 2015 (UNFCCC, 2015), countries have agreed to limit emissions of GHGs to the atmosphere with the goal of minimizing the impact of human activities on the global climate system. In pursuit of this objective, they have agreed:

TABLE 1 A brief summary of the emission estimates highlighted in this section.

	Approach	Energy AD	EF/	Gridded?	Reference	Source
IEA	Sectoral and reference	IEA	IPCC	No	IEA (2020), CO_2 emissions from fuel combustion: overview	https://www.iea.org/data-and-statistics
CDIAC	Reference	UNSD	Gilfillan and Marland (2021) and Marland & Rotty, 1984	Yes	Gilfillan and Marland (2021)	https://energy.appstate.edu/research/work-areas/cdiac-appstate
EDGAR	Sectoral	IEA	IPCC	0.1	Janssens-Maenhout et al. (2019)	https://edgar.jrc.ec.europa.eu/
bp	Reference	bp	IPCC	No	Statistical review of world energy (70th edition)	https://www.bp.com/en/global/corporate/energy-economics/statistical-review-of-world-energy.html
GCP-GCB	Mixed	Various	Mixed	No	Friedlingstein et al. (2020)	https://www.globalcarbonproject.org/
GridFED	EDGAR scaled annually to GCP. Seasonality corrected annually based on heating/cooling degree days	Inherits from EDGAR	Inherits from EDGAR	0.1	Jones et al. (2021)	https://zenodo.org/record/4277267#.YN18wBNKj0o

CO₂ and other GHGs from energy and industrial process Chapter | 2   **33**

(1) to establish nationally determined contributions (NDCs) in pursuit of "holding the increase in the global average temperature to well below 2°C above preindustrial levels and pursuing efforts to limit the temperature increase to 1.5°C above preindustrial levels,"

(2) to report GHG emissions regularly, and

(3) "to communicate a nationally determined contribution every five years," as part of the Global Stocktake, with the objective "to assess progress and inform further action" and with the "view to enhancing its level of ambition" in reducing emissions.

In this context, it is important to quantify GHG emissions from human activities with accuracy and precision. We need to accurately evaluate the current emissions while anticipating the future emissions, assess progress in reducing emissions to provide a basis for further action, and be prepared for independent verification of GHG accounting. Further, GHG emissions from human activities influence global biogeochemical cycling of these gases, thus a comprehensive understanding of biogeochemical cycles requires complete and accurate accounting of all components.

This chapter is focused on the basic concepts of estimating and evaluating GHG emissions from their largest anthropogenic sources: fossil fuel-based energy systems and industrial processes. The chapter has a special focus on carbon dioxide (CO_2) since the fundamental approach used for emission calculations is consistent across other gases. We know, for example, that CO_2 is emitted when fossil fuels are oxidized and when certain industrial processes are carried out, and we evaluate the magnitudes of these processes. In this chapter, we do not deal with the human role in forest degradation and land-use change (these are addressed in Chapter 3).

To date, commitments to reduce emissions have been largely at the national level; however, efforts are also coming from states, cities, and corporations. There are important issues at multiple spatial and temporal scales. Monitoring, reporting, and verification (MRV) depends on observations and data at multiple levels: nations, individual facilities, and perhaps even communities or households. We recognize that global corporations have multinational emissions footprints, and that there are critical issues of equity and fairness in reducing global total emissions. Emissions in one country, for example, may be from producing products that will be consumed in another country (Peters, Davis, & Andrew, 2012). Global solutions to climate change will be facilitated by emissions inventories at spatial scales from continents to communities and individual facilities and neighborhoods. We need to understand emissions at the temporal scales of data collection—years for many survey methods, hours for space-based observations. Inventories need to have well-defined spatial and temporal boundaries and these will depend on the question being posed.

Emissions inventories are now carried out in many countries, as required in the UNFCCC, and there are several independent organizations that estimate emissions globally or regionally. Some emission estimates are made as a part of research projects. Many of the inventories are in different forms and/or with different boundaries. Given the challenge of covering all the emission estimates and their variants, we focus on a subset of the emission estimates and products to give an overview of those different elements. We believe these examples capture the diversity of the emission estimates and products and provide basic information for those who produce or use emissions inventories. Users of specific emission estimates and data products should refer to the suggested documentation.

This chapter provides a short overview of the general inventory approach, introduces some major sources of emission estimates (see Table 1)—such as the International Energy Agency (IEA), the US Department of Energy/Energy Information Administration (US DOE/EIA), the Carbon Dioxide Information and Analysis Center (CDIAC), and the Emissions Database for Global Atmospheric Research (EDGAR)—which have been recognized as primary sources of emissions data and are frequently seen in major reports and scientific studies as the principal primary international datasets on global CO_2 emissions. They differ in elements such as the sources of fuel consumption data; the time period of coverage; the emissions factors used for converting fuel consumption to CO_2 emissions; the reporting of emissions by fuel or economic sector; the inclusion of industrial processes, international transport fuels, and gas flaring; and spatial and temporal coverage. While these datasets differ in detail and emphasis, they are generally complementary and provide a way to quantify uncertainty. The goal of this chapter is to provide an overview of these data sources so that users can discern which data product is best for their research needs. In addition, this chapter discusses the uncertainties associated with CO_2 emissions estimates.

2 Overview of inventory approaches

2.1 Emission estimation from energy statistics

CO_2 emission estimates from different sources differ in objectives and details, but they all adhere to the basic principle that CO_2 emissions (from fossil fuels, for example) can be estimated as the product of the amount of fuel burned and the carbon content of that fuel—since CO_2 is a necessary and stable product of fossil fuel combustion (e.g., Andres et al., 2012; Eggleston, 2006; Gilfillan & Marland, 2021; Marland & Rotty, 1984). This can be described in a general form (e.g., IPCC Guidance for National Greenhouse Gas Inventories, 2006—see Eggleston, 2006) as

$$E = AD \times EF$$

where AD is the activity data, which indicate the amount of fuels combusted and EF is the emission factor that converts the fuel amount into the emissions of

CO$_2$ and other GHGs from energy and industrial process Chapter | 2   **35**

interest. A value for *AD* and a value for *EF* are required for each different fuel. In the calculation of CO$_2$ emissions from the energy system (fuel combustion), *AD* should be a good estimate of the fuel amount oxidized and capture all the CO$_2$ from the potential carbon sources in the system of interest. Thus, *AD* defines the spatial and time extent of the resulting emission estimates. For example, Marland and Rotty (1984) proposed the method to calculate global, annual total CO$_2$ emissions using fuel production statistics. The basic assumption is that all of the fuels produced in the year of interest are oxidized in that year or can be accounted for as changes in the stocks on hand. Any other portions of the fuels that are not oxidized and the carbon not released as CO$_2$ into the atmosphere (because of incomplete combustion or inclusion into long-lived products) are factored into the *EF*.

The IPCC Good Practice Guidance, which provides general but extensive guidance to the compilation of national annual emission inventories, recommends fuel statistics (e.g., IPCC, 2006) as AD and calculates annual emissions assuming the fuels are oxidized in the year of interest. Due to the economic incentives, reasonable domestic fuel statistics are available for most countries, and the import/export of fuels is also reported regularly. Annual emissions from a specific sector can be estimated using domestic (or sectorally disaggregated) fuel statistics or energy balances. For example, the IPCC guidelines recommend the use of the fuel-sale statistics for estimating emissions from the transportation sector. Emission estimation can be done in the same way for smaller regions, such as for provinces or states (e.g., EPA), or for cities (Global Carbon Project (GCP) protocol). However, the collection of data on AD that matches the system boundaries can be challenging as the system boundaries are not as clear for other designations as for countries. For smaller system boundaries or time intervals, one could do an emission estimation using a region-specific *EF* and potentially do detailed calculations for smaller entities.

A challenge for policy-relevant emission inventories is the time required to compile the requisite activity data. The United Nations energy statistics, for example, are based on questionnaires submitted to individual countries and these in turn are based on in-country surveys of fuel producers and consumers. The consequence is that UN energy data tend to lag 2–3 years behind the current calendar year. Also, as countries accumulate and refine energy data, there are routinely extensive revisions that are reported in years following the initial report. For example, Marland, Hamal, and Jonas (2009) show that for Austria, data are refined for many years after initial reporting as activity data are refined. Different countries, and different compilations of energy data, are more inclined toward subsequent revisions and/or use estimation techniques to provide informed estimates of more recent time periods. The bp Statistical Review of World Energy, for example, provides annual estimates of energy consumption that reports more recent estimates than most of the other international compilations. In recent years, various groups have been reducing this lag, including

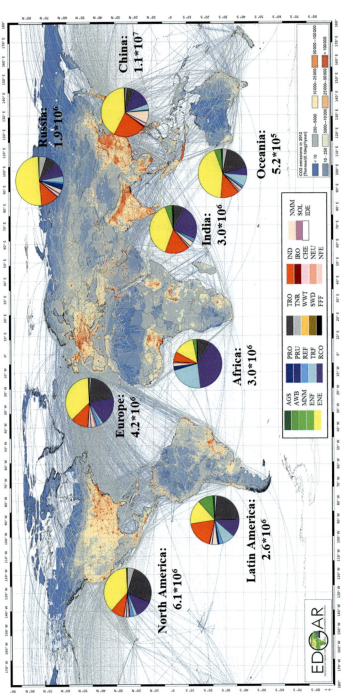

FIG. 1 EDGAR, as an example of global, gridded estimates of CO_2 emissions. Gridding here is at a scale of 0.1×0.1 degree resolution. CO_2 emission grid map and relative contribution of the EDGAR sectors in world regions (pie charts) for 2012. The legend for the pie charts relates to the EDGAR sectors is: *AGS*, agricultural soils; *AWB*, agricultural waste burning; *CHE*, chemicals; *ENE*, power industry; *ENF*, enteric fermentation; *FFF*, fossil fuel fires; *IDE*, indirect emissions; *IND*, manufacturing industry; *IRO*, iron and steel; *MNM*, manure management; *NEU*, nonenergy use; *NFE*, nonferrous metals; *NMM*, nonmetallic minerals; *PRO*, fuel production; *PRU*, production and use of products; *RCO*, residential; *REF*, oil refineries; *SOL*, solvents; *SWD*, solid waste disposal; *TNR*, nonroad transport; *TRF*, transformation industry; *TRO*, road transport; *WWT*, waste water. The represented CO_2 emissions also include those from short-cycle carbon (i.e., of e.g., biofuel combustion and agricultural waste burning).

CO₂ and other GHGs from energy and industrial process **Chapter | 2** **37**

the publication of estimates for the current year before it has finished (e.g., Friedlingstein et al., 2020; IEA, 2021a, 2021b).

Beyond the amount of CO_2 emitted from the oxidation of fossil fuels, the largest source of human-driven CO_2 emissions (not including biomass burning) is from the industrial decomposition of carbonate minerals. The largest portion is generated during the manufacture of cement, where calcium carbonate ($CaCO_3$) is calcined at high temperature to release CO_2 and yields calcium oxide (CaO), or clinker, which is the primary component of cement. Estimates of CO_2 emissions thus sometimes include emissions from cement manufacture (e.g., EDGAR, CDIAC, and GCP's Global Carbon Budget) and sometimes stay with only emissions related to fossil fuels (e.g., IEA and US DOE/EIA).

2.2 Activity data

The spatial extent or spatial scale (also the time scale) of emission inventories is established largely by the activity data available. Furthermore, the quality of the available AD is the key for obtaining reliable estimates. Data completeness and quality vary across countries and other spatial determinants. The IPCC (2006) provides good practice guidance to help countries compile and report reasonably reliable and comparable inventories regardless of the challenges.

However, data collection across time and countries by a single organization requires some uniformity. The IEA and UN have been the two major sources of global energy data (IPCC, 2006). The global reference for energy statistics definitions and methodologies is now the International Recommendations for Energy Statistics (IRES), a foundation that was adopted by several agencies in 2011. The IPCC and UNFCCC have also been part of the group of organizations (InterEnerStat) that have conducted a 10-year consultation to agree on definitions of energy products and flows so that data can be comparable across agencies, countries, and time. The IRES has been adopted by the United Nations Statistical Commission and is available online. The IRES ensures that the different formats and definitions used by countries are harmonized to produce internationally comparable energy balances, to the extent that the requisite data exist at the national level and are shared with data agencies such as the IEA and the UN.

A particular focus of the IEA is on capacity building in countries with weak statistical management for energy statistics (e.g., methodological work like manuals; training events; webinars; etc.). The IEA, like the UN, welcomes feedback on energy statistics data quality from users and data providers, constantly striving to improve the quality of global data. In terms of international cooperation, the IEA cooperates with the IPCC on data-related tasks including preparation of computation guidelines, running workshops on data management and emission factors, sharing data, etc. The IEA works with the United Nations Framework Convention on Climate Change (UNFCCC) by reviewing energy statistics submitted in official national reports of GHG emissions.

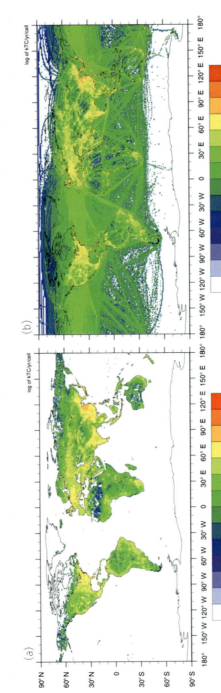

FIG. 2 CDIAC emissions (part A) disaggregated within countries using data on population density (Andres et al., 1996; Andres & Boden, 2016) and (part B) using satellite-observed night lights plus data on large point sources and other proxy data (Oda, Maksyutov, & Andres, 2018). Note the CDIAC (part A) does not include emissions from international bunker fuels (i.e., international transport). *(From Oda, T., Maksyutov, S., & Andres, R.J. (2018). A global monthly fossil fuel CO_2 gridded emissions data product for tracer transport simulations and surface flux inversions. Earth System Science Data, 10(1), 87–107. https://doi.org/10.5194/essd-10-87-2018.)*

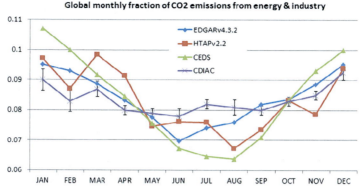

FIG. 3 Comparison of the monthly fractions of CO_2 emissions from energy and industry applied by EDGAR v4.3.2 (Janssens-Maenhout et al., 2019), with those derived by Andres et al. (2011) for CDIAC, Hoesly et al. (2018) for CEDS, and Janssens-Maenhout et al. (2015) for HTAPv2.2.

The United Nations Statistics Division (UNSD) collects data on energy production, processing, and use for all countries, with data back to 1950. The data are updated annually with a number of revisions and additions for previous years. The UNSD also provides country-year-specific conversion factors for solid fuels to convert to energy units. The UNSD data collection is global and their focus is on energy production, while the IEA tends toward a focus on energy consumption.

The annual publication of bp's Statistical Review of World Energy has been a widely used dataset, particularly due to its timeliness, i.e., the inclusion of the two most recent years that are not often covered by other international data sources. Their updates on the energy data are often used to provide emissions data estimates that bring other time series data up essentially to the present.

2.3 Emission factors

Emission factors (EFs) are conversion factors to obtain CO_2 emission estimates from activity values indicated by AD. EFs can be decomposed into the oxidation ratio (i.e., what fraction of the carbon in the fuel was oxidized and released as CO_2) and the carbon content (i.e., how much carbon is included in the fuel). The IPCC's EFs for the many different fuels are stored at the Emission Factor Database (EFDB, https://www.ipcc-nggip.iges.or.jp/EFDB/main.php). The emissions factor for a given fuel can vary slightly over time or from country to country, for example, as the composition of coal varies, but relatively constant default values can be derived when the default value, for fuels, is expressed in units of mass of carbon per unit of energy released.

2.4 Spatial and temporal emission disaggregation

The original scope of the calculations and IPCC guidelines was at the global and country levels on annual time scales. However, with appropriate consideration of spatial and temporal boundaries, the approach can be applicable at other scales such as provinces or cities. Some emissions estimates are provided by estimating the subsystem-level spatial and temporal distributions and this can be done in a gridded and/or vector data form. However, these emission estimates are often downscaled in space and/or time from the primary country/annual scale estimates. Especially for atmospheric modeling applications, this is often a necessary step. For convenience in this chapter, we define these downscaled estimates to be emission products.

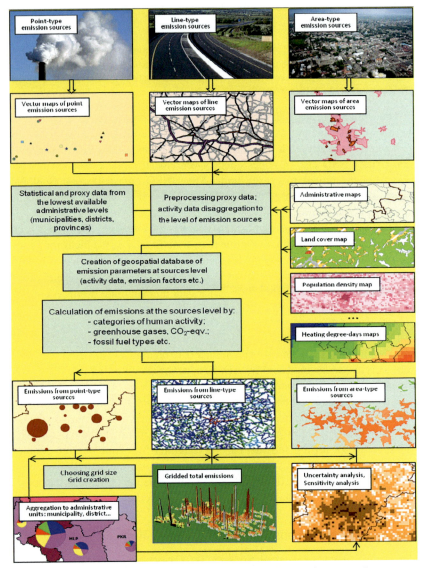

FIG. 4 The GESAPU modeling framework. Represented here are the main stages and components of a bottom-up estimation of GHG emissions at a fine spatial scale.

For example, Andres, Marland, Fung, and Matthews (1996) created estimates of the spatial distribution of emissions at 1-degree latitude by 1-degree longitude resolution by assuming a correlation between human population density and CO_2 emissions. Shown in Fig. 1, the EDGAR (Janssens-Maenhout et al., 2019) distributes emissions on a fine spatial grid, but employs a variety of proxy spatial data that correlate with CO_2 emissions, such as power plant and industrial facilities locations, road network, rural and urban population distribution, etc. (refer to Table S4b of Janssens-Maenhout et al., 2019). Emissions from nonstationary sources and from line and area sources such as highways and residential areas are difficult to allocate spatially but can be done with data such as highway traffic counts and census data. Spatial emission downscaling of sector-specific emissions (emission spatial disaggregation) is a common way to estimate the fine-scale spatial emission distribution within a system boundary. Population, satellite-observed night lights, GDP, census data, etc. are commonly used for the spatial disaggregation of emission estimates. In order to estimate human emissions at high spatial and temporal resolution, a combination of proxy geospatial data is often used. See Fig. 2 for proxy decomposition. For a list of inventories mentioned in this chapter along with AD and EF sources, refer to Table 1.

Similarly, annual emissions are often disaggregated in time in order to achieve emissions at shorter time scales, such as monthly (e.g., Andres, Gregg, Losey, Marland, & Boden, 2011; Andrew, 2020b), weekly, and hourly time scales (e.g., Crippa et al., 2020; Nassar et al., 2013). The temporal changes in CO_2 emissions are closely related to patterns of human activities. For example, emissions from domestic heating are higher in cold months. The energy demand for cooling can cause an emission peak in the summer time in hot areas. Road transport emissions are expected to show diurnal and weekly cycles that are closely related to commuting cycles.

An early estimate of the monthly cycle of CO_2 emissions in the United States was prepared by Blasing, Broniak, and Marland (2005) using monthly data on energy consumption, but most estimates at scales less than a year have required using approximations from limited data or proxy data to represent the temporal variability. Andres et al. (2011) collected data from major fuel-use patterns in 20 countries to approximate the annual pattern of CO_2 emissions globally. Andrew (2020b) developed estimates of India's CO_2 emissions with low temporal lag using detailed monthly activity data. Nassar et al. (2013) created a global gridded dataset of the scaling factor to derive weekly and hourly emissions estimates. The sectoral emission temporal profiles were taken from the Vulcan US data (Gurney et al., 2009) and the sectoral profiles were allocated using the EDGAR sectoral composition. Recently, Crippa et al. (2020) created extensive high time-resolution profiles for air pollutants and greenhouse gases coemitted by anthropogenic sources within the EDGAR datasets in support of atmospheric modeling, Earth observation communities, and decision makers (Crippa et al., 2020; JRC, 2020). Compared to other global inventories, the key novelties of the EDGAR temporal profiles are the development of

42 Section | B Methods

country/region- and sector-specific yearly profiles for all sources, time-dependent yearly profiles for sources with interannual variability of their seasonal pattern, and country-specific weekly and daily profiles to represent hourly emissions. In particular, regional yearly profiles are defined mainly based on climate zones, heating degree days since weather conditions strongly affect the energy consumption and emissions, and ecological zones (i.e., the temporal variation of activities and emissions can also be different among various ecological zones, especially for agriculture and biomass burning). A comparison of the EDGAR v4.3.2 monthly profiles (Janssens-Maenhout et al., 2019) and those used for the CDIAC (Andres et al., 2011), CEDS (Hoesly et al., 2018), and HTAP (Janssens-Maenhout et al., 2015) is given in Fig. 3.

2.5 Other approaches

While the large-scale emission estimation is more or less the same, subnational emission estimation has more diversity in the methodologies used. For example, traffic emissions can be estimated using the vehicle miles traveled (VMT; also known as VKT, vehicle kilometers traveled) and the appropriate emission factors (Gately, Hutyra, & Sue Wing, 2015). This would not get the benefit of economic intensives, but VMT data should provide the direct regional or local estimates of emissions. This might be a better direct emission estimate for a shorter time scale and smaller areas, as the data come with spatial information. Another way is to use the measurement of associated species such as CO to estimate CO_2 emissions (Gurney et al., 2009).

Especially at higher temporal scales, hybrid approaches are useful to reduce the degree of disaggregation by directly calculating emissions at the location of sources (e.g., Bun et al., 2018). For example, in the model GESAPU emissions data, separate sources are calculated using both statistical and proxy data. Statistical data (the amount of fossil fuels used for industrial production, etc.) from the lowest available administrative level (ideally municipality) decrease the depth of disaggregation and the errors associated with this procedure. Depending on emission category, proxy data might include the capacity of electricity generation and heat production, population density, access to energy sources, heating degree days, the gross value of production in the industry sector, the numbers of cars and road categories (see, for example, Bun et al., 2018), etc. At the final stage, the emission data from all emission categories are combined to calculate total emissions. A vector grid map is used, where each cell is represented as a polygon feature. Any grid size can be chosen as long as it is larger than the lowest resolution of area-type emission sources. As a case study, Bun et al. (2018) and Charkovska et al. (2018) calculated the emissions of CO_2, CH_4, N_2O, SO_2, CF_4, C_2F_6, NO_x, and NMVOC according to this approach from all point, linear, and area sources in Poland separately for more than 100 categories of human activity covered by the IPCC guidelines (Fig. 4).

CO₂ and other GHGs from energy and industrial process **Chapter | 2** **43**

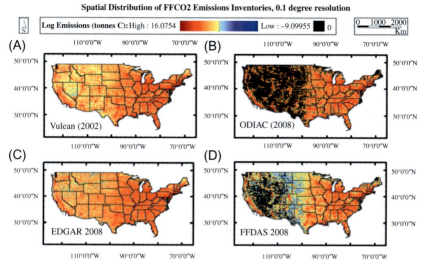

FIG. 5 Presented here are four spatial distributions of fossil fuel emissions across the United States. Selected inventories and years include: (A) Vulcan, 2002, (B) ODIAC, 2008, (C) EDGAR, 2008, and (D) FFDAS, 2008. All inventories are distributed at 0.1° resolution with units expressed as the natural log of tons of carbon. Zero values are presented in *black*. *(From Hutchins, M. G., Colby, J. D., Marland, G., & Marland, E. (2016). A comparison of five high-resolution spatially-explicit, fossil-fuel, carbon dioxide emission inventories for the United States. Mitigation and Adaptation Strategies for Global Change, 22(6), 947–972. https://doi.org/10.1007/s11027-016-9709-9.)*

3 Uncertainty

3.1 Emission estimation

Understanding uncertainty is important in the study of climate change. While significant uncertainty exists, much is known about the climate system and much of the uncertainty can be managed if we are well informed. We can reduce uncertainty by carefully defining the system, cooperating with other parties to make sure common system boundaries are used, and taking responsibility for reducing leakage in the system. We also can use our knowledge of the various components of uncertainty to target study areas where uncertainty can be reduced as efficiently as possible. The conservative approach is to tend toward over estimation of uncertainty (i.e., to underestimate our ability to measure and evaluate).

In general, due to the simple calculation and the nature of CO_2 emitted by fossil fuels (FFCO$_2$), the uncertainty associated with annual emissions can be relatively small. For example, the percentage uncertainty associated with the

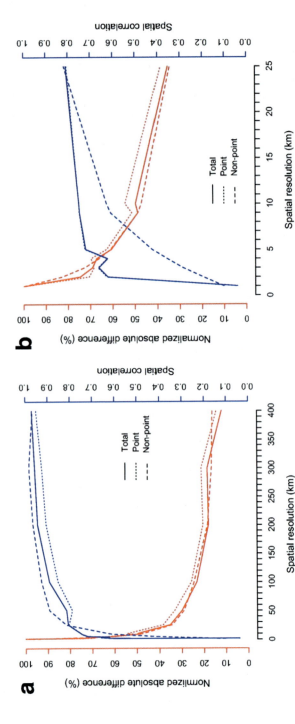

FIG. 6 Difference between ODIAC and GESAPU as a function of spatial resolution. Data are from Ukraine. *(From Oda, T., Bun, R., Kinakh, V., Topylko, P., Halushchak, M., Marland, G., et al. (2019). Errors and uncertainties in a gridded carbon dioxide emissions inventory. Mitigation and Adaptation Strategies for Global Change, 24(6), 1007–1050. https://doi.org/10.1007/s11027-019-09877-2.)*

global total of CO_2 emissions has been estimated at 8.4% (2σ, Andres, Boden, & Higdon, 2014). The percent uncertainty associated with the estimates is calculated by following equation:

$$U = \sqrt{U_{AD}^2 + U_{EF}^2}$$

Here is a caveat. The emission calculation is just a multiplication of two or three terms. Thus, the evaluation of the AD and EF is extremely important to reduce the potential uncertainties. The uncertainty calculation may not capture the errors and/biases or uncertainties associated with AD or EF; therefore, more advanced techniques, such as Monte Carlo simulations, may be used to incorporate additional uncertainties.

3.1.1 Uncertainty in AD

To understand the uncertainty in emissions, we first deal with system boundaries. A careful description of what it is that we are measuring is needed and special care to make sure we do not double count or miss any components must be taken. For anthropogenic emissions, we need to account for the flow of all of the carbon in the defined system—all of the possible ways that carbon enters the system and all of the possible ways that carbon can leave the system. This becomes particularly important when goods are traded between accounting parties. If petroleum is produced in Country 1, final products are produced in Country 2, and those products are sold to Country 3, it is important to track the flow of carbon. Note that the uncertainties outlined here are based on creating an accurate inventory. Determining accountability and attributing credit for the release and sequestration of the carbon is a separate consideration.

In pursuing emissions reductions over time, there has been a marked increase in emissions inventories at smaller and smaller scales, in both time and space. Pushing toward smaller scales presents many of the same challenges that we face at the larger scales—carefully define the system and keep careful track of carbon entering and leaving the system. While we might start with a national-level uncertainty of 2%–3%, the uncertainty for a city or county might reach values over 100% (Gurney et al., 2020). If we want to understand emissions in a city, we need to consider things like where people live, where people buy gas, and where the electric power is produced. Several projects have pursued careful development of city-scale inventories, but these are expensive (Gurney et al., 2019; Kona et al., 2021). Proxy data such as population density or the intensity of satellite-observed night lights are often used to stand-in for energy consumption or CO_2 emissions. These approximations are likely reasonable, but the conservatively calculated uncertainty is greater.

46 Section | B Methods

At the country scale, data collection is often very good and the uncertainties well defined (especially where taxable transactions occur). The import and export of carbon containing products is tracked carefully. At smaller spatial scales, data collection and analysis become increasingly labor and cost expensive. For these targeted programs, it is possible to estimate CO_2 savings even though a full inventory can be very complex.

But the consistency needs to be checked in quality. In China, for example, the difference between the sum of the province-level emissions estimates and the country-level emission estimates has been significant and this has been interpreted to suggest data biases (Guan, Liu, Geng, Lindner, & Hubacek, 2012).

3.1.2 Uncertainty in EF

Given the diversity of the emission estimation methods and the data used, it is critical that the emission developers should perform careful quality checks and uncertainty analyses (e.g., IPCC, 2006). Here, we discuss the uncertainties along with the methodologies discussed earlier, see Liu et al. (2013).

3.2 Spatial and temporal modeling

Since the spatial and temporal emission disaggregation is an independent procedure from the overall emission calculation, these processes will be additional sources of uncertainty. Errors are unique to the proxy data and/or modeling approaches used (see Fig. 5). Currently, it is difficult to determine which inventories are close to "truth" due to the lack of evaluation data (Andres & Boden, 2016; Oda et al., 2018). Thus, the error assessment is often done just using the spread of multiple estimates as a proxy for the emission uncertainty. An exception is Bun et al. (2018). CO_2 spatial estimates are characterized by multiple sources of uncertainty: geolocation uncertainty of emission sources/sinks; aggregated statistical data uncertainty; uncertainty of proxy data magnitude and location; uncertainty in the proxy data representation; and uncertainty of the emission factors (Fig. 6).

For point sources, Woodard et al. (2015) noted that uncertainties in magnitude and location were of concern. They estimated that the US eGRID dataset geolocation error is about 0.8 km. The errors could be mitigated by additional attention to data collection. Oda et al. (2019) suggest that the point source geolocation errors can be due to the errors or lack of completeness in the underlying power plant databases. The uncertainty connected with the geolocation of large point sources plays an important role in some applications (Hogue, Marland, Andres, Marland, & Woodard, 2016; Hogue, Roten, Marland, Marland, & Boden, 2017). For inventories at all scales, detailed data on emissions from these sources can be useful. Since large point source emissions values are reported every year and incorporate a large fraction of total emissions, accurate locational data on these sources could be a relatively inexpensive way to reduce overall uncertainty.

Similarly, errors in temporal modeling could be evaluated. As with spatial differences, temporal differences could also be used as an additional proxy for uncertainty in large-scale comparisons. However, the examination of submonth temporal emission changes across different sectors is difficult, mainly due to the lack of evaluation data. The level of the uncertainty should correlate with that of AD. The higher temporal resolution, then the more uncertainty.

3.3 Other sources of uncertainty

As mentioned earlier, the uncertainty assessment method suggested by the IPCC may not fully capture all possible sources of uncertainties. While the IPCC methodology allows for capturing all of the sometimes subtle sources of uncertainty, they may not be used consistently or detailed transparently in the reporting. For example, a detailed study by Quick and Marland reviews the difference in uncertainty in power plant emissions depending on the method of measuring, whether by stock change or stack measurement (Quick & Marland, 2019). Other uncertainties might arise from the lack of homogeneity of the quality of coal being mined in a certain region. Some emissions inventories are prone to systematic errors due to aspects such as self-reporting misunderstandings, typos, or even translation errors (Oda et al., 2019). The use of atmospheric measurements can be very important to resolving those uncertainties. While this is challenging for CO_2, there has been some success for CH_4 and N_2O (Calvo Buendia et al., 2019).

4 Examples of emission estimates and products

Here, we discuss the sources of emission estimates that were mentioned in this chapter. Since emissions reported to the UNFCCC do not fully cover the entire world, the emission estimates listed here are often used for whatever analyses need to be done. Table 1 shows the brief summary of the emission estimates introduced here. See also Andrew (2020a) for a recent comparison of these and other datasets.

4.1 IEA

The IEA produces global estimates of energy-related CO_2 emissions based on its own data from national energy balances and the 2006 IPCC Guidelines for estimating CO_2 emissions, Tier 1 methodology (IEA, 2020). The IEA's energy data are sourced both directly from countries and, for a number of smaller countries, from the UN. Emissions are reported by economic flows and detailed energy products. Emission estimates are not official national figures, but they provide consistent values for comparing across countries and over time.

The IEA has been collecting energy statistics since its foundation in 1974, and their data now cover energy supply and demand for over 150 countries/regions worldwide; emissions were first reported in 1997 (Andrew, 2020a). The IEA data collection is based partly on direct reporting requirements for OECD member countries, individual contacts with country data providers, and other access to national reporting. Sources are listed in the documentation file of the World Energy Balances (IEA, 2020).

4.2 bp

The Statistical Review of World Energy published annually by energy company bp p.l.c. has been another source of global statistical data for CO_2 emissions estimates (bp, 2021). The data include fuel consumption and production as well as key variables useful for fossil emission estimations since 1965. Energy data are sourced directly from countries, although sources are not publicly documented. First produced in 1952 the Review introduced estimates of CO_2 emission for total oxidation of energy-related fossil fuel use using the default EF values provided by the IPCC and following the IPCC guidelines. The IEA's nonenergy fractions are used to remove nonenergy uses of fossil fuels, which mean that most oxidation emissions are included (such as iron and steel), but some are excluded (such as decomposition of carbon anodes in aluminum smelting) (pers. comm., bp, March 2021). However, these national estimates also include emissions from bunker fuels, in conflict with the IPCC guidelines (Andrew, 2020a). Emissions from flaring were introduced in the 2021 edition. Due to the timely availability of data—usually in June reporting up to the previous year—the energy data are often used to project emission estimates for the most recent past year (e.g., Friedlingstein et al., 2020; Myhre, Alterskjaer, & Lowe, 2009; Oda et al., 2018).

4.3 CDIAC-FF

The Carbon Dioxide Information and Analysis Center (CDIAC) at the US Department of Energy's Oak Ridge National Laboratory (ORNL) has maintained estimates of CO_2 emissions since 1984. The CDIAC estimates of global total emissions have been used as a primary estimate for carbon budget analyses, such as the Global Carbon Project's annual carbon balance (Friedlingstein et al., 2020) and atmospheric inversions (Gurney et al., 2002). The CDIAC estimates are based on activity data from the UNSD, using fuel production data to estimate global emissions and data on apparent consumption to estimate national emissions. The methodology was developed by Marland and Rotty (1984), and has recently been updated (Gilfillan & Marland, 2021) after the data management migrated to the Appalachian State University. The two notable changes to the methodology are the inclusion of stock changes

CO$_2$ and other GHGs from energy and industrial process **Chapter | 2** **49**

in global calculations of CO$_2$ emissions and new assumptions concerning the amount of clinker contained in cement.

The dataset includes emissions estimates by country and fuel beginning in 1751. A gridded product has been produced by Andres et al. (1996) and Andres and Boden (2016) using population density as a proxy for the distribution within countries. The CDIAC first developed the monthly FFCO$_2$ emissions estimates based on the analysis of energy data for the United States (Blasing et al., 2005) and then globally by Andres et al. (2011). The uncertainty has been estimated at 8.4% for the global total (Andres et al., 2014). The data from the ORNL CDIAC are currently archived by the DOE's Environmental Systems Science Data Infrastructure for a Virtual Ecosystem (ESS-DIVE) data repository as well as the newer estimates developed at the Appalachian State University. There is also a high-resolution downscaled emission product called ODIAC (https:// db.cger.nies.go.jp/dataset/ODIAC/) that uses data on large point sources and within-country spatial distributions based on night lights data (Oda & Maksyutov, 2011; Oda et al., 2018).

4.4 EDGAR

The Emissions Database for Global Atmospheric Research (EDGAR) is a global emission inventory of GHGs and air pollutants developed by the Joint Research Centre (JRC) of the European Commission (EC) (https://edgar.jrc. ec.europa.eu/index.php). The EDGAR includes a time series of emissions esti-mates, starting in 1970 and up to most recent years (i.e., the current year minus one for CO$_2$, as presented by Crippa et al., 2020, and the current year minus four for non-CO$_2$ GHGs and air pollutants, as presented by Crippa et al. (2018). The EDGAR applies the same methodology for the emission computation, at the country and annual level, providing consistent and comparable estimates for all world countries. The EDGAR emissions estimates are computed using activ-ity data from international statistics (e.g., fuel balances from the International Energy Agency (IEA), agriculture statistics from Food and Agriculture Orga-nization (FAOSTAT, 2021), cement data from the US Geological Survey data (USGS, 2020), etc.)) and following the IPCC guidelines (IPCC, 2006) for the GHGs (Janssens-Maenhout et al., 2019) and the EMEP/EEA guidebook (European Environment Agency, 2019) for the air pollutants (Crippa et al., 2018), covering all anthropogenic emission sectors with the exception of Land Use, Land Use Change and Forestry (LULUCF). GHG emission uncertainty is addressed by Solazzo et al. (2020) following the IPCC guidelines (IPCC, 2006). Once the emission database is compiled for all countries, sectors, and pollutants, annual emission data are disaggregated to monthly emissions applying sector and country-specific yearly emission profiles (Crippa et al., 2020). In addition, monthly emissions are spatially distributed over global grid maps with a reso-lution of 0.1 degree $\times$ 0.1 degree making use of sector-specific spatial proxies

(Crippa, Guizzardi, et al., 2021; Janssens-Maenhout et al., 2019) as input for atmospheric modelers.

The EDGAR data are used as the default emission inventory in air quality modeling projects (e.g., Hemispheric Transport of Air Pollution (HTAP_v1 and HTAP_v2) (Janssens-Maenhout et al., 2015), FP7 PEGASOS (Crippa et al., 2016), etc.) and model intercomparisons (e.g., HTAP, AQMEII, EURO-DELTA, etc. (Solazzo, Bianconi, et al., 2017; Solazzo, Hogrefe, Colette, Garcia-Vivanco, & Galmarini, 2017)). The EDGAR emission grids have also been used by the Global Carbon Project as a priori fluxes to run inverse atmospheric modeling (Le Quéré et al., 2018). The EDGAR is widely used in support of policy design and treaty compliance, by the Intergovernmental Panel on Climate Change (IPCC AR5 and AR6), and for emission verification (http://verify.lsce.ipsl.fr/). The latest development of the EDGAR includes the EDGAR-FOOD global database (Crippa, Solazzo, et al., 2021), providing a picture of how an evolving world food system has responded to the evolution of world population in the last decades, together with changes in dietary habits and food-related technology. The dataset is being used in the Sixth Assessment Report of the IPCC.

4.5 Global carbon project

The Global Carbon Project (GCP) was established in 2001 with the aim "to work with the international science community to establish a common and mutually agreed knowledge base to support policy debate and action to slow down and ultimately stop the increase of greenhouse gases in the atmosphere" (GCP, n.d.). The GCP produces an annual Global Carbon Budget (GCB), which presents estimates of both sources of carbon (fossil and land-use change) sources and sinks (atmosphere, terrestrial, ocean). The fossil CO_2 component of the GCB began as a direct replication of the CDIAC extended by bp for the final years, but has since developed to include national official estimates as well as specific country-level improvements, with the CDIAC still forming the core dataset over which data assumed to be superior are written (Andrew, 2020a, 2020b; Friedlingstein et al., 2020). Further, to create time series more useful to current countries, some countries that came into existence in recent years—such as the countries of the former Soviet Union—are disaggregated out of their united origins. Data are published for the period 1750 to the year before publication, for eight categories including "other" for noncement carbonate decomposition (e.g., lime, glass, and ceramics production).

4.6 GCP gridded fossil emissions dataset

The GCP Gridded Fossil Emissions Dataset (GCP-GridFED, Jones et al., 2021) is a 0.1 degree × 0.1 degree gridded global dataset of monthly CO_2 emissions resulting from fossil fuel oxidation and the calcination of limestone during cement production since 1959. Emissions estimates are provided for five source classes, such as coal, oil, gas, international bunkers, and cement. The GCP-GridFED grids are consistent with the national and source-specific emissions inventories compiled by the GCP in its annual assessment of the global carbon budget (GCB, e.g., Friedlingstein et al., 2020).

GCP-GridFED is explicitly designed to support the contribution of inversion models to the annual GCB assessment. The inversion models require a prior constraint on fossil CO_2 emissions, and they correspondingly estimate the sinks of CO_2 to the land and oceans by accounting for atmospheric transport and observations of CO_2 concentrations. The key benefits of using GCP-GridFED as a prior in inversion models are compliance with the GCP's national emission estimates and its long time series, which means that the inversion model estimates of the land and ocean sinks can be compared directly with estimates based on other approaches (e.g., process-based models) over six decades. Following the introduction of GCP-GridFED in the 2020 edition of the GCB, the spread of the land and ocean sink estimates was reduced among the inversion models in comparison to previous years (Friedlingstein et al., 2020).

GCP-GridFED scales monthly gridded emissions for the year 2010, from EDGAR v4.3.2 (Janssens-Maenhout et al., 2019), to the national annual emissions estimates compiled by GCP (Friedlingstein et al., 2020). In all, 18 activity sectors of the EDGAR dataset are included, collectively representing fossil fuel combustion, the noncombustion use of fossil fuels, and cement production. The EDGAR emissions estimates were summed to the five source classes used by GCP (coal, oil, gas, international bunkers, and cement), and thereafter summed within national borders and scaled to the GCP estimates of national emissions for each year since 1959.

GCP-GridFED also adopts the monthly distribution of annual emissions from EDGAR v4.3.2 at the grid cell level, with modifications in each year based on meteorological reanalyzes of heating/cooling degree days. Relationships between monthly emissions from a subset of the EDGAR activity sectors and monthly heating/cooling degree days are derived for each grid cell in the year 2010 (Harris, Osborn, Jones, & Lister, 2020; Spinoni et al., 2017). The monthly distribution of annual emissions is then adjusted in all other years based on reanalysis of monthly heating/cooling degree days.

GCP-GridFED also includes gridded uncertainty estimates that respect differences in uncertainty across country groups and across sectors in national emission inventories. The reported uncertainties include uncertainty in total fossil CO_2 emissions of 5% for Annex I nations and 10% for others (Friedlingstein et al., 2020), while the uncertainty in specific sectors is larger and sums in quadrature to the total uncertainty at the national level (Marshall & Ramirez, 2019). In GCP-GridFED, like most gridded emission inventories, the uncertainty estimates do not include uncertainties associated with the spatial or temporal (monthly) disaggregation of national emissions to the cell level.

5 Summary

This chapter provides an overview of the inventory approaches used for CO_2 emissions from the energy sector, and discusses some data sources as examples. While this chapter focuses on CO_2, the calculations for the same emission sectors for CH_4 and N_2O are done in the same way (but with appropriate EFs). Due to the nature of $FFCO_2$, calculation of the annual country $FFCO_2$ can yield robust estimates in general with the simple, yet established calculation method. However, the subnational emission estimations are more challenging and could be subject to errors and uncertainties due to the less robust values for AD and the demands for establishing clear accounting boundaries. The emission estimates can be available in a gridded form that is convenient for atmospheric modeling applications. Emission disaggregation is a common approach to provide country emissions in a gridded form, but it introduces errors and uncertainties that are unique to the data and/or approaches used. The errors and uncertainties associated with the emission estimates are difficult to evaluate due to the lack of independent evaluation data.

References

Andres, R., Marland, G., Fung, I., & Matthews, E. (1996). A 1° by 1° distribution of carbon dioxide emissions from fossil fuel consumption and cement manufacture, 1950-1990. *Global Biogeochemical Cycles, 10,* 419–429. https://doi.org/10.1029/96GB01523.

Andres, R. J., & Boden, T. A. (2016). *Annual fossil-fuel CO_2 emissions: Uncertainty of emissions gridded by one degree latitude by one degree longitude (1950-2013) (V. 2016).* United States. https://doi.org/10.3334/CDIAC/FFE.ANNUALUNCERTAINTY.2016.

Andres, R. J., Boden, T. A., Bréon, F.-M., Ciais, P., Davis, S., Erickson, D., et al. (2012). A synthesis of carbon dioxide emissions from fossil-fuel combustion. *Biogeosciences, 9*(5), 1845–1871. https://doi.org/10.5194/bg-9-1845-2012.

Andres, R. J., Boden, T. A., & Higdon, D. (2014). A new evaluation of the uncertainty associated with CDIAC estimates of fossil fuel carbon dioxide emission. *Tellus B: Chemical and Physical Meteorology, 66*(1), 23616. https://doi.org/10.3402/tellusb.v66.23616.

Andres, R. J., Gregg, J. S., Losey, L., Marland, G., & Boden, T. A. (2011). Monthly, global emissions of carbon dioxide from fossil fuel consumption. *Tellus B, 63*(3). https://doi.org/10.3402/tellusb.v63i3.16211.

Andrew, R. M. (2020a). A comparison of estimates of global carbon dioxide emissions from fossil carbon sources. *Earth System Science Data, 12*(2), 1437–1465. https://doi.org/10.5194/essd-12-1437-2020.

Andrew, R. M. (2020b). Timely estimates of India's annual and monthly fossil CO_2 emissions. *Earth System Science Data, 12*(4), 2411–2421. https://doi.org/10.5194/essd-12-2411-2020.

Blasing, T., Broniak, C., & Marland, G. (2005). The annual cycle of fossil-fuel carbon dioxide emissions in the United States. *Tellus B, 57*(2), 107–115. https://doi.org/10.1111/j.1600-0889.2005.00136.x.

bp. (2021). *Statistical review of world energy (70th ed.).* bp p.l.c https://www.bp.com/en/global/corporate/energy-economics/statistical-review-of-world-energy.html.

CO_2 and other GHGs from energy and industrial process Chapter | 2 **53**

Bun, R., Nahorski, Z., Horabik-Pyzel, J., Danylo, O., See, L., Charkovska, N., et al. (2018). Development of a high resolution spatial inventory of GHG emissions for Poland from stationary and mobile sources. *Mitigation and Adaptation Strategies for Global Change*, 24(6), 853–880. https://doi.org/10.1007/s11027-018-9791-2.

Calvo Buendia, E., Tanabe, K., Kranjc, A., Baasansuren, J., Fukuda, M., Ngarize, S., ... Federici, S. (2019). *Vol. I. 2019 Refinement to the 2006 IPCC Guidelines for National Greenhouse Gas Inventories*. Switzerland: IPCC.

Charkovska, N., Horabik-Pyzel, J., Bun, R., Danylo, O., Nahorski, Z., Jonas, M., et al. (2018). High-resolution spatial distribution and associated uncertainties of greenhouse gas emissions from the agricultural sector. *Mitigation and Adaptation Strategies for Global Change*, 24(6), 881–905. https://doi.org/10.1007/s11027-017-9779-3.

Crippa, M., Guizzardi, D., Muntean, M., Schaaf, E., Dentener, F., van Aardenne, J. A., et al. (2018). Gridded emissions of air pollutants for the period 1970–2012 within EDGAR v4.3.2. *Earth System Science Data*, 10(4), 1987–2013. https://doi.org/10.5194/essd-10-1987-2018.

Crippa, M., Guizzardi, D., Pisoni, E., Solazzo, E., Guion, A., Muntean, M., et al. (2021). Global anthropogenic emissions in urban areas: Patterns, trends, and challenges. *Environmental Research Letters*, 16(7). https://doi.org/10.1088/1748-9326/ac00e2, 074033.

Crippa, M., Janssens-Maenhout, G., Dentener, F., Guizzardi, D., Sindelarova, K., Muntean, M., et al. (2016). Forty years of improvements in European air quality: Regional policy-industry interactions with global impacts. *Atmospheric Chemistry and Physics*, 16(6), 3825–3841. https://doi.org/10.5194/acp-16-3825-2016.

Crippa, M., Solazzo, E., Guizzardi, D., Monforti-Ferrario, F., Tubiello, F. N., & Leip, A. (2021). Food systems are responsible for a third of global anthropogenic GHG emissions. *Nature Food*, 2(3), 198–209. https://doi.org/10.1038/s43016-021-00225-9.

Crippa, M., Solazzo, E., Huang, G., Guizzardi, D., Koffi, E., Muntean, M., et al. (2020). High resolution temporal profiles in the emissions database for global atmospheric research. *Scientific Data*, 7(1). https://doi.org/10.1038/s41597-020-0462-2.

Eggleston, S. (2006). *IPCC guidelines for national greenhouse gas inventories*. IPCC. https://www.ipcc-nggip.iges.or.jp/public/2006gl/.

European Environment Agency. (2019). *EMEP/EEA air pollutant emission inventory guidebook 2019: Technical guidance to prepare national emission inventories*. Publications Office.

FAOSTAT. (2021). *Statistics division of the food and agricultural organisation of the UN, live animal numbers, crop production, total nitrogen fertiliser consumption statistics*. http://www.fao.org/faostat/en/#home.

Friedlingstein, P., O'Sullivan, M., Jones, M. W., Andrew, R. M., Hauck, J., Olsen, A., et al. (2020). Global carbon budget 2020. *Earth System Science Data*, 12(4), 3269–3340. https://doi.org/10.5194/essd-12-3269-2020.

Gately, C. K., Hutyra, L. R., & Sue Wing, I. (2015). Cities, traffic, and CO_2: A multidecadal assessment of trends, drivers, and scaling relationships. *Proceedings of the National Academy of Sciences*, 112(16), 4999–5004. https://doi.org/10.1073/pnas.1421723112.

Woodard, D., Branham, M., Buckingham, G., Hogue, S., Hutchins, M., Gosky, R., ... Marland, E. (2015). A spatial uncertainty metric for anthropogenic CO_2 emissions. *Greenhouse Gas Measurement and Management*, 4(2–4), 139–160. https://doi.org/10.1080/20430779.2014.1000793.

GCP n.d. About GCP, https://www.globalcarbonproject.org/about/index.htm [Accessed 20 July 2021].

54 Section | B Methods

Gilfillan, D., & Marland, G. (2021). CDIAC-FF: Global and national CO_2 emissions from fossil fuel combustion and cement manufacture: 1751–2017. *Earth System Science Data*, *13*(4), 1667–1680. https://doi.org/10.5194/essd-13-1667-2021.

Guan, D., Liu, Z., Geng, Y., Lindner, S., & Hubacek, K. (2012). The gigatonne gap in China's carbon dioxide inventories. *Nature Climate Change*, *2*(9), 672–675. https://doi.org/10.1038/nclimate1560.

Gurney, K., Law, R., Denning, A., Rayner, P., Baker, D., Bousquet, P., ... Yuen, C.-W. (2002). Towards robust regional estimates of CO_2 sources and sinks using atmospheric transport models. *Nature*, *415*. https://doi.org/10.1038/415626a.

Gurney, K. R., Liang, J., O'Keeffe, D., Patarasuk, R., Hutchins, M., Huang, J., et al. (2019). Comparison of global downscaled versus bottom-up fossil fuel CO_2 emissions at the urban scale in four U.S. urban areas. *Journal of Geophysical Research: Atmospheres*, *124*(5), 2823–2840. https://doi.org/10.1029/2018jd028859.

Gurney, K. R., Liang, J., Patarasuk, R., Song, Y., Huang, J., & Roest, G. (2020). The vulcan version 3.0 high-resolution fossil fuel CO_2 emissions for the United States. *Journal of Geophysical Research: Atmospheres*, *125*(19). https://doi.org/10.1029/2020jd032974.

Gurney, K. R., Mendoza, D. L., Zhou, Y., Fischer, M. L., Miller, C. C., Geethakumar, S., et al. (2009). High resolution fossil fuel combustion CO_2 emission fluxes for the United States. *Environmental Science & Technology*, *43*(14), 5535–5541. https://doi.org/10.1021/es900806c.

Harris, I., Osborn, T. J., Jones, P., & Lister, D. (2020). Version 4 of the CRU TS monthly high-resolution gridded multivariate climate dataset. *Scientific Data*, *7*(1). https://doi.org/10.1038/s41597-020-0453-3.

Hoesly, R. M., Smith, S. J., Feng, L., Klimont, Z., Janssens-Maenhout, G., Pitkanen, T., et al. (2018). Historical (1750–2014) anthropogenic emissions of reactive gases and aerosols from the community emission data system (CEDS). *Geoscientific Model Development*, *11*(1), 369–408. https://doi.org/10.5194/gmd-2017-43.

Hogue, S., Marland, E., Andres, R. J., Marland, G., & Woodard, D. (2016). Uncertainty in gridded CO_2 emissions estimates. *Earth's Future*, *4*(5), 225–239. https://doi.org/10.1002/2015ef000343.

Hogue, S., Roten, D., Marland, E., Marland, G., & Boden, T. A. (2017). Gridded estimates of CO_2 emissions: Uncertainty as a function of grid size. *Mitigation and Adaptation Strategies for Global Change*, *24*(6), 969–983. https://doi.org/10.1007/s11027-017-9770-z.

IEA. (2020). *CO_2 emissions from fuel combustion: Overview*. Paris: IEA. https://www.iea.org/reports/co2-emissions-from-fuel-combustion-overview.

IEA. (2021, May 5). *World energy statistics and balances April 2021 edition (for 1960–2019)*. https://www.iea.org/data-and-statistics/data-product/world-energy-statistics-and-balances.

IEA. (2021b). *Global energy review: CO_2 emissions in 2020. International energy agency*. Available at: https://www.iea.org/articles/global-energy-review-co2-emissions-in-2020. (Accessed 2 March 2021).

Janssens-Maenhout, G., Crippa, M., Guizzardi, D., Dentener, F., Muntean, M., Pouliot, G., et al. (2015). HTAP_v2.2: A mosaic of regional and global emission grid maps for 2008 and 2010 to study hemispheric transport of air pollution. *Atmospheric Chemistry and Physics*, *15*(19), 11411–11432. https://doi.org/10.5194/acp-15-11411-2015.

Janssens-Maenhout, G., Crippa, M., Guizzardi, D., Muntean, M., Schaaf, E., Dentener, F., et al. (2019). EDGAR v4.3.2 global Atlas of the three major greenhouse gas emissions for the period 1970–2012. *Earth System Science Data*, *11*(3), 959–1002. https://doi.org/10.5194/essd-11-959-2019.

CO$_2$ and other GHGs from energy and industrial process **Chapter | 2 55**

Jones, M. W., Andrew, R. M., Peters, G. P., Janssens-Maenhout, G., De-Gol, A. J., Ciais, P., et al. (2021). Gridded fossil CO$_2$ emissions and related O2 combustion consistent with national inventories 1959–2018. *Scientific Data, 8*(1). https://doi.org/10.1038/s41597-020-00779-6.

JRC. (2020). *New high resolution temporal profiles in EDGAR*. https://edgar.jrc.ec.europa.eu/overview.php?v=temp_profile. (Accessed 20 July 2021).

Kona, A., Bertoldi, P., Monforti-Ferrario, F., Baldi, M. G., Lo Vullo, E., Kakoulaki, G., et al. (2021). *Global covenant of mayors, a dataset of GHG emissions for 6,200 cities in Europe and the southern mediterranean*. Copernicus GmbH. https://doi.org/10.5194/essd-2021-67.

Le Quéré, C., Andrew, R. M., Friedlingstein, P., Sitch, S., Hauck, J., Pongratz, J., et al. (2018). Global carbon budget 2018. *Earth System Science Data, 10*(4), 2141–2194. https://doi.org/10.5194/essd-10-2141-2018.

Liu, M., Wang, H., Wang, H., Oda, T., Zhao, Y., Yang, X., et al. (2013). Refined estimate of China's CO$_2$ emissions in spatiotemporal distributions. *Atmospheric Chemistry and Physics, 13*(21), 10873–10882. https://doi.org/10.5194/acp-13-10873-2013.

Marland, G., Hamal, K., & Jonas, M. (2009). How uncertain are estimates of CO$_2$ emissions? *Journal of Industrial Ecology, 13*(1), 4–7. https://doi.org/10.1111/j.1530-9290.2009.00108.x.

Marland, G., & Rotty, R. M. (1984). Carbon dioxide emissions from fossil fuels: A procedure for estimation and results for 1950-1982. *Tellus B, 36B*(4), 232–261. https://doi.org/10.1111/j.1600-0889.1984.tb00245.x.

Marshall, J., & Ramirez, T. (2019). *Attribution problem configurations (Ver. 4.1). CO2 human emissions project*. https://www.che-project.eu/sites/default/files/2020-01/CHE-D4-3-V4-1.pdf.

Myhre, G., Alterskjaer, K., & Lowe, D. (2009). A fast method for updating global fossil fuel carbon dioxide emissions. *Environmental Research Letters, 4*(034012). https://doi.org/10.1088/1748-9326/4/3/034012.

Nassar, R., Napier-Linton, L., Gurney, K. R., Andres, R. J., Oda, T., Vogel, F. R., et al. (2013). Improving the temporal and spatial distribution of CO$_2$ emissions from global fossil fuel emission data sets. *Journal of Geophysical Research: Atmospheres, 118*(2), 917–933. https://doi.org/10.1029/2012JD018196.

Oda, T., Bun, R., Kinakh, V., Topylko, P., Halushchak, M., Marland, G., et al. (2019). Errors and uncertainties in a gridded carbon dioxide emissions inventory. *Mitigation and Adaptation Strategies for Global Change, 24*(6), 1007–1050. https://doi.org/10.1007/s11027-019-09877-2.

Oda, T., & Maksyutov, S. (2011). A very high-resolution (1 km × 1 km) global fossil fuel CO$_2$ emission inventory derived using a point source database and satellite observations of nighttime lights. *Atmospheric Chemistry and Physics, 11*(2), 543–556. https://doi.org/10.5194/acp-11-543-2011.

Oda, T., Maksyutov, S., & Andres, R. J. (2018). The open-source data inventory for anthropogenic CO$_2$, version 2016 (ODIAC2016): A global monthly fossil fuel CO$_2$ gridded emissions data product for tracer transport simulations and surface flux inversions. *Earth System Science Data, 10*(1), 87–107. https://doi.org/10.5194/essd-10-87-2018.

Peters, G. P., Davis, S. J., & Andrew, R. (2012). A synthesis of carbon in international trade. *Biogeosciences, 9*(8), 3247–3276. https://doi.org/10.5194/bg-9-3247-2012.

Quick, J. C., & Marland, E. (2019). Systematic error and uncertain carbon dioxide emissions from U. S. power plants. *Journal of the Air & Waste Management Association, 69*(5), 646–658. https://doi.org/10.1080/10962247.2019.1578702.

Solazzo, E., Bianconi, R., Hogrefe, C., Curci, G., Tuccella, P., Alyuz, U., et al. (2017). Evaluation and error apportionment of an ensemble of atmospheric chemistry transport modeling systems: Multivariable temporal and spatial breakdown. *Atmospheric Chemistry and Physics, 17*(4), 3001–3054. https://doi.org/10.5194/acp-17-3001-2017.

56 Section | B Methods

Solazzo, E., Crippa, M., Guizzardi, D., Muntean, M., Choulga, M., & Janssens-Maenhout, G. (2020). *Uncertainties in the EDGAR emission inventory of greenhouse gases.* Copernicus GmbH. https://doi.org/10.5194/acp-2020-1102.

Solazzo, E., Hogrefe, C., Colette, A., Garcia-Vivanco, M., & Galmarini, S. (2017). Advanced error diagnostics of the CMAQ and Chimere modelling systems within the AQMEII3 model evaluation framework. *Atmospheric Chemistry and Physics, 17*(17), 10435–10465. https://doi.org/10.5194/acp-17-10435-2017.

Spinoni, J., Vogt, J. V., Barbosa, P., Dosio, A., McCormick, N., Bigano, A., et al. (2017). Changes of heating and cooling degree-days in Europe from 1981 to 2100. *International Journal of Climatology, 38*, e191–e208. https://doi.org/10.1002/joc.5362.

UNFCCC. (2015). *Paris agreement.* United Nations. https://unfccc.int/sites/default/files/english_paris_agreement.pdf.

USGS. (2020). *Data of cement, lime, ammonia and ferroalloys of the USGS commodity statistics (June 2020).* https://minerals.usgs.gov/minerals/pubs/commodity/.

Further reading

Arsanjani, J. J., Zipf, A., Mooney, P., & Helbich, M. (Eds.). (2015). *Openstreetmap in giscience; experiences, research, and applications: experiences, research, and applications.* https://doi.org/10.1007/978-3-319-14280-7.

Baldasano, J. M., Güereca, L. P., López, E., Gassó, S., & Jiménez-Guerrero, P. (2008). Development of a high-resolution (1km × 1km, 1h) emission model for Spain: The high-elective resolution modelling emission system (HERMES). *Atmospheric Environment, 42*(31), 7215–7233. https://doi.org/10.1016/j.atmosenv.2008.07.026.

Büttner, G., Kosztra, B., Maucha, G., & Pataki, R. (2012). *Implementation and achievements of CLC2006.* Institute of Geodesy, Cartography and Remote Sensing (FÖMI). https://www.eea.europa.eu/data-and-maps/data/corine-land-cover-3/clc-final-report/clc-final-report.

Crisp, D., Atlas, R. M., Breon, F.-M., Brown, L. R., Burrows, J. P., Ciais, P., et al. (2004). The orbiting carbon observatory (OCO) missions. *Advances in Space Research, 34*(4), 700–709. https://doi.org/10.1016/j.asr.2003.08.062.

European Commission. Joint Research Centre. (2019). *Fossil CO_2 and GHG emissions of all world countries: 2019 report.* Publications Office. https://doi.org/10.2760/687800.

European Commission. Joint Research Centre. (2020). *Fossil CO_2 and GHG emissions of all world countries: 2020 report.* Publications Office. https://doi.org/10.2760/143674.

Ganesan, A. L., Rigby, M., Lunt, M. F., Parker, R. J., Boesch, H., Goulding, N., et al. (2017). Atmospheric observations show accurate reporting and little growth in India's methane emissions. *Nature Communications, 8*(1). https://doi.org/10.1038/s41467-017-00994-7.

Hutchins, M. G., Colby, J. D., Marland, G., & Marland, E. (2016). A comparison of five high-resolution spatially-explicit, fossil-fuel, carbon dioxide emission inventories for the United States. *Mitigation and Adaptation Strategies for Global Change, 22*(6), 947–972. https://doi.org/10.1007/s11027-016-9709-9.

Jones, M. W., Andrew, R. M., Peters, G. P., Janssens-Maenhout, G., De-Gol, A. J., Ciais, P., et al. (2020). *Gridded fossil CO_2 emissions and related O_2 combustion consistent with national inventories 1959-2019 (GCP-GridFEDv2020.1) [data set].* Zenodo. https://doi.org/10.5281/ZENODO.4277267.

Kühlwein, J., Wickert, B., Trukenmüller, A., Theloke, J., & Friedrich, R. (2002). Emission modelling in high spatial and temporal resolution and calculation of pollutant concentrations for

CO$_2$ and other GHGs from energy and industrial process Chapter | 2   **57**

comparisons with measured concentrations. *Atmospheric Environment*, *36*, 7–18. https://doi. org/10.1016/s1352-2310(02)00209-1.

Lenhart, L., & Friedrich, R. (1995). European emission data with high temporal and spatial resolution, water, air. *Soil Pollution*, *85*(4), 1897–1902. https://doi.org/10.1007/BF01186111.

Schaap, M., Timmermans, R. M. A., Roemer, M., Boersen, G. A. C., Builtjes, P. J. H., Sauter, F. J., et al. (2008). The LOTOS EUROS model: Description, validation and latest developments. *International Journal of Environment and Pollution*, *32*(2), 270. https://doi.org/10.1504/ IJEP.2008.017106.

Simpson, D., Benedictow, A., Berge, H., Bergström, R., Emberson, L. D., Fagerli, H., et al. (2012). The EMEP MSC-W chemical transport model—Technical description. *Atmospheric Chemistry and Physics*, *12*(16), 7825–7865. https://doi.org/10.5194/acp-12-7825-2012.

Solazzo, E., & Galmarini, S. (2015). Comparing apples with apples: Using spatially distributed time series of monitoring data for model evaluation. *Atmospheric Environment*, *112*, 234–245. https://doi.org/10.1016/j.atmosenv.2015.04.037.

Terrenoire, E., Bessagnet, B., Rouïl, L., Tognet, F., Pirovano, G., Létinois, L., et al. (2015). High-resolution air quality simulation over Europe with the chemistry transport model CHIMERE. *Geoscientific Model Development*, *8*(1), 21–42. https://doi.org/10.5194/gmd-8-21-2015.

Thiruchittampalam, B. (2014). *Entwicklung und Anwendung von Methoden und Modellen zur Berechnung von räumlich und zeitlich hochaufgelösten Emissionen in Europa*. PhD dissertation Universität Stuttgart.

Verstraete, J. (2017). The spatial disaggregation problem: Simulating reasoning using a fuzzy inference system. *IEEE Transactions on Fuzzy Systems*, *25*(3), 627–641. https://doi.org/10.1109/ tfuzz.2016.2567452.

Yokota, T., Yoshida, Y., Eguchi, N., Ota, Y., Tanaka, T., Watanabe, H., et al. (2009). Global concentrations of CO$_2$ and CH$_4$ retrieved from GOSAT: First preliminary results. *SOLA*, *5*, 160–163. https://doi.org/10.2151/sola.2009-041.

Yoshida, Y., Ota, Y., Eguchi, N., Kikuchi, N., Nobuta, K., Tran, H., et al. (2011). Retrieval algorithm for CO$_2$ and CH$_4$ column abundances from short-wavelength infrared spectral observations by the greenhouse gases observing satellite. *Atmospheric Measurement Techniques*, *4*(4), 717–734. https://doi.org/10.5194/amt-4-717-2011.

Zhou, Y., & Gurney, K. (2010). A new methodology for quantifying on-site residential and commercial fossil fuel CO2 emissions at the building spatial scale and hourly time scale. *Carbon Management*, *1*(1), 45–56. https://doi.org/10.4155/cmt.10.7.

Zhu, D., Tao, S., Wang, R., Shen, H., Huang, Y., Shen, G., et al. (2013). Temporal and spatial trends of residential energy consumption and air pollutant emissions in China. *Applied Energy*, *106*, 17–24. https://doi.org/10.1016/j.apenergy.2013.01.040.

Chapter 3

Bottom-up approaches for estimating terrestrial GHG budgets: Bookkeeping, process-based modeling, and data-driven methods

Benjamin Poulter[a], Ana Bastos[b], Josep G. Canadell[c], Philippe Ciais[d], Deborah Huntzinger[e], Richard A. Houghton[f], Werner Kurz[g], A.M. Roxana Petrescu[h], Julia Pongratz[i,j], Stephen Sitch[k], and Sebastiaan Luyssaert[h]

[a]Biospheric Sciences Laboratory, NASA Goddard Space Flight Center, Greenbelt, MD, United States, [b]Department of Biogeochemical Integration, Max Planck Institute for Biogeochemistry, Jena, Germany, [c]Global Carbon Project, Climate Science Centre, CSIRO Oceans and Atmosphere, Canberra, ACT, Australia, [d]Laboratoire des Sciences du Climat et de l'Environnement LSCE CEA CNRS UVSQ, Gif sur Yvette Cedex, France, [e]School of Earth and Sustainability, Northern Arizona University, Flagstaff, AZ, United States, [f]Woodwell Climate Research Center, Falmouth, MA, United States, [g]Natural Resources Canada, Canadian Forest Service, Victoria, British Columbia, Canada, [h]Department of Ecological Sciences, Vrije Universiteit Amsterdam, Amsterdam, The Netherlands, [i]Ludwig-Maximilians-Universität Munich, München, Germany, [j]Max Planck Institute for Meteorology, Hamburg, Germany, [k]University of Exeter, Exeter, United Kingdom

1 Introduction to bottom-up (BU) approaches

Human activities have transformed the biogeochemistry of the Earth system, leading to increases in greenhouse gas concentrations in the atmosphere (IPCC, 2013). Increases in atmospheric carbon dioxide (CO_2), methane (CH_4), and nitrous oxide (N_2O) concentrations have led to an overall anthropogenic radiative forcing of $3\,W\,m^{-2}$ since preindustrial times (von Schuckmann et al., 2020; WMO, 2020). Changes in atmospheric CO_2 concentration is mainly from the combustion of fossil fuels used for energy production and fossil carbonates, and from land-use and related land-cover change, with the equivalent of $\sim 45\%$ of emitted CO_2 remaining in the atmosphere, as the land and ocean

Balancing Greenhouse Gas Budgets. https://doi.org/10.1016/B978-0-12-814952-2.00010-1
Copyright © 2022 Elsevier Inc. All rights reserved.

60 Section | B Methods

remove the rest (Friedlingstein et al., 2020). Atmospheric CH_4 concentration increases result from roughly equal contributions from oil and gas exploration and from agricultural activities linked to livestock (enteric fermentation and waste management) and rice cultivation, with the atmospheric hydroxyl sink buffering part of this increase (Saunois et al., 2020). Nitrous oxide emissions are mostly linked to agricultural activities such as the addition of nitrogen-based urea fertilizer and manure waste management practices (Tian et al., 2020). Efforts to understand the sources of these gases, their spatial patterns, seasonal, interannual, and decadal dynamics, and their removal from the atmosphere (Hong et al., 2021), are central to our predictive understanding of the Earth system as well as for informing policies on climate mitigation and air pollution reduction. Greenhouse gas budgeting is the science of accounting for these sources and has a long history in informing scientific syntheses from regional to global scales (see Chapter 1).

Greenhouse gas (GHG) budgeting activities can be carried out following one or more independent methodologies and grouped into either "top-down" or "bottom-up" approaches. Top-down approaches use atmospheric inversions (see Chapter 4) where monitoring of atmospheric concentrations of a particular gas provide a direct observational constraint for estimating likely emission sources and their uncertainties for a limited set of sectors. In contrast, bottom-up approaches use a combination of activity data and emission factors (in the case of inventories), process-based models, or data-driven (i.e., remote sensing) approaches to estimate GHG fluxes and the net balance of emissions over time. GHG budgets that use both top-down and bottom-up methodologies are complementary because to some extent, the difference between the two approaches can be used to quantify uncertainties, which can then be used to identify sectoral emission sources or sinks for further detailed observation and understanding (Ciais et al., 2014). This chapter presents the background for the different bottom-up approaches that include bookkeeping, process-based models and data-driven methods presented in the topical chapters.

One of the main objectives of greenhouse gas budget reporting[a] is to quantify the net emissions of trace (e.g., GHG) gases as the sum of their gross sources and sinks, where sinks represent either natural or human-driven removal of gases from the atmosphere or the accumulation of CO_2 in the atmosphere, land, and ocean (Luyssaert et al., 2012; Pan et al., 2011; Petrescu et al., 2020). Quantifying the net emissions of GHGs to the atmosphere is relevant for estimating the radiative forcing responsible for causing climate change, and for addressing scientific questions or for informing policies related to monitoring, reporting,

---

a. Here we defined reporting as the informal summation of GHG fluxes, rather than using the formal 'Reporting' and 'Accounting' definitions of the UNFCCC, where 'Reporting' is the reporting of anthropogenic sources and sinks via the National Greenhouse Gas Inventories (NGHGI) and 'Accounting' is related to Kyoto Protocol guidance.

and verification (MRV) of emission reductions. Greenhouse gas accounting thus tracks the magnitude and the changes over time of gases such as carbon dioxide, methane, or nitrous oxide, as well as hydrofluorocarbons (HFCs) and other short-lived climate pollutants like black carbon. Non-GHG budgets are also carried out using similar approaches, such as for carbon monoxide (CO) that serves an important role in determining the lifetime of atmospheric hydroxyl radicals (Zheng et al., 2019) but also oxidizes to CO_2 (Ciais, Bastos, et al., 2020; Ciais, Gasser, et al., 2020). The final summary of all the individual sources and sinks provides the context for a "budget" that can be used to evaluate net emissions and their changes, rank relative contributions of gases, or analyze gases by their groupings into anthropogenic, biogenic or thermogenic categories. The following discussion in this chapter of bookkeeping, process-based modeling, and data-driven approaches can be applied to understanding the biogeochemistry of ecosystems as well as be applied to policy-related questions. The methods can be linked to the various "Tier" levels of reporting outlined by the IPCC Good Practice Guidelines (2006, 2019), and the themes central to the Paris Agreement, i.e., Article 4.1 *"to achieve a balance between anthropogenic emissions by sources and removals by sinks of greenhouse gases."*

The range of options available for carrying out bottom-up (BU) approaches is more diverse than top-down methods (TD). Compared with TD methods, BU approaches can take advantage of a diverse set of data sources and methods for estimating emissions, whereas atmospheric inversions rely mainly on networks of atmospheric concentrations derived from the surface monitoring station, aircraft, or satellite observations (Crisp et al., 2004) that are integrated within atmospheric chemistry transport models. Some atmospheric inversion systems now include the capability for using land "observations" as statistical *priors* that come from remote sensing, such as solar-induced fluorescence (SIF, e.g., Miller, Michalak, Yadav, & Tadić, 2018). In contrast, BU approaches rely on (i) direct observations of fluxes using eddy covariance or chamber-based methods, (ii) indirect observations derived from remote-sensing surface reflectance data, (iii) simulation models to estimate fluxes, or (iv) from statistical reporting of energy use or land use based on activity data and their corresponding emission factors. Some of the BU methods, aside from statistics reported at the country level, distinguish themselves from TD methods by providing higher-resolution gridded estimates of GHG fluxes that are linked directly to specific geographic locations, e.g., fossil fuel emissions from nightlight activity that can be linked to cities, towns, and transportation networks (Oda, Maksyutov, & Andres, 2020) or the Emission Database for Global Atmospheric Research (Janssens-Maenhout et al., 2019) or anthropogenic CH_4 emissions (Maasakers et al., 2016).

In addition, BU methods can be applied to a wide range of, if not all, individual emission sectors or land-use activities producing GHGs and then these sectors aggregated later for total trace gas emission fluxes expressed as their own mass fluxes or as carbon dioxide equivalents (CO_2eq). These individual sectoral emissions and their temporal dynamics are important to track as they may include "legacy" effects, i.e., lagged effects of deforestation and degradation on coarse woody debris and soil carbon emissions, or CO_2 fertilization effects, which cannot be measured directly and require ecosystem models to elucidate. The accuracy of BU approaches is limited mainly by the quality and quantity of observations that inform spatial and temporal scaling and inform model development through parameter estimation that allows deriving theoretical relationships. The accuracy of the models is in part limited by the present-day understanding of the processes the models aim to simulate. However, the comparison between BU approaches can be problematic if different definitions are used in the reporting of GHG fluxes from managed and unmanaged systems, or if there is overlap in the definitions that might lead to double counting of emissions (see next section).

The Intergovernmental Panel on Climate Change Guidelines (IPCC GL, 2006), and their revisions in 2019 (IPCC, 2019 = Refinement), rely on BU methodologies for estimating emissions from energy, industrial processes, agriculture, forestry and other land-use activities (AFOLU) and waste. The IPCC GL (2006) document further ranks BU methods by their accuracy using a tiered system (Petrescu et al., 2020), where Tier 1 estimates of GHG emissions use global emission factors and are associated with high uncertainty, Tier 2 uses regional or country-level emission factors, and Tier 3 estimates have the lowest uncertainties using local emission factors and/or parameterized models with detailed remote-sensing observations. The accuracy in the reporting of GHG fluxes is particularly important from a policy perspective, where the fulfillment of country commitments to the Paris Agreement will be evaluated every 5 years, starting in 2023, via the Global "Stocktake" described in Article 14, as well as from an economic perspective, where financial markets trade on GHG credits that require verification.

This chapter describes how different BU approaches are applied to estimate biogenic GHG fluxes from natural and managed ecosystems (Fig. 1), whereas Chapter 2 focuses on estimating industrial emissions, including those on farms from transportation and electric consumption, from their respective inventories. We focus here on the three main categories of BU methods, (i) bookkeeping and inventory, (ii) forward, or process-based, biogeochemical modeling, and (iii) data-driven (remote-sensing)-based models, and contrast these methods in terms of their data requirements and uncertainties. Some of the questions that BU approaches address include the following: What are the net or gross emissions of a particular source or sink category of trace gases? Where are the emissions located and how are the emissions changing at diurnal to seasonal to annual and decadal time scales? How do trace gases respond to different management and disturbance regimes, climatic conditions, and atmospheric CO_2 concentrations?

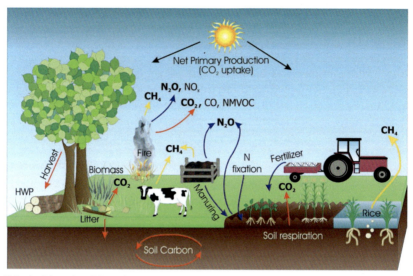

FIG. 1 Biogenic AFOLU fluxes as described by the IPCC GPG (2019). This chapter focuses on how bottom-up methods are used to quantify these fluxes, with a focus on carbon dioxide. *(Reproduced with permission of IPCC.)*

1.1 Definitions and discrepancies

Recent work has highlighted the importance that definitions and terminology have in explaining uncertainties in GHG budgets (Grassi et al., 2018; Kondo et al., 2019). Differences between TD and BU accounting have been explored in several studies that have evaluated the role of lateral fluxes, i.e., the movement of dissolved and particulate carbon through streams and rivers to the coastal estuaries, shelf, and deep ocean as well as the trade-induced movement of crops and harvested wood around the globe (Ciais, Gasser, et al., 2020). To some extent, the differences in TD and BU estimates can be corrected by introducing riverine flux adjustments (0.45–0.78 PgC) to the atmospheric inversions (e.g., Le Quéré et al., 2018), and recommendations for how to include the contribution to atmospheric CO_2 from the oxidation of reduced carbon gases (CO, CH_4, and BVOCs) have recently been proposed (Ciais, Bastos, et al., 2020; IPCC, 2019).

Systematic differences in FOLU fluxes from bookkeeping and process-based dynamic global vegetation models (DGVMs) have been explored in detail (Houghton, 2020; Pongratz, Reick, Houghton, & House, 2014). The main differences have been attributed to (1) how bookkeeping models rely on contemporary stock estimates that exclude climate and atmospheric CO_2 effects, and, related to this, (2) how the process-based models include "loss of additional sink capacity (LASC)" in their FOLU estimate. The LASC term represents the loss in potential carbon storage from CO_2 fertilization, and for some geographic regions, climate change, that is included implicitly in the use of a baseline "no land-use change" simulation and can be up to a 0.4–0.7 PgC additional carbon flux (Gasser et al., 2020; Obermeier et al., 2021).

In addition, differences between country-level National Greenhouse Gas Inventory (NGHGI) reporting to the United Nations Framework Convention on Climate Change (UNFCCC) and scientific-based GHG budgets used in the IPCC Assessment Reports have recently been reconciled by paying particular attention to definitions of managed and unmanaged lands (Grassi et al., 2018; IPCC SRCCL, 2019). For example, UNFCCC reporting of AFOLU fluxes refers strictly to managed lands, whereas estimates from DGVM models consider both managed and unmanaged lands, and includes both indirect effects of anthropogenic activities (climate, nitrogen deposition, and atmospheric CO_2 fertilization) and direct effects from management. Globally, the difference between the two approaches leads to a ~ 1.1 PgC yr^{-1} larger FOLU estimates reported by IPCC versus the sum of country-level reporting (Grassi et al., 2018).

The BU approaches described in this chapter are relevant for estimating natural carbon exchange processes from photosynthesis, autotrophic and heterotrophic respiration and disturbance, and fluxes from LULUCF for FOLU accounting. The methodologies described are used in NGHGI [e.g., annual National Inventory Reporting (NIR) and ancillary documents], reporting to the UNFCCC (i.e., "stock change" or "stock difference"), and also include those used in scientific studies as reported in the IPCC Assessment Reports (mainly bookkeeping estimates and DGVMs). We also mention non-CO_2 gases, CH_4 and N_2O that complement the full GHG reporting used for AFOLU accounting. Within FOLU, we describe deforestation and afforestation and other land-use-induced changes in natural vegetation, wood harvest, peatland burning and drainage, and changes in soil carbon from cropland or grassland management (tillage and grazing). Cropland and grassland management is typically included implicitly in bookkeeping approaches and sometimes included explicitly by DGVM models (Friedlingstein et al., 2020).

2 Bottom-up methodologies

2.1 Stock-change versus flux-based accounting

Emissions can be estimated by comparing the change in "stock" between two time periods, or by measuring the flux directly and integrating this flux over some time period. The two approaches are referred to as "stock-change" and "flux-based" approaches, respectively, and will give the same emissions estimate for the same time period under perfectly known conditions. A third approach, "gain-loss," applies activity data to some baseline year of carbon stocks (see Harris et al., 2021). The bookkeeping, process-based, and data-driven approaches combine stock-change and flux-based methods where

Bottom-up approaches for estimating terrestrial GHG budgets **Chapter | 3** **65**

uncertainties might be lower for stocks (in the case of remote-sensing approaches, Baccini et al., 2012) or where processes might be better known, i.e., leaf-level photosynthesis. Reporting requirements may also specify when and how to use stock-change versus flux-based approaches. For example, the IPCC GL (2006, 2019) is based on activity data, e.g., land-cover change, combined with emission factors or stock estimates, to estimate gross or net emissions.

Based on the work by Ciais, Bastos, et al. (2020) and Ciais, Gasser, et al. (2020), the regional carbon stock-change approach, ΔC_{region}, can be represented mathematically by changes in the different pools as.

$$\Delta C_{region} = \Delta C_{forest} + \Delta C_{crop} + \Delta_{grass} + \Delta_{woodharvest} + \Delta_{cropharvest} + \Delta C_{peat}$$
$$+ \Delta C_{burial} + \Delta C_{litho} \tag{1}$$

$$\Delta C_{forest} = \Delta C_{biomass} + \Delta C_{litter} + \Delta C_{soil} \tag{2}$$

$$\Delta C_{croplands} = \Delta C_{biomass} + \Delta C_{litter} + \Delta C_{soil} \tag{3}$$

$$\Delta C_{grasslands} = \Delta C_{biomass} + \Delta C_{litter} + \Delta C_{soil} \tag{4}$$

where the total stock change can be further separated into changes in biomass, litter and woody debris, and soil (mineral and organic) carbon for forests, croplands, and grasslands. These pools change in response to climate, atmospheric CO_2, and management, i.e., wood harvest, tillage, crop harvest. The regional stock change also includes changes in the import and export of wood and crop products and their decay in landfills. Changes in peatland carbon, from harvest, fire, or drainage can also be accounted for, as well as the burial of carbon in lakes, ponds, and reservoirs. In some regions, lithogenic carbon, or carbon released through weathering processes that consume atmospheric CO_2, can be included, but it is more relevant at global scales and long time periods, i.e., multiannual.

The measurements of stocks are typically made through repeated national inventories or bookkeeping models that combine information on activity data (area of forest, etc.) with corresponding carbon density. Some forest biomass national inventories however do not use a stock change approach but rather consider the area of forests subject to harvest and other disturbances and use growth curves for subsequent carbon gains, which do not integrate the impacts of environmental factors. The frequency and time interval between inventories may differ by country and their reporting commitments. For example, the Food and Agriculture Organization (FAO) assembles country-level forest data every 5 years for publication and analysis in the Global Forest Resources Assessment (FRA), with the latest version covering the period 1990–2020. NGHGI reporting by Annex 1 countries to the UNFCCC National Inventory Reports (NIR) is annual and includes emissions of GHGs [CO_2, CH_4, N_2O, perfluorocarbons (PFCs), hydrofluorocarbons (HFCs), sulfur hexafluoride (SF6), and nitrogen trifluoride (NF$_3$) from five sectors (energy; industrial processes and product

66 Section | B Methods

use; agriculture; land use, land-use change and forestry (LULUCF); and waste] from the base year up to 2 years before the inventory is due.

Compared to the stock-change approach, where the stock components integrate a variety of fluxes, the flux-based estimates of emissions must consider each flux separately. Based on the work of Ciais, Bastos, et al. (2020) and Ciais, Gasser, et al. (2020), the flux-based equation for CO_2 can be written as

$$NEE = \left(NPP + F_{weathering\ CO2\ uptake}\right) - \left(F_{weathering\ CO2\ outgas} - F_{lithoriverinput}\right)$$
$$- \left(F_{lakes+res.outgassing} - F_{rivers\ C\ outgas}\right) - F_{bio.riverinput}$$
$$- \left(SHR - F_{LUC} - F_{Land\ Management} - F_{fires} - F_{insect} - F_{reduced} - F_{grazing}\right)$$
$$- \left(F_{crop\ harvest} - F_{wood\ harvest} - F_{crop\ products} - F_{wood\ products\ decay}\right.$$
$$\left. - F_{wood\ products\ burning} - F_{peat\ use}\right) - \left(F_{rivers\ export} - F_{wood\ trade} - F_{crop\ trade}\right)$$

$$(5)$$

where NEE is Net Ecosystem Exchange, following the definition of Hayes (2012), extended to include nonecosystem, but natural geologic fluxes. This definition of NEE includes all carbon fluxes that are observed by top-down atmospheric inversions, including the oxidation of reduced gases (CH_4, CO, and BVOCs) in the $F_{reduced}$ term. As such, this definition reflects what the atmosphere integrates from the land (excluding fossil fuel and cement production emissions). Because of the inclusion of geologic terms, aquatic fluxes, and trade fluxes, the definition is not exactly the same as used by Schulze, Wirth, and Heimann (2000) for Net Biome Production, NBP, or the definition proposed by Randerson, Chapin III, Harden, Neff, and Harmon (2002) for Net Ecosystem Production, NEP, or the Net Ecosystem Carbon Balance term proposed by Chapin et al. (2006), that were each proposed to integrate ecosystem carbon processes over larger temporal and spatial scales rather than to be aligned or reconciled with the atmospheric perspective (Ciais, Bastos, et al., 2020; Ciais, Gasser, et al., 2020).

Presently, there is no single bottom-up measurement or model that can provide an estimate of NEE, and so each component flux must be estimated separately. Ideally, multiple, quasi-independent estimates (e.g., using model ensembles) can be used to provide the range of uncertainty in the resulting flux. A variety of approaches can be used to estimate each flux, which is discussed below in the bookkeeping, process-based, and data-driven sections. Because of the complexity in estimating the individual component flux terms, bottom-up budgets are prone to "double counting" of emissions, whereby overlap in flux estimates, interpretation of definitions, and spatial resolution of data all can lead to biases and uncertainties in the estimated flux.

2.2 Bookkeeping methodology

The bookkeeping approach was developed in the early 1980s by Houghton et al. (1983) to estimate carbon emissions from land-use and related land-cover

Bottom-up approaches for estimating terrestrial GHG budgets Chapter | 3 **67**

change. At that time, anthropogenic CO_2 emissions were roughly half (i.e., $\sim$5 PgC yr^{-1}) of what they are today in 2020, and the atmospheric growth rate (2–3 PgC yr^{-1}) and ocean uptake (1.8–2.4 PgC yr^{-1}) meant that the global carbon budget was largely balanced. The additional carbon flux from net land-use and land-cover change activities of 1.3 ± 0.7 PgC yr^{-1} (for the 1980s) meant that there was an additional "missing" sink, most likely on land, that was absorbing the CO_2 that did not accumulate in the atmosphere or ocean. Bookkeeping models were developed to more accurately estimate the land-use and land-cover change flux, and currently, three approaches are in use that inform the annual Global Carbon Budget (Friedlingstein et al., 2020) and IPCC activities. The three approaches are referred to as BLUE, "bookkeeping of land-use emissions" (Hansis, Davis, & Pongratz, 2015), HandN2017 (Houghton & Nassikas, 2017), and the reduced-form Earth system model, OSCAR (Gasser et al., 2020). The bookkeeping approaches share similarities in how land-area data and carbon densities (i.e., above and belowground biomass) are integrated and tracked over time, but use different data sources, time frames, management practices, and regional groupings. A main difference is that HandN2017 used forest area from FAO and regional estimates before 1960, whereas BLUE used the LUH2 land use reconstruction where forest area change result from a potential vegetation map and the historical simulated expansion of agricultural land from the HYDE model, constrained by FAO agricultural area reports at national scale. As a result of these different drivers, forest area decreased in HandN2017, leading to higher net land use emissions, whereas it remained approximately stable in LUH2 driving the BLUE model, leading to decreasing emissions in the last decade.

Bookkeeping models are used to estimate carbon fluxes from each of the individual land-use and land management activities, including those from wood harvest and forest degradation, shifting cultivation, cropland harvest, crop management (nitrogen fertilizer, tillage, irrigation), pasture grazing and mowing, peat fires and peatland drainage, fire suppression, product pools, and the regrowth of secondary forests. Because the bookkeeping approaches track each LULCC component, the LULCC flux can be presented as the gross emissions and removals (from the atmosphere), or as the net flux. Crop and grassland management practices are included implicitly through the use of observed carbon densities. The long-term trends and interannual variability from climate change, atmospheric CO_2 fertilization, and nitrogen deposition on natural carbon densities or stocks, are not considered in BLUE or HandN2017, with both approaches using fixed carbon densities as well as decay and regrowth curves. The bookkeeping models do not include non-CO_2 gases, i.e., CH_4 and N_2O, and so their application is relevant for FOLU accounting rather than AFOLU (Hong et al., 2021).

Compared to DGVM models, an important distinction is that bookkeeping models estimate LULCC directly, rather than as the difference between a "no-land cover and land-use change scenario" and a "with land-cover and

68 Section | B Methods

land-use change scenario." Another distinction is that bookkeeping models focus on land management and direct anthropogenic land cover change rather than natural land cover changes (Pongratz et al., 2017). This means that the bookkeeping models do not include the loss of additional sink capacity (LASC) term and that the approach is not directly comparable with DGVM models without adjusting the approaches to either include or exclude this flux. The decision on how to incorporate the LASC flux has important policy implications because countries are required to track LULCC fluxes during a "commitment period" with specific treatment of historical carbon fluxes (Hansis et al., 2015). The bookkeeping approaches quantify what is referred to as "legacy fluxes," i.e., fluxes from slow processes such as soil carbon adjustment, forest regrowth and delayed fluxes from products (pulp, paper, and lumber), as well as the "instantaneous fluxes" released within about a year such as clearing through fire. In contrast, remote-sensing and stock-change-based approaches estimate "committed fluxes" (e.g., Davis, Burney, Pongratz, & Caldeira, 2014; Ramankutty et al., 2007) via their approach that compares land cover states, or biomass maps, between time periods (Baccini et al., 2017; Harris et al., 2012, 2021; Petrescu, Abad-Vinas, Janssens-Maenhout, Blujdea, & Grassi, 2012). The remote-sensing approaches assume that the carbon stocks before and after a LULCC event are at equilibrium (Hansis et al., 2015) because they do not consider time-dynamics for delayed processes, thus providing estimates of committed rather than instantaneous LULCC emissions. Because of these differences, and due to the fact that bookkeeping approaches include degradation, shifting cultivation, wood harvest, and other processes, comparisons with remote-sensing-based approaches show large differences, with remote-sensing-based estimates having lower LULCC flux (e.g., Harris et al., 2012).

At their core, the bookkeeping approaches estimate LULCC fluxes by combining data on the land cover area or "states" with data on carbon densities for biomass and soil. The removal of biomass due to a change from one land cover type to another, or from management, is tracked over time, taking into account how carbon decays to the atmosphere, or how carbon is removed from the atmosphere during the regrowth of natural vegetation. Because of the importance of legacy fluxes, the start date of when the bookkeeping model begins is important, with models initializing with land-area data from as early as the year 1500 or 1700 to avoid having to assume instantaneous or committed emissions. The HandN2017 approach represents carbon densities for 20 ecosystem types, BLUE represents cropland and grassland conversions with 11 natural plant functional types, and OSCAR calibrates carbon densities to DGVM estimates. The bookkeeping models represent the LULCC processes on an annual timestep, and each year update the following four carbon pools: biomass (above and belowground), slash, products (fuelwood and charcoal, pulp and paper, and lumber), and soil carbon. The four pools are represented by the age since when they were first created by a land-cover transition, either by deforestation or regrowth. Peatland degradation processes are not explicitly included in the

Bottom-up approaches for estimating terrestrial GHG budgets **Chapter | 3 69**

models but calculated off-line with estimates for peat fire from van der Werf et al. (2010) and drainage effects on oxidation from Hooijer et al. (2010) or FAOSTAT (Conchedda & Tubiello, 2020).

Three main sources of data are used for the bookkeeping models. First, data on land-use and land-cover change processes, including land cover area and wood harvest. These data are provided from the FAO-FRA, FAOSTAT (for wood harvest), and remote sensing (for mapping grasslands and forests). HandN2017 approach is carried out at the country level or regional level (10 regions) and uses the FAO-FRA reporting. In contrast, BLUE uses a spatially explicit approach, based on the Land-Use Harmonization (LUH) gridded information on land states, transitions, and management (Hurtt et al., 2020). OSCAR is used at the biome level (5 biomes). Data for carbon density come from literature estimates provided by the FAO for biomass, and from literature meta-analyses for soil carbon (Schlesinger, 1984; Zinke, 1986). Comparisons of literature-based biomass summaries with remote-sensing observations have shown higher LULCC fluxes by 21% when using remotely sensed biomass (Baccini et al., 2012). The decay and regrowth curves are also used from FAO and literature-based estimates, with uncertainties assessed through evaluating minimum and maximum bounds (Hansis et al., 2015). Because the estimates are fixed in time, they do not include explicit effects of climate change, meteorological variability, and atmospheric CO_2 fertilization.

The three different approaches are useful for assessing uncertainties, with cumulative LULCC emissions from 1850 to 2019 ranging from 150 PgC (HandN2017), 200 PgC (OSCAR), and 275 PgC (BLUE). The large difference is partly due to differences in rates of land-use and land-cover change, with OSCAR and BLUE using LUH and HandN2017 and OSCAR using FAO. HandN2017 did not include shifting cultivation, but it did include an alternative land use to account for the net loss of tropical forests that exceeded increases in areas under permanent croplands and pastures. Decadal LULCC fluxes, from 2010 to 2019, were 0.97 PgC, 1.86 PgC, and 2.56 PgC, for HandN2017, OSCAR, and BLUE, respectively. For the past decade, the three bookkeeping approaches do not agree with whether there has been an increase or decrease in global LULCC fluxes, with HandN2017 suggesting a downward trend and OSCAR and BLUE showing upward trends (Fig. 2). For the same decade, the gross emissions (source) were 4.1 ± 1.2 PgC yr^{-1} and gross uptake (sink from regrowth) was -2.7 ± 1.1 PgC yr$-^{1}$, showing the importance of LULCC processes relative to fossil fuel emissions.

Reducing uncertainties and advancing bookkeeping models are needed to improve LULCC carbon flux estimates. Uncertainties can be reduced through the integration of remote-sensing observations, i.e., using IPCC land cover categories that can track land-cover change over time with improved biomass estimates from radar or lidar satellite missions. New processes to be considered include separating management practices (such as irrigation) and intensities, and also lateral fluxes related to erosion and redeposition of carbon, woody

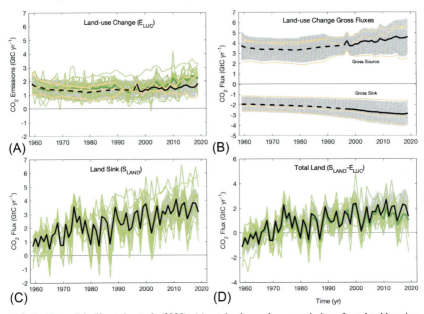

FIG. 2 From Friedlingstein et al. (2020), (a) net land-use change emissions from bookkeeping models (black, dark green, yellow lines) and from dynamic global vegetation models (TRENDY v9, light green lines), (b) gross fluxes from LULCC processes for the three bookkeeping models, (c) land carbon uptake from the TRENDY v9 DGVMs, (d) total land carbon uptake, including losses from LUCC.

encroachment of grasslands, and refining carbon densities, carbon decay, and regrowth models using more geographically representative observations. Approaches that directly ingest national forest inventory data directly are being expanded to other countries, i.e., the Canadian Biosphere Model (CBM, Kurz et al., 2009; Kurz et al., 2018). Advancing current bookkeeping methods to include CH_4 and N_2O accounting from agriculture would provide a single, integrated model for AFOLU accounting.

2.3 Process-based methodology

Dynamic global vegetation models (DGVMs) represent the carbon cycle (CO_2, CH_4, and increasingly, the nitrogen cycle) by coupling biophysics, biogeochemistry, and biogeography using a "process-based" approach (Fig. 3, Fisher, Huntzinger, Schwalm, & Sitch, 2014; Prentice et al., 2007). Compared with bookkeeping approaches, DGVMs use "first principles" to represent ecological processes through a combination of physical and semiempirical equations derived from theory, experiments, and observations. This process-based approach provides estimates for a range of carbon flux components and carbon stock pools under varying climate, atmospheric CO_2, nitrogen

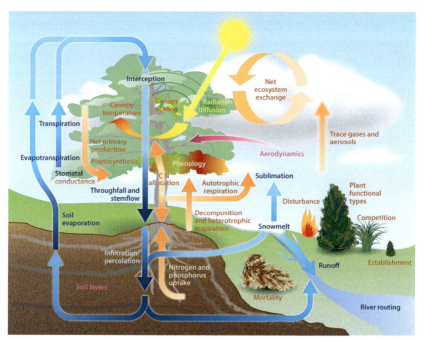

FIG. 3 Schematic representation of processes represented within the DGVM process-based accounting. *(Reproduced from Fisher, J., Huntzinger, D., Schwalm, C. R., & Sitch, S. 2014. Modeling the terrestrial biosphere.* Annual Review of Environment and Resources, *39, 91–123.)*

deposition, land management (i.e., irrigation), and land-cover change scenarios. Process-based approaches also provide flexibility for evaluating the carbon cycle under nonanalog conditions (e.g., paleo-and future climate) or for attributing model responses to individual drivers. While bookkeeping approaches have the advantage in that they use data-driven, verified statistical reporting of land area and changes over time, they so far use fairly simple look-up table approaches for estimating changes in carbon stocks. In contrast, DGVMs can provide information for a range of global, climatic conditions, using either a fully prognostic mode, with a minimal set of driver data (meteorology, soils, atmospheric CO_2, N-deposition) or in a diagnostic mode. Diagnostic mode means combining a variety of remote-sensing observations to initialize model state, i.e., plant functional type distribution (Poulter, Ciais, et al., 2011; Poulter, Frank, et al., 2011) or disturbance history (Bellassen, Le Maire, Dhote, Ciais, & Viovy, 2010) or optimize model parameters (e.g., phenology, MacBean, Peylin, Chevallier, Scholze, & Schürmann, 2016). Diagnostic mode typically limits the model to the period of time that observations are available.

DGVMs simulate a full range of carbon fluxes, allocate carbon to various storage organs, i.e., pools and represent the turnover of carbon within these "pools" due to mortality, disturbance, decomposition, or management processes

72   Section | B Methods

(Prentice et al., 2007). With their heritage partly in forest gap models (Botkin, Janak, & Wallis, 1972), DGVMs simulate the establishment, growth and mortality of managed and natural ecosystems. Beginning with photosynthesis, or gross primary production (GPP), DGVMs use either an enzyme-kinetic approach (Farquhar, O'Leary, & Berry, 1982) or a light-use efficiency model (Monteith, 1977). The enzyme-kinetic approach estimates photosynthesis from biochemical colimitations between maximum rate of carboxylation, or Rubisco limited rate (Vcmax), the maximum rate of electron transport, light-limited rate (Jmax), and export or sucrose limited assimilation rate (Js). Leaf internal CO_2 concentration (Ci) is solved iteratively to account for stomatal and leaf-boundary layer conductance (gs) effects on the diffusion of atmospheric CO_2 (Ca) using either the Ball-Berry model (Collatz, Ball, Grivet, & Berry, 1991) or derivatives coupled to soil moisture (e.g., Sitch et al., 2003). Leaf respiration is typically estimated as a function of Vcmax (Collatz et al., 1991) to estimate net photosynthesis (Anet). In contrast, light-use efficiency approaches integrate photosynthetic active radiation with a conversion efficiency parameter to estimate carbon uptake that is modified by a series of limitations related to temperature and moisture. Autotrophic respiration is the sum of maintenance respiration [represented by an Arrhenius function (Lloyd & Taylor, 1994) and applied to leaf, sapwood, and root carbon pools] and growth respiration (as a fixed 25% of GPP and autotrophic respiration). In some modeling approaches, mainly those using light-use efficiency models to solve for GPP, NPP is simply a fixed fraction of GPP of around 45%–47% (Waring, Landsberg, & Williams, 1998).

Process-based models estimate above and belowground carbon stocks, and soil carbon, from the allocation of NPP and turnover to various plant storage organs. These storage organs include leaf biomass, heartwood, and sapwood biomass, and root biomass, and in some cases, nonstructural carbohydrates (Fatichi, Leuzinger, & Korner, 2013). The rules governing the allocation of NPP to storage organs follow either fixed ratios (i.e., Landsberg & Waring, 1997), or are driven by a resource-based approach (Friedlingstein, Joel, Field, & Fung, 1999), or by functional constraints, i.e., the "pipe" model of Shinozaki, Yoda, Hozumi, and Kira (1964) that links leaf area and sapwood area. The summation of the biomass pools are equal to total live biomass, and the partitioning to above and below ground components needs to be explicitly defined. The turnover of live biomass is determined by leaf phenology and longevity, fine-root turnover or mortality (from drought, NPP deficit or carbon starvation, fire, windthrow, pests, temperature extremes), with some modeling approaches including a branch turnover term. The carbon from the biomass pools first enters the litter, with some fraction returning to the atmosphere as CO_2, and then a fraction of the litter pool entering the soil mineral carbon pools, which are defined by their own turnover times and heterotrophic respiration response functions. The number of soil carbon pools can range from just two-bulk carbon pools, i.e., CENTURY (Parton, 1996) to more discretized pools as in the case of the Yasso model (Tuomi, Vanhala, Karhu, Fritze, & Liski, 2008). Disturbance, in the case of fire, converts live biomass and litter carbon to CO_2, based on a combination of area burned, fire intensity, and resiliency to fire (Thonicke, Spessa, Prentice, Harrison, & Carmona-Moreno, 2010). Land-use and land-cover change processes include transitions from natural land cover to managed states (i.e., crop or pasture) and the harvesting of wood from forested land cover states or crop or pasture biomass in the case of agriculture or grazing (Arneth et al., 2017; Calle & Poulter, 2020).

Process-based models require a fairly extensive set of driver data to estimate establishment, growth, and mortality-related carbon fluxes. For example, photosynthesis requires information on air temperature, shortwave radiation, relative humidity, precipitation (to determine soil moisture), and atmospheric CO_2 concentrations. Internal hydrologic schemes require information on soil physical structure, i.e., sand, silt, and clay content, that determines infiltration rates, soil-water holding capacity, subsurface drainage, and soil depth (van Genuchten, 1980; Cosby, Hornberger, Clapp, & Ginn, 1984). Disturbance subroutines may require information on lightning strikes, or on population density and its relation to human-driven fire ignitions (Thonicke et al., 2010). Land-use and land-cover change processes require information on land cover states, transitions, and management for both wood harvest and croplands (i.e., cropland type, irrigation, fertilizer use). The most common meteorological datasets used are those by gridding observational station data, i.e., the Climate Research Unit (Harris, Jones, Osborn, & Lister, 2013), that covers the period 1901-present data in monthly time step. Alternatively, "Reanalysis" products, which combine diverse sets of observations from in situ and spaceborne measurements with general circulation models, are used to produce hourly meteorological information, starting around 1980 (when observation density is high enough) to the present day. Hybridized observation-reanalysis products are used when DGVM models require subdaily, moderate-spatial resolution information, i.e., CRU-NCEP or CRU-JRA (Sitch et al., 2015; Wei et al., 2014). Soil data are typically used from globally consistent datasets, such as the Harmonized World Soils Database, which provides information on soil depth and physical and chemical constituents, for the global at 1-km resolution (Nachtergaele & van Velthuisen, 2008). Atmospheric CO_2 is used from globally averaged observations from the NOAA ESRL network, using sites representative of well-mixed marine boundary layer conditions (Conway, Tans, & Waterman, 1994).

Compared to bookkeeping models, there are a much larger number of DGVMs, and these have been used in various model intercomparison project (MIP) activities to understand and quantify epistemological uncertainties. For example, the MsTMIP model intercomparison activity brought together modeling groups (including land-surface models without vegetation dynamics) to understand North American and global carbon dynamics (Huntzinger et al., 2017). Similarly, the annual TRENDY model intercomparison is used to assess the status of the land carbon sink on an annual time step (Friedlingstein et al., 2020). The Climate Model Intercomparison Project (CMIP) activities, which incorporate DGVMs, are used to inform the IPCC assessments for a variety of time scales, climate, and land-use change scenarios (Anav et al., 2013).

The MIP activities rely on an ensemble of models that are often associated with various scenarios that control for whether climate, CO_2, land cover are changing simultaneously or individually. These factorial model experiments allow for attribution of modeled fluxes and stocks to various drivers through factor separation (Stein & Alpert, 1993) or for the explicit calculation for the loss of additional sink capacity (Pongratz et al., 2014) or for the CO_2-fertilization effect (Frank et al., 2015).

Because of the complexity of DGVM modeling approaches and their goal to incorporate feedbacks, model ensembles have a large spread of estimates for component fluxes and for carbon stocks (Luo et al., 2012). The mean of ensemble output is often a more useful indicator of carbon fluxes and states. Uncertainties are due to model processes, model parameters, and model driver data, and these sources have been extensively assessed through perturbation and driver-ensemble analysis (i.e., Poulter, Ciais, et al., 2011; Poulter, Frank, et al., 2011). Data assimilation and the use of diagnostic methods by combining remote-sensing or field data constraints is one successful approach to reduce these uncertainties (MacBean et al., 2016), but these approaches are limited to the time period of available observations or may be biased toward the observational datasets and their own inherent uncertainties. The use of process-based approaches for greenhouse gas accounting is gaining appreciation (i.e., Petrescu et al., 2020) and as scientific-based accounting becomes more synergistic with policy-driven requirements, i.e., in the case of the Regional Carbon Cycle and Processes study (RECCAP, Canadell, 2012, and described in Chapter 1).

2.4 Data-driven methodologies

Data-driven methods use statistical models, trained on observations, to produce spatially and temporally continuous carbon flux and carbon stock estimates. In the case of carbon fluxes, towers using the eddy covariance method have been used to "upscale" point measurements to globally continuous products, i.e., for GPP, FLUXCOM (Jung et al., 2011) or FluxSat (Joiner et al., 2018; Yoshida et al., 2015), for soil respiration Hashimoto et al. (2015), for NEE, the VPRM model (Mahadevan et al., 2008). Similar upscaling techniques have been applied to soil chamber measurement networks to develop soil respiration products (Bond-Lamberty et al., 2020; Bond-Lamberty & Thomson, 2010; Warner, Bond-Lamberty, Jian, Stell, & Vargas, 2019). The development of carbon stock products combines optical remote-sensing metrics, like NDVI, with discontinuous observations of lidar using ICESat (a spaceborne lidar) returns (Baccini et al., 2012; Saatchi et al., 2011). In the future though, global coverage of lidar is expected from combining ICESAat-2 and GEDI (Duncanson et al., 2020), similar to how global coverage of L, X, and C-band radar currently exists (e.g., GLOBBIOMASS, Santoro et al., 2020). In comparison to bookkeeping or process-based models, data-driven approaches rely heavily on remote-sensing observations and other driver-data (i.e., meteorology, typically Reanalysis such as MERRA-2 (Gelaro et al., 2017) or ERA (Dee et al., 2011) as covariates in the statistical models, hence limiting their approach to the observational era.

Data-driven approaches are particularly useful for handling large datasets through the use of neural networks (NN), machine learning (ML), deep learning (DL), and artificial intelligence algorithms (AI) that can be used to extract complex nonlinear relationships (Reichstein et al., 2019). However, these approaches should be distinguished from diagnostic modeling approaches that combine remote-sensing data with semiempirical models to estimate GPP or NPP, i.e., the MODID GPP/NPP Product (MOD17, Running et al., 2004) or see Zhang et al. (2017) for alternative integration of MODIS surface reflectance with light-use efficiency modeling. These diagnostic approaches use surface reflectance, typically red and near-infrared reflectance, to estimate the normalized difference vegetation index (Tucker, 1979), or derivatives such as NIRv, which is the product of NDVI and NIR (Badgley, Field, & Berry, 2017), to relate photosynthesis to light availability and other growth limitations, including CO_2 (Stocker et al., 2020). Efforts to upscale flux tower data using solar-induced fluorescence (SIF) from TROPOMI, GOME, OCO-2 satellites are underway (see Hu, Liu, Guo, Du, & Liu, 2018).

Fig. 4 provides an example of how an upscaling workflow is implemented, using FLUXCOM (Jung et al., 2019). First, the flux tower data are processed,

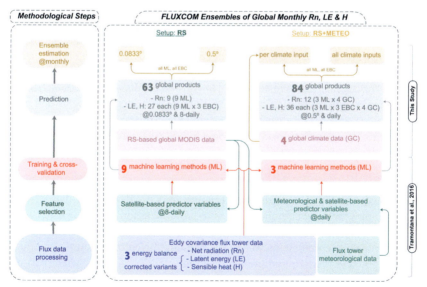

FIG. 4 The schematic representation shows how data from flux towers are combined with remote-sensing and meteorological information to train statistical models that can be used to generate global "upscaled" products. *(Reproduced from Jung, M., Koirala, S., Weber, U., Ichii, K., Gans, F., Camps-Valls, G., Papale, D., Schwalm, C., Tramontana, G., & Reichstein, M. 2019. The FLUXCOM ensemble of global land-atmosphere energy fluxes. Scientific Data, 6, 74.)*

including applying quality control filters, gap-filling, and standardization (units, time step, etc.). Predictor data from meteorological and/or satellite surface information are selected as covariates and then used in machine learning methods to develop predictive models. The predictive models are applied to global, gridded time series of the meteorological and remote-sensing surface data to produce global products. Given the computational efficiency of machine learning techniques, multiple tree ensembles can be developed to understand uncertainties (i.e., by withholding data used in training the model), and to evaluate the role of the predictor data in model performance. For example, some upscaling models use both meteorological and remote-sensing surface information whereas other modeling approaches find a high correlation with observations by using remote-sensing surface information alone, which to some degree, incorporates meteorological information implicitly (Joiner et al., 2018).

Data-driven approaches provide useful independent estimates for gross fluxes that can be compared with prognostic or diagnostic process-based modeling approaches. For example, Beer et al. (2010) used an ensemble-based approach to constrain global primary production to $123 + 8$ PgC yr^{-1}. More recent upscaling modeling approaches, e.g., FluxSat, suggest that meteorology can be excluded as a driver in estimating GPP, with surface reflectance variability containing equivalent information (Joiner et al., 2018). Global heterotrophic respiration has considerable uncertainty, given measurement challenges and partitioning autotrophic versus heterotrophic components. Recent upscaling work suggests a lower flux (40–50 Pg C yr^{-1}, Ciais, Gasser, et al., 2020) than process modeling approaches, yielding insights into the importance of lateral fluxes in removing carbon off-site. Modeling of net ecosystem fluxes and scaling to regional or global totals, has proven difficult because of sampling biases in flux tower networks (to mid-age, more productive ecosystems), and the representation of disturbance histories and lateral fluxes (Jung et al., 2011). Efforts to reconcile regional and global NEE estimates with process models and top-down inversions are currently underway as new information on disturbance history, forest age (Poulter et al., 2018), and lateral flux modeling are developed.

3 Relevance to Stock-Change and flux-based accounting

Combined, the three methods, bookkeeping, process-based, and data-driven, provide the necessary carbon variables required to complete the stock-change (Eq. 1) or flux-based (Eq. 5) approaches. The stock-change and the "gain-loss" methods used to complete the UNFCCC National Inventory Reports. The flux-based approach is more traditionally used in scientific studies, where various hypotheses on how the carbon cycle responds to climate, atmospheric CO_2, and other drivers can be tested. More recently, interest in combining the three approaches to reduce uncertainties has led to the process-based methods being considered for in UNFCCC reporting as described in the 2019 IPCC Refinement (2019). For example, the bookkeeping approaches have advanced to use activity data from country-reporting or from spaceborne measurements of forest gain and forest loss. Additionally, the bookkeeping approaches can now incorporate biomass stocks and biomass recovery curves derived from lidar satellite missions, moving beyond regional emission factors to fine-scale and spatially varying emission factors. In a similar way, process-based modeling approaches are also being adapted so that they can use lower uncertainty remote-sensing biomass products (Dubayah et al., 2020) but still provide model-based estimates for respiration and GPP carbon fluxes (Joetzjer et al., 2017). While complex, a comprehensive bottom-up accounting methodology is becoming increasingly necessary to help understand and partition GHG fluxes derived from top-down approaches (Chapter 4).

3.1 Uncertainties

Currently, the three methods require continued development to be accurate enough to estimate the GHG budget from regional to country and global scales. For example, considerable uncertainties remain in estimating sectoral emission sources and sinks that make tracking of emission changes over time difficult to detect in the context of natural background variability (Peters et al., 2017). For example, for CO_2 the imbalance of the global carbon budget ranges from -0.4 to $+0.5$ PgC yr^{-1} from the 1960s to the present (Friedlingstein et al., 2020). For CH_4, the TD and BU estimates of the total global methane sources differ by $\sim$150 Tg CH_4 yr^{-1}, most likely due to double counting of vegetated wetland and inland water methane emissions (Saunois et al., 2020). The COVID19 pandemic in 2020 and 2021 provided a case study to evaluate how well carbon monitoring networks can detect anomalies in anthropogenic emissions given natural variability (Diffenbaugh et al., 2020; Le Quéré et al., 2020). These COVID19 studies showed that the GHG satellite accuracy could detect XCO_2 (where "X" indicates the atmospheric column average) anomalies $\ll 1$ ppm but that ecosystem fluxes and synoptic scale meteorology impose a large and important signature (Buchwitz et al., 2021).

Moving forward, GHG budgets will have to combine all three approaches to be comprehensive (i.e., solve Eq. 1 and 5), and the budgets should also include ensemble-based and independent approaches as a way to quantify uncertainties (e.g., BU and TD comparisons). Ensemble-based approaches have proven important in identifying models that might be outliers, or using the ensemble mean as a representative, and often a more accurate source of information. Data-driven approaches, such as Fluxcom (Jung et al., 2019), can be used as a benchmark to evaluate process-based estimates, but data-driven uncertainties and their underlying model biases needs to be included in such comparisons. There are many gaps in observations and monitoring that span site to global scales (Ciais et al., 2014; Luyssaert et al., 2012). To some extent, new remote-sensing missions will help contribute wall-to-wall and time-varying information on forest structure (i.e., GEDI, ICESat-2), on solar-induced fluorescence (SIF, see OCO-2, TROPOMI), on canopy chemistry, using imaging spectroscopy [i.e., DESIS, CHIME, Surface Biology and Geology (SBG)]. However, direct field measurements are needed to provide continuity in ground-based observations, as well as for calibration and validation of remote-sensing products.

3.2 Comparisons between approaches

The methods described here have been used in numerous synthesis studies on greenhouse gases. In addition to the global carbon, methane, and nitrous oxide budgets coordinated by the Global Carbon Project (Friedlingstein et al., 2020; Saunois et al., 2020; Tian et al., 2020), these approaches have been used to complete national scientific assessments such as the United States Global Change Research Program's Second State of the Carbon Cycle Report (SOCCR-2, Hayes et al., 2012), and regional assessments, such as H2020 VERIFY project that covers European emission sources and sinks (Petrescu et al., 2020). These assessments have shown the value of including multiple methodologies as a way to understand and reconcile uncertainties. As discussed in Section 1.1, definitions have emerged as a large source of uncertainty between methods, and that harmonizing approaches to include the same areas, fluxes, and boundary conditions can greatly improve estimates (Grassi et al., 2018).

4 Conclusions

Here we have provided an overview of three methods that can be used to estimate emission sources and sinks to solve for the stock-change or flux-based approach. The methods are primarily used in scientific studies of greenhouse gas budgets and do not necessarily follow the strict criteria outlined by the IPCC Guidelines for National Greenhouse Gas Inventories (2006, 2019). The chapter points out areas of overlap where the methods have similarities with the IPCC guidelines, and also suggests areas of research for how these methods can be used to move toward higher tier, lower uncertainty GHG budgets. The topical chapters in the book (5–13) describe how these methods have been applied to develop GHG budgets for different land, ocean, and managed systems.

References

Anav, A., Friedlingstein, P., Kidston, M., Bopp, L., Ciais, P., Cox, P., et al. (2013). Evaluating the land and ocean components of the global carbon cycle in the CMIP5 earth system models. *Journal of Climate*, 26(18), 6801–6843.

Arneth, A., Sitch, S., Pongratz, J., Stocker, B. D., Ciais, P., Poulter, B., et al. (2017). Historical carbon dioxide emissions caused by land-use changes are possibly larger than assumed. *Nature Geoscience*, 10, 79–84.

Bottom-up approaches for estimating terrestrial GHG budgets **Chapter | 3 79**

Baccini, A., Goetz, S. J., Walker, W. S., Laporte, N. T., Sun, M., Sulla-Menashe, D., et al. (2012). Estimated carbon dioxide emissions from tropical deforestation improved by carbon-density maps. *Nature Climate Change, 2,* 182–185.

Baccini, A., Walker, W., Carvalho, L., Farina, M., Sulla-Menashe, D., & Houghton, R. A. (2017). Tropical forests are a net carbon source based on aboveground measurements of gain and loss. *Science, 358,* 230–234.

Badgley, G., Field, C. B., & Berry, J. A. (2017). Canopy near-infrared reflectance and terrestrial photosynthesis. *Science Advances, 3*(3), e1602244.

Beer, C., Reichstein, M., Tomelleri, E., Ciais, P., Jung, M., Carvalhais, N., et al. (2010). Terrestrial gross carbon dioxide uptake: Global distribution and covariation with climate. *Science, 329,* 834–838.

Bellassen, V., Le Maire, G., Dhote, J. F., Ciais, P., & Viovy, N. (2010). Modelling forest management within a global vegetation model—Part 1: Model structure and general behaviour. *Ecological Modelling, 221,* 2458–2474.

Bond-Lamberty, B., Christianson, D. S., Malhotra, A., Pennington, S. C., Sihi, D., AghaKouchak, A., et al. (2020). COSORE: A community database for continuous soil respiration and other soil-atmosphere greenhouse gas flux data. *Global Change Biology, 26,* 7268–7283.

Bond-Lamberty, B., & Thomson, A. (2010). A global database of soil respiration data. *Biogeosciences, 7,* 1915–1926.

Botkin, D. B., Janak, J. F., & Wallis, J. R. (1972). Some ecological consequences of a computer model of forest growth. *Journal of Ecology, 60,* 849–872.

Buchwitz, M., Reuter, M., Noël, S., Bramstedt, K., Schneising, O., Hilker, M., … Crisp, D. (2021). Can a regional-scale reduction of atmospheric CO_2 during the COVID-19 pandemic be detected from space? A case study for East China using satellite XCO_2 retrievals. *Atmospheric Measurement Techniques, 14,* 2141–2166.

Calle, L., & Poulter, B. (2020). Ecosystem age-class dynamics and distribution in the LPJwsl v2.0 global ecosystem model. *Geoscientific Model Development Discussions, 14*(5), 2575–2601.

Canadell, J. G. (2012). An international effort to quantify regional carbon fluxes. *EOS, Transactions, American Geophysical Union, 92,* 81–88.

Chapin, F. S., Woodwell, G. M., Randerson, J. T., Rastetter, E. B., Lovett, G. M., Baldocchi, D., et al. (2006). Reconciling carbon-cycle concepts, terminology, and methods. *Ecosystems, 9,* 1041–1050.

Ciais, P., Bastos, A., Chevallier, F., Lauerwald, R., Poulter, B., Canadell, P., et al. (2020). Definitions and methods to estimate regional land carbon fluxes for the second phase of the REgional Carbon Cycle Assessment and Processes Project (RECCAP-2). *Geoscientific Model Development Discussions.* https://doi.org/10.5194/gmd-2020-259 (in review, 2020).

Ciais, P., Dolman, A. J., Bombelli, A., Duren, R., Peregon, A., Rayner, P. J., et al. (2014). Current systematic carbon cycle observations and needs for implementing a policy-relevant carbon observing system. *Biogeosciences, 11,* 3547–3602.

Ciais, P., Gasser, T., Luaerwald, R., Peng, S., Raymond, P., Wang, Y., et al. (2020). Observed regional carbon budgets imply reduced soil heterotrophic respiration. *National Science Reviews, 8*(2), nwaa145.

Collatz, G. J., Ball, J. T., Grivet, C., & Berry, J. A. (1991). Physiological and environmental regulation of stomatal conductance, photosynthesis and transpiration: A model that includes a laminar boundary layer. *Agricultural and Forest Meteorology, 54,* 107–136.

Conchedda, G., & Tubiello, F. N. (2020). Drainage of organic soils and GHG emissions: validation with country data. *Earth System Science Data, 12,* 3113–3137.

80 Section | B Methods

Conway, T. J., Tans, P. P., & Waterman, L. S. (1994). Atmospheric CO_2 records from sites in the NOAA/CMDL air sampling network. In T. Boden, D. P. Kaiser, R. J. Sepanski, & F. W. Stoss (Eds.), *Trends '93: A compendium of data on global change*. Tenn., U.S.: Carbon Dioxide Information Analysis Center, Oak Ridge National Laboratory, U.S. Department of Energy, Oak Ridge.

Cosby, B. J., Hornberger, G. M., Clapp, R. B., & Ginn, T. R. (1984). A statistical exploration of the relationships of soil moisture characteristics to the physical properties of soils. *Water Resources Research, 20*, 682–690.

Crisp, D., Atlas, R. M., Breon, F.-M., Brown, L. R., Burrows, J. P., Ciais, P., et al. (2004). The orbiting carbon observatory (OCO) mission. *Trace Constituents in the Troposphere and Lower Stratosphere, 34*, 700–709.

Davis, S. J., Burney, J. A., Pongratz, J., & Caldeira, K. (2014). Methods for attributing land-use emissions to products. *Carbon Management, 5*(2), 233–245.

Dee, D. P., Uppala, S. M., Simmons, A. J., Berrisford, P., Poli, P., Kobayashi, S., et al. (2011). The ERA-interim reanalysis: Configuration and performance of the data assimilation system. *Quarterly Journal of the Royal Meteorological Society, 137*, 553–597.

Diffenbaugh, N. S., Field, C. B., Appel, E. A., Azevedo, I. L., Baldocchi, D. D., Burke, M., et al. (2020). The COVID-19 lockdowns: A window into the earth system. *Nature Reviews Earth & Environment, 1*, 470–481.

Dubayah, R., Blair, J. B., Goetz, S., Fatoyinbo, L., Hansen, M., Healey, S., et al. (2020). The global ecosystem dynamics investigation: High-resolution laser ranging of the Earth's forests and topography. *Science of Remote Sensing, 1*, 100002.

Duncanson, L., Neuenschwander, A., Hancock, S., Thomas, N., Fatoyinbo, T., Simard, M., et al. (2020). Biomass estimation from simulated GEDI, ICESat-2 and NISAR across environmental gradients in Sonoma County, California. *Remote Sensing of Environment, 242*, 111779.

Farquhar, G. D., O'Leary, M. H., & Berry, J. A. (1982). On the relationship between carbon isotope discrimination and the intercellular carbon dioxide concentration in leaves. *Australian Journal of Plant Physiology, 9*, 121–137.

Fatichi, S., Leuzinger, S., & Korner, C. (2013). Moving beyond photosynthesis: From carbon source to sink-driven vegetation modeling. *New Phytologist, 201*, 1086–1095.

Fisher, J., Huntzinger, D., Schwalm, C. R., & Sitch, S. (2014). Modeling the terrestrial biosphere. *Annual Review of Environment and Resources, 39*, 91–123.

Frank, D., Poulter, B., Saurer, M., Esper, J., Huntingford, C., Helle, G., et al. (2015). Water use efficiency and transpiration across European forests during the Anthropocene. *Nature Climate Change, 5*, 579–583.

Friedlingstein, P., Joel, G., Field, C. B., & Fung, I. Y. (1999). Toward an allocation scheme for global terrestrial carbon models. *Global Change Biology, 5*, 755–770.

Friedlingstein, P., O'Sullivan, M., Jones, M. W., Andrew, R. M., Hauck, J., Olsen, A., et al. (2020). Global carbon budget 2020. *Earth System Science Data, 12*, 3269–3340.

Gasser, T., Crepin, L., Quilcaille, Y., Houghton, R. A., Ciais, P., & Obersteiner, M. (2020). Historical CO_2 emissions from land use and land cover change and their uncertainty. *Biogeosciences, 17*, 4075–4101.

Gelaro, R., McCarty, W., Suárez, M. J., Todling, R., Molod, A., Takacs, L., et al. (2017). The modern-era retrospective analysis for research and applications, version 2 (MERRA-2). *Journal of Climate, 30*, 5419–5454.

Grassi, G., House, J., Kurz, W. A., Cescatti, A., Houghton, R. A., Peters, G. P., et al. (2018). Reconciling global-model estimates and country reporting of anthropogenic forest CO_2 sinks. *Nature Climate Change, 8*, 914–920.

Bottom-up approaches for estimating terrestrial GHG budgets Chapter | 3 **81**

Hansis, E., Davis, S. J., & Pongratz, J. (2015). Relevance of methodological choices for accounting of land use change carbon fluxes. *Global Biogeochemical Cycles, 29*(8), 1230–1246.

Harris, I., Jones, P. D., Osborn, T. J., & Lister, D. H. (2013). Updated high-resolution grids of monthly climatic observations—The CRU TS3.10 dataset. *International Journal of Climatology, 34*, 623–642.

Harris, N. L., Brown, S., Hagen, S., Saatchi, S. S., Petrova, S., Salas, W., et al. (2012). Baseline map of carbon emissions from deforestation in tropical regions. *Science, 336*, 1573–1576.

Harris, N. L., Gibbs, D. A., Baccini, A., Birdsey, R. A., de Bruin, S., Farina, M., et al. (2021). Global maps of twenty-first century forest carbon fluxes. *Nature Climate Change, 11*, 234–240.

Hashimoto, S., Carvalhais, N., Ito, A., Migliavacca, M., Nishina, K., & Reichstein, M. (2015). Global spatiotemporal distribution of soil respiration modeled using a global database. *Biogeosciences, 12*, 4121–4132.

Hayes, D. (2012). The need for "Apples-to-Apples" comparisons of carbon dioxide source and sink estimates. *EOS, Transactions, American Geophysical Union, 93.*

Hayes, D. J., Turner, D. P., Stinson, G., McGuire, A. D., Wei, Y., West, T. O., et al. (2012). Reconciling estimates of the contemporary north American carbon balance among terrestrial biosphere models, atmospheric inversions, and a new approach for estimating net ecosystem exchange from inventory-based data. *Global Change Biology, 18*, 1282–1299.

Hong, C., Burney, J. A., Pongratz, J., Nabel, J. E. M. S., Mueller, N. D., Jackson, R. B., et al. (2021). Global and regional drivers of land-use emissions in 1961–2017. *Nature, 589*, 554–561.

Hooijer, A., Page, S., Canadell, J. G., Sivius, M., Kwadijk, J., Wosten, H., et al. (2010). Current and future CO_2 emissions from drained peatlands in Southeast Asia. *Biogeosciences, 7*, 1505–1514.

Houghton, R. A. (2020). Terrestrial fluxes of carbon in GCP carbon budgets. *Global Change Biology, 26*, 3006–3014.

Houghton, R. A., Hobbie, J. E., Melillo, J. M., Moore, B., Peterson, B. J., Shaver, G. R., et al. (1983). Changes in the carbon content of terrestrial biota and soils between 1860 and 1980: A net release of CO'_2 to the atmosphere. *Ecological Monographs, 53*, 235–262.

Houghton, R., & Nassikas, A. A. (2017). Global and regional fluxes of carbon from land use and land cover change 1850–2015. *Global Biogeochemical Cycles, 31*, 456–472.

Hu, J., Liu, L., Guo, J., Du, S., & Liu, X. (2018). Upscaling solar-induced chlorophyll fluorescence from an instantaneous to daily scale gives an improved estimation of the gross primary productivity. *Remote Sensing, 10*(10), 1663.

Huntzinger, D. N., Michalak, A. M., Schwalm, C., Ciais, P., King, A. W., Fang, Y., et al. (2017). Uncertainty in drivers of terrestrial carbon sink undermines carbon-climate feedback predictions. *Scientific Reports, 7*, 4765.

Hurtt, G. C., Chini, L., Sahajpal, R., Frolking, S., Bodirsky, B. L., Calvin, K., et al. (2020). Harmonization of global land-use change and management for the period 850–2100 (LUH2) for CMIP6. *Geoscientific Model Development, 13*, 5425–5464.

IPCC. (2006). In S. Eggleston, L. Buendia, K. Miwa, T. Ngara, & K. Tanabe (Eds.), *IPCC guidelines for national greenhouse gas inventories.* Hayama, Japan: (Prepared by the National Greenhouse Gas Inventory Programme), published by the Institute for Global Environmental Strategies, Hayama, Japan, IPCC-TSUNGGIP,IGES.

IPCC. (2013). In T. F. Stocker, D. Qin, G.-K. Plattner, M. Tignor, S. K. Allen, J. Boschung, … P. M. Midgley (Eds.), *Climate change 2013: The physical science basis. Contribution of working group I to the fifth assessment report of the intergovernmental panel on climate change.* Cambridge, United Kingdom/New York, NY, USA: Cambridge University Press.

IPCC. (2019). Chapter 4: Forest land. In D. Blain, F. Agus, M. A. Alfaro, & H. Vreuls (Eds.), *Refinement to the 2006 IPCC guidelines for National Greenhouse Gas Inventories (Vol. 4): Agriculture, forestry and other land use* (p. 68). IPCC (advance version).

82   Section | B Methods

IPCC SRCCL. (2019). In P. R. Shukla, J. Skea, E. C. Buendia, V. Masson-Delmotte, H.-O. Pörtner, D. C. Roberts, … J. Malley (Eds.), *Climate change and land: An IPCC special report on climate change, desertification, land degradation, sustainable land management, food security, and greenhouse gas fluxes in terrestrial ecosystems.*

Janssens-Maenhout, G., Crippa, M., Guizzardi, D., Muntean, M., Schaaf, E., Dentener, F., et al. (2019). EDGAR v4.3.2 Global atlas of the three major greenhouse gas emissions for the period 1970–2012. *Earth System Science Data, 11*(3), 959–1002.

Joetzjer, E., Pillet, M., Barbier, N., Barichivich, J., Chave, J., Herault, B., et al. (2017). Assimilating satellite-based canopy height within an ecosystem model to derive above ground forest biomass. *Geophysical Research Letters, 44*(13), 6823–6832.

Joiner, J., Yoshida, Y., Zhang, Y., Duveiller, G., Jung, M., Lyapustin, A., et al. (2018). Estimation of terrestrial global gross primary production (GPP) with satellite data-driven models and Eddy covariance flux data. *Remote Sensing, 10*(9), 1346.

Jung, M., Koirala, S., Weber, U., Ichii, K., Gans, F., Camps-Valls, G., et al. (2019). The FLUXCOM ensemble of global land-atmosphere energy fluxes. *Scientific Data, 6*, 74.

Jung, M., Reichstein, M., Margolis, H., Cescatti, A., Richardson, A. D., Arain, M. A., et al. (2011). Global patterns of land-atmosphere fluxes of carbon dioxide, latent heat, and sensible heat derived from eddy covariance, satellite, and meteorological observations. *Journal of Geophysical Research, 116*. https://doi.org/10.1029/2010JG001566.

Kondo, M., Patra, P. K., Sitch, S., Friedlingstein, P., Poulter, B., Chevallier, F., et al. (2019). State of the science in reconciling top-down and bottom-up approaches for terrestrial CO_2 budget. *Global Change Biology, 26*(3), 1068–1084.

Kurz, W. A., Dymond, C. C., White, T. M., Stinson, G., Shaw, C. H., Rampley, G. J., et al. (2009). CBM-CFS3: A model of carbon-dynamics in forestry and land-use change implementing IPCC standards. *Ecological Modelling, 220*, 480–504.

Kurz, W. A., Hayne, S., Fellows, M., MacDonald, J. D., Metsaranta, J. M., Hafer, M., et al. (2018). Quantifying the impacts of human activities on reported greenhouse gas emissions and removals in Canada's managed forest: Conceptual framework and implementation. *Canadian Journal of Forest Research, 48*, 1227–1240.

Landsberg, J. J., & Waring, R. H. (1997). A generalised model of forest productivity using simplified concepts of radiation-use efficiency, carbon balance and partitioning. *Forest Ecology and Management, 95*, 209–228.

Le Quéré, C., Andrew, R. M., Friedlingstein, P., Sitch, S., Pongratz, J., Manning, A. C., et al. (2018). Global carbon budget 2017. *Earth System Science Data, 10*, 405–448.

Le Quéré, C., Jackson, R. B., Jones, M. W., Smith, A. J. P., Abernethy, S., Andrew, R. M., et al. (2020). Temporary reduction in daily global CO2 emissions during the COVID-19 forced confinement. *Nature Climate Change, 10*, 647–653.

Lloyd, J., & Taylor, J. A. (1994). On the temperature dependence of soil respiration. *Functional Ecology, 8*, 315–323.

Luo, Y., Randerson, J. T., Abramowitz, G., Bacour, C., Blyth, E., Carvalhais, N., et al. (2012). A framework for benchmarking land models. *Biogeosciences, 9*(10), 3857–3874.

Luyssaert, S., Abril, G., Andres, R., Bastviken, D., Bellassen, V., Bergamaschi, P., et al. (2012). The European land and inland water CO_2, CO, CH_4 and N_2O balance between 2001 and 2005. *Biogeosciences, 9*, 3357–3380.

Maasakers, J. D., Jacob, D. J., Sulprizio, M. P., Turner, A. J., Weitz, M., Wirth, T., et al. (2016). Gridded National Inventory of U.S. methane emissions. *Environmental Science and Technology, 50*(23), 13123–13133.

Bottom-up approaches for estimating terrestrial GHG budgets **Chapter | 3** **83**

MacBean, N., Peylin, P., Chevallier, F., Scholze, M., & Schürmann, G. (2016). Consistent assimilation of multiple data streams in a carbon cycle data assimilation system. *Geoscientific Model Development*, *9*, 3569–3588.

Mahadevan, P., Wofsy, S. C., Matross, D. M., Xiao, X., Dunn, A. L., Lin, J. C., et al. (2008). A satellite-based biosphere parameterization for net ecosystem CO_2 exchange: Vegetation photosynthesis and respiration model (VPRM). *Global Biogeochemical Cycles*, *22*(2).

Miller, S. M., Michalak, A. M., Yadav, V., & Tadić, J. M. (2018). Characterizing biospheric carbon balance using CO_2 observations from the OCO-2 satellite. *Atmospheric Chemistry and Physics*, *18*, 6785–6799.

Monteith, J. L. (1977). Climate and the efficiency of crop production in Britain. *Philosophical Transactions of the Royal Society B: Biological*, *281*, 277–294.

Nachtergaele, F., & van Velthuisen, H. (2008). *The harmonized world soils database (HWSD)*. *ISRIC report 2016/02*.

Obermeier, W. A., Nabel, J. E. M. S., Loughran, T., Hartung, K., Bastos, A., Havermann, F., et al. (2021). Modelled land use and land cover change emissions—A spatio-temporal comparison of different approaches. *Earth System Dynamics*, *12*, 635–670.

Oda, T., Maksyutov, S., & Andres, R. J. (2020). The Open-source Data Inventory for Anthropogenic CO_2, version 2016 (ODIAC2016): A global monthly fossil fuel CO_2 gridded emissions data product for tracer transport simulations and surface flux inversions. *Earth System Science Data*, *10*, 87–107.

Pan, Y., Birdsey, R., Fang, J., Houghton, R. A., Kauppi, P. E., Kurz, W., et al. (2011). A large and persistent carbon sink in the World's forests. *Science*, *333*, 988–993.

Parton, W. J. (1996). The CENTURY model. In D. S. Powlson, P. Smith, & J. U. Smith (Eds.), *Evaluation of soil organic matter models* (pp. 283–291). Berlin Heidelberg: Springer.

Peters, G. P., Le Quéré, C., Andrew, R. M., Canadell, J. G., Friedlingstein, P., Ilyina, T., et al. (2017). Towards real-time verification of CO_2 emissions. *Nature Climate Change*, *7*, 848–850.

Petrescu, A. M. R., Abad-Vinas, R., Janssens-Maenhout, G., Blujdea, V. N. B., & Grassi, G. (2012). Global estimates of carbon stock changes in living forest biomass: EDGARv4.3—Time series from 1990 to 2010. *Biogeosciences*, *9*, 3437–3447.

Petrescu, A. M. R., Peters, G. P., Janssens-Maenhout, G., Ciais, P., Tubiello, F. N., Grassi, G., et al. (2020). European anthropogenic AFOLU greenhouse gas emissions: a review and benchmark data. *Earth System Science Data*, *12*, 961–1001.

Pongratz, J., Dolman, H., Don, A., Erb, K.-H., Fuchs, R., Herold, M., et al. (2017). Models meet data: Challenges and opportunities in implementing land management in earth system models. *Global Change Biology*, *24*(4), 1470–1487.

Pongratz, J., Reick, C. H., Houghton, R. A., & House, J. I. (2014). Terminology as a key uncertainty in net land use and land cover change carbon flux estimates. *Earth System Dynamics*, *5*, 177–195.

Poulter, B., Aragão, L., Andela, N., Bellassen, V., Ciais, P., Kato, T., … Shivdenko, A. (2018). The global forest age dataset (GFADv1.0), link to NetCDF file. NASA National Aeronautics and Space Administration, PANGAEA. https://doi.org/10.1594/PANGAEA.889943.

Poulter, B., Ciais, P., Hodson, E., Lischke, H., Maignan, F., Plummer, S., et al. (2011). Plant functional type mapping for earth system models. *Geoscientific Model Development*, *4*, 1–18.

Poulter, B., Frank, D. C., Hodson, E., & Zimmermann, N. (2011). Impacts of land cover and climate data selection on understanding terrestrial carbon dynamics and the CO_2 airborne fraction. *Biogeosciences*, *8*, 2027–2036.

Prentice, I. C., Bondeau, A., Cramer, W., Harrison, S. P., Hickler, T., Lucht, W., et al. (2007). Dynamic global vegetation modeling: Quantifying terrestrial ecosystem responses to large-scale environmental change. In P. Canadell, D. E. Pataki, & L. F. Pitelka (Eds.), *Terrestrial ecosystems in a changing world* (pp. 175–192). Berlin, Heidelberg, DE: Springer-Verlag.

Ramankutty, N., Gibbs, H. K., Achard, F., Defries, R., Foley, J. A., & Houghton, R. A. (2007). Challenges to estimating carbon emissions from tropical deforestation. *Global Change Biology*, *13*, 51–66.

Randerson, J. T., Chapin, F. S., III, Harden, J. W., Neff, J. C., & Harmon, M. E. (2002). Net ecosystem production: A comprehensive measure of net carbon accumulation by ecosystems. *Ecological Applications*, *12*, 937–947.

Reichstein, M., Camps-Valls, G., Stevens, B., Jung, M., Denzler, J., Carvalhais, N., et al. (2019). Deep learning and process understanding for data-driven earth system science. *Nature*, *566*, 195–204.

Running, S. W., Nemani, R. R., Heinsch, F. A., Zhao, M., Reeves, M., & Hashimoto, H. (2004). A continuous satellite-derived measure of global terrestrial primary production. *Bioscience*, *54*, 547–560.

Saatchi, S. S., Harris, N. L., Brown, S., Lefsky, M. A., Mitchard, E. T. A., Salas, W., et al. (2011). Benchmark map of forest carbon stocks in tropical regions across three continents. *Proceedings of the National Academy of Science*, *108*, 9899–9904.

Santoro, M., Cartus, O., Carvalhais, N., Rozendaal, D., Avitabilie, V., Araza, A., et al. (2020). The global forest above-ground biomass pool for 2010 estimated from high-resolution satellite observations. *Earth System Science Data Discussions*, *2020*, 1–38.

Saunois, M., Stavert, A. R., Poulter, B., Bousquet, P., Canadell, J. G., Jackson, R. B., et al. (2020). The global methane budget 2000–2017. *Earth System Science Data*, *12*(3), 1561–1623.

Schlesinger, W. H. (1984). Soil organic matter: A source of atmospheric CO_2. In G. M. Woodwell (Ed.), *The role of terrestrial vegetation in the global carbon cycle: Measurement by remote sensing* John Wiley & Sons, Ltd.

Schulze, E. D., Wirth, C., & Heimann, M. (2000). Managing forests after Kyoto. *Science*, *289*, 2058–2059.

Shinozaki, K., Yoda, K., Hozumi, K., & Kira, T. (1964). A quantitative analysis of plant form—The pipe model theory. I. Basic analyses. *Japanese Journal of Ecology*, *14*, 97–105.

Sitch, S., Friedlingstein, P., Gruber, N., Jones, S., Murray-Tortarolo, G., Ahlström, A., … Myneni, R. (2015). Recent trends and drivers of regional sources and sinks of carbon dioxide. *Biogeosciences*, *12*, 653–679.

Sitch, S., Smith, B., Prentice, I. C., Arneth, A., Bondeau, A., Cramer, W., et al. (2003). Evaluation of ecosystem dynamics, plant geography and terrestrial carbon cycling in the LPJ dynamic global vegetation model. *Global Change Biology*, *9*, 161–185.

Stein, U., & Alpert, P. (1993). Factor separation in numerical simulations. *Journal of Atmospheric Sciences*, *50*, 2107–2115.

Stocker, B. D., Wang, H., Smith, N. G., Harrison, S. P., Keenan, T. F., Sandoval, D., et al. (2020). P-model v1.0: An optimality-based light use efficiency model for simulating ecosystem gross primary production. *Geoscientific Model Development*, *13*, 1545–1581.

Thonicke, K., Spessa, A., Prentice, I. C., Harrison, S. P., & Carmona-Moreno, C. (2010). The influence of vegetation, fire spread and fire behaviour on biomass burning and trace gas emissions: Results from a process-based model. *Biogeosciences*, *7*, 1991–2011.

Thonicke, K., Spessa, A., Prentice, I. C., Harrison, S. P., Dong, L., & Carmona-Moreno, C. (2010). The influence of vegetation, fire spread and fire behaviour on biomass burning and trace gas emissions: Results from a process-based model. *Biogeosciences*, *7*(6), 1991–2011.

Tian, H., Xu, R., Canadell, J. G., Thompson, R. L., Winiwarter, W., Suntharalingam, P., et al. (2020). A comprehensive quantification of global nitrous oxide sources and sinks. *Nature*, *586*, 248–256.

Tucker, C. J. (1979). Red and photographic infrared linear combinations for monitoring vegetation. *Remote Sensing of Environment, 8*, 127–150.

Tuomi, M., Vanhala, P., Karhu, K., Fritze, H., & Liski, J. (2008). Heterotrophic soil respiration—comparison of different models describing its temperature dependence. *Ecological Modelling, 211*(1–2), 182–190.

van der Werf, G. R., Randerson, J. T., Giglio, L., Collatz, G. J., Mu, M., Kasibhatla, P. S., et al. (2010). Global fire emissions and the contribution of deforestation, savanna, forest, agricultural, and peat fires (1997–2009). *Atmospheric Chemistry Physics, 10*, 11707–11735.

van Genuchten, M. T. (1980). A closed-form equation for predicting the hydraulic conductivity of unsaturated soils. *Soil Science Society of America Journal, 44*, 892–898.

von Schuckmann, K., Cheng, L., Palmer, M. D., Hansen, J., Tassone, C., Aich, V., et al. (2020). Heat stored in the earth system: Where does the energy go? *Earth System Science Data, 12*, 2013–2041.

Waring, R. H., Landsberg, J. J., & Williams, M. (1998). Net primary productivity of forests: A constant fraction of gross primary production? *Tree Physiology, 18*, 129–134.

Warner, D. L., Bond-Lamberty, B., Jian, J., Stell, E., & Vargas, R. (2019). Spatial predictions and associated uncertainty of annual soil respiration at the global scale. *Global Biogeochemical Cycles, 33*, 1733–1745.

Wei, Y., Liu, S., Huntzinger, D. N., Michalak, A. M., Viovy, N., Post, W. M., et al. (2014). The north American carbon program multi-scale synthesis and terrestrial model Intercomparison project—Part 2: Environmental driver data. *Geoscientific Model Development, 7*, 2875–2893.

WMO. (2020). *United in science 2020*. World Meteorological Organization.

Yoshida, Y., Joiner, J., Tucker, C., Berry, J., Lee, J.-E., Walker, G., et al. (2015). The 2010 Russian drought impact on satellite measurements of solar-induced chlorophyll fluorescence: Insights from modeling and comparisons with parameters derived from satellite reflectances. *Remote Sensing of Environment, 166*, 163–177.

Zhang, Y., Xiao, X., Wu, X., Zhou, S., Zhang, G., Qin, Y., & Dong, J. (2017). A global moderate resolution dataset of gross primary production of vegetation for 2000–2016. *Scientific Data, 4*, 170165.

Zheng, B., Chevallier, F., Yin, Y., Ciais, P., Fortems-Cheiney, A., Deeter, M. N., et al. (2019). Global atmospheric carbon monoxide budget 2000–2017 inferred from multi-species atmospheric inversions. *Earth System Science Data, 11*, 1411–1436.

Zinke. (1986). *Global organic soil carbon and nitrogen*.

Chapter 4

Top-down approaches

Rona L. Thompson[a], Frédéric Chevallier[b], Shamil Maksyutov[c], Prabir K. Patra[d], and Kevin Bowman[e]

[a]NILU—Norsk Institutt for Luftforskning, Kjeller, Norway, [b]Laboratoire des Sciences du Climat et de l'Environnement, Gif sur Yvette, France, [c]National Institute for Environmental Studies, Tsukuba, Japan, [d]Center for Environmental Remote Sensing, Chiba University, Chiba, Japan, [e]Jet Propulsion Laboratory, California Institute of Technology, CA, United States

Key messages

- Top-down approaches reconcile fluxes of greenhouse gases with atmospheric observations and thus provide a mass-balance constraint on estimates derived from other methods.
- Top-down approaches can be applied at global as well as regional scales using the same methodology and can be applied to any relatively long-lived atmospheric species.
- Top-down approaches are most reliable on large spatiotemporal scales, owing to the correlation of errors, which is in contrast (and complementary) to the majority of bottom-up approaches, which are most accurate at small scales. Similarly, top-down approaches are most reliable for the total (or net) flux but provide only limited information on contribution from different sources and sinks.
- In top-down approaches, random errors are relatively straightforward to estimate, however, systematic error is more elusive. The use of model ensembles can help quantify transport error and provide a more comprehensive uncertainty estimate.

1 Introduction

Top-down approaches involve proceeding from a general picture to the details. In this chapter, we will see how atmospheric measurements of greenhouse gases can provide a general picture of the emissions and sinks and how, with the help of an atmospheric transport model, these can be used to gain more detailed information about where and when the emissions and sinks occur and how large they are.

Balancing Greenhouse Gas Budgets. https://doi.org/10.1016/B978-0-12-814952-2.00008-3
Copyright © 2022 Elsevier Inc. All rights reserved.

88 Section | B Methods

We can think of the atmosphere as a box in which all emissions that enter it are slowly mixed. Since the mixing is incomplete, gradients in concentration occur, and it is precisely these gradients that are used to establish the pattern of emissions. In the simplest case, we can imagine a single point source, such as a factory, emitting some species into the atmosphere creating a plume. If the concentration of this species is measured continuously at several points around the factory, and knowing the wind direction, we can estimate its location. With further information, such as the wind speed and the buoyancy of the air, we can also estimate how much the factory is emitting. Establishing the emissions of the major greenhouse gases (CO_2, CH_4, and N_2O), however, is vastly more complicated, as there are many sources, both point and diffuse, as well as sinks. To extract information on the pattern of emissions in this case, we need to use an atmospheric transport model and apply statistical methods to determine the most likely solution. Atmospheric transport models relate emissions to atmospheric concentrations. In this case, however, the emissions may be unknown or at best, only poorly known, while atmospheric concentrations can be measured. Thus, to solve the problem the atmospheric transport model needs to be effectively *inverted*, i.e., to relate the concentrations to the emissions. Hence, this approach is referred to as *inverse modeling*.

Inverse modeling is used widely in earth sciences, e.g., in seismology to locate the epicenter of an earthquake, or in atmospheric sciences to determine temperature profiles from satellite measurements. In our context, the method was first used in the late 1980s to determine the global pattern of sources and sinks of CO_2 (Tans, Conway, & Nakazawa, 1989) and in 1990 to determine the emissions of N_2O (Prinn et al., 1990). Since then, the method has grown enormously in use and is now widely applied for estimating emissions of numerous atmospheric species on global and as well regional scales.

This chapter introduces the reader to the principles of inverse modeling and its application to estimating emissions of CO_2, CH_4, and N_2O. Since this is a very broad subject itself, it is not possible to go into fine details, however, suggestions for additional reading are given throughout. Before embarking further on top-down approaches, it is necessary to define some terminology. The term *emissions* will be used from hereon to describe fluxes from the surface to the atmosphere (this is a *positive* flux) while the term *flux* will be used in the case the fluxes are in both directions (i.e., positive and negative). The term *source* will be used to refer to a specific type of emission, e.g., emissions of CO_2 from the combustion of fossil fuels, and similarly for the term *sink*, e.g., the uptake of CO_2 through photosynthesis or, for CH_4, the oxidation is by hydroxyl radicals in the atmosphere.

In the following sections, the reader will be acquainted with the types of atmospheric observations available and how they are made (Section 2), introduced to the concepts of atmospheric chemistry and transport modeling (Section 3), and learn about optimization methods and how these are implemented (Section 4). Sections 5–8, present applications of inverse modeling to estimating fluxes of CO_2, CH_4, and other greenhouse gases and, lastly, Sections 9 and 10 address the uncertainties associated with top-down methods and ways to validate the results.

2 Measurements of greenhouse gases in the atmosphere

The longest-running time series of atmospheric measurements are the CO_2 records from La Jolla Pier, California, from Mauna Loa, Hawaii, and from South Pole Observatory. These records were started in the late 1950s by Professor Keeling (1960) and provided the first evidence of rising CO_2 as a result of the combustion of fossil fuels and land-use changes. The curve at Mauna Loa is now referred to as the "Keeling Curve." Since then, atmospheric measurements of CO_2, and other greenhouse gases, have proliferated and include ground-based networks as well as remote sensing by satellites.

2.1 Ground-based measurements

Ground-based measurements refer to those made close to the Earth's surface, typically from a mast or tower at a height of a few meters to over a hundred meters above ground, or from a ship, aircraft, or even a balloon. These measurement samples are taken from the air in the surface or mixed layer of the atmosphere and, in the case of aircraft and balloons, the measurements are taken from the free troposphere (Fig. 1) either by discrete samples, such as flasks or continuously by an in situ instrument. Both types of measurements involve sampling of ambient air through an inlet using a pump to create an underpressure and thus a flow of air. The air is filtered for particles and dried by cooling and trapping condensation (often used for continuous sampling) or through chemical drying (often used for discrete samples). The flow rate and pressure of the air are controlled to ensure the reproducibility of the sampling procedure. Atmospheric measurements of gases are reported as dry air mole fractions $(\text{mol}\,\text{mol}^{-1})$ or equivalently as volume mixing ratios given in, e.g., parts per million (ppm). Although the term *concentration* is often used in reference to atmospheric gas measurements, strictly speaking, it is not correct. Concentration is a measure of the ratio of the mass (or moles) of a species to volume $(\text{mol}\,\text{V}^{-1})$. From hereon, we will use the term *mixing ratio* when discussing atmospheric measurements.

There are numerous analytical techniques used to measure air samples. For CO_2, spectroscopic methods are most widely used taking advantage of the strong absorption band in the near-infrared at $\sim 4.26\,\mu\text{m}$ (e.g., Crosson, 2008). For CH_4 and N_2O, the absorption bands are not as strong or as distinct as for CO_2, therefore, these species are predominantly measured using gas chromatographs equipped with a flame ionization detector for CH_4 and an election capture detector for N_2O (Weiss, 1981). However, developments in spectroscopic techniques over the past couple of decades, including a greatly increased effective absorption path length and the use of lasers to provide a narrow range of light frequency to target specific absorption bands, have enabled CH_4 and

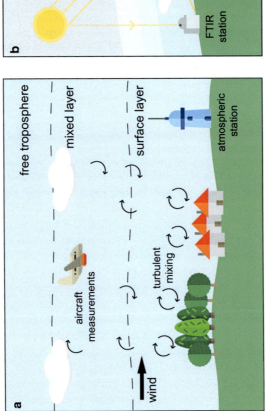

FIG. 1 Atmospheric observing systems. (A) Ground-based observations from a tower sampling air in the surface or mixed layer. Additional observations from, e.g., aircraft provide information on the vertical profile and mixing ratios in the mixed layer and free troposphere. (B) Satellite observations from Nadir and Limb viewing. Also shown are observations from FTIR and aircraft, which are used to validate satellite observations.

N_2O to be measured with sufficient precision for atmospheric monitoring. Thus gas chromatographs are now slowly being replaced by modern spectroscopic methods, such as integrated cavity output spectroscopy (ICOS) (O'Keefe, 1998) and cavity ring-down spectroscopy (CRDS) (Zalicki & Zare, 1995). These methods typically use the CH_4 absorption band at $\sim$1.65 µm (Rella, Hoffnagle, He, & Tajima, 2015) and the N_2O band at $\sim$4.55 µm (Harris et al., 2014).

Regardless of the measurement technique used, calibration is necessary to ensure consistency in time and among instruments. For each species, there are established calibration scales defined by a set of primary gas standards (Aoki, Nakazawa, Murayama, & Kawaguchi, 1992; Dlugokencky et al., 2005; Hall, Sutton, & Elkins, 2007; Zhao, Tans, & Thoning, 1997). The scales are propagated to each instrument through secondary gas standards, which have been calibrated against the primaries. To make atmospheric measurements comparable among laboratories worldwide, regular exchanges of gas standards are made, such as within the Round Robin Comparison Experiment organized by the World Meteorological Organization (WMO).

A number of programs exist for operating long-term atmospheric measurement networks and observatories. Among some of the most important networks used for estimating emissions of greenhouse gases are the Global Greenhouse Gas Reference Network operated by the National Oceanic and Atmospheric Administration (NOAA ESRL), the Advanced Global Atmospheric Gases Experiment (AGAGE), and the Integrated Carbon Observation System (ICOS). The NOAA network consists of discrete sampling at more than 50 sites at approximately weekly intervals, continuous measurements at 15 sites, and sampling from aircraft at 16 sites every 2–3 weeks (https://www.esrl.noaa.gov/gmd/ccgg/). AGAGE is the longest operating network with measurements starting in 1978 and currently consists of 13 sites with continuous measurements (http://agage.mit.edu). ICOS is a European infrastructure project primarily for measurements of CO_2 and CH_4 and consists of more than 25 sites in Europe and the Arctic with continuous measurements (https://www.icos-ri.eu/home). (Information on how the data can be obtained is available from the links provided.)

2.2 Satellite measurements

Remote sensing measurements exploit the wavelength-dependent molecular absorption, emission, and scattering of radiation to infer greenhouse gases concentrations and their distribution (Chance & Martin, 2017). From spectroscopy, the molecular dipole structure of CO_2 and CH_4 lead to absorption at longwave and shortwave infrared wavelengths. For example, CO_2 strongly absorbs in a wavelength band around 15 µm, which is also the range where thermal emission peaks from the Earth's surface. It is this absorption that makes changes in CO_2 concentrations, the primary driver of radiative forcing and consequently, climate change (Myhre et al., 2013). CO_2 also strongly absorbs reflected solar radiation at shorter wavelengths such as 1.6 and 2.06 µm. In conjunction with

92 Section | B Methods

knowledge of total column pressure, which can be obtained from well-known molecular oxygen abundances, the total column of CO_2, i.e., the total number of CO_2 molecules in an atmospheric column can be inferred. These basic physical principles are the foundation of satellite measurements of greenhouse gases.

In order to use these principles with remote sensing measurements, the details of instrument location, effects, and noise must be incorporated. To that end, the relationship between the observed radiances at a satellite and the atmospheric state, including trace gases, can be described by an additive noise model:

$$\mathbf{y} = F(\mathbf{c}) + \mathbf{n} \tag{1}$$

where $\mathbf{y} \in \mathbb{R}^m$ is a vector whose m elements contain discretized radiance measurements, $\mathbf{n} \in \mathbb{R}^m$ accounts for measurement error, which may be both random and systematic, and $\mathbf{c} \in \mathbb{R}^n$ is a vector whose n elements is the geophysical state generally discretized in altitude or pressure also including surface characterization. The forward model, $F: \mathbb{R}^n \rightarrow \mathbb{R}^m$ projects the atmospheric state to synthetic radiances incorporating surface, atmospheric, and instrumental effects. The forward model incorporates models of the atmosphere, surface, and instrument along with a mechanism to propagate radiation between them. That mechanism, radiative transfer, is a critical component of the forward model that accounts for the absorption and scattering of radiation with matter (e.g., Spurr, 2006). As it is a relatively expensive part of the calculation, significant effort is put into fast algorithms (e.g., Spurr, Natraj, Lerot, Van Roozendael, & Loyola, 2013). The inference of greenhouse gas (GHG) mixing ratios from these satellite measurements is known as *retrieval*, which can be described generally as an inverse model (Tarantola, 2005):

$$\hat{\mathbf{c}} = G(\mathbf{y}) \tag{2}$$

where $G: \mathbb{R}^m \rightarrow \mathbb{R}^n$ and $\hat{\mathbf{c}}$ is the estimated geophysical state. There are several methods to implement a retrieval algorithm such as Markov Chain Monte Carlo or Machine Learning (Crevoisier et al., 2009; Tamminen, 2004). However, the most widespread approach in the atmospheric composition is based on optimal estimation (e.g., Bowman et al., 2006; Deeter, Emmons, Edwards, Drummond, & Gille, 2004; O'Dell et al., 2012; Rodgers, 2000), which incorporates important a priori statistical information about the atmosphere and computes critical error diagnostics needed for quantitative comparisons and assimilation.

The retrieval process starts with an initial geophysical state, which describes the vertical distribution of the atmospheric state including the target species (i.e., CO_2 or CH_4), temperature, aerosols, and water vapor along with surface properties, e.g., albedo. A radiative transfer algorithm, which accounts for the absorption and scattering of radiation through the atmosphere, is used to compute radiation leaving the atmosphere. Once instrumental effects are included, this synthetic radiation is compared to a satellite-measured radiance. The geophysical state is adjusted through nonlinear optimization techniques until the measured and computed radiation match within the precision of the instrument. This updated state, e.g., CO_2, is known as the *retrieved* state. With these techniques, the GHG estimate is influenced by both the radiance measurements and a priori information is used to *regularize* the estimate (Engl, Hanke, & Neubauer, 1996).

A fundamental diagnostic is a relationship between the retrieved and the true, but unknown, state. Within an optimal estimation framework, this relationship is given by

$$\hat{\mathbf{c}} = \mathbf{c_a} + \mathbf{A}(\mathbf{c} - \mathbf{c_a}) + \mathbf{n} \tag{3}$$

where $\mathbf{c_a}$ is the a priori vector and $\mathbf{A}$ is the averaging kernel matrix defined as

$$\mathbf{A} = \frac{\partial \hat{\mathbf{c}}}{\partial \mathbf{c}} \tag{4}$$

which is the sensitivity of the retrieved state to the actual or true state. It is important to note that, in general, the optimal estimate $\hat{\mathbf{c}}$ includes other parameters that impact the measured radiances, e.g., aerosols, in addition to the GHG mixing ratios. The uncertainties and biases in the GHG mixing ratio are correlated with the uncertainties in the other parts of the geophysical state (e.g., Worden et al., 2004). Given the relatively small variations in GHG abundances compared to the relatively large variations in other geophysical parameters, the impact of bias in the estimate has become an important area of focus.

Eq. (3) also serves another purpose: it is the basis of a retrieval observation operator (Jones et al., 2003; Kaminski & Mathieu, 2017; Migliorini, Piccolo, & Rodgers, 2008; Rodgers, 2000), which is represented as

$$H(\cdot) = \mathbf{c_a} + \mathbf{A}(\cdot - \mathbf{c_a}) \tag{5}$$

where $H(\cdot)$ is an operator that maps a modeled mixing ratio to the corresponding retrieved mixing ratio. With this operator, a synthetic retrieval can be computed given a model prediction of the atmospheric state. This synthetic retrieval can be directly related to a satellite retrieval allowing for an "apples-to-apples" comparison by explicitly describing the retrieval process. Observation operators for each satellite retrieval are incorporated into GHG assimilation systems (see Section 4).

Thermal infrared instruments such as the Tropospheric Emission Spectrometer (TES) (Beer, 2006) and Atmospheric Infrared Sounder (AIRS) (Aumann et al., 2003) directly measure the greenhouse gas effect of CO_2 in the 10–15 μm band. However, they are primarily sensitive to free tropospheric CO_2, which is relatively well-mixed and consequently has been challenging to use (Chevallier et al., 2009; Engelen & Stephens, 2004; Kulawik et al., 2010; Nassar et al., 2011). Carbon dioxide measurements from the near-infrared were pioneered by the SCIAMACHY satellite (Buchwitz et al., 2005) followed by the Japan Aerospace Exploration Agency (JAXA) Greenhouse gases Observing Satellite "IBUKI" (GOSAT) and the National Space and Aeronautics Administration (NASA) Orbital Carbon Observatory (OCO-2) satellites (Crisp, Miller,

94 Section | B Methods

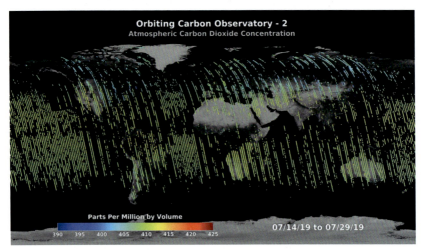

FIG. 2 An example of atmospheric column measurements of CO_2 (XCO_2) from the OCO-2 satellite between June 1 and July 2, 2015. *(Courtesy of D. Crisp, OCO-2/NASA.)*

& DeCola, 2008; Kuze, Suto, Nakajima, & Hamazaki, 2009). As shown in Fig. 2, these measurements are usually defined in terms of column-averaged CO_2 dry air mole fraction X_{CO2}. These measurements must be both accurate and precise in order to capture small spatiotemporal gradients (<2.5%) imposed by regional carbon fluxes (O'Dell et al., 2012). Consequently, validation of these data is a central focus, the pillar of which has been a suite of ground-based high-resolution uplooking Fourier transform spectrometers known as the Total Carbon Column Observing Network (TCCON) (Wunch et al., 2011). Comparisons of these to OCO-2 V8 data have shown correlations greater than 0.9 and mean differences between 0.2 and 0.3 ppm (O'Dell et al., 2018). The differences have been further used to assess predicted uncertainties and to tease out systematic and correlated errors. The calculation of the uncertainties in the retrievals can be computed directly from optimal estimation and compared with errors computed from independent measurements. If the errors exceed the expected uncertainties, then additional, systematic errors can be quantified and investigated. Kulawik et al. (2016) compared successively averaged GOSAT data around TCCON sites with theoretical reductions in precision to show that correlated errors were 0.8 ppm and random errors were 1.6 ppm. However, Worden et al. (2017) showed using OCO-2 data collected in small neighborhoods over 100 km that the precision and accuracy of OCO-2 land data were 0.75 and 0.65 ppm, respectively, indicating that systematic errors are significant compared to the calculated measurement and interference errors of 0.36 and 0.2 ppm. Retrieval characterization and improvement is an ongoing activity as the accuracies demanded to infer more local fluxes increases.

Satellite carbon observations will increase in the future forming a constellation that will play an important role for both carbon cycle science and carbon mitigation (Crisp, 2018). In addition to GOSAT, GOSAT-2 was launched in October 2018, in a 613 km, sun-synchronous orbit with a 6-day ground track repeat period and a mean local time of 1300 ± 15 min. GOSAT-2 includes an additional CO channel to aid in identifying carbon sources of combustion. Similarly, OCO-3 was launched in April 2019, on the International Space Station (ISS). While maintaining a similar footprint as OCO-2, it can map out $100 \text{ km} \times 100 \text{ km}$ areas to characterize the XCO_2 distribution within a large urban area. These satellites are in low earth orbit (LEO), which permits global measurements but poor temporal resolution. New geostationary (GEO) sounders such as GeoCarb will measure CO_2, CH_4, and CO with spatial resolutions similar to OCO-2 but multiple times of day contiguously over a region (Polonsky et al., 2014). A constellation of GEO and LEO sounders anchored by surface observations could play a critical role in support of carbon mitigation strategies (Ciais et al., 2014).

3 Atmospheric modeling

3.1 Modeling atmospheric transport and chemistry

Atmospheric measurements of greenhouse gases integrate information about the sources and sinks of these gases in time and space. Interpretation of the measurements requires a numerical model describing the relationship between the sources and sinks and the measured atmospheric mixing ratios. In other words, the model has to represent the evolution of the plumes of greenhouse gas and their augmentation or depletion due to sources and sinks, as well as be capable of representing the mixing ratios as seen by measurements. Depending on the measurement type and location, and on the available computing resources, models of varying complexity and with varying degrees of detail can be used. The atmosphere may be represented by a single box (Craig, 1957), by several boxes that exchange mass with each other (Cunnold et al., 2002), or up to tens of millions of boxes (Agusti-Panareda, Diamantakis, Bayona, Klappenbach, & Butz, 2017). These types of models are known as Eulerian models. The boxes can be ordered in a single dimension in the latitudinal or the vertical dimension (Bolin & Keeling, 1963), or along both latitudinal and vertical dimensions (Enting & Mansbridge, 1989), or along the three dimensions of space (Hartley & Prinn, 1993). Alternatively, atmospheric transport may be described by an ensemble of virtual particles that follow wind fields (Uliasz, 1993). These types of models are known as Lagrangian models. Transport models may represent the global atmosphere or only a region, actual dates or only a climatology. Moreover, transport models may be run either forward in time, i.e., from the sources/

96 Section | B Methods

sinks to a measurement point (or receptor), or backward[a] in time from a receptor to the sources/sinks (Uliasz & Pielke, 1991).

A transport model, regardless of the number of boxes, must solve the continuity equation of mass conservation, which stems from fundamental physical principles. The continuity equation for mass conservation simply expresses the fact that the temporal evolution of the mixing ratio in an elemental volume of the atmosphere results from the balance between exchanges with the outside of the volume by transport and the sources and sinks inside the volume or at the boundary, including chemistry and deposition. With a single box, the source/sink estimation becomes a simple mass-balance problem based on measurements inside and, if the box does not cover the whole globe, around the box (e.g., Crevoisier et al., 2006).

The discretization of the continuity equation in space (along grid boxes) means that this equation is integrated into the box volume. It prevents explicitly resolving the transport processes initiated at smaller scales that still contribute significantly to the evolution of the mixing ratio within a box. For most models, this concerns turbulent (chaotic) processes within the planetary boundary layer and around unstable clouds. Sub-box-scale transport processes cannot be directly described by general physical laws and their overall effect is, therefore, parameterized empirically. This effect is usually described as a function of tracer mass fluxes and vertical turbulent exchange coefficients. The sub-box-scale parameterizations are the most uncertain component of the models. At all scales, they induce systematic errors that partly corrupt the source/sink to receptor relationship.[b] Locally, they hamper the assimilation of nighttime surface measurements in the models, when conditions are very heterogeneous in the vertical, with little turbulence close to the surface, which occurs intermittently. By modulating the local exchange between the planetary boundary layer and the free troposphere, their effect can also be seen at a large scale, for instance in the modeling of the exchange between low and high latitudes (Stephens et al., 2007).

Various techniques allow the continuity equation to be applied on discrete time steps while ensuring, in principle, important properties of the advection scheme (i.e., mass conservation, stability, positivity, monotonicity) and limiting numerical diffusion (Cabannes & Temam, 1973). However, even modeling

---

a. Backward computations may simply use the time symmetry of fluid transport (Hourdin & Talagrand, 2006) or be done through adjoint codes of the forward model, where numerical sensitivities are analytically computed backward in time (Errico, 1997; see Section 4.2.2).

b. As an example, release CT2013B of NOAA's CarbonTracker inversion system (https://www.esrl.noaa.gov/gmd/ccgg/carbontracker/) states the following: "A significant improvement was made recently to the atmospheric transport model used in CarbonTracker. As a result of this "convective flux fix" (see relevant documentation), CT2013B surface flux estimates are significantly different than those of previous releases. While global-scale fluxes remain basically unchanged, differences in regional fluxes are pronounced (...). This is the most significant revision to CarbonTracker fluxes in the history of our program." (https://www.esrl.noaa.gov/gmd/ccgg/carbontracker/CT2013B/, access 6 December 2018).

box-scale advection remains challenging with equivocal technical solutions and with nonnegligible uncertainty in the meteorological fields used as input (Prather, Zhu, Strahan, Steenrod, & Rodriguez, 2008).

This weakness of transport models explains why the atmospheric inversion community has, since the early 1990s, primarily put their collaborative effort into the comparison of transport models and the evaluation of the impact of their differences for the estimation of surface fluxes (Law et al., 1996). This effort is referred to as the Atmospheric Tracer Transport Model Intercomparison Project. The mathematical model that links sources/sinks to a receptor also includes the relevant atmospheric chemistry and deposition processes. Chemical reactions involve one or several species at once and transform them. They conserve mass but may absorb or emit energy in the form of heat or radiation. They depend on external factors like temperature, pressure or light, occur at variable speeds and compete with each other. The relevant chemical and photochemical mechanisms are parameterized with production and loss rates and deposition is parameterized using deposition velocities (Müller & Brasseur, 1995).

The main greenhouse gases, which are our focus, are only directly involved in a few chemical or photochemical reactions, e.g., photolysis and oxidation by $O(^1D)$ of N_2O, and oxidation of CH_4 by the OH, Cl, and $O(^1D)$ radicals. However, the OH radical itself is highly reactive with many pollutants in the troposphere that are, therefore, linked to methane. Its concentrations result from a poorly known equilibrium. They are low and difficult to measure (Prinn et al., 1995).

3.2 Types of atmospheric transport models

The previous subsection has mentioned the diversity of model formulations: from 0D to full 3D models, global, regional, Eulerian (with a fixed frame of reference with respect to the Earth), and Lagrangian (with a frame of reference attached to moving particles) models. In practice, the diversity is enhanced by the choice of the meteorological data, which drive the continuity equation. Outside specific local studies with dense or outside the mainstream meteorological observations (Lauvaux et al., 2016), the most reliable source of information comes from Numerical Weather Prediction (NWP) centers that assimilate many meteorological observations into their forecasting model, in operational or reanalysis mode. They provide estimates of a large set of atmospheric variables close to each point of space and time, but with a standard resolution that rarely fits the configuration of a particular study. For efficient global simulations, the NWP resolution is usually too fine, because air parcels need to be traced back far away in time over the globe, especially for long-lived tracers like CH_4 and N_2O or for CO_2, which does not have a defined lifetime (Tans, 1997). Alternatively, it may be too coarse if the study is focused on a very local domain. In both cases, some interpolation scheme can be used, but changing the resolution implies changing the distinction between what is sub-box-scale and what is not, since sub-box-scale variables are empirical, this evolution adds another

98 Section | B Methods

layer of empiricism. The problem can be circumvented by running an intermediate model at the target resolution, forced in some way by the NWP data, in order to simulate atmosphere dynamics as a whole, and not just tracer transport processes. Disaggregation with the help of a higher resolution model allows complementing the original NWP data with more detailed orography or land surface characteristics while reducing numerical diffusion. Aggregation with the help of a coarser resolution model directly guarantees some physical consistency in the description of the atmosphere at the target resolution. However, the empiricism of the combination with NWP (relaxation of some of the meteorological variables toward the NWP ones, forcing of the domain boundary, etc.) necessarily limits the benefit of this solution. We usually call a model "offline" if it simply simulates chemistry and transport based on the dynamical information provided by a parent "online" model, whether from NWP or not.

3.3 Relating surface fluxes to atmospheric mixing ratios

This section has so far dealt only with forward modeling of atmospheric transport, i.e., simulating atmospheric mixing ratios given a prior estimate of surface fluxes. However, we have not yet discussed how an atmospheric transport model can be used to find the optimal fluxes given a set of observations. In essence, there are two approaches for how a transport model can be used for this purpose, first by effectively *inverting* the transport model to relate the observations to surface fluxes, and second by using *ensembles*.

The first method involves linearizing the transport model. This can be achieved through a step-by-step process in which forward runs (as many runs as there are flux regions or grid cells to solve for) are made each with a perturbation in a different flux region. Each of these runs relates to a given flux region to a change in the whole mixing ratio field, and thus gives one row of a *Jacobian* matrix (Enting, 2002; Fung et al., 1991; Rayner, Enting, Francey, & Langenfelds, 1999). This Jacobian matrix is a linearized transport operator relating fluxes to mixing ratios (see Fig. 3). To use this approach, however, any chemistry needs to be linear, i.e., zero order. Jacobian matrices can also be derived from Lagrangian models (Seibert & Frank, 2004). With a Jacobian matrix, the inversion can be solved analytically and is discussed in Section 4.1. Alternatively, the transport can be linearized by calculating line by line the tangent linear of each equation to give the *tangent-linear model*. By taking the transpose of each line of the tangent-linear model and reversing the order of equations, the *adjoint model* can be derived, which is equivalent to the transpose of the Jacobian matrix (Errico, 1997). With an adjoint model, the inversion can be solved using gradient methods and is discussed in Section 4.2.3.

The second method uses the transport model as it is, but it is run many times each with a different perturbation to the prior fluxes to build an ensemble of mixing ratios. This approach is discussed in Section 4.2.3.

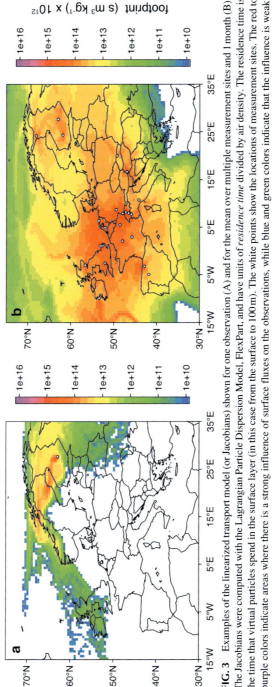

FIG. 3 Examples of the linearized transport model (or Jacobians) shown for one observation (A) and for the mean over multiple measurement sites and 1 month (B). The Jacobians were computed with the Lagrangian Particle Dispersion Model, FlexPart, and have units of *residence time* divided by air density. The residence time is the time that virtual particles spend in the surface layer (in this case from the surface to 100m). The white points show the locations of measurement sites. The red to purple colors indicate areas where there is a strong influence of surface fluxes on the observations, while blue and green colors indicate that the influence is weak.

100 Section | B Methods

4 Inversion concepts

An atmospheric transport model coupled to flux estimates can predict the outcome of measurements, the so-called *forward problem*. However, we want to use the measurements to determine the fluxes, this constitutes the so-called *inverse problem*. While the forward problem has a unique solution, the inverse problem does not—it is underdetermined in the mathematical sense. Therefore, a probabilistic approach is necessary. In this approach, *prior*[c] information about the fluxes is represented by a probability distribution that is transformed into a posterior probability distribution by incorporating a model (relating the measurements to the fluxes) and the measurements themselves.

We will use the concept of conditional probability and Bayes' theorem to derive the expressions for solving the inverse problem. Note that these are not the only concepts on which inverse theory may be based, but the most straightforward and widely used in the community. For a more complete understanding of inverse theory, we recommend Tarantola (2005).

The inverse problem can be broken down into three steps:

1. *Defining the variables to be optimized:* these could be the fluxes themselves discretized in space and time or a number of parameters that describe the fluxes. Independently of how the variables are defined, we can think of them as a vector, $\mathbf{x}$, in an abstract space, which we will call the control space, $\mathcal{X}$, such that $\mathbf{x} \in \mathcal{X}$.
2. *Forward modeling:* describes the relationship between the variables $\mathbf{x}$ and the observations. We can think of the observations, $\mathbf{y}$, as a vector in another abstract space, which we will call the observation space, such that $\mathbf{y} \in \mathcal{Y}$ and the forward model as a mapping from the control to the observation space, $\mathcal{X} \to \mathcal{Y}$.
3. *Inverse modeling:* uses the actual observations to infer the optimal values of $\mathbf{x}$.

The second step, *forward modeling*, is dealt with in Section 3 and the third step, *inverse modeling* will be the focus of this section. (An overview of the nomenclature for this section is given in Table 1.)

4.1 Bayes' theorem and its application to optimizing fluxes

Bayes' theorem describes the probability of an outcome, given some observations and some prior knowledge. In this case the posterior probability of the fluxes, given their prior probability, $\rho(\mathbf{x})$, the probability of the observations $\rho(\mathbf{y})$ and the conditional probability, $\rho(\mathbf{y}|\mathbf{x})$, i.e., the likelihood of observing $\mathbf{y}$ given $\mathbf{x}$. This is stated as

$$\rho(\mathbf{x}|\mathbf{y}) = \frac{\rho(\mathbf{y}|\mathbf{x})\rho(\mathbf{x})}{\rho(\mathbf{y})} \tag{6}$$

---

c. *prior* means *before the observations are known.*

Top-down approaches **Chapter | 4** **101**

TABLE 1 Overview of the nomenclature used in inverse modeling, where n is number of control variables and m is the number of observations.

Variable	Dimension	Description
$\mathcal{X}$	$\mathcal{X} \subset \mathbb{R}$	Control space
$\mathcal{Y}$	$\mathcal{Y} \subset \mathbb{R}$	Observation space
$\mathbf{x}$	$n \times 1$	Control vector
$\mathbf{x_b}$	$n \times 1$	Prior control vector
$\mathbf{y}$	$m \times 1$	Observation vector
$\mathbf{H}$	$m \times n$	Transport operator
H		Transport model
H^*		Adjoint transport model
$\mathbf{B}$	$n \times n$	Prior error covariance
$\mathbf{R}$	$m \times m$	Observation error covariance
$\mathbf{P}$	$n \times n$	Posterior error covariance (in Kalman Filter methods this is the update error covariance matrix)
$\mathbf{Q}$	$n \times n$	Error covariance of the forecast model in Kalman Filter methods
$\mathbf{F}$		Forecast model operator in Kalman Filter methods
$\mathbf{K}$	$n \times m$	Kalman Gain matrix
$\mathbf{M}$	$w \times n$	Mapping operator to reduce dimensionality of problem

If the probability distribution functions (*pdf*) $\rho(\mathbf{x})$ and $\rho(\mathbf{y}|\mathbf{x})$ are Gaussian, and the forward model is linear, then the *pdf* $\rho(\mathbf{y})$ is also Gaussian. Here we will assume that $\rho(\mathbf{x})$ has a Gaussian distribution so that taking the logarithm of $\rho(\mathbf{x})$ we can derive the following expression:

$$-2\ln\rho(\mathbf{x}) = (\mathbf{x} - \mathbf{x_b})^{\mathrm{T}}\mathbf{B}^{-1}(\mathbf{x} - \mathbf{x_b}) + c_1 \tag{7}$$

where c_1 is a constant, $\mathbf{x_b}$ is the prior value of $\mathbf{x}$ and $\mathbf{B}$, its associated error covariance:

$$\mathbf{B} = \varepsilon\left\{(\mathbf{x_t} - \mathbf{x_b})(\mathbf{x_t} - \mathbf{x_b})^{\mathrm{T}}\right\} \tag{8}$$

where ε means is the *expected value* and $\mathbf{x_t}$ is the true value of $\mathbf{x}$. We can derive a similar expression to Eq. (7) for the conditional probability $\rho(\mathbf{y}|\mathbf{x})$:

$$-2\ln\rho(\mathbf{y}|\mathbf{x}) = (\mathbf{y} - H(\mathbf{x}))^{\mathrm{T}}\mathbf{R}^{-1}(\mathbf{y} - H(\mathbf{x})) + c_2 \tag{9}$$

102 Section | B Methods

where c_2 is again a constant, H is a model that maps $\mathcal{X} \rightarrow \mathcal{Y}$, and $\mathbf{R}$ is the observation error covariance defined as

$$\mathbf{R} = \varepsilon\left\{(\mathbf{y} - H(\mathbf{x_t}))(\mathbf{y} - H(\mathbf{x_t}))^{\mathrm{T}}\right\} \tag{10}$$

Note that the true values of $\mathbf{B}$ and $\mathbf{R}$ cannot be known as this would require knowing $\mathbf{x_t}$ before solving the inverse problem. Instead, $\mathbf{B}$ and $\mathbf{R}$ must be approximated, and approaches to do this are discussed in Section 4.4.

By substituting Eqs. (7) and (9) into Eq. (6), and given that $\rho(\mathbf{y})$ is a constant (and can be ignored), we can derive an expression for the posterior *pdf* of $\mathbf{x}$:

$$-2\ln\rho(\mathbf{y}|\mathbf{x}) = (\mathbf{y} - H(\mathbf{x}))^{\mathrm{T}}\mathbf{R}^{-1}(\mathbf{y} - H(\mathbf{x})) + (\mathbf{x} - \mathbf{x_b})^{\mathrm{T}}\mathbf{B}^{-1}(\mathbf{x} - \mathbf{x_b}) + c_3 \tag{11}$$

Eq. (11) gives us the expression for the cost function based on a Gaussian distribution:

$$2J(\mathbf{x}) = (\mathbf{x} - \mathbf{x_b})^{\mathrm{T}}\mathbf{B}^{-1}(\mathbf{x} - \mathbf{x_b}) + (\mathbf{y} - H(\mathbf{x}))^{\mathrm{T}}\mathbf{R}^{-1}(\mathbf{y} - H(\mathbf{x})) \tag{12}$$

We can see that Eq. (12) has a quadratic form and thus will have a minimum where the first-order derivative is equal to zero:

$$2J'(\mathbf{x}) = \mathbf{B}^{-1}(\mathbf{x} - \mathbf{x_b}) + (H'(\mathbf{x}))^{\mathrm{T}}\mathbf{R}^{-1}(H(\mathbf{x}) - \mathbf{y}) = 0 \tag{13}$$

Thus, the optimal value of $\mathbf{x}$, i.e., the one that minimizes the cost function can be found by solving Eq. (13). If the *pdf* is Gaussian, this is equivalent to finding the maximum probability solution of $\mathbf{x}$. It is also the minimum of the least-squares misfit function and is also called the *least-squares estimator*. If the model $H(\mathbf{x})$ can be defined as a matrix operator, $\mathbf{H}$, then Eq. (13) can be solved directly by using some algebra to derive the expressions:

$$\mathbf{x} = \left(\mathbf{H}^{\mathrm{T}}\mathbf{R}^{-1}\mathbf{H} + \mathbf{B}^{-1}\right)^{-1}\left(\mathbf{H}^{\mathrm{T}}\mathbf{R}^{-1}\mathbf{y} + \mathbf{B}^{-1}\mathbf{x_b}\right) \tag{14}$$

$$\mathbf{x} = \mathbf{x_b} + \left(\mathbf{H}^{\mathrm{T}}\mathbf{R}^{-1}\mathbf{H} + \mathbf{B}^{-1}\right)^{-1}\mathbf{H}^{\mathrm{T}}\mathbf{R}^{-1}(\mathbf{y} - \mathbf{H}\mathbf{x_b}) \tag{15}$$

$$\mathbf{x} = \mathbf{x_b} + \mathbf{B}\mathbf{H}^{\mathrm{T}}\left(\mathbf{H}\mathbf{B}\mathbf{H}^{\mathrm{T}} + \mathbf{R}\right)^{-1}(\mathbf{y} - \mathbf{H}\mathbf{x_b}) \tag{16}$$

Eqs. (14)–16) are equivalent. Note that in (15) the matrix $(\mathbf{H}^{\mathrm{T}}\mathbf{R}^{-1}\mathbf{H} + \mathbf{B}^{-1})$ is inverted, which is a square matrix with the dimensions of the control vector $\mathbf{x}$, while in (16) the matrix $(\mathbf{H}\mathbf{B}\mathbf{H}^{\mathrm{T}} + \mathbf{R})$ is inverted which is a square matrix with the dimensions of the observation vector $\mathbf{y}$. The computation time of large problems can be reduced by choosing the expression in which the smaller matrix is inverted.

The posterior error covariance, $\mathbf{P}$, can be found from the inverse of the Hessian matrix of the cost function in Eq. (12) (Fisher & Courtier, 1995):

$$\mathbf{P} = (J''(\mathbf{x}))^{-1} = \left(\mathbf{H}^{\mathrm{T}}\mathbf{R}^{-1}\mathbf{H} + \mathbf{B}^{-1}\right)^{-1} \tag{17}$$

The simplest interpretation of $\mathbf{P}$ is to use the square root of its diagonal elements as the error bars on the posterior fluxes, $\mathbf{x}$. However, the off-diagonal elements contain useful information as well, especially if one examines the correlations (Tarantola, 2005):

$$C_{ij} = \frac{v_{ij}}{\sqrt{v_{ii}}\sqrt{v_{jj}}} \quad (18)$$

where c_{ij} is the correlation and v_{ij} is the variance between the ith and jth variables. The correlation falls within the interval -1 to $+1$. Obviously, if c_{ij} is close to zero then the variables are nearly independent of each other, and conversely, if c_{ij} is close to $+1$ or -1, then they are strongly positively or negatively correlated meaning that they are not independently resolved.

4.2 Introduction to different optimization methods

In the case that $H(\mathbf{x})$ cannot be defined as a matrix operation, i.e., if the matrix $\mathbf{H}$ would be (1) too big to store in memory, (2) require too many computations of the forward matrix to generate all the elements of $\mathbf{H}$, or (3) if the model is nonlinear, then other methods must be used to find the minimum of (13). Here we introduce a few methods that are commonly used in atmospheric sciences. Fig. 4 shows schematically the generic steps involved in an atmospheric inversion system and the input data required.

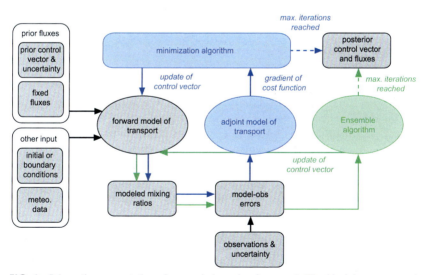

FIG. 4 Schematic representation of a generic inversion framework. The *black boxes* represent input/output and steps required in any atmospheric inversion. The *blue arrows* represent the steps involved in a gradient method inversion and are performed in an iterative loop around forward and adjoint model runs and the minimization algorithm. The *green arrows* represent an ensemble of control vectors, model simulations, and model-observation errors as used in an ensemble method. There is also an iterative loop between the forward model runs and the ensemble algorithm, e.g., the Kalman Ensemble filter update step, but in this case it iterates over the assimilation time windows.

104 Section | B Methods

4.2.1 Kalman filters

Kalman Filters can be used when the matrix $\mathbf{H}$ cannot be stored in memory, as in case (1). Instead of solving the problem in one step, using all observations simultaneously, Kalman Filters update the control vector, and its uncertainty, using the observations sequentially (Kalman, 1960). The Kalman Filter is a recursive algorithm in which only the estimated state from the previous time step and the observations in the current time step are needed to compute the update. This can be seen in the pseudo-code example:

Pseudo-code **Kalman Filter**

INPUT: $\mathbf{y} \in \mathcal{Y}$, $\mathbf{x_b} \in \mathcal{X}$, $\mathbf{R}$, $\mathbf{B}$, $\mathbf{F}$, $\mathbf{Q}$, τ

$\mathbf{x}_{0|0} = \mathbf{x_b}$ (i)

$\mathbf{P}_{0|0} = \mathbf{B}$ (ii)

do $t = 1$ to τ

    Forecast step:

    $\mathbf{x}_{t|t-1} = \mathbf{F}\mathbf{x}_{t-1|t-1} + \mathbf{q}$, where $\mathbf{q} \in N(0, \mathbf{Q})$ (iii)

    $\mathbf{P}_{t|t-1} = \mathbf{F}\mathbf{P}_{t-1|t-1}\mathbf{F}^{\mathsf{T}} + \mathbf{Q}$ (iv)

    Update step:

    $\mathbf{K}_t = \mathbf{P}_{t|t-1}\mathbf{H}_t^{\mathsf{T}}\left(\mathbf{H}_t\mathbf{P}_{t|t-1}\mathbf{H}_t^{\mathsf{T}} + \mathbf{R}_t\right)^{-1}$ (v)

    $\mathbf{x}_{t|t} = \mathbf{x}_{t|t-1} + \mathbf{K}_t\left(\mathbf{y}_t - \mathbf{H}_t\mathbf{x}_{t|t-1}\right)$ (vi)

    $\mathbf{P}_{t|t} = (\mathbf{I} - \mathbf{K}_t\mathbf{H}_t)\mathbf{P}_{t|t-1}$ (vii)

end do

---

where $\mathbf{x}_{t|t-1}$ is the control vector at time step t based on observations up to and including time step $t-1$, $\mathbf{F}$ is a forecast operator that projects the value of $\mathbf{x}_{t-1|t-1}$ (that is the control vector from the previous time step) to the current time step. In step (iii), $\mathbf{q}$ is a random perturbation vector drawn from a normal distribution with zero mean and covariance $\mathbf{Q}$, where $\mathbf{Q}$ describes the error in the forecast operator, $\mathbf{F}$. The matrix $\mathbf{P}_{t|t-1}$ is the error covariance matrix corresponding to $\mathbf{x}_{t|t-1}$. Lastly, the matrix $\mathbf{K}$ is the so-called *Kalman Gain*, which projects the model-observation error, $\mathbf{y}_t - \mathbf{H}_t\mathbf{x}_{t|t-1}$ to the control space and updates the value of $\mathbf{x}$ in step (vi). We can see the similarity of step (vi) to Eq. (16) if the expression for $\mathbf{K}_t$ in (v) is substituted into (vi). Furthermore, once all observations have been used, the Kalman Filter is mathematically equivalent to the direct solution given by Eqs. (14)–(16) if the sensitivity to the control variables (described by $\mathbf{H}$) is limited to the assimilation time window. Note that at any time step, t, only the part of the matrix $\mathbf{H}$ corresponding to t is needed (that is $\mathbf{H}_t$), thus avoiding the need to store the whole matrix in memory. This has the effect though, that mass may not be conserved.

Top-down approaches **Chapter | 4** **105**

There are a number of variants of the Kalman Filter, including the Extended Kalman Filter (Jazwinski, 1970) and the Ensemble Kalman Filter (Evensen, 1994) (discussed in Section 4.2.3). The Ensemble Kalman Filter can be used for cases when the model cannot be defined as a matrix operator.

4.2.2 Gradient methods

Gradient methods can be used to find the minimum in cases (1,2) listed above, and some approaches can also be used in case (3). They involve numerical approximations to find the minimum of Eq. (13) and avoid the need for $\mathbf{H}$ and $\mathbf{H}^T$. These approaches are iterative and involve finding the direction of descent toward the minimum, a new estimate of $\mathbf{x}$, and recalculation of the gradient, $\mathbf{g}$, at each iteration. Some algorithms also require the cost, c, to be calculated. A simplified pseudo-code for the procedure is:

Pseudo-code **Gradient methods**

INPUT: $\mathbf{y} \in \mathcal{Y}$, $\mathbf{x_b} \in \mathcal{X}$, $\mathbf{R}$, $\mathbf{B}$, *maxiter*
$i = 0$

$\mathbf{x}_i = \mathbf{x_b}$ (i)

do while $i \leq$ *maxiter*

$\chi = \mathbf{B}^{-1/2}(\mathbf{x} - \mathbf{x_b})$ (ii)

$\mathbf{g} = \chi + \mathbf{B}^{-1/2} H^* \mathbf{R}^{-1} (H(\mathbf{x}) - \mathbf{y})$ (iii)

$c = \dfrac{1}{2}\chi^T \chi + \dfrac{1}{2}(H(\mathbf{x}) - \mathbf{y})^T \mathbf{R}^{-1}(H(\mathbf{x}) - \mathbf{y})$ (iv)

$\chi \leftarrow$ descent algorithm using $\chi, \mathbf{g}, c$ (v)

$\mathbf{x} = \mathbf{B}^{1/2}\chi + \mathbf{x_b}$ (vi)

$i = i + 1$

end do

In practice, the gradient methods do not optimize $\mathbf{x}$ directly, but a preconditioned value of it, χ, where the conversion of $\mathbf{x}$ to χ is given in step (ii) (in this case using the square root of the prior error covariance matrix). Preconditioning increases the convergence rate of the optimization (Fisher & Courtier, 1995). In this pseudo example, *maxiter* is given as the stopping criteria, but in practice other stopping criteria, such as the reduction in the norm of the gradient can also be used. We can see that the gradient calculation in step (iii) can be derived by substituting $\mathbf{x}$ for χ and $(H'(\mathbf{x}))^T$ for H^* in Eq. (13). This step requires running the forward model on the latest estimate of $\mathbf{x}$ to derive the model-observation error $H(\mathbf{x}) - \mathbf{y}$. The operator H^* in step (iii) is the so-called adjoint of H or *adjoint model*. While H is a mapping from $\mathcal{X} \to \mathcal{Y}$, H^* is the mapping from $\mathcal{Y} \to \mathcal{X}$. The adjoint model is applied to the vector $\mathbf{R}^{-1}(H(\mathbf{x}) - \mathbf{y})$ (in the observation space, $\mathcal{Y}$) and gives the contribution to the gradient from the observation-

106 Section | B Methods

model mismatch. Adjoint models are derived from the forward models and are mathematically equivalent to $(H'(\mathbf{x}))^T$. For a detailed explanation of adjoint models see Errico (1997) and for how to derive them see Giering and Kaminski (1998).

In step (v) a descent algorithm is used to find the direction of descent and to update the estimate of χ. The conjugate gradient algorithm can be used if $H(\mathbf{x})$ is linear (Paige & Saunders, 1975). The Lanczos algorithm, which is closely related to the conjugate gradient algorithm, provides a method to solve a system of linear equations and simultaneously gives an estimate of the Hessian matrix, $J''(\mathbf{x}) = \mathbf{B}^{-1} + \mathbf{H}^T\mathbf{R}^{-1}\mathbf{H}$, the inverse of which is the posterior error covariance, $\mathbf{P}$ (Fisher & Courtier, 1995; Golub & Van Loan, 1996). If $H(\mathbf{x})$ is slightly nonlinear, then a quasi-Newton algorithm may be used. One such method is the Broyden-Fletcher-Goldfarb-Shanno (BFGS) algorithm (Broyden, 1970), which can also be used to estimate the posterior error covariance (Fisher & Courtier, 1995).

Note that the linearity of any model can be assessed using the chi-test statistic:

$$\delta\mathbf{y} = H(\mathbf{x}) - H(\mathbf{x} + \delta\mathbf{x}) - H^{TL}(\delta\mathbf{x}) \qquad (19)$$

$$\chi = \delta\mathbf{y}^T\mathbf{R}^{-1}\delta\mathbf{y} \qquad (20)$$

where H^{TL} is the tangent linear model of H and $\delta\mathbf{x}$ is an increment of $\mathbf{x}$ (Rodgers, 2000). If the model is linear, χ will be close to 1.

4.2.3 Ensemble methods

Ensemble methods are particularly useful when the model $H(\mathbf{x})$ is nonlinear, case (3) above, as well as in cases where the *pdf* is non-Gaussian or in hierarchical Bayesian models (see Section 4.4.3). Another advantage of these methods is that they do not require the adjoint model of transport, H^*. Some of these methods are relatively easy to code making them also popular alternatives to the methods already discussed. There are many ensemble algorithms, but here we will discuss two that are often used in atmospheric sciences: Markov Chain Monte Carlo (MCMC) and the Ensemble Kalman Filter (EnKF).

4.2.3.1 Markov chain Monte Carlo

Markov Chain Monte Carlo methods generate a sequence of probability distributions. At each step in the sequence, a *candidate* variable (or in multidimensional problems, a vector) is drawn from the distribution and is either accepted or rejected based on some criteria—if accepted it is used to generate the next distribution in the sequence. As the number of steps in the sequence increases, the probability distribution will converge to the *target* distribution, $P(\mathbf{x})$, i.e., the posterior distribution. One of the most versatile MCMC algorithms is the Metropolis-Hastings algorithm (Chib & Greenberg, 2012) and will be discussed here.

The Metropolis-Hastings algorithm has the advantage that the target distribution, $P(\mathbf{x})$, need not be calculated but only a function, $f(\mathbf{x})$, which is proportional to $P(\mathbf{x})$ (Hastings, 1970). This is particularly useful, as the normalizing factor is often unknown. The acceptance/rejection of a candidate value is determined by the *acceptance probability*, A. The algorithm is demonstrated in the pseudo-code:

Pseudo-code **Metropolis-Hastings algorithm**

INPUT: $\mathbf{y} \in \mathcal{Y}$, $\mathbf{x_b} \in \mathcal{X}$, $\mathbf{R}$, $\mathbf{B}$, *maxiter*
initialize: $\mathbf{x}_{(0)} = \mathbf{x_b}$
do $i = 1$ to *maxiter*

$$\mathbf{x}'_i \sim Q((\mathbf{x}'_i | \mathbf{x}_{i-1})) \tag{i}$$

$$A(\mathbf{x}'_i, \mathbf{x}_{i-1}) = \min\left\{1, \frac{f(\mathbf{x}'_i) Q(\mathbf{x}_{i-1} | \mathbf{x}'_i)}{f(\mathbf{x}_{i-1}) Q(\mathbf{x}'_i | \mathbf{x}_{i-1})}\right\} \tag{ii}$$

$$u \sim \text{uniform}(0, 1) \tag{iii}$$

if $u \le A(\mathbf{x}'_i, \mathbf{x}_{i-1})$ **then**

$$\mathbf{x}_i = \mathbf{x}'_i \tag{iv}$$

else
$$\mathbf{x}_i = \mathbf{x}_{i-1} \tag{v}$$

end if
end do

In step (i), the distribution, $Q(\mathbf{x}'_i | \mathbf{x}_{i-1})$, or *proposal density*, proposes a new distribution on $\mathbf{x}_i$ (based on the previous state $\mathbf{x}_{i-1}$) and a new candidate, $\mathbf{x}'_i$, is drawn from this distribution. Usually, $Q(\mathbf{x}'_i | \mathbf{x}_{i-1})$ is defined as a *random walk*:

$$\mathbf{x}'_i = \mathbf{x}_{i-1} + \varepsilon \tag{21}$$

where $\varepsilon \in \mathcal{N}(0, \sigma)$ and σ is the uncertainty of $\mathbf{x}$. Step (ii) calculates the acceptance probability, A, in which the function $f(\mathbf{x})$, in our case, is the conditional *pdf* derived in Eqs. (6)–(11):

$$P(\mathbf{x}) \propto f(\mathbf{x}) = \exp\left(-\frac{1}{2}(\mathbf{y} - H(\mathbf{x}))^{\mathrm{T}} \mathbf{R}^{-1}(\mathbf{y} - H(\mathbf{x})) - \frac{1}{2}(\mathbf{x} - \mathbf{x_b})^{\mathrm{T}} \mathbf{B}^{-1}(\mathbf{x} - \mathbf{x_b})\right) \tag{22}$$

The acceptance probability ensures that the condition of *detailed balance* is met, i.e., for every pair of states the probability of being in state $\mathbf{x}_{i-1}$ and transitioning to state $\mathbf{x}_i$ must be equal to the probability of being in state $\mathbf{x}_i$ and transitioning to state $\mathbf{x}_{i-1}$ (Hastings, 1970). Thus, the acceptance probability, A, ensures that the following equation is balanced:

$$P(\mathbf{x}_{i-1})Q(\mathbf{x}_i | \mathbf{x}_{i-1})A(\mathbf{x}_i, \mathbf{x}_{i-1}) = P(\mathbf{x}_i)Q(\mathbf{x}_{i-1} | \mathbf{x}_i)A(\mathbf{x}_{i-1}, \mathbf{x}_i) \tag{23}$$

108 Section | B Methods

A new candidate, $\mathbf{x}'_i$, however, is only accepted if $u \leq A(\mathbf{x}_i, \mathbf{x}_{i-1})$, where u is a random number drawn from a uniform distribution over $(0,1)$. If the candidate is rejected, then the previous state, $\mathbf{x}_{i-1}$, is used in the next iteration.

In a special case of the Metropolis-Hastings algorithm, when Q has a symmetric distribution, e.g., Gaussian, the acceptance probability simplifies to:

$$A(x_i, x_{i-1}) = \min\left\{1, \frac{f(\mathbf{x}'_i)}{f(\mathbf{x}_{i-1})}\right\} \tag{24}$$

This special case constitutes the Metropolis algorithm (Chib & Greenberg, 2012; Metropolis, Rosenbluth, Rosenbluth, Teller, & Teller, 1953). From Eq. (24), we can easily see that if the candidate $\mathbf{x}'_i$ leads to a higher probability density, i.e., $f(\mathbf{x}'_i)/f(\mathbf{x}_{i-1}) > 1$, then the transition is always accepted, otherwise, it is accepted only if $f(\mathbf{x}'_i)/f(\mathbf{x}_{i-1}) \geq u$. In MCMC algorithms, all the states, $\mathbf{x}_1, \mathbf{x}_2, \ldots, \mathbf{x}_N$ define the target distribution, $P(\mathbf{x})$.

There are a couple of disadvantages of the Metropolis-Hastings algorithm (and MCMC methods in general). First, the initial samples of the Markov chain may follow a very different distribution to the target distribution, especially if the starting point is in a region of low probability density. Therefore, a *burn-in* period is often necessary for which an initial number of samples are discarded. Second, the samples are sequentially correlated although overall they follow the target distribution. This means that a set of nearby samples correlate with each other but will not correctly reflect the distribution. If we want a set of independent samples, we have to throw away a number of samples and only take every nth sample, where n is typically determined by examining the autocorrelation between adjacent samples. Lastly, the number of iterations required depends on the accuracy desired, the dimensionality of $\mathbf{x}$, and the relationship between $f(\mathbf{x})$ and $P(\mathbf{x})$.

4.2.3.2 Ensemble Kalman filter

The ensemble Kalman Filter (EnKF) is a Monte Carlo implementation of the Kalman Filter developed by Evensen (1994). EnKF is suited to solving inverse problems with very large numbers of variables and has the advantage that it does not require the full error covariance matrix to be updated at each time step and it does not require an adjoint model of transport. There are several variants of EnKF as discussed in the review papers by Evensen (2003) and Houtekamer and Zhang (2016). Here we will present one variant of EnKF, i.e., the *stochastic filter*, which is one of the more straightforward variants to describe. EnKF proceeds as the Kalman Filter but, instead of just updating one control vector, an ensemble of vectors is updated. The mean and spread of the ensemble are used to approximate the error covariance matrix, $\mathbf{P}$, thus avoiding the need to ever specify $\mathbf{P}$ fully. To start, an ensemble of q members is generated based on the prior estimate of the control vector:

$$\mathbf{X}_{0|0} = \left[\mathbf{X}^1_{0|0}, \mathbf{X}^2_{0|0}, \ldots, \mathbf{X}^q_{0|0}\right], \text{ where } \mathbf{X}^i_{0|0} = \mathbf{x_b} + \mathbf{w}^i \tag{25}$$

where $\mathbf{X}_{0|0}$ is an $n \times q$ matrix with each column equal to one control vector, and $\mathbf{w}^i$ is a random perturbation vector with zero mean and covariance $\mathbf{B}$. Similarly, an ensemble of observations is generated:

$$\mathbf{Y} = [\mathbf{y}^1, \mathbf{y}^2, ..., \mathbf{y}^q], \text{ where } \mathbf{y}^i = \mathbf{y} + \mathbf{u}^i \tag{26}$$

where $\mathbf{Y}$ is an $m \times q$ matrix with each column equal to the observation vector plus a random perturbation, $\mathbf{u}^i$ with zero mean and covariance $\mathbf{R}$. The algorithm then proceeds similarly to the Kalman Filter with a *forecast step* and an *update step*:

Pseudo-code **Ensemble Kalman Filter**

INPUT: $\mathbf{y} \in \mathcal{Y}$, $\mathbf{x_b} \in \mathcal{X}$, $\mathbf{R}$, $\mathbf{B}$, $\mathbf{F}$, $\mathbf{Q}$, τ

$$\mathbf{X}_{0|0} = \left[\mathbf{x}_{0|0}^1, \mathbf{x}_{0|0}^2, ..., \mathbf{x}_{0|0}^q\right] \tag{i}$$

$$\mathbf{Y} = [\mathbf{y}^1, \mathbf{y}^2, ..., \mathbf{y}^q] \tag{ii}$$

do $t = 1$ to τ

    Forecast step:

$$\mathbf{x}_{t|t-1}^i = \mathbf{F}\mathbf{x}_{t-1|t-1}^i + \mathbf{q}^i, \text{ for } i = 1, ..., q \tag{iii}$$

    Update step:

$$\overline{\mathbf{x}}_{t|t-1} = \frac{1}{q-1} \sum_{i=1}^{q} \mathbf{x}_{t|t-1}^i \tag{iv}$$

$$\mathbf{P}_{t|t-1} = \frac{1}{q-1} \sum_{i=1}^{q} \left(\mathbf{x}_{t|t-1}^i - \overline{\mathbf{x}}_{t|t-1}\right)\left(\mathbf{x}_{t|t-1}^i - \overline{\mathbf{x}}_{t|t-1}\right)^{\mathsf{T}} \tag{v}$$

$$\mathbf{K}_t = \mathbf{P}_{t|t-1}\mathbf{H}_t^{\mathsf{T}}\left(\mathbf{H}_t\mathbf{P}_{t|t-1}\mathbf{H}_t^{\mathsf{T}} + \mathbf{R}_t\right)^{-1} \tag{vi}$$

$$\mathbf{x}_{t|t}^i = \mathbf{x}_{t|t-1}^i + \mathbf{K}_t\left(\mathbf{y}_t^i - \mathbf{H}_t\mathbf{x}_{t|t-1}^i\right), \text{ for } i = 1, ..., q \tag{vii}$$

end do

The loop begins with step (iii) adding a random perturbation vector, $\mathbf{q}$, which is drawn from a normal distribution with zero mean and covariance $\mathbf{Q}$, where $\mathbf{Q}$ describes the error in the forecast operator, $\mathbf{F}$. In step (v) the error covariance matrix, $\mathbf{P}_{t|t-1}$ is represented by the spread of the ensemble at time step t, where $\overline{\mathbf{x}}_{t|t-1}$ is the ensemble mean.

In this pseudo-code example for the stochastic filter, an ensemble of observations is created by adding random perturbations. This is necessary because without perturbation of the observations the analysis ensemble would not include the uncertainty due to observation uncertainty and thus would underestimate the analysis uncertainty (Tippett, Anderson, Bishop, Hamill, & Whitaker, 2003). However, perturbing the observations also introduces sampling issues and thus *deterministic filters* were developed, such as the Square

Root Filter (SRF), which is more commonly used in atmospheric sciences (for details about the SRF see Tippett et al., 2003).

EnKF may also be used when the model, $H(\mathbf{x})$, cannot be expressed as a matrix, $\mathbf{H}$, and when there is no adjoint model. In this case the matrices $\mathbf{P}_{t|t-1}\mathbf{H}_t^{\mathrm{T}}$ and $\mathbf{H}_t\mathbf{P}_{t|t-1}\mathbf{H}_t^{\mathrm{T}}$ are approximated by:

$$\mathbf{P}_{t|t-1}\mathbf{H}_t^{\mathrm{T}} = \frac{1}{q-1} \sum_{i=1}^{q} \left(\mathbf{x}_{t|t-1}^i - \overline{\mathbf{x}}_{t|t-1}\right) \left(H\left(\mathbf{x}_{t|t-1}^i\right) - \overline{H\left(\mathbf{x}_{t|t-1}\right)}\right)^{\mathrm{T}}, \text{ where } \overline{H\left(\mathbf{x}_{t|t-1}\right)}$$

$$= \frac{1}{q-1} \sum_{i=1}^{q} H\left(\mathbf{x}_{t|t-1}^i\right)$$

$$(27)$$

$$\mathbf{H}_t\mathbf{P}_{t|t-1}\mathbf{H}_t^{\mathrm{T}} = \frac{1}{q-1} \sum_{i=1}^{q} \left(H\left(\mathbf{x}_{t|t-1}^i\right) - \overline{H\left(\mathbf{x}_{t|t-1}\right)}\right) \left(H\left(\mathbf{x}_{t|t-1}^i\right) - \overline{H\left(\mathbf{x}_{t|t-1}\right)}\right)^{\mathrm{T}}$$

$$(28)$$

From the equation for $\mathbf{P}$ in step (v) we can see that $\mathbf{P}$ will have a maximum rank of $q-1$, thus the information from the observations is projected onto $q-1$ directions in the control space. Since owing to practical limitations $q \ll n$ (where n is the number of control variables) $\mathbf{P}$ has a greatly reduced rank compared to the full covariance matrix in the standard Kalman Filter. For EnKF to work well, the ensemble size must be large enough to accurately represent the mean and covariance, otherwise, the optimization may over or underfit the observations and/or introduce spurious correlations between control variables. For a full discussion of these problems and methods to address them, see Houtekamer and Zhang (2016). Another important consideration is the assimilation time window: EnKF requires a sufficiently short assimilation window so that the ensemble perturbations [in step (iii)] evolve linearly and remain Gaussian. However, too short assimilation windows have the disadvantage that they may miss the influence of the control variables at the current time step on observations beyond this time step (through long-range transport), and they limit the amount of information from the observations that are available to update the control vector (Liu, Bowman, & Lee, 2016). Choosing an appropriate assimilation time window is challenging and requires consideration of the density of observations, as well as the characteristics of the control variables and model, and different optimization systems for CO_2 fluxes have used time windows varying from a few hours (Kang, Kalnay, Miyoshi, Liu, & Fung, 2012) to a few weeks (Michalak, 2008; Peters et al., 2005).

4.3 Ensembles for estimating the posterior uncertainty

Ensembles can also be used to estimate the posterior uncertainty and may be needed if the optimization method does not provide an estimate of the error

Top-down approaches Chapter | 4 **111**

covariance, $\mathbf{P}$, such as some descent algorithms used in the gradient methods, or if the computation of $\mathbf{P}$ is too costly, for instance, too many iterations are required to arrive at a sufficiently accurate estimate of $\mathbf{P}$. In such cases, the posterior uncertainty can be estimated as the standard deviation of an ensemble of inversions (Chevallier, Bréon, & Rayner, 2007). The prior fluxes and observations used in the ensemble, however, must have the same statistical characteristics as the prior flux and observation uncertainties. To achieve this, a random error is added to the prior fluxes (observations) in each member of the ensemble such that the standard deviation over the whole ensemble approaches the prior flux (observation) uncertainty. The perturbed flux for the ith ensemble member is calculated as (Chevallier et al., 2007):

$$\mathbf{x}_i = \mathbf{x_b} + \mathbf{B}^{1/2}\mathbf{r}_i \tag{29}$$

where $\mathbf{r}$ is a random vector of the same length as $\mathbf{x_b}$ and $\mathbf{B}^{1/2}$ is the square root of the prior error covariance matrix. A similar expression is used to perturb the observations in each ensemble member (Chevallier et al., 2007):

$$\mathbf{y}_i = \mathbf{y} + \mathbf{R}^{1/2}\mathbf{r}_i \tag{30}$$

Computing an ensemble of inversions for large problems can be costly but provides a means to estimate $\mathbf{P}$ where other methods are not possible or feasible.

4.4 Estimating prior flux and observation uncertainties

In Section 4.1, we have seen the definitions of the prior and observation error covariance matrices, $\mathbf{B}$ and $\mathbf{R}$, respectively. In practice, however, the prior and observation uncertainties, and their correlations, are difficult to estimate. In this section, we will see how these matrices can be approximated and present a way to assess the approximations.

4.4.1 Prior error covariance

Defining the prior error covariance, $\mathbf{B}$, poses a dilemma in that we cannot know the true uncertainty before we know the true fluxes. Therefore, the flux uncertainties and their correlations must be estimated based on the knowledge available about the prior flux model and/or, if direct observations exist, the comparison of the modeled with the observed fluxes.

The first approximation is, usually, to assume that the prior flux uncertainty is linearly proportional to the prior flux itself. A lower limit may be chosen as well to ensure a nonzero value for the uncertainty where the prior flux is zero. A simple and common choice for the error correlation is exponential decay over distance and time with some chosen scale lengths, D, and T, respectively. This is a convenient choice as it leads to a positive definite matrix, which is a

112 Section | B Methods

requirement for the error covariance matrices. With this model, the elements of $\mathbf{B}$ can then be calculated as

$$b_{ij} = \sigma_i \sigma_j \exp\left(-\frac{d_{ij}}{D}\right) \exp\left(-\frac{t_{ij}}{T}\right) \tag{31}$$

where σ_i is the uncertainty of the ith variable, and d_{ij} and t_{ij} are the distance and time intervals between the ith and jth variables. The correlation scale length, D, may be based on considerations of the extent of a particular land cover or ecosystem type and/or the extent of a weather system and range from a few tens to a few hundreds of kilometers over land (and longer over the ocean). Similarly, T may be based on considerations of the duration of a weather system or the seasonality of the flux considered and range from a few days to months.

Since the number of elements of $\mathbf{B}$ is equal to the square of the length of the control vector, this matrix may be very large and it potentially exceeds the available memory. To deal with this, we can decompose $\mathbf{B}$ into its spatial and temporal correlations, $\mathbf{C}^S$ and $\mathbf{C}^T$:

$$\mathbf{B} = \mathbf{C}^T \otimes \mathbf{C}^S \circ \boldsymbol{\sigma}^T \boldsymbol{\sigma} \tag{32}$$

where $\otimes$ is the Kronecker product and $\circ$ is the Hadamard product, and $\boldsymbol{\sigma}$ is a vector of the flux uncertainties corresponding to one-time step. This assumes that the spatial correlations and flux uncertainties are the same for each time step, and thus $\mathbf{C}^S \circ \boldsymbol{\sigma}^T \boldsymbol{\sigma}$ can be saved as the spatial error covariance matrix $\mathbf{B}^S$ (with the dimensions of the control vector for one-time step). If $\mathbf{B}^S$ is also large, then its eigen-decomposition may be saved instead with truncation of the smallest eigenvalues and eigenvectors, e.g., truncation of those for which the eigenvalues are $<1\%$ of the largest eigenvalue. In case the flux uncertainties are different for each time step, then $\mathbf{B}$ can be decomposed block-wise:

$$\mathbf{B}_{ij} = c_{ij}^T \mathbf{B}_{ij}^S \tag{33}$$

where $\mathbf{B}_{ij}$ is the block of $\mathbf{B}$ for the ith and jth time steps, $\mathbf{B}_{ij}^S = \mathbf{C}^S \circ \boldsymbol{\sigma}_i^T \boldsymbol{\sigma}_j$ and c_{ij}^T is the temporal correlation of the ith and jth time steps (from $\mathbf{C}^T$). All calculations involving $\mathbf{B}$ ($\mathbf{B}^{-1}$ or $\mathbf{B}^{1/2}$) can be made row-wise and column-wise using its decomposition.

4.4.2 Observation error covariance

In Eq. (10), we saw that error covariance in the observation space, $\mathbf{R}$, is the expected value of the outer product of model-observation error, $\{(\mathbf{y} - H(\mathbf{x_t}))(\mathbf{y} - H(\mathbf{x_t}))^T\}$. A gross approximation is to estimate $\mathbf{R}$ by substituting $\mathbf{x_t}$ with $\mathbf{x_b}$ in Eq. (5). However, $\mathbf{y} - H(\mathbf{x_b})$ is the signal used to correct the prior fluxes and using this to calculate $\mathbf{R}$ overestimates the uncertainty in the observation space since it also includes the error of the prior with respect to the true fluxes. Instead, we need to estimate the uncertainty in the observation space independently of $\mathbf{x_b}$. We can think of the uncertainty in the observation space as the

Top-down approaches **Chapter | 4** **113**

error that would remain even if the true fluxes were known. If this were the case, the uncertainty in the observation space still needs to account for the uncertainty in the measurements, the representation error of our model, i.e., the error arising from the finite resolution (the so-called *aggregation error*), and the errors inherent in the model, such as the transport and, if applicable, chemistry.

The measurement uncertainty is the most straightforward of these to obtain, as it depends only on the measurement method, sampling, and calibration and is usually reported with the observations. Model transport and chemistry errors, however, are much more challenging to estimate. The only way to estimate these errors is to assess the model performance in a controlled way. This can be done, e.g., by tracer release experiments in which a known amount of a tracer (for which there are no other sources in the vicinity of the experiment) is released and measured at one or more locations downstream—the error between the observed and simulated mixing ratios gives the model error. The problem is that these experiments, and hence the errors, are not representative of all locations and meteorological conditions. Alternatively, one can examine a species for which the fluxes are widespread and relatively well known, e.g., [222]Radon (Karstens, Schwingshackl, Schmithüsen, & Levin, 2015). The transport error assessed this way will only be an approximation, as the actual error is also dependent on the magnitude and distribution of the actual fluxes to be optimized. (Methods to deal with the uncertainty in atmospheric chemistry, relevant for CH_4 and N_2O, are dealt with in Sections 7 and 8, respectively.)

The last type of error to consider is the aggregation error. This error arises because the solution for the control vector depends on how its variables are sampled in space in time, or in other words, how they are aggregated (Trampert & Snieder, 1996). This problem can be simply expressed as

$$\mathbf{H}_H \mathbf{x}_H \neq \mathbf{H}_L \mathbf{x}_L \text{ where } \mathbf{H}_L = \mathbf{H}_H \mathbf{M}^\mathrm{T} \text{ and } \mathbf{x}_L = \mathbf{M} \mathbf{x}_H \tag{34}$$

where $\mathbf{H}_H$ and $\mathbf{H}_L$ are the transport operators at high and low resolution, respectively, and similarly for the control vector, and $\mathbf{M}$ is a mapping operator from high to low resolution. In practice, a downgrading in resolution may arise when the control vector used in the inversion has a coarser resolution than the transport model. Trampert and Snieder (1996) showed that the error associated with the downgrading of the resolution from $\mathbf{x}_H$ to $\mathbf{x}_L$ can be projected into the observation space as

$$\mathbf{R}_A = \mathbf{H}_H \mathbf{G}_- \mathbf{B}_H \mathbf{G}_-^\mathrm{T} \mathbf{H}_H^\mathrm{T} \tag{35}$$

where $\mathbf{R}_A$ is the aggregation error covariance (with the same dimensions as $\mathbf{R}$), $\mathbf{B}_H$ is the prior error covariance for the control vector at high resolution, and $\mathbf{G}_-$ is a projection matrix defined as

$$\mathbf{G} = \sum_{i=1}^{} \mathbf{m}_i^T \mathbf{m}_i \text{ and } \mathbf{G}_- = \mathbf{I} - \mathbf{G} \tag{36}$$

where $\mathbf{m}_i$ is a row vector of the mapping operator, $\mathbf{M}$, and $\mathbf{I}$ is the identity matrix. The matrices $\mathbf{H}_H$, $\mathbf{G}_-$, and $\mathbf{B}_H$ are potentially very large and methods

114 Section | B Methods

to calculate $\mathbf{R}_A$ in practice are given in Kaminski, Rayner, Heimann, and Enting (2001) and Thompson and Stohl (2014). Note that $\mathbf{R}_A$ requires $\mathbf{H}_H$ that can be stored in memory, which is one of the reasons that $\mathbf{R}_A$ is often ignored in inverse calculations.

The diagonal elements of $\mathbf{R}$, that is the variances $\sigma^2_{ii,R}$, are calculated as the quadratic sum of the contributions from measurement, transport, and, if available, aggregation errors:

$$\sigma^2_{ii,R} = \sigma^2_{ii,meas} + \sigma^2_{ii,trans} + \sigma^2_{ii,agg} \tag{37}$$

In many inverse calculations, $\mathbf{R}$ is assumed to be diagonal—an assumption that simplifies the calculations in the gradient methods discussed above. This assumption may be fine if the observations are made with some time and/or space interval between them—an assumption that can be tested by examining the autocorrelation between observations. If the correlation is considerable, then error inflation methods can be used to account for this while retaining a diagonal $\mathbf{R}$ matrix (Chevallier, 2007).

4.4.3 Assessing choices for the error covariance matrices

The specification of the matrices $\mathbf{B}$ and $\mathbf{R}$ can have a significant impact on the inversion results, especially if there are only a few observations.

A simple answer to the question of the appropriateness of the choice for $\mathbf{B}$ and $\mathbf{R}$ within a defined theoretical framework is to calculate the reduced chi-squared statistic, χ^2. In a perfectly tuned system with the correct statistical hypothesis (e.g., unbiased and linear), the value of the cost function (Eq. 12) at its minimum, normalized by the number of degrees of freedom, is close to 1 (i.e., $\chi^2 \approx 1$) (e.g., Rayner et al., 1999). The number of degrees of freedom, d, is the number of independent pieces of information used in the inversion, i.e., the number of observations plus the number of prior flux variables (in the case of diagonal $\mathbf{B}$ and $\mathbf{R}$), minus the number of variables optimized in the inversion—thus d is generally equal to the number of observations. It is important to highlight the fact that χ^2 close to 1 is an internal diagnostic only; it is neither a sufficient nor a necessary condition for optimality of the inversion system (Talagrand, 2014). Wrong covariance values in $\mathbf{B}$ and $\mathbf{R}$ can compensate for wrong statistical hypotheses in terms of statistical optimality while yielding an apparently ok χ^2. Practical examples are also given in a study by Chevallier et al. (2007).

In the absence of reliable information to assign $\mathbf{B}$ and $\mathbf{R}$, the covariance matrices may still be scaled until this reduced chi-squared criterion is met. However, the χ^2 statistic does not contain any information about the relative weighting between the model-observation mismatch (determined by $\mathbf{R}$) and the prior flux estimates (determined by $\mathbf{B}$). One approach is to find the maximum likelihood for $\mathbf{B}$ and $\mathbf{R}$ within the given statistical hypotheses. If $\mathbf{B}$ and $\mathbf{R}$ are diagonal, this is equivalent to finding the optimal values for their variances. We are interested in the joint *pdf* of $\mathbf{B}$ and $\mathbf{R}$, which we shall call $\boldsymbol{\theta}$.

Top-down approaches **Chapter | 4** **115**

Michalak et al. (2005) showed that the maximum likelihood (or optimal) estimate of **B** and **R** can be obtained by solving:

$$2J(\boldsymbol{\theta}) = \ln\left|\mathbf{HBH}^{\mathrm{T}} + \mathbf{R}\right| + (\mathbf{y} - \mathbf{Hx_b})^{\mathrm{T}}\left(\mathbf{HBH}^{\mathrm{T}} + \mathbf{R}\right)^{-1}(\mathbf{y} - \mathbf{Hx_b}) \tag{38}$$

where | | denotes the matrix determinant. We can see from Eq. (38) that larger variances in **B** and **R** will increase the value of the first term but decrease the value of the second term. To solve this equation we need to use iterative methods such as quasi-Newton algorithms (see Section 4.2.2) or ensemble methods such as MCMC (see Section 4.2.3).

An alternative approach is to include the parameters of **B** and **R**, along with the prior fluxes, in the optimization, the so-called *hierarchical Bayesian* approach (Gelman & Hill, 2006). In this approach, we want to find the posterior *pdf* of the fluxes, **x**, and uncertainties, $\boldsymbol{\theta}$, given the observations, **y**. From Baye's theorem we can derive:

$$\rho(\mathbf{x}, \boldsymbol{\theta}|\mathbf{y}) \propto \rho(\mathbf{y}|\mathbf{x}, \boldsymbol{\theta})\rho(\mathbf{x}, \boldsymbol{\theta}) \tag{39}$$

Note that we give the equation a proportionality without the denominator, $\rho(\mathbf{y})$, which is constant. Using the chain rule of probability, we can expand the joint probability, $\rho(\mathbf{x},\boldsymbol{\theta})$:

$$\rho(\mathbf{x}, \boldsymbol{\theta}) \propto \rho(\mathbf{x}|\boldsymbol{\theta})\rho(\boldsymbol{\theta}) \tag{40}$$

and substituting Eq. (40) into Eq. (39) we obtain:

$$\rho(\mathbf{x}, \boldsymbol{\theta}|\mathbf{y}) \propto \rho(\mathbf{y}|\mathbf{x}, \boldsymbol{\theta})\rho(\mathbf{x}|\boldsymbol{\theta})\rho(\boldsymbol{\theta}) \tag{41}$$

In the simple case, where **B** and **R** are diagonal matrices defined by the uncertainties, $\boldsymbol{\sigma_B}$ and $\boldsymbol{\sigma_R}$, respectively, then the full equation for the hierarchical posterior probability is

$$\rho(\mathbf{x}, \boldsymbol{\sigma_B}, \boldsymbol{\sigma_R}|\mathbf{y}) \propto \rho(\mathbf{y}|\mathbf{x}, \boldsymbol{\sigma_R})\rho(\mathbf{x}|\boldsymbol{\sigma_B})\rho(\boldsymbol{\sigma_B})\rho(\boldsymbol{\sigma_R}) \tag{42}$$

If a Gaussian *pdf* is assumed for each term in Eq. (42) then the maximum likelihood solution is that which minimizes the cost function:

$$2J(\mathbf{x}, \boldsymbol{\sigma_B}, \boldsymbol{\sigma_R}) = (\mathbf{y} - \mathbf{Hx})^{\mathrm{T}}\mathbf{R}^{-1}(\mathbf{y} - \mathbf{Hx}) + (\mathbf{x} - \mathbf{x_b})^{\mathrm{T}}\mathbf{B}^{-1}(\mathbf{x} - \mathbf{x_b})$$
$$+ (\boldsymbol{\sigma_B} - \boldsymbol{\sigma_{B,b}})^{\mathrm{T}}\mathbf{S}_{\mathbf{B}}^{-1}(\boldsymbol{\sigma_B} - \boldsymbol{\sigma_{B,b}}) + (\boldsymbol{\sigma_R} - \boldsymbol{\sigma_{R,b}})^{\mathrm{T}}\mathbf{S}_{\mathbf{R}}^{-1}(\boldsymbol{\sigma_R} - \boldsymbol{\sigma_{R,b}})$$
$$\tag{43}$$

where $\boldsymbol{\sigma_{B,b}}$ and $\boldsymbol{\sigma_{R,b}}$ are the prior estimates for the prior and observation uncertainties, and $\mathbf{S_B}$ and $\mathbf{S_R}$ are the error covariance matrices for the uncertainty on the uncertainties $\boldsymbol{\sigma_{B,b}}$ and $\boldsymbol{\sigma_{R,b}}$. Eq. (43) may be solved for **x**, $\boldsymbol{\sigma_B}$, and $\boldsymbol{\sigma_B}$ using, e.g., MCMC methods.

4.5 Boundary conditions

We have seen how Bayes' theorem can be applied to derive a cost function for optimizing fluxes and reviewed a number of optimization methods commonly

116   Section | B Methods

used in atmospheric sciences. However, we have not yet considered the *boundary conditions* of the problem, i.e., the conditions (or parameters) at the boundary of the space and/or time domains. In our case, these are the atmospheric mixing ratios at the initial time step of a forward or inverse model run (the so-called *initial mixing ratios*), or for regional inversions, they may be the mixing ratios at the edges of the domain. Since the GHGs CO_2, CH_4, and N_2O are all long-lived species (with atmospheric lifetimes of more than a few years), their atmospheric mixing ratios represent the sum of all fluxes and atmospheric sinks over many years. At the start of a model run, a realistic representation of the mixing ratios at the initial time step is required, otherwise, the modeled mixing ratios will be biased, even if the true value of the fluxes and sinks are known. A corollary of this is that the optimized fluxes from an inversion will also be biased.

Initial mixing ratios can be estimated from integrations of the forward model starting with approximate mixing ratios (e.g., a homogeneous atmosphere with a value close to the observed value) with fluxes approximately balanced with the sink (no net positive flux). This is known as a model *spin-up*. Obviously, the estimated mixing ratios will still have some error compared to the true mixing ratios. One method to deal with this is to include correction factors for the initial conditions in the control vector and optimize these along with the fluxes (e.g., Pison, Bousquet, Chevallier, Szopa, & Hauglustaine, 2009). An alternative and novel method is to integrate the forward model from a starting point a few years before the first inversion time step using prior estimates for the fluxes, and *nudge* the modeled mixing ratios within a given distance around each observation toward the observed value (Groot Zwaaftink et al., 2018). In this way, the simulated mixing ratios will remain close to the past observations even if there are errors in the prior fluxes and/or atmospheric sink.

In regional inversions, the model is only calculating the contribution to the observed mixing ratios from fluxes (and atmospheric sinks) inside the domain. Therefore, the boundary conditions need to account for all the contributions to the observations from outside the domain. Boundary conditions may be estimated for short time periods (a few days to weeks) and for small domains by sampling the mixing ratios at the domain boundaries using, e.g., aircraft (e.g. Miller et al., 2016). However, for longer periods and/or large domains this is not feasible and the boundary conditions must be modeled. One approach to do this is the two-step scheme described by Rödenbeck, Gerbig, Trusilova, and Heimann (2009). In this scheme, the first step involves running a global inversion (typically using a different dataset and coarser model resolution than the regional inversion), setting the optimized fluxes within the regional domain to zero, and integrating the forward model with the adjusted fluxes to calculate the mixing ratios. The resulting mixing ratios (also known as the background) are the contribution from fluxes outside the regional domain and, if relevant, the atmospheric sink. The second step involves running the regional inversion having corrected the observations for the outside domain influence (or background) calculated in step one. The Rödenbeck et al. (2009) scheme ensures that the regional budget is consistent with the global one and the general approach may be adapted to use with a regional Eulerian or Lagrangian model.

Alternatively, for a Lagrangian model, the background may be estimated by coupling the model output to 4D (latitude, longitude, altitude, time) fields of mixing ratios from an optimized global model. This can be done by either by (1) coupling in spatial domain using an average of the mixing ratios at the times and locations where the virtual particles exit the regional domain as in Nevison et al. (2018), or (2) coupling in the time domain using a weighted average of the mixing ratios at the termination times and locations of the virtual particles and adding the contribution along the particle trajectories outside the regional domain as in Thompson and Stohl (2014). Again, in both these approaches, correction factors for the background can be included in the control vector to reduce the error from these in the optimized fluxes.

5 Application to land biosphere CO_2 fluxes (NEE)

Traditional CO_2 atmospheric inversions estimate the net CO_2 exchange of the Earth's surface with the atmosphere. This net flux can be decomposed in various ways. Over land, the distinction between fossil fuel fluxes and biosphere fluxes may seem obvious, but both of these fluxes can be disaggregated further according to emission processes. Biosphere fluxes include photosynthesis and respiration from vegetation, respiration from soils, humans and animals, vegetation disturbance (e.g., combustion), lateral fluxes of organic carbon (e.g., harvest), and numerous diverse minor contributions (see Fig. 5). Extracting the biosphere flux from the net flux requires either extending the observation vector, y, with appropriate observations, like isotopic measurements or complementary tracers, or using some prior information, like inventories of fossil fuel emissions, in order to subtract a "reliable" estimate of the nonbiosphere contributions from the net flux. These topics are further explored in the complementary Section 6 about fossil fuel emissions.

Similarly, within the biosphere flux, inversion users may wish to extract some specific components, like the net ecosystem exchange (NEE) or the gross primary production (GPP) that can be more easily compared with land-surface models, but, again, additional measurements (e.g., tracers of GPP such as fluorescence or carbonyl sulfide and tracers of biomass burning such as CO) or prior information have to be used for the disaggregation (Bowman et al., 2017; Liu et al., 2017).

At first glance, the estimation of land biosphere CO_2 fluxes from inverse modeling appears to have progressed little over the last three decades. By compiling existing knowledge about the carbon surface fluxes (i.e., by selecting some prior information x_b with an appropriate matrix B) and with the help of a 3D transport operator H, Tans, Fung, and Takahashi (1990) identified a large terrestrial carbon sink at temperate latitudes that were needed to balance the carbon budget and to match the north–south gradient of atmospheric

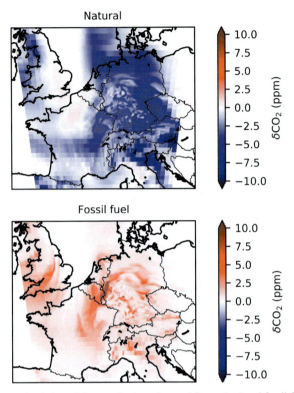

FIG. 5 Regional simulation of the contribution of natural fluxes (top) and fossil fuel emissions (bottom) within the simulation domain to the CO_2 mixing ratio at 400 m above ground level on June 2, 2016, at 24:00. The domain is centered over the Benelux region (courtesy from E. Potier, LSCE).

CO_2 over the remote oceans. More than 25 years later, that claim is not clear anymore since some of the modern inversion results agree with it, while others relocate some of the sinks to tropical lands (Le Quere et al., 2018). In addition, there was hope that the progressively increasing density of ground-based measurement networks would allow longitudinal gradients in the fluxes to also be resolved. In this aspect, initial results were not consistent across inversions, such as the importance of the carbon sink in North America (Bousquet et al., 2000; Fan et al., 1998).

If we consider geographical Europe, a region that has been the focus of extensive scientific research by the carbon cycle community, the last major inversion intercomparison exercise, using surface measurements of CO_2, concluded that the European terrestrial biosphere is a moderate carbon sink in absolute terms (Peylin et al., 2013). Reuter et al. (2014) reevaluated this sink by increasing it several fold by assimilating satellite XCO_2 retrievals, but their finding seems to be contradicted by the use of more recent satellite data (Crowell et al., 2019). Similar disagreement among inversion results can be seen in the analysis of the African carbon budget during the 2015–16 El-Niño (Gloor et al., 2018; Liu et al., 2017).

Diverging conclusions for some features of the land biosphere CO_2 fluxes are paralleled by converging ones, for instance, the distribution of the carbon budget within Asia (Thompson et al., 2016), and by agreement of some flux features with correlated observations. Specifically, the ability of some inversions to document the link between the land biosphere CO_2 fluxes and influencing variables such as temperature and rainfall anomalies (Patra, Ishizawa, Maksyutov, Nakazawa, & Inoue, 2005; Peng et al., 2013; Rödenbeck, Zaehle, Keeling, & Heimann, 2018), sea-level pressure variability (Bastos et al., 2016), or others (e.g., Fernández-Martínez et al., 2019), or to describe flux changes during extreme climate events in various parts of the globe, e.g., Detmers et al. (2015) for Australia; Alden et al. (2016), Bowman et al. (2017), Gatti et al. (2014), van der Laan-Luijkx et al. (2015) for the Amazon; Liu et al. (2018), Wolf et al. (2016) for North America, and Bastos et al. (2018), for the Tropics, is remarkable (see Fig. 6). An important explanation behind these successes lies in the greater robustness of flux temporal variations inferred by atmospheric inversions compared to the absolute long-term values. This is due to some recurrent errors in transport models (e.g., biases in winter boundary

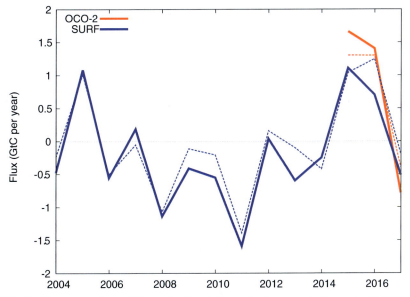

FIG. 6 Time series of annual fluxes (positive indicates to the atmosphere) over tropical land between 2004 and 2017, estimated by two global atmospheric inversions driven by ground-based observations (blue lines) and two global atmospheric inversions driven by satellite observations (red lines). For each given observation dataset, the two inversions use different atmospheric transport models. The temporal variations are all in phase and show some coherence with the occurrence of El Nino (2015–16) and La Nina (2008, 2011) events.

layer heights) and prior error models (e.g., biases in prior error variances in winter) from year to year, which partially cancel when computing anomalies. The heterogeneous technological level among current inversion systems, and their underlying atmospheric transport models, may also artificially increase the spread among inversion results. In this hypothesis, the best inversion systems would be more robust than inter-comparison exercises suggest.

Ideally, accurate independent measurements of the land biosphere CO_2 fluxes would allow measuring the skill of inversion systems and of their components (transport models H, error models B and R). In the specific case of global inverse systems, the spatial resolution of existing flux observations (in the order of kilometers) is much smaller than the spatial resolution of global transport models (in the order of a couple of degrees). Regional models can reach a higher resolution that approaches that of flux observations (made with the eddy covariance technique). A few regional studies have demonstrated good skill in inferring land biosphere fluxes at the local scale (a $300 \times 300\,km^2$ domain in southwest France, (Lauvaux et al., 2009); or the "Corn Belt" region of the United States, (Lauvaux et al., 2012), national scale (The Netherlands, Meesters et al., 2012) or even larger scales by aggregating flux observations over similar ecosystems in western Europe (Broquet et al., 2013). In addition, independent aircraft measurements of CO_2 have been used to assess the accuracy of inversion results within ensembles, based on the fit of the posterior modeled mixing ratios to measured vertical gradients (Stephens et al., 2007) or latitudinal gradients (Houweling et al., 2015). They have also demonstrated the additional value of some new measurements in inversion systems (Kim et al., 2017). These various and encouraging results help establish the credibility of atmospheric inversions for studying land biosphere CO_2 fluxes. They have justified the important role that global inversions have played in assessing regional carbon budgets over most subcontinents within the first REgional Carbon Cycle Assessment and Processes (RECCAP, https://www.biogeosciences.net/special_issue107.html) of the Global Carbon Project. It is expected that the development of the remote sensing of CO_2 from space (Section 2.2) will reinforce this role by contributing to flux estimates at higher spatial resolution and with greater robustness than is currently possible.

6 Application to fossil fuel emissions of CO_2

The estimation of the fossil fuel emissions from atmospheric measurements is now considered one of the top priorities for atmospheric GHG observation programs. However, the level of uncertainty in fossil CO_2 emissions at the national and regional scales is already as low as 5%–10%, according to Marland, Hamal, and Jonas (2009). The already low level of uncertainty makes improving estimates of fossil fuel CO_2 (FFCO$_2$) emissions more challenging than for some of the other more uncertain components of the carbon cycle. That said, the uncertainty in fossil fuel, at least in some regions, is still large enough to be a significant source of uncertainty when deriving NEE from atmospheric inversions, which requires accurate prior estimates of FFCO$_2$ (Thompson et al., 2016; Saeki & Patra, 2017). The need for very accurate FFCO$_2$ emission estimates has led to a strong interest in developing ways to estimate these emissions from atmospheric observations and has advanced the understanding of the problems, limitations, and viable solutions to do this. Several examples of ongoing studies are available to introduce the methodologies available for estimating FFCO$_2$ emissions.

Many studies of FFCO$_2$ emissions are at a city scale since cities are responsible for about 70% of the global emissions. Such ongoing studies are made in Indianapolis, IN, in the framework of the INFLUX experiment (The Indianapolis Flux Experiment) (Davis et al., 2017; Turnbull et al., 2015), Boston, MA (Gately, Hutyra, & Wing, 2015), Salt Lake City, UT, (Patarasuk et al., 2016), the Los Angeles Basin, CA (Newman et al., 2016), and Paris, France (Staufer et al., 2016). At the city scale, the primary approach is to combine monitoring of atmospheric CO$_2$ (surface, aircraft, and remote sensing data) with high-resolution atmospheric transport models and inversions to quantify FFCO$_2$ emissions and resolve the sources in space and time within the urban domain. These inversions rely on prior information in the form of gridded FFCO$_2$ emissions from inventories. In contrast to the global scale, the construction of city-level gridded FFCO$_2$ emissions is based on direct estimates of fuel consumption or fuel-consuming activities, supported by location and time-of-activity data. The high-resolution transport modeling is based on wind fields produced by mesoscale model simulations at 1–10 km resolution, and is implemented with either Lagrangian particle plume dispersion models (e.g., FlexPart), off-line regional transport models (e.g., CHIMERE) or online transport (e.g., WRF-Chem). Bayesian inversion schemes at city scale are often set up to optimize scaling factors for the prior emission patterns (grouped, e.g., by source type) in order to reduce the numbers of unknowns.

One example of a city scale study is the INFLUX experiment, which installed 13 continuous analyzers measuring CO$_2$, CH$_4$, and CO, and complemented these with aircraft flights, meteorological instruments, discrete flask sampling, and eddy covariance tower measurements (Figs. 7 and 8). The high-density network of observations was designed to quantify the anthropogenic and biogenic components of the greenhouse gas budget over the 12th biggest metropolitan area of the United States (Davis et al., 2017). The inversion system, supported by the Hestia CO$_2$ emission inventory, a building-level emission dataset (Gurney et al., 2012), produced 5-day 1-km resolution maps of CO$_2$ emissions with a 20% uncertainty over 6 months (Lauvaux et al., 2016). Sensitivity tests showed that spatial structures in the prior emissions influenced the spatial pattern in the optimized emissions and the total carbon emissions over the entire area by 15%. At the same time, the total emission estimate was less sensitive to the CO$_2$ boundary conditions and to the different prior emissions, such as the Open-source Data Inventory for Anthropogenic CO$_2$ (ODIAC) (Oda, Maksyutov, & Andres, 2018). Owing to the sensitivity of high-resolution inversions to errors in the transport models, in the INFLUX

122 Section | B Methods

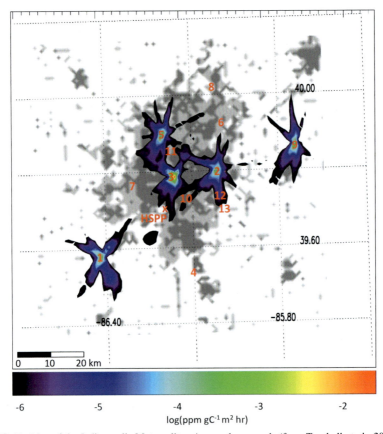

FIG. 7 Map of the Indianapolis Metropolitan Area and surrounds (from Turnbull et al., 2015). Gray shading indicates urban density from low (white) to high (dark gray) from the 2006 US National Land Cover Database. Locations of the INFLUX towers and the Harding Street Power Plant (HSPP) are shown in red. The modeled aggregate surface influence functions for mid-afternoon in October 2012 for Towers 1, 2, 3, 5, and 9 are overlaid using the color scale shown. All wind directions are included in the modeled aggregate influence functions. The "star" shape of the influence functions is a consequence of the dominant wind directions over the 1-month period.

study a preoptimization of the wind fields, used in the transport model, was performed. In this study, the winds were improved using a regional meteorological data assimilation system ingesting surface, aircraft and lidar observations of wind, temperature, and mixing height. Lauvaux et al. (2020) further refined the modeling set-up and found their top-down $FFCO_2$ emissions matched the bottom-up inventory to within 3% of the annual city total, although there were differences in the spatial and sectoral breakdown of emissions between the inverse model and the inventory estimates.

Hedelius et al. (2018) developed another inversion system based on the Hybrid Single-Particle Lagrangian Integrated Trajectory (HYSPLIT) model and ODIAC to estimate the net CO_2 flux from the South Coast Air Basin

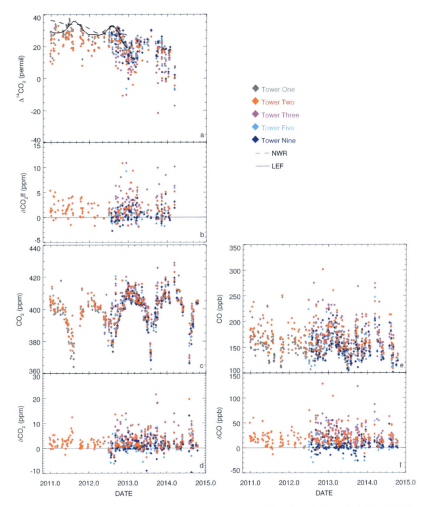

FIG. 8 Trace gas measurements from the INFLUX towers (from Turnbull et al., 2015). (A) The observed $\Delta^{14}CO_2$ at each tower along with the smoothed curve $\Delta^{14}CO_2$ values from Niwot Ridge, CO (NWR) and Park Falls, WI (LEF) stations. (B) The calculated $\delta CO_2 ff$ relative to Tower 1. (C) The observed CO_2 mole fractions. (D) CO_2 enhancements are relative to Tower 1. (E) The observed CO mole fractions. (F) CO enhancements are relative to Tower 1.

(SoCAB), Los-Angeles. They used North American Mesoscale Forecast System (NAM) data at 12 km and 3-hourly resolution from the NOAA data archive as the primary model input. To constrain the emissions, they used Total Carbon Column Observing Network (TCCON) and Orbiting Carbon Observatory-2 (OCO-2) observations. Using TCCON data, they estimated the direct net CO_2 flux from SoCAB to be 104 ± 26 TgCO$_2$ yr^{-1} for the study period of July 2013–August 2016 and obtained an estimate of 120 ± 30 TgCO$_2$ yr^{-1} using OCO-2 data. To determine the uncertainty of the net flux, they performed

124 Section | B Methods

sensitivity tests to estimate the contribution from errors in the prior and in the covariance, as well as from different inversion schemes and a coarser dynamical model. Overall, the uncertainty was estimated to be 25%, with the largest contribution from the dynamical model.

Staufer et al. (2016) tested the ability of a Bayesian atmospheric inversion to quantify the Paris region's fossil fuel CO_2 emissions on a monthly basis based on a network of three surface stations operating for 1 year as part of the CO_2-MEGAPARIS experiment (August 2010–July 2011). The inversion uses the CHIMERE transport model run with a zoom region over Paris at 2 km × 2 km horizontal resolution (Bréon et al., 2015). Prior information in the inversion includes the spatial distribution of $FFCO_2$ emissions from a local 6-hourly and 1 km × 1 km resolution inventory (AIRPARIF, http://www.airparif.asso.fr) and the spatial distribution of biogenic CO_2 fluxes from the C-TESSEL land surface model. Differences in the hourly CO_2 atmospheric mixing ratios between the near-ground monitoring sites located at the northeastern and southwestern edges of the urban area (i.e., the CO_2 gradients across Paris) were used to optimize the 6-hourly mean $FFCO_2$ emissions. The MEGAPARIS experiment found that a stringent selection of CO_2 gradients was necessary for getting reliable inversion results, due to large modeling uncertainties. The most robust data selection analyzed in the study uses only mid-afternoon gradients if wind speeds are larger than $3 \, \text{ms}^{-1}$ and if the modeled wind at the upwind site is within $\pm 15°$ of the transect between downwind and upwind sites. This data selection removed 92% of the hourly observations. Even though this left few remaining data to constrain the emissions, the inversion system diagnosed that their assimilation significantly reduced the uncertainty in monthly emissions: by 50% in October 2010 and 9% in November 2010. The results showed significant sensitivity to the spatial distribution in the prior emissions used in the inversion system demonstrating the need to rely on high-resolution local inventories augmented with accurate temporal emission profiles (Lian et al., 2021). Although the inversion constrains emissions through the assimilation of CO_2 gradients, the results were sensitive to estimates of CO_2 fluxes outside the local domain when air masses originated from urban areas northeast of Paris. The large uncertainties in the transport model and boundary conditions meant that strict data selection criteria were needed, which limited the ability to monitor $FFCO_2$ emissions from Paris. This was especially true for September and November 2010, when the fluxes were poorly constrained by too few CO_2 measurements. These limitations could be overcome through an improvement of the transport model and boundary conditions and through the expansion of the observation network.

For modeling highly heterogeneous $FFCO_2$ emissions, the effects of the prior inventory, inverse model and wind field resolution are important. Feng et al. (2016) studied the impacts of emission inventory and wind field resolution and concluded that the use of high resolution (1.3 km) rather than 10 km resolution emission inventory is important for the simulations and inverse modeling in the LA region, while similar performance can be achieved with 1.3 and 4 km WRF-Chem simulations. For the INFLUX experiment, Deng et al. (2017)

evaluated the impact of assimilating different meteorological observations on the optimization of CO_2 fluxes using observed CO_2 mixing ratios over Indianapolis for the 2-month period. Assimilating surface meteorological measurements improved the simulated wind speed and direction, but their impact on the simulated planetary boundary layer (PBL) was limited. Simulated PBL wind statistics improved significantly when assimilating upper-air observations from commercial airline flights and continuous ground-based Doppler lidar wind observations. The inversion results indicated that the spatial distribution of optimized CO_2 fluxes was affected by the model, but the performance overall changed little across WRF simulations when the fluxes were aggregated over the entire domain. Results show that assimilating observations of PBL winds is a powerful tool for increasing the precision of urban meteorological reanalyzes, but that the impact on flux estimates from inversions is dependent on the specific urban environment.

On the national scale, the problem of separating $FFCO_2$ emissions from land biosphere CO_2 fluxes is more difficult to address than on city scale where the relative contribution from fossil fuel emissions is higher. Radiocarbon ($\Delta^{14}CO_2$), CO and NOx observations are considered to become useful tools to address the problem of estimating $FFCO_2$ emissions separately from land biosphere fluxes (Beirle, Boersma, Platt, Lawrence, & Wagner, 2011; Levin, Hammer, Eichelmann, & Vogel, 2011). In addition, Basu, Miller, and Lehman (2016) argued that dual tracer (CO_2 and $\Delta^{14}CO_2$) inversions would be able to detect and minimize biases in estimates of the land biosphere CO_2 flux that would otherwise arise in a traditional CO_2-only inversion when prescribing fossil fuel emissions. To demonstrate a practical implementation of the radiocarbon-based technique, $FFCO_2$ emissions were estimated at a national scale for the United States for the first time by Basu et al. (2020). They used $^{14}CO_2$ observations to estimate US $FFCO_2$ emissions for 2010 and studied the impact of multiple factors to understand possible biases in the model estimates. Their model estimated uncertainty, of 2%–3% for the annual total $FFCO_2$ emission, is well in the range of practical needs and the national monthly emission estimates agreed well with the high-resolution inventory, Vulcan v3 (Gurney et al., 2012), for most of the year. However, the emissions exceeded the level of global and Environmental Protection Agency (EPA) inventories (the latter did not account for international bunker fuel use). Among several directions for reducing the errors and biases in $FFCO_2$ emission estimates, several authors point to a need to reduce the uncertainty of radiocarbon emissions from soil respiration. Model estimates suggest that measurement programs, similar in scale to that recommended by Pacala et al. (2010), can help to extend estimates of $FFCO_2$ emissions to the regional scale. Recent studies with both real atmospheric measurements of $\Delta^{14}CO_2$, and using Observing System Simulation Experiments (OSSEs) for a network of sites, have shown that atmospheric $\Delta^{14}CO_2$ can be used to estimate monthly mean $FFCO_2$ emissions for California with posterior uncertainties of 5%–8% (Fischer et al., 2017) or 15% (Brophy et al., 2019).

126 Section | B Methods

Some improvements in estimating FFCO$_2$ can also be achieved via the optimal design of the inversion set-up. Yadav, Michalak, Ray, and Shiga (2016) proposed a geostatistical inverse modeling scheme for disaggregating winter-time fluxes based on their unique temporal and spatial flux patterns. The application of the method is demonstrated with one synthetic and two real data prototypical inversions using in situ CO$_2$ flux measurements from eddy covariance flux towers over North America. Inversions are performed for January since in other months the predominance of land biosphere CO$_2$ fluxes over FFCO$_2$ and observational limitations preclude the disaggregation of the fluxes. The quality of disaggregation is assessed primarily through the examination of the a posteriori covariance between disaggregated FFCO$_2$ and land biosphere fluxes at regional scales. Findings indicate that the proposed method is able to robustly disaggregate fluxes regionally for the study period with a posteriori cross covariance lower than $0.15\,\mu\text{mol}\,\text{m}^{-2}\,\text{s}^{-1}$ between FFCO$_2$ and land biosphere fluxes.

7 Application to CH$_4$ fluxes

Changes in atmospheric CH$_4$ result from an imbalance between the CH$_4$ sources and sinks on the Earth's surface and loss in the atmosphere. Global atmospheric chemistry transport models (ACTMs) indicate a loss of 518–579 Tg CH$_4$ yr^{-1} over the period of 1990–99 due to the chemical reactions with hydroxyl radicals (OH), chlorine radicals (Cl), and oxygen radicals (in the state O^{1}D) (Kirschke et al., 2013; Patra et al., 2011).

About 90% of the CH$_4$ loss is due to the reaction with OH. Although the atmospheric abundance of OH globally is fairly stable, small variations in OH influence the amount of CH$_4$ loss. Since OH is a highly reactive species, it is not possible to measure directly but attempts have been made to infer the OH abundance using measurements of methyl chloroform (CH$_3$CCl$_3$). Methyl chloroform has only industrial emissions (which are generally easier to estimate accurately than biological emissions) and, like CH$_4$, is primarily lost in the troposphere by reaction with OH. Thus assuming the emissions of CH$_3$CCl$_3$ are well known, changes in its atmospheric mixing ratio can be used to derive changes in OH (Krol & Lelieveld, 2003; Montzka et al., 2011; Patra et al., 2014; Prinn et al., 2005). Since the atmospheric abundance of CH$_3$CCl$_3$ has now reached very low levels, new species for determining the OH abundance are being sought, such as hydroflurocarbons (HFCs) and hydrochloroflurocarbons (HCFCs), which also have only industrial sources (Liang et al., 2017).

Methane has a large number of anthropogenic and natural sources. The diversity of sources and their uncertainty makes it challenging to compile inventory and model estimates (i.e., from bottom-up approaches) that are consistent with the observed atmospheric change in CH$_4$ and the estimated global atmospheric loss. The global total CH$_4$ emission during 2000–09 is estimated to be in the range of 583–861 Tg yr^{-1} based on bottom-up approaches; this range is much wider and nonoverlapping with the top-down estimate of 535–566 Tg yr^{-1} (Saunois et al., 2016). About 25% of the emission is thought to be due to agriculture (specifically, enteric fermentation, manure management, and rice cultivation), another 30% from industrial activities (specifically coal mining as well as natural gas production, storage, and transport) and the remaining 45% from natural sources (specifically, wetlands, ocean, termites, mud volcanoes, and biomass burning). Due to the large spatial and temporal variability of CH_4 fluxes, and its atmospheric loss, observed CH_4 mixing ratios show large spatial gradients in the troposphere and stratosphere, as well as large temporal variations on hourly, monthly, and annual timescales (Aoki et al., 1992; Dlugokencky, Nisbet, Fisher, & Lowry, 2011; Khalil & Rasmussen, 1993; Prinn et al., 2000).

Atmospheric CH_4 mixing ratios can be modeled by ACTMs using prior emission estimates, which may be based on inventories (specifically for anthropogenic sources) and empirically upscaled observed fluxes or ecosystem models (for natural sources) (Crutzen & Gidel, 1983; Fung et al., 1991; Patra et al., 2009; Warwick, Bekki, Law, Nisbet, & Pyle, 2002). The ACTMs use parameterizations of atmospheric chemistry to calculate CH_4 loss due to OH, Cl, and O^1D radicals. In the TransCom-CH_4 model intercomparison, Patra et al. (2011) showed that using the best available bottom-up emission estimates the decadal variability in global atmospheric CH_4 can be reproduced reasonably well over the period 1992–2007. They also found that the differences between the simulated CH_4 gradients simulations arose largely from the modeled transport and to a lesser extent from the modeled chemistry. Patra et al. (2011) suggest that about 40% of the difference between models in the meridional CH_4 gradient is due to model transport differences (see also Houweling et al., 2017). The TransCom model intercomparison highlighted the need for evaluating the ACTMs using independent atmospheric observations before conducting CH_4 inversions.

Atmospheric measurements of CH_4 are available from the late 1970s and inversions of CH_4 have been conducted since the late 1990s (Hein, Crutzen, & Heimann, 1997) (Bousquet et al., 2006; Fraser et al., 2013; Houweling, Kaminski, Dentener, Lelieveld, & Heimann, 1999; Patra et al., 2016). Currently, there are still fewer than 100 sites globally where measurements of CH_4 have been made uninterrupted for more than a decade. Long and consistent time series of CH_4 mixing ratio are important for studying its interannual and interdecadal evolution in the atmosphere and the causes of these variations, that is whether it is due to changes in the sources, the atmospheric sink, or both, and where these changes are occurring. Satellite remote sensing data are improving the global observational coverage, but the accuracy of satellite measurements and the long-term consistency of remote sensing data, e.g., across satellite instruments, remains an issue (Frankenberg et al., 2008).

Still, the optimization of CH_4 fluxes, as with all the applications addressed in this chapter, remains an underdetermined problem, Furthermore, the optimization of CH_4 emissions by source sector is even more poorly constrained than the total flux and the attribution of CH_4 emissions to different sources is very uncertain regionally and even globally (Kirschke et al., 2013; Saunois et al., 2017). In order to improve the constraint on the source attribution of CH_4 emissions, and to some extent the sink, further information in the form of measurements of the ratios of the stable isotopes of C and H have been included in inversions (Ghosh et al., 2015; Mikaloff Fletcher, Tans, Bruhwiler, Miller, & Heimann, 2004; Monteil et al., 2011; Rice et al., 2016; Rigby et al., 2017; Schwietzke et al., 2016; Thompson et al., 2018; Turner, Frankenberg, Wennberg, & Jacob, 2017). Briefly, the isotopic ratios, $^{13}C:^{12}C$ and D:H, are influenced by kinetic processes, an effect that is known as the Kinetic Isotope Effect (KIE), and thus CH_4 from different sources has different isotopic ratios. The relationship between these ratios and the source sector is shown in Fig. 9. The isotopic ratios are expressed relative to a standard ratio as defined by Vienna Pee Dee Belemnite (VPDB)[d] in parts per mil (‰):

$$\delta^{13}C = \left(\frac{\left(\frac{^{13}C}{^{12}C}\right)_{sample}}{\left(\frac{^{13}C}{^{12}C}\right)_{standard}} - 1 \right) \times 1000 \qquad (44)$$

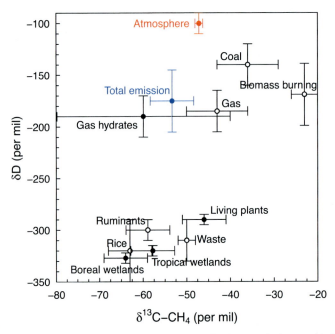

FIG. 9 Typical carbon and hydrogen isotopic signatures of anthropogenic (open symbol) and natural (solid symbol) CH_4 sources (based on Fischer et al., 2008; Quay et al., 1999; Rice et al., 2016; Whiticar, 1993). The *red* point indicates the global atmospheric mean and the blue point indicates the value of global total emission.

---

d. Pee Dee Belemnite (PDB) is based on a Cretaceous marine fossil, from the Pee Dee Formation in South Carolina. Since the original sample is no longer available it has been replaced by the VPDB standard (Miller and Wheeler, 2012, p. 186).

Top-down approaches **Chapter | 4 129**

and similarly, for δD. KIEs also occur in the atmospheric sinks due to OH, Cl, and O^1D and strongly influence the $\delta^{13}C$ and δD values in atmospheric CH. The atmospheric sinks result in enrichment of ^{13}C in atmospheric CH_4. This effect is seen in the fact that the observed tropospheric $\delta^{13}C$ value is approximately −47.2‰ (Rice et al., 2016; Thompson et al., 2018) although the $\delta^{13}C$ value of the global total source is approximately −53.5‰. The isotopic values for various sources are shown in Fig. 9; microbial sources, such as wetlands, have more negative $\delta^{13}C$ and δD values (i.e., they are strongly depleted in ^{13}C) and pyrogenic sources, such as biomass burning, have less negative values. The information provided by isotopic measurements, however, is imperfect as there is considerable overlap in the $\delta^{13}C$ and δD values of the different source sectors. A discussion of the constraint that the isotopic measurements can provide is given by Rigby, Manning, and Prinn (2012). Despite the limitations, isotopic measurements have indicated some important inconsistencies between inventory estimates of CH_4 sources compared to the atmospheric record of $\delta^{13}C$ (Saunois et al., 2017; Schwietzke et al., 2016; Thompson et al., 2018).

Since 2007, there has been a significant increase in global atmospheric CH_4, departing from a period of relatively stable mixing ratio from the late 1990s (Nisbet et al., 2016; Rigby et al., 2008). Numerous inversion studies have tried to determine the cause of the increase using various additional constraints, such as $\delta^{13}C$, δD, CH_3CCl_3, and ethane (Rice et al., 2016; Rigby et al., 2017; Schwietzke et al., 2016; Thompson et al., 2018; Turner et al., 2017; Worden et al., 2017). However, these studies have attributed the increase to various causes, including an increase in one or more source terms and a decrease in the atmospheric sink, illustrating the large uncertainties that remain in attributing changes in atmospheric CH_4 to its sources and sinks.

While globally there are still large uncertainties in the estimation of methane's sources, progress is being made at regional levels. One good example of this is the top-down determination of East Asian emissions of CH_4, which have increased continuously during the period 2000–2012. While inventories and inversions are in agreement that the emissions have increased, numerous inversions consistently indicate that this increase has been at about half the rate suggested by global inventories (Bergamaschi et al., 2013; Janssens-Maenhout et al., 2017; Patra et al., 2016; Peng et al., 2016; Thompson et al., 2015) (see Fig. 10). In the inversion study of Patra et al. (2016), the emission increase from the fossil fuel industry in East Asia (specifically China) during 2002–2012 is reduced by half in comparison with their prior (bottom-up) emissions. On the other hand, Patra et al. (2016) found that the rate of increase in the bottom-up emission due to enteric fermentation and manure management (based on FAOSTAT[e]) was strongly supported by the inversions. However,

---

e. FAOSTAT is the Food and Agricultural Organization of the United Nations, http://www.fao.org/.

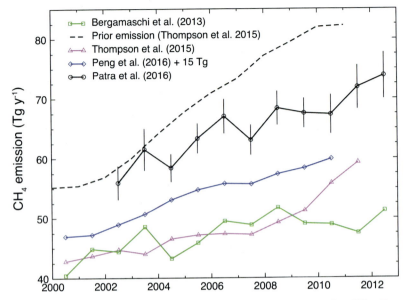

FIG. 10 Time evolution of CH_4 emissions from the East Asian region, including China, Japan, North and South Korea, and Mongolia (as in Patra et al., 2016), estimated by the prior emissions (based on Emissions Database for Global Atmospheric Research, EDGAR-v4.2 FT2010, broken *black line* and *blue line*) and inversions (*black*, *magenta*, and *green lines*). Note that the regional boundaries and the included sectors do not match exactly between studies. For example, the emissions from Peng et al. (2016) are for China only and do not include natural sectors such as termites and wetlands [therefore, an offset of 15 Tg was added to the emission estimate from Peng et al., 2016]. Despite these differences, the large discrepancy between the prior and posterior emissions and trends can still be seen.

there are still large differences in the estimated emissions arising from the differences in modeled transport and chemistry.

8 Application to other GHG fluxes

Top-down methods are also used to estimate emissions of N_2O (e.g., Hirsch et al., 2006; Huang et al., 2008; Thompson, Chevallier, et al., 2014; Wells et al., 2018) and various human-made species, including HFCs, HCFCs (Brunner et al., 2017; Stohl et al., 2010), and SF_6 (Fang et al., 2014). While the inversion principles are the same as for CO_2 and CH_4, there are some special considerations for these species. Here, we will just discuss N_2O but many of the considerations for this species also apply to HFCs, HCFCs, and SF_6.

N_2O has a long atmospheric lifetime of 116 ± 9 years (Prather et al., 2015) and the only atmospheric sinks are photolysis and oxidation, by O^1D, in the stratosphere (Minschwaner, Salawitch, & McElroy, 1993). Despite the long lifetime, the atmospheric sink still needs to be accounted for in global inversions as the amount of N_2O lost amounts to approximately 13 TgN yr^{-1} or 77% of the estimated global source (Prather et al., 2015). Since N_2O is lost only in the stratosphere and the source is predominantly at the surface (only $\sim$3% is produced by atmospheric chemistry) a large gradient in N_2O is induced across the tropopause. The long lifetime and strong vertical gradient means that a long spin-up time is essential to establish realistic initial mixing ratios for the forward and inverse model runs (see Section 4.5). If the initial mixing ratios are not close to steady state, then the source will strongly decrease (or increase) for the first few months to years of the inversion, while the opposite trend will be seen in the sink.

For global inversions, over multiple years, specific attention also needs to be paid to the strength of the sink, as this will directly determine the strength of the source needed to match the atmospheric growth rate. As the sink strength is dependent on the model's atmospheric abundance, the lifetime (i.e., the abundance divided by the sink rate) is a better metric to determine if the sink strength is appropriate. The modeled lifetime should be consistent with independent estimates such as that of Prather et al. (2015).

N_2O variability in the troposphere is influenced by the transport of air, relatively depleted in N_2O, from the stratosphere into the troposphere, i.e., stratosphere to troposphere transport (STT). Seasonal variations in STT has a significant impact on the seasonal cycle of N_2O in the lower troposphere, especially in the high latitudes (Nevison, Mahowald, Weiss, & Prinn, 2007; Thompson, Patra, et al., 2014), and thus should be represented as accurately as possible to avoid incurring errors in the optimized fluxes, in particular, in the seasonality.

In contrast to CO_2 and CH_4, the gradients in N_2O observed in the troposphere are relatively small compared to the measurement accuracy—the current accuracy is $\sim$0.2 ppb compared to an inter-hemispheric gradient of $\sim$1.4 ppb and a seasonal cycle amplitude of $\sim$0.9 ppb in the high latitudes (where it is the largest). Particular care needs to be taken when using measurements from different laboratories to correct these for any calibration scale differences as these can be on the same order of magnitude as the signals to be resolved.

Lastly, unlike CH_4, there is a very limited possibility to resolve N_2O emissions by source sector. The sources of N_2O on land include agriculture, uncultivated (or natural) soils, industry, energy, and waste treatment. Distinguishing natural from agricultural emissions in most regions of the world is not possible for the following reasons: (1) there is often a mosaic of uncultivated and cultivated land, emissions from which cannot be resolved at the scales of inversions, (2) dispersion of anthropogenic nitrogen through air and water to uncultivated land means N_2O emissions may be enhanced over the natural or preindustrial levels, and (3) land that is used for agriculture may have also emitted some N_2O naturally, i.e., without human nitrogen additions in the form of fertilizer or manure. In summary, even if these regions can be resolved spatially, the emission sources may not be able to be resolved. On the other hand, high-resolution inversions may be able to pick up emissions from large industrial

132 Section | B Methods

or waste-treatment sources. Moreover, there are no atmospheric tracers that can be used to separate the emissions by source type, as the underlying processes producing N_2O (i.e., nitrification and denitrification) are the same for all microbial sources (i.e., agriculture, waste treatment, and natural emissions). Recent advances in measuring isotopomers of N_2O ($\delta^{15}N$ and $\delta^{18}O$) do indicate, however, a global decreasing trend in $\delta^{15}N$ and $\delta^{18}O$ (Park et al., 2012). For $\delta^{15}N$ the enzyme kinetics favor ^{14}N when N-substrate is plentiful, and thus the N_2O emitted is lighter from well-fertilized agricultural soils (Park et al., 2012; Pérez et al., 2001). However, owing to the absence of a clear source signature for agricultural emissions, it is not clear how this information could be used to constrain sources at regional scales in inversions.

9 Sources of error

Inversion estimates of GHG fluxes are subject to random but also to a number of systematic errors. While the random error is given by the posterior error covariance matrix, or in the case that this cannot be calculated from the use of Monte Carlo ensembles, the systematic error is more difficult to determine. In this section, we will look into sources of systematic error that are common to all atmospheric inversion frameworks. This includes *model representation errors*, including transport and aggregation errors.

Model representation errors concern how well the model would be able to represent the measurements even if the *true* fluxes were known. In other words, how accurately the model maps the control space into the observation space, $\mathcal{X} \rightarrow \mathcal{Y}$. There are two considerations involved in this, first how well the processes are represented in the model, in our case, this is the atmospheric transport and chemistry, and second, the spatial and temporal resolution of the control variables, which determine the *aggregation error*, which was introduced in Section 4.2.2.

9.1 Transport errors

Errors in modeled atmospheric chemistry and transport are largely due to (1) limitations of the meteorological data driving the model due to its finite resolution but also its accuracy, which is determined by, e.g., the number of meteorological observations assimilated, (2) the resolution of the transport model itself (specifically for Eulerian models), and (3) assumptions made in the parameterizations of the physics describing the transport and, where relevant, the chemistry (note that errors due to chemistry as they pertain to CH_4 and other GHGs are discussed in Sections 7 and 8).

The resolution of the model (and the driving meteorological data) affects the accuracy of the transport, and thus the atmospheric observations can be represented, in a number of ways. First, the resolution of topography determines how well complex terrain is represented. In a coarse model, the topography will be significantly smoothed meaning that a mountain site, if it is located at the surface in the model, may be at a much lower altitude compared to the reality, but on the other hand, if it is placed at an altitude in the model that is comparable to the reality, it will be above and more decoupled from the surface (Geels et al., 2007). In both cases, the site may not be well represented and the best results might be obtained with the site located somewhere in between these two scenarios. Second, mountain sites are affected by local thermally induced circulations, i.e., upslope winds over sunlit slopes during the day and downslope winds during the night. An accurate representation of up/downslope winds requires sufficient resolution of the topography and in the meteorological data. A third problem, also directly related to the model resolution, is the representation of coastal sites. In a coarse resolution model, the coordinates of the site may in fact fall in a grid cell that is classified as an ocean and not land, thus care should be taken to check and, if needed, adjust, the model location of coastal sites. Furthermore, local land-sea breezes can have a strong impact on the observations at coastal sites but may not be resolved in coarse models.

Low altitude (flat terrain) sites are generally better represented by atmospheric transport models than mountain sites and continental sites provide the most information about land-atmosphere fluxes. However, how well these sites are represented is strongly determined by the accuracy of the mixed layer or planetary boundary layer (PBL) height. This is not only linked to the horizontal resolution of the model but also to the vertical resolution and the representation of vertical mixing. The PBL height determines the volume into which a gas emitted at the surface can efficiently be mixed and hence, what impact this will have on the mixing ratio in the volume. For instance, if the model overestimates the PBL height it will underestimate the change in mixing ratio from a given surface flux and vice versa and obviously, this will lead to bias errors in the fluxes estimated from atmospheric inversions. The modeled PBL height can be validated against radiosonde data and the error in the height can be used to estimate the error in the modeled mixing ratio (Gerbig, Körner, & Lin, 2008). While Gerbig et al. (2008) found the systematic error (or bias) in the daytime mixing heights to be negligible, at least in summer in temperate latitudes, the errors in the nocturnal mixing heights were large and in some cases exceeded 50%.

For CO_2, the impact of mixing height errors is particularly important as the PBL height covaries with CO_2 flux. During the day when the CO_2 flux is negative (due to photosynthesis), the mixed layer is deep and dilutes the effect of the flux on the mixing ratios, but during the night when the CO_2 flux is positive (due to respiration), the mixed layer is shallow and the effect on the mixing ratios is large (Denning, Randall, Collatz, & Sellers, 1996; Denning, Takahashi, & Friedlingstein, 1999) (Fig. 11). This is known as the diurnal rectifier effect in reference to electrical rectifiers, as the different processes "beat" on the same frequency. Similarly, there is also a seasonal rectifier effect. The diurnal and seasonal rectifiers effectively trap high CO_2 air (from respiration) close to the surface while depleted CO_2 air (from photosynthesis) is aloft. This also leads to an enhanced gradient in the surface layer from the coast to inland

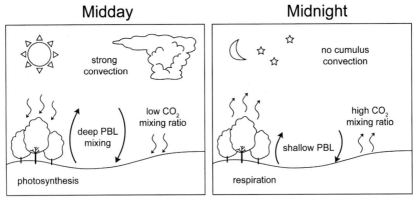

FIG. 11 Schematic illustration of the diurnal covariance between the land-atmosphere CO_2 flux (from photosynthesis and respiration) and the change in the planetary boundary layer (PBL) height.

and an enhanced latitudinal gradient from the south to the north (Denning et al., 1999). If the rectifier effect is too strong in the model, resulting in too strong a gradient, then a spuriously large sink in the northern midlatitudes will be necessary to match the simulated mixing ratios with the observations, and vice versa (Denning et al., 1999).

The last problem we consider here is the effect of resolution on the modeling of subgrid variability and point sources. Mixing ratios simulated by Eulerian models are subject to an error arising from the assumption that the influence of fluxes is mixed instantaneously into the grid cell volume, while in reality, observations at a site are sensitive to the subgrid scale variability of the fluxes and the wind direction. A proxy method for how to estimate this error for each observation is given by Bergamaschi et al. (2010). Their method considers the influence on the mixing ratio from a change in PBL height over the time interval needed for air to pass through the grid-cell volume (where the observation is located), where this time interval is determined by horizontal advection (see Bergamaschi et al., 2010 for details). In addition, a proxy method to estimate the horizontal transport error in Eulerian models is provided by Rödenbeck, Houweling, Gloor, and Heimann (2003). This is proxy based on the spatial and temporal mixing ratio gradients between the grid-cell where an observation is located and the neighboring grid cells, using the argument that the error in the modeled mixing ratio will be larger the more heterogeneous the mixing ratio field is around that location.

In summary, we give the following recommendations to minimize model transport errors [the first three are based on those given by Geels et al., 2007]:

1. Low altitude sites (flat terrain) are generally better represented in models, but if mountain sites are included then the model layers that most accurately represent the sites should be chosen.

Top-down approaches **Chapter | 4 135**

2. At low altitude sites, afternoon values are less susceptible to errors in the PBL height and are more representative of sources and sinks on large scales.
3. For low altitude sites, observations at $\sim100\,$m or more above the surface (i.e., from tall towers) will be more representative and more reproducible in transport models.
4. Including estimates of model transport error in the observation uncertainty will reduce the impact that these have on the posterior fluxes.

9.2 Aggregation errors

Aggregation errors arise from averaging (or aggregating) the control variables leading to a loss of information from small-scale structures. In reality, the observations may be sensitive to these small-scale structures but are not represented by the forward model. Discretization to some level is necessary and averaging the control variables from the resolution of the transport model (the native resolution) to a coarser one in many cases is unavoidable since the full resolution may be too computationally demanding. Fig. 12 shows schematically how the mixing ratio at an observation site depends on the way the control variables are averaged in space and time. The temporal aggregation error arises when the sensitivities to the source (described by the corresponding elements of the Jacobian matrix) are different between time steps t_1 and t_2. If the sensitivity does not change, then averaging the sources over these time steps will not impact the mixing ratio. Similarly, the spatial aggregation error arises when the elements of the Jacobian matrix have different sensitivities to the sources s_1 and s_2. From this, we can see that the aggregation error depends not only on the resolution of the control variables but also on how they are sampled, or in other words, how sensitive the observations are to the loss of small-scale structure, as described by the Jacobian matrix.

The problem of aggregation errors was first raised by Trampert and Snieder (1996) as it pertains to tomography, and then by Kaminski et al. (2001) who described the spatial aggregation error in the application to inversions of CO_2. Trampert and Snieder (1996) derived a method to project the loss of information from averaging/aggregating the control variables into the observation space so that it can be included in the observation uncertainty estimate. This method gives less weight to observations that are strongly affected by the loss of information and are described in Section 4.4.2. However, since this method requires the transport model at the native resolution to be formulated as a matrix operator, it is not always possible let alone practical to implement. Moreover, while maintaining or increasing the resolution helps to minimize the aggregation error there is a trade-off not only in computational efficiency but also in the dependence of the solution on the regularization or prior information used (for a discussion on this see Bocquet, 2009). Here, we discuss strategies to minimize the aggregation error by choosing appropriate averaging of the control variables.

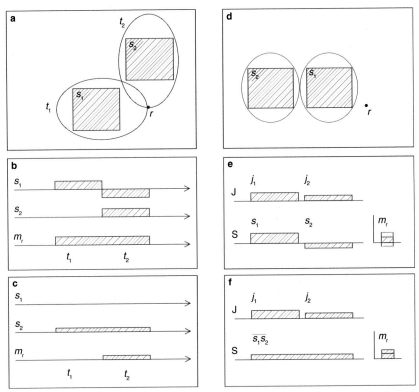

FIG. 12 Schematic illustration demonstrating the temporal (panels A–C) and spatial (panels D–F) aggregation errors. (A) Map of a receptor site, r, and two sources, s_1 and s_2, which are sampled by atmospheric transport at times t_1 and t_2, respectively, where the Jacobians (indicated by the ellipses) have the same magnitude. (B) The magnitude of the sources at times t_1 and t_2 and the mixing ratio, m_r simulated at the receptor at each time (note s_1 is sampled at t_1 and s_2 is sampled at t_2). (C) Shows the same as (B) but where the sources are averaged over t_1 and t_2, and the effect this has on the mixing ratio, m_r. The difference in the mixing ratio between cases (B) and (C) is the effect of temporal aggregation error. (D) Map of two sources, s_1 and s_2, are sampled at the same time. (E) The values of the Jacobian, j_1 and j_2 corresponding to s_1 and s_2, respectively, and the magnitudes of the sources and the simulated mixing ratio, m_r. (F) Shows the same as (E) but where s_1 and s_2 are averaged into one source and the effect this has on the simulated mixing ratio, m_r. Here the difference in mixing ratio between cases (E) and (F) is the effect of spatial aggregation error.

Since the aggregation error arises only when the observations are sensitive to the loss of small-scale structure, aggregating the control variables where the observations have little sensitivity to the fluxes is a practical solution. These are the variables that lie in, or are close to, the null space of the Jacobian matrix, **H**. This approach is used by Stohl et al. (2009) who aggregate variables (at the native resolution) so that the value of the corresponding aggregated elements of the Jacobian matrix is greater than some threshold. Aggregating in this way transforms the variables onto an irregular grid where the resolution is finest

Top-down approaches **Chapter | 4** **137**

close to observations and decreases with distance, where the sensitivity to the fluxes is lower. Bocquet (2009) uses a similar but more formal approach and shows how an optimal representation of the control variables can be found using some criterion, such as maximizing the trace of the posterior confidence matrix (i.e., the inverse of the posterior error covariance matrix, $\mathbf{P}$), which is also known as the Fisher Information (FI):

$$FI = tr\left(\mathbf{P}^{-1}\right) = tr\left(\mathbf{B}^{-1} + \mathbf{H}^T\mathbf{R}^{-1}\mathbf{H}\right) \quad (45)$$

Alternatively, the aggregation error can be minimized by examining the off-diagonal elements of the error covariance matrix and aggregating variables that have correlated errors. Although the true error correlations are usually not known, some idea of the correlation is used to construct the prior error covariance matrix $\mathbf{B}$. For example, in the optimization of NEE, the errors may be correlated within biomes and on a scale comparable to that of weather systems. The inversion framework, CarbonTracker, used by Peters et al. (2010) to estimate NEE, uses this type of approach. In this example, the control variables are scalars of the prior fluxes and are defined for each biome. This is equivalent to assuming that the prior fluxes within each biome are fully correlated as well as the prior model correctly represents the biomes and that the fluxes within each biome respond in the same way to a given forcing, or at least in a way that is consistent with the prior model.

The choice of aggregation method will determine the mapping operator, $\mathbf{M}$, needed to map from the higher to lower resolution control vector (see Eq. 34). Regardless of the aggregation method, an appropriate prior error covariance for the low-resolution control vector needs to be used. The low-resolution prior error covariance matrix can be found as:

$$\mathbf{B}_L = \mathbf{MBM}^T \quad (46)$$

For a full discussion about a priori information and the choice of inversion resolution, see Rodgers (2000), Chapter 10.

10 Validation of flux estimates from inversions

In a well set up inversion, control variables that have no observational constraint (i.e., they lie in the null space of the Jacobian matrix $\mathbf{H}$) should remain at their prior value—if this is not the case there is a good reason to revise the inversion! Moreover, the dependence of the posterior solution on the prior information can be written as

$$\mathbf{x} = \mathbf{x_b} + \mathbf{G}(\mathbf{y} - \mathbf{Hx_b}), \text{ where } \mathbf{G} = \left(\mathbf{H}^T\mathbf{R}^{-1}\mathbf{H} + \mathbf{B}^{-1}\right)\mathbf{H}^T\mathbf{R}^{-1} \quad (47)$$

$$= \mathbf{x_b} + \mathbf{G}(\mathbf{H}(\mathbf{x_t} - \mathbf{x_b}) + \boldsymbol{\varepsilon}) \quad (48)$$

$$= \mathbf{x_b} + \mathbf{A}(\mathbf{x_t} - \mathbf{x_b}) + \mathbf{G}\boldsymbol{\varepsilon}, \text{ where } \mathbf{A} = \mathbf{GH} \quad (49)$$

$$= (\mathbf{I} - \mathbf{A})\mathbf{x_b} + \mathbf{Ax_t} + \mathbf{G}\boldsymbol{\varepsilon} \quad (50)$$

where the matrix $\mathbf{A}$ is known as the averaging kernel and $\boldsymbol{\varepsilon}$ is the observation error. The averaging kernel is conceptually the same as in Eqs. (3) and (4) for atmospheric retrievals, but here it is in terms of fluxes. The matrix $\mathbf{I} - \mathbf{A}$ represents the fraction of the prior information that appears in the posterior solution (Rodgers, 2000). From Eq. (50) we can see that the posterior solution is effectively a weighting between the prior and true states, which provides a means to check how well constrained the solution is by the observations versus the prior. In an inversion system with no systematic error, this would already give us a good idea about how accurate the posterior solution is. (Note though, for very large inversions, or where $\mathbf{H}$ cannot be defined as a matrix, the terms in Eq. (50) cannot be computed explicitly.) Unfortunately, systematic errors are unavoidable (as we have seen in the previous section). Therefore, validation of the posterior solution using independent observations is highly recommendable. Independent observations are those that were not used in the inversion and, ideally, should not be too closely correlated with observations that were included, but should still be sensitive to the fluxes in the inversion domain. These could be atmospheric measurements (from ground-based sites or aircraft) or measurements of the fluxes themselves.

If the posterior fluxes are nonbiased (and closer to the true fluxes) then the simulated mixing ratios from a forward model run should be closer to the independent atmospheric observations than those simulated using the prior fluxes. Comparing simulated versus observed vertical profiles, such as from aircraft measurements, is a particularly useful diagnostic since this may help identify errors in the modeled vertical mixing, which would bias the optimized fluxes (see Section 9.1). This diagnostic is used in the study of Bergamaschi et al. (2018) who compare the simulated vertical profiles from five inversions for CH_4 with observations from aircraft. For more quantitative analysis, Bergamaschi et al. (2018) also compared the enhancement of CH_4 mixing ratio above the background and integrated this enhancement vertically throughout the boundary layer. The rationale behind this is that the impact of fluxes is strongest on mixing ratios within the boundary layer and, therefore, differences between the modeled and observed integrated enhancements should reflect biases in the posterior fluxes. A similar approach is used by Pickett-Heaps et al. (2011) who evaluate the posterior NEE from four inversions by running forward simulations of all the NEE fields with an independent atmospheric transport model and comparing the simulated CO_2 mixing ratios with those from aircraft profiles. Since the bias errors in the posterior fluxes are model-dependent, a forward model simulation with the same model as used for the inversion will compensate to a certain extent for these errors in the simulated mixing ratios. For this reason, Pickett-Heaps et al. (2011) perform the forward simulation with an independent transport model. They found that all models had a significant negative bias for June to August in the lower troposphere at a northern hemisphere continental site indicating systematic errors in modeling vertical transport (Pickett-Heaps et al., 2011). A comparison of the prior and

Top-down approaches **Chapter | 4 139**

posterior simulated mixing ratios versus the observations, however, does not tell us about where the posterior fluxes have improved. Liu and Bowman (2016) formulated a method to project the prior and posterior model-observation errors to spatial and temporal differences between the prior and posterior fluxes. They further show mathematically that the posterior fluxes are more accurate than the prior ones over the regions that contribute to the changes in mixing ratios when the posterior mixing ratios are more accurate (i.e., closer to the observations).

For high-resolution (typically regional) inversions, direct flux measurements, such as from eddy covariance towers may also be used for validation. However, even with relatively high spatial resolution, the inversion grid is still typically much larger than the footprint of an eddy covariance tower, which has a radius in the order of 100 m (Baldocchi, 1997). Although eddy covariance measurements are much more sensitive to local fluxes, they can still be used for validation by filtering out the high-frequency component. Validation of an inversion for NEE against eddy covariance measurements is made by Kountouris et al. (2018) who compare the seasonal cycle from their inversion for NEE at 50 km resolution with that from eddy covariance measurements.

In this section, we have mentioned studies using multiple inversions, but we have not yet discussed the use of inversion ensembles. Although inversion ensembles cannot be used for validation in the strict sense, they can help estimate the overall accuracy of the posterior fluxes. An ensemble of inversions, each with independent atmospheric transport models, can provide clues to the accuracy of the posterior fluxes by taking into account (at least to some extent) systematic errors from transport. A number of recent studies have looked at inversion ensembles to establish an uncertainty range. Peylin et al. (2013) examined an ensemble of 11 global inversion frameworks (each with different prior information and atmospheric transport models) optimizing CO_2 fluxes over land and ocean. They found that the fossil fuel emission estimates used in the inversions varied greatly and that the difference between these showed up in the adjustments to the nonfossil fuel fluxes, contributing an additional source of systematic error.

11 Summary and conclusions

This chapter has introduced the top-down approach, specifically atmospheric inversions, to estimate surface-atmosphere fluxes of greenhouse gases (GHGs). We have seen how Bayesian statistics can be applied to derive equations to optimize the fluxes using the model-observation mismatch. Although there are a vast number of optimization algorithms, we have examined the principal ones used in the atmospheric sciences and for estimating GHG fluxes. We have also introduced the reader to the practical considerations needed in inverse modeling, such as the estimation of the prior and observation uncertainties, and error covariance matrices, as well as methods for assessing the appropriateness of these choices.

140 Section | B Methods

We have presented the application of atmospheric inversion to estimate the fluxes of the principle GHGs, i.e., CO_2, CH_4, and N_2O. We have seen for CO_2, there are in fact two distinct problems that can be addressed with top-down approaches: first, improving estimates of the net land-biosphere CO_2 fluxes (or NEE) and second, improving estimates of the fossil fuel emissions. For the first problem, which is the more traditional use of inverse modeling, we have seen that progress has been made since the late 1980s as shown by the coherent patterns of interannual variability in NEE and meteorological conditions found by a number of inverse models, as well as by the coherent results at regional scales among inversion ensembles. For the second problem, we have looked at the importance of improving fossil fuel emission estimates using top-down approaches, particularly at the city scale since cities are responsible for about 70% of global emissions. We have presented a number of examples of city-scale inversion studies and shown how these can lead to improvements in city emission estimates. For CH_4, we have seen the importance of atmospheric chemistry in the CH_4 budget, in particular, oxidation by the OH radical. We have also seen that the CH_4 budget is still poorly constrained owing to the difficulty to uniquely quantify the sources and atmospheric sink of CH_4 given the current observation networks. While additional tracers for constraining the OH sink (e.g., CH_3CCl_3 and various HFC and HCFC species) and the emission sources (e.g., stable isotopes of C and H in CH_4) can help, the problem remains poorly constrained. Moreover, we have shown that the same principles of atmospheric inversion can be applied to any long-lived GHG, such as N_2O as well as many of the industrially produced species, such as HFCs, HCFCs, and SF_6.

Finally, this chapter looked at the principal sources of systematic error in atmospheric inversions, such as model representation error, and how the effects of these may be minimized, and we have provided guidance on how the results of an atmospheric inversion may be assessed and validated.

Acknowledgments

K. Bowman's contribution was supported by the Jet Propulsion Laboratory, California Institute of Technology, under a contract with the National Aeronautics and Space Administration (80NM0018D0004).

References

Agusti-Panareda, A., Diamantakis, M., Bayona, V., Klappenbach, F., & Butz, A. (2017). Improving the inter-hemispheric gradient of total column atmospheric CO_2 and CH_4 in simulations with the ECMWF semi-Lagrangian atmospheric global model. *Geoscientific Model Development*, *10*(1), 1–18. https://doi.org/10.5194/gmd-10-1-2017.

Alden, C. B., Miller, J. B., Gatti, L. V., Gloor, M. M., Guan, K., Michalak, A. M., et al. (2016). Regional atmospheric CO_2 inversion reveals seasonal and geographic differences in Amazon net biome exchange. *Global Change Biology*, *22*(10), 3427–3443. https://doi.org/10.1111/gcb.13305.

Aoki, S., Nakazawa, T., Murayama, S., & Kawaguchi, S. (1992). Measurements of atmospheric methane at the Japanese Antarctic Station, Syowa. *Tellus Series B: Chemical and Physical Meteorology, 44*(4), 273–281. https://doi.org/10.1034/j.1600-0889.1992.t01-3-00005.x.

Aumann, H. H., Chahine, M. T., Gautier, C., Goldberg, M. D., Kalnay, E., McMillin, L. M., et al. (2003). AIRS/AMSU/HSB on the aqua mission: Design, science objectives, data products, and processing systems. *IEEE Transactions on Geoscience and Remote Sensing, 41*(2), 253–264. https://doi.org/10.1109/TGRS.2002.808356.

Baldocchi, D. (1997). Flux footprints within and over forest canopies. *Boundary-Layer Meteorology, 85*(2), 273–292. https://doi.org/10.1023/A:1000472717236.

Bastos, A., Friedlingstein, P., Sitch, S., Chen, C., Mialon, A., Wigneron, J.-P., et al. (2018). Impact of the 2015/2016 El Niño on the terrestrial carbon cycle constrained by bottom-up and top-down approaches. *Philosophical Transactions of the Royal Society, B: Biological Sciences, 373* (1760), 20170304. https://doi.org/10.1098/rstb.2017.0304.

Bastos, A., Janssens, I. A., Gouveia, C. M., Trigo, R. M., Ciais, P., Chevallier, F., et al. (2016). European land CO_2 sink influenced by NAO and East-Atlantic Pattern coupling. *Nature Communications, 7*(1), 10315. https://doi.org/10.1038/ncomms10315.

Basu, S., Lehman, S. J., Miller, J. B., Andrews, A. E., Sweeney, C., Gurney, K. R., et al. (2020). Estimating US fossil fuel CO_2 emissions from measurements of ^{14}C in atmospheric CO_2. *Proceedings of the National Academy of Sciences of the United States of America*, 201919032.

Basu, S., Miller, J. B., & Lehman, S. (2016). Separation of biospheric and fossil fuel fluxes of CO_2 by atmospheric inversion of CO_2 and $^{14}CO_2$ measurements: Observation system simulations. *Atmospheric Chemistry and Physics, 16*(9), 5665–5683.

Beer, R. (2006). TES on the aura mission: Scientific objectives, measurements, and analysis overview. *IEEE Transactions on Geoscience and Remote Sensing, 44*(5), 1102–1105. https://doi.org/10.1109/TGRS.2005.863716.

Beirle, S., Boersma, K. F., Platt, U., Lawrence, M. G., & Wagner, T. (2011). Megacity emissions and lifetimes of nitrogen oxides probed from space. *Science, 333*. https://doi.org/10.1126/science.1207824.

Bergamaschi, P., Houweling, S., Segers, A., Krol, M., Frankenberg, C., Scheepmaker, R. A., et al. (2013). Atmospheric CH_4 in the first decade of the 21st century: Inverse modeling analysis using SCIAMACHY satellite retrievals and NOAA surface measurements. *Journal of Geophysical Research-Atmospheres, 118*, 7350–7369. https://doi.org/10.1002/jgrd.50480.

Bergamaschi, P., Karstens, U., Manning, A. J., Saunois, M., Tsuruta, A., Berchet, A., et al. (2018). Inverse modelling of European CH_4 emissions during 2006-2012 using different inverse models and reassessed atmospheric observations. *Atmospheric Chemistry and Physics, 18*(2), 901–920. https://doi.org/10.5194/acp-18-901-2018.

Bergamaschi, P., Krol, M., Meirink, J. F., Dentener, F., Segers, A., van Aardenne, J., et al. (2010). Inverse modeling of European CH_4 emissions 2001-2006. *Journal of Geophysical Research, 115*(D22). https://doi.org/10.1029/2010jd014180, D22309.

Bocquet, M. (2009). Toward optimal choices of control space representation for geophysical data assimilation. *Monthly Weather Review, 137*(7), 2331–2348. https://doi.org/10.1175/2009mwr2789.1.

Bolin, B., & Keeling, C. D. (1963). Large-scale atmospheric mixing as deduced from the seasonal and meridional variations of carbon dioxide. *Journal of Geophysical Research-Atmospheres, 68* (13), 3899–3920. https://doi.org/10.1029/JZ068i013p03899.

Bousquet, P., Ciais, P., Miller, J. B., Dlugokencky, E. J., Hauglustaine, D. A., Prigent, C., et al. (2006). Contribution of anthropogenic and natural sources to atmospheric methane variability. *Nature, 443*(7110), 439–443. https://doi.org/10.1038/nature05132.

142 Section | B Methods

Bousquet, P., Peylin, P., Ciais, P., Le Quere, C., Friedlingstein, P., & Tans, P. P. (2000). Regional changes in carbon dioxide fluxes of land and oceans since 1980. *Science, 290*(5495), 1342–1346. https://doi.org/10.1126/science.290.5495.1342.

Bowman, K. W., Liu, J., Bloom, A. A., Parazoo, N. C., Lee, M., Jiang, Z., et al. (2017). Global and Brazilian carbon response to El Niño Modoki 2011–2010. *Earth and Space Science, 4*(10), 637–660. https://doi.org/10.1002/2016EA000204.

Bowman, K. W., Rodgers, C. D., Kulawik, S. S., Worden, J., Sarkissian, E., Osterman, G., et al. (2006). Tropospheric emission spectrometer: Retrieval method and error analysis. *IEEE Transactions on Geoscience and Remote Sensing, 44*(5), 1297–1307. https://doi.org/10.1109/TGRS.2006.871234.

Bréon, F. M., Broquet, G., Puygrenier, V., Chevallier, F., Xueref-Remy, I., Ramonet, M., et al. (2015). An attempt at estimating Paris area CO_2 emissions from atmospheric concentration measurements. *Atmospheric Chemistry and Physics, 15*(4), 1707–1724. https://doi.org/10.5194/acp-15-1707-2015.

Brophy, K., Graven, H., Manning, A. J., White, E., Arnold, T., Fischer, M. L., et al. (2019). Characterizing uncertainties in atmospheric inversions of fossil fuel CO_2 emissions in California. *Atmospheric Chemistry and Physics, 19*(5), 2991–3006. https://doi.org/10.5194/acp-19-2991-2019.

Broquet, G., Chevallier, F., Bréon, F. M., Alemanno, M., Apadula, F., Hammer, S., et al. (2013). Regional inversion of CO_2 ecosystem fluxes from atmospheric measurements: Reliability of the uncertainty estimates. *Atmospheric Chemistry and Physics, 13*(3), 9039–9056. https://doi.org/10.5194/acp-13-9039-2013.

Broyden, C. G. (1970). The convergence of a class of double-rank minimization algorithms 1. General considerations. *IMA Journal of Applied Mathematics, 6*(1), 76–90. https://doi.org/10.1093/imamat/6.1.76.

Brunner, D., Arnold, T., Henne, S., Manning, A., Thompson, R. L., Maione, M., et al. (2017). Comparison of four inverse modelling systems applied to the estimation of HFC-125, HFC-134a, and SF_6 emissions over Europe. *Atmospheric Chemistry and Physics, 17*(17), 10651–10674. https://doi.org/10.5194/acp-17-10651-2017.

Buchwitz, M., Beek, R., Burrows, J. P., Bovensmann, H., Warneke, T., Notholt, J., et al. (2005). Atmospheric methane and carbon dioxide from SCIAMACHY satellite data: Initial comparison with chemistry and transport models. *Atmospheric Chemistry and Physics, 5*, 941–962.

Cabannes, H., & Temam, R. (Eds.). (1973). *Vol. 18. Proceedings of the Third International Conference on Numerical Methods in Fluid Mechanics* (pp. 163–168). Berlin, Heidelberg: Springer Berlin Heidelberg. https://doi.org/10.1007/BFb0118659.

Chance, K., & Martin, R. V. (2017). *Spectroscopy and Radiative Transfer of Planetary Atmospheres.* Oxford: Oxford University Press.

Chevallier, F. (2007). Impact of correlated observation errors on inverted CO_2 surface fluxes from OCO measurements. *Geophysical Research Letters, 34*(24). https://doi.org/10.1029/2007gl030463.

Chevallier, F., Bréon, F.-M., & Rayner, P. J. (2007). Contribution of the orbiting carbon observatory to the estimation of CO_2 sources and sinks: Theoretical study in a variational data assimilation framework. *Journal of Geophysical Research-Atmospheres, 112*(D9). https://doi.org/10.1029/2006jd007375.

Chevallier, F., Engelen, R. J., Carouge, C., Conway, T. J., Peylin, P., Pickett-Heaps, C., et al. (2009). AIRS-based versus flask-based estimation of carbon surface fluxes. *Journal of Geophysical Research, 114*(D20), 1342. https://doi.org/10.1029/2009JD012311.

Chib, S., & Greenberg, E. (2012). Understanding the metropolis-hastings algorithm. *The American Statistician, 49*(4), 327–335. https://doi.org/10.1080/00031305.1995.10476177.

Top-down approaches **Chapter | 4** **143**

Ciais, P., Dolman, A. J., Bombelli, A., Duren, R., Peregon, A., Rayner, P. J., et al. (2014). Current systematic carbon-cycle observations and the need for implementing a policy-relevant carbon observing system. *Biogeosciences, 11*, 3547–3602. https://doi.org/10.5194/bg-11-3547-2014.

Craig, H. (1957). The natural distribution of radiocarbon and the exchange time of carbon dioxide between atmosphere and sea. *Tellus, 9*(1), 1–17. https://doi.org/10.3402/tellusa.v9i1.9078.

Crevoisier, C., Chédin, A., Matsueda, H., Machida, T., Armante, R., & Scott, N. A. (2009). First year of upper tropospheric integrated content of CO_2 from IASI hyperspectral infrared observations. *Atmospheric Chemistry and Physics, 9*(14), 4797–4810. https://doi.org/10.5194/acp-9-4797-2009.

Crevoisier, C., Gloor, M., Gloaguen, E., Horowitz, L. W., Sarmiento, J. L., Sweeney, C., et al. (2006). A direct carbon budgeting approach to infer carbon sources and sinks. Design and synthetic application to complement the NACP observation network. *Tellus Series B: Chemical and Physical Meteorology, 58*(5), 366–375. https://doi.org/10.1111/j.1600-0889.2006.00214.x.

Crisp, D. (2018). *A Constellation Architecture for Monitoring Carbon Dioxide and Methane From Space.* Retrieved from http://ceos.org/ourwork/virtual-constellations/acc/.

Crisp, D., Miller, C. E., & DeCola, P. L. (2008). NASA orbiting carbon observatory: Measuring the column averaged carbon dioxide mole fraction from space. *Journal of Applied Remote Sensing, 2*(1). https://doi.org/10.1117/1.2898457, 023508.

Crosson, E. R. (2008). A cavity ring-down analyzer for measuring atmospheric levels of methane, carbon dioxide, and water vapor. *Applied Physics B, 92*(3), 403–408. https://doi.org/10.1007/s00340-008-3135-y.

Crowell, S., Baker, D., Schuh, A., Basu, S., Jacobson, A. R., Chevallier, F., et al. (2019). The 2015-2016 carbon cycle as seen from OCO-2 and the global *in situ* network. *Atmospheric Chemistry and Physics Discussions, 2019*, 1–79. https://doi.org/10.5194/acp-2019-87.

Crutzen, P. J., & Gidel, L. T. (1983). A two-dimensional photochemical model of the atmosphere: 2. The tropospheric budgets of the anthropogenic chlorocarbons CO, CH_4, CH_3Cl and the effect of various NOx sources on tropospheric ozone. *Journal of Geophysical Research, 88*(C11), 6641–6661. https://doi.org/10.1029/JC088iC11p06641.

Cunnold, D. M., Steele, L. P., Fraser, P. J., Simmonds, P. G., Prinn, R. G., Weiss, R. F., et al. (2002). In situ measurements of atmospheric methane at GAGE/AGAGE sites during 1985-2000 and resulting source inferences. *Journal of Geophysical Research Atmospheres (1984-2012), 107* (D14). https://doi.org/10.1029/2001JD001226, ACH–20.

Davis, K. J., Deng, A., Lauvaux, T., Miles, N. L., Richardson, S. J., Sarmiento, D. P., et al. (2017). The Indianapolis Flux Experiment (INFLUX): A test-bed for developing urban greenhouse gas emission measurements. *Elementa: Science of the Anthropocene, 5*, 21. https://doi.org/10.1525/elementa.188.

Deeter, M. N., Emmons, L. K., Edwards, D. P., Drummond, J. R., & Gille, J. C. (2004). Vertical resolution and information content of CO profiles retrieved by MOPITT. *Geophysical Research Letters, 31*(15). https://doi.org/10.1029/2004GL020235, L15112.

Deng, A., Lauvaux, T., Davis, K. J., Gaudet, B. J., Miles, N., Richardson, S. J., et al. (2017). Toward reduced transport errors in a high resolution urban CO_2 inversion system. *Elementa: Science of the Anthropocene, 5*, 20. https://doi.org/10.1525/elementa.133.

Denning, A. S., Randall, D. A., Collatz, G. J., & Sellers, P. J. (1996). Simulations of terrestrial carbon metabolism and atmospheric CO_2 in a general circulation model. *Tellus Series B: Chemical and Physical Meteorology, 48*(4), 543–567. https://doi.org/10.1034/j.1600-0889.1996.t01-1-00010.x.

Denning, A. S., Takahashi, T., & Friedlingstein, P. (1999). Can a strong atmospheric CO_2 rectifier effect be reconciled with a "reasonable" carbon budget? *Tellus Series B: Chemical and Physical Meteorology, 51*(2), 249–253. https://doi.org/10.1034/j.1600-0889.1999.t01-1-00010.x.

Detmers, R. G., Hasekamp, O., Aben, I., Houweling, S., van Leeuwen, T. T., Butz, A., et al. (2015). Anomalous carbon uptake in Australia as seen by GOSAT. *Geophysical Research Letters, 42* (19), 8177–8184. https://doi.org/10.1002/2015GL065161.

Dlugokencky, E. J., Myers, R. C., Lang, P. M., Masarie, K. A., Crotwell, A. M., Thoning, K. W., et al. (2005). Conversion of NOAA atmospheric dry air CH_4 mole fractions to a gravimetrically prepared standard scale. *Journal of Geophysical Research, 110*(D18). https://doi.org/10.1029/2005JD006035, D18306.

Dlugokencky, E. J., Nisbet, E. G., Fisher, R., & Lowry, D. (2011). Global atmospheric methane: Budget, changes and dangers. *Philosophical Transactions of the Royal Society A: Mathematical, Physical and Engineering Sciences, 369*(1943), 2058–2072. https://doi.org/10.1098/rsta.2010.0341.

Engelen, R. J., & Stephens, G. L. (2004). Information content of infrared satellite sounding measurements with respect to CO_2. *Journal of Applied Meteorology, 43*(2), 373–378. https://doi.org/10.1175/1520-0450(2004)043<0373:ICOISS>2.0.CO;2.

Engl, H. W., Hanke, M., & Neubauer, A. (1996). *Regularization of inverse problems*. Dordrecht, The Netherlands: Kluwer Academic Publishers.

Enting, I. G. (2002). *Inverse Problems in Atmospheric Constituent Transport*. New York: Cambridge University Press, Cambridge.

Enting, I. G., & Mansbridge, J. V. (1989). Seasonal sources and sinks of atmospheric CO_2 direct inversion of filtered data. *Tellus B, 41*(2), 111–126. https://doi.org/10.1111/j.1600-0889.1989.tb00129.x.

Errico, R. M. (1997). What is an adjoint model? *Bulletin of the American Meteorological Society, 78* (11), 2577–2591. https://doi.org/10.1175/1520-0477(1997)078<2577:WIAAM>2.0.CO;2.

Evensen, G. (1994). Sequential data assimilation with a nonlinear quasi-geostrophic model using Monte Carlo methods to forecast error statistics. *Journal of Geophysical Research, 99*(C5), 10143–10162.

Evensen, G. (2003). The Ensemble Kalman Filter: Theoretical formulation and practical implementation. *Ocean Dynamics, 53*(4), 343–367. https://doi.org/10.1007/s10236-003-0036-9.

Fan, S., Gloor, M., Mahlman, J., Pacala, S., Sarmiento, J., Takahashi, T., et al. (1998). A large terrestrial carbon sink in North America implied by atmospheric and oceanic carbon dioxide data and models. *Science, 282*, 442–446. https://doi.org/10.1126/science.282.5388.442.

Fang, X., Thompson, R. L., Saito, T., Yokouchi, Y., Kim, J., Li, S., et al. (2014). Sulfur hexafluoride (SF_6) emissions in East Asia determined by inverse modeling. *Atmospheric Chemistry and Physics, 14*(9), 4779–4791. https://doi.org/10.5194/acp-14-4779-2014.

Feng, S., Lauvaux, T., Newman, S., Rao, P., Ahmadov, R., Deng, A., et al. (2016). LA megacity: A high-resolution land-atmosphere modelling system for urban CO_2 emissions. *Atmospheric Chemistry and Physics, 16*, 9019–9045. https://doi.org/10.5194/acp-2016-143.

Fernández-Martínez, M., Sardans, J., Chevallier, F., Ciais, P., Obersteiner, M., Vicca, S., et al. (2019). Global trends in carbon sinks and their relationships with CO_2 and temperature. *Nature Climate Change, 9*(1), 73–79.

Fischer, H., Behrens, M., Bock, M., Richter, U., Schmitt, J., Loulergue, L., et al. (2008). Changing boreal methane sources and constant biomass burning during the last termination. *Nature, 452*, 864–867.

Fischer, M. L., Parazoo, N., Brophy, K., Cui, X., Jeong, S., Liu, J., et al. (2017). Simulating estimation of California fossil fuel and biosphere carbon dioxide exchanges combining in-situ tower and satellite column observations. *Journal of Geophysical Research-Atmospheres*. https://doi.org/10.1002/2016JD025617, 2016JD025617.

Fisher, M., & Courtier, P. (1995). *Estimating the covariances matrices of analysis and forecast error in variational data assimilation (No. 220)*. European Centre for Medium-Range Weather Forecasts.

Frankenberg, C., Bergamaschi, P., Butz, A., Houweling, S., Meirink, J. F., Notholt, J., et al. (2008). Tropical methane emissions: A revised view from SCIAMACHY onboard ENVISAT. *Geophysical Research Letters*, *35*(15). https://doi.org/10.1029/2008GL034300.

Fraser, A., Palmer, P. I., Feng, L., Boesch, H., Cogan, A., Parker, R., et al. (2013). Estimating regional methane surface fluxes: The relative importance of surface and GOSAT mole fraction measurements. *Atmospheric Chemistry and Physics*, *13*(11), 5697–5713. https://doi.org/10.5194/acp-13-5697-2013.

Fung, I., John, J., Lerner, J., Matthews, E., Prather, M., Steele, L. P., et al. (1991). Three-dimensional model synthesis of the global methane cycle. *Journal of Geophysical Research*, *96*(D7), 13033–13065. https://doi.org/10.1029/91JD01247.

Gately, C. K., Hutyra, L. R., & Wing, I. S. (2015). Cities, traffic, and CO_2: A multidecadal assessment of trends, drivers, and scaling relationships. *Proceedings of the National Academy of Sciences of the United States of America*, *112*(16), 201421723–201425004. https://doi.org/10.1073/pnas.1421723112.

Gatti, L. V., Gloor, M., Miller, J. B., Doughty, C. E., Malhi, Y., Domingues, L. G., et al. (2014). Drought sensitivity of Amazonian carbon balance revealed by atmospheric measurements. *Nature*, *506*(7486), 76–80. https://doi.org/10.1038/nature12957.

Geels, C., Gloor, M., Ciais, P., Bousquet, P., Peylin, P., Vermeulen, A. T., et al. (2007). Comparing atmospheric transport models for future regional inversions over Europe—Part 1: Mapping the atmospheric CO_2 signals. *Atmospheric Chemistry and Physics*, *7*, 3461–3479.

Gelman, A., & Hill, J. (2006). *Data analysis using regression and multilevel/hierarchical models* (pp. 1–651). Cambridge: Cambridge University Press.

Gerbig, C., Körner, S., & Lin, J. C. (2008). Vertical mixing in atmospheric tracer transport models: Error characterization and propagation. *Atmospheric Chemistry and Physics*, *8*(3), 591–602. https://doi.org/10.5194/acp-8-591-2008.

Ghosh, A., Patra, P. K., Ishijima, K., Umezawa, T., Ito, A., Etheridge, D. M., et al. (2015). Variations in global methane sources and sinks during 1910–2010. *Atmospheric Chemistry and Physics*, *15*(5), 2595–2612. https://doi.org/10.5194/acp-15-2595-2015.

Giering, R., & Kaminski, T. (1998). Recipes for adjoint code construction. *ACM Transactions on Mathematical Software*, *24*(4), 437–474. https://doi.org/10.1145/293686.293695.

Gloor, E., Wilson, C., Chipperfield, M. P., Chevallier, F., Buermann, W., Boesch, H., et al. (2018). Tropical land carbon cycle responses to 2015/16 El Niño as recorded by atmospheric greenhouse gas and remote sensing data. *Philosophical Transactions of the Royal Society, B: Biological Sciences*, *373*(1760), 20170302. https://doi.org/10.1098/rstb.2017.0302.

Golub, G. H., & Van Loan, C. F. (1996). *Matrix computations*. Baltimore: The Johns Hopkins University Press.

Groot Zwaaftink, C. D., Henne, S., Thompson, R. L., Dlugokencky, E. J., Machida, T., Paris, J.-D., et al. (2018). Three-dimensional methane distribution simulated with FLEXPART 8-CTM-1.1 constrained with observation data. *Geoscientific Model Development*, *11*(11), 4469–4487. https://doi.org/10.5194/gmd-11-4469-2018.

Gurney, K. R., Razlivanov, I., Song, Y., Zhou, Y., Benes, B., & Abdul-Massih, M. (2012). Quantification of fossil fuel CO_2 emissions on the building/street scale for a large U.S. city. *Environmental Science & Technology*, *46*(21), 12194–12202. https://doi.org/10.1021/es3011282.

Hall, B. D., Sutton, G. S., & Elkins, J. W. (2007). The NOAA nitrous oxide standard scale for atmospheric observations. *Journal of Geophysical Research*, *112*, D09305. https://doi.org/10.1029/2006JD007954.

Harris, E., Nelson, D. D., Olszewski, W., Zahniser, M., Potter, K. E., McManus, B. J., et al. (2014). Development of a spectroscopic technique for continuous online monitoring of oxygen and site-

146 Section | B Methods

specific nitrogen isotopic composition of atmospheric nitrous oxide. *Analytical Chemistry, 86* (3), 1726–1734. https://doi.org/10.1021/ac403606u.

Hartley, D., & Prinn, R. (1993). Feasibility of determining surface emissions of trace gases using an inverse method in a three-dimensional chemical transport model. *Journal of Geophysical Research-Atmospheres, 98*(D3), 5183–5197. https://doi.org/10.1029/92JD02594.

Hastings, W. K. (1970). Monte Carlo sampling methods using Markov chains and their applications. *Biometrika, 57*(1), 97–109.

Hedelius, J. K., Liu, J., Oda, T., Maksyutov, S., Roehl, C. M., Iraci, L. T., et al. (2018). Southern California megacity CO_2, CH_4, and CO flux estimates using ground- and space-based remote sensing and a Lagrangian model. *Atmospheric Chemistry and Physics, 18*(22), 16271–16291. https://doi.org/10.5194/acp-18-16271-2018.

Hein, R., Crutzen, P. J., & Heimann, M. (1997). An inverse modeling approach to investigate the global atmospheric methane cycle. *Global Biogeochemical Cycles, 11*(1), 43–76. https://doi.org/10.1029/96GB03043.

Hirsch, A. I., Michalak, A. M., Bruhwiler, L. M., Peters, W., Dlugokencky, E. J., & Tans, P. P. (2006). Inverse modeling estimates of the global nitrous oxide surface flux from 1998–2001. *Global Biogeochemical Cycles, 20*, GB1008. https://doi.org/10.1029/2004gb002443.

Hourdin, F., & Talagrand, O. (2006). Eulerian backtracking of atmospheric tracers. I: Adjoint derivation and parametrization of subgrid-scale transport. *Quarterly Journal of the Royal Meteorological Society, 132*(615), 567–583. https://doi.org/10.1256/qj.03.198.A.

Houtekamer, P. L., & Zhang, F. (2016). Review of the ensemble kalman filter for atmospheric data assimilation. *Monthly Weather Review, 144*(12), 4489–4532. https://doi.org/10.1175/MWR-D-15-0440.1.

Houweling, S., Baker, D., Basu, S., Boesch, H., Butz, A., Chevallier, F., et al. (2015). An intercomparison of inverse models for estimating sources and sinks of CO_2 using GOSAT measurements. *Journal of Geophysical Research-Atmospheres*. https://doi.org/10.1002/2014JD022962, 2014JD022962.

Houweling, S., Bergamaschi, P., Chevallier, F., Heimann, M., Kaminski, T., Krol, M., et al. (2017). Global inverse modeling of CH_4 sources and sinks: An overview of methods. *Atmospheric Chemistry and Physics, 17*(1), 235–256. https://doi.org/10.5194/acp-17-235-2017.

Houweling, S., Kaminski, T., Dentener, F., Lelieveld, J., & Heimann, M. (1999). Inverse modeling of methane sources and sinks using the adjoint of a global transport model. *Journal of Geophysical Research-Atmospheres, 104*(D21), 26137–26160. https://doi.org/10.1029/1999jd900428.

Huang, J., Golombek, A., Prinn, R., Weiss, R., Fraser, P., Simmonds, P., et al. (2008). Estimation of regional emissions of nitrous oxide from 1997 to 2005 using multinetwork measurements, a chemical transport model, and an inverse method. *Journal of Geophysical Research, 113.* https://doi.org/10.1029/2007JD009381, D17313.

Janssens-Maenhout, G., Crippa, M., Guizzardi, D., Muntean, M., Schaaf, E., Dentener, F., et al. (2017). EDGAR v4.3.2 Global Atlas of the three major Greenhouse Gas Emissions for the period 1970–2012. *Earth System Science Data Discussions*, 1–55. https://doi.org/10.5194/essd-2017-79.

Jazwinski, A. A. (1970). Stochastic and filtering theory. In *Mathematics in Sciences and Engineering Series, Series* (p. 64). Retrieved from https://www.elsevier.com/books/stochastic-processes-and-filtering-theory/jazwinski/978-0-12-381550-7.

Jones, D. B. A., Bowman, K. W., Palmer, P. I., Worden, J. R., Jacob, D. J., Hoffman, R. N., et al. (2003). Potential of observations from the Tropospheric Emission Spectrometer to constrain continental sources of carbon monoxide. *Journal of Geophysical Research, 108*(D24), 4789. https://doi.org/10.1029/2003JD003702.

Kalman, R. E. (1960). A new approach to linear filtering and prediction problems. *Journal of Basic Engineering, 82*(1), 35–45. https://doi.org/10.1115/1.3662552.

Kaminski, T., & Mathieu, P.-P. (2017). Reviews and syntheses: Flying the satellite into your model: On the role of observation operators in constraining models of the Earth system and the carbon cycle. *Biogeosciences, 14*(9), 2343–2357. https://doi.org/10.5194/bg-14-2343-2017.

Kaminski, T., Rayner, P. J., Heimann, M., & Enting, I. G. (2001). On aggregation errors in atmospheric transport inversions. *Journal of Geophysical Research, 106*(D5), 4703–4715. https://doi.org/10.1029/2000JD900581.

Kang, J.-S., Kalnay, E., Miyoshi, T., Liu, J., & Fung, I. (2012). Estimation of surface carbon fluxes with an advanced data assimilation methodology. *Journal of Geophysical Research, 117*(D24). https://doi.org/10.1029/2012JD018259, D24101.

Karstens, U., Schwingshackl, C., Schmithüsen, D., & Levin, I. (2015). A process-based ^{222}radon flux map for Europe and its comparison to long-term observations. *Atmospheric Chemistry and Physics, 15*(22), 12845–12865. https://doi.org/10.5194/acp-15-12845-2015.

Keeling, C. D. (1960). The concentration and isotopical abundance of carbon dioxide in the atmosphere. *Tellus, 12*, 200–203.

Khalil, M. A. K., & Rasmussen, R. A. (1993). Decreasing trend of methane: Unpredictability of future concentrations. *Chemosphere, 26*(1-4), 803–814. https://doi.org/10.1016/0045-6535(93)90462-E.

Kim, J., Kim, H. M., Cho, C. H., Boo, K. O., Jacobson, A. R., Sasakawa, M., et al. (2017). Impact of Siberian observations on the optimization of surface CO_2 flux. *Atmospheric Chemistry and Physics, 17*(4), 2881–2899. https://doi.org/10.5194/acp-17-2881-2017.

Kirschke, S., Bousquet, P., Ciais, P., Saunois, M., Canadell, J. G., Dlugokencky, E. J., et al. (2013). Three decades of global methane sources and sinks. *Nature Geoscience, 6*(10), 813–823. https://doi.org/10.1038/ngeo1955.

Kountouris, P., Gerbig, C., Rödenbeck, C., Karstens, U., Koch, T. F., & Heimann, M. (2018). Atmospheric CO_2 inversions on the mesoscale using data-driven prior uncertainties: Quantification of the European terrestrial CO_2 fluxes. *Atmospheric Chemistry and Physics, 18*(4), 3047–3064. https://doi.org/10.5194/acp-18-3047-2018.

Krol, M., & Lelieveld, J. (2003). Can the variability in tropospheric OH be deduced from measurements of 1,1,1-trichloroethane (methyl chloroform)? *Journal of Geophysical Research-Atmospheres, 108*(D3). https://doi.org/10.1029/2002JD002423.

Kulawik, S. S., Jones, D. B. A., Nassar, R., Irion, F. W., Worden, J. R., Bowman, K. W., et al. (2010). Characterization of tropospheric emission spectrometer (TES) CO_2 for carbon cycle science. *Atmospheric Chemistry and Physics, 10*(12), 5601–5623. https://doi.org/10.5194/acp-10-5601-2010.

Kulawik, S. S., Wunch, D., O'Dell, C., Frankenberg, C., Reuter, M., Oda, T., et al. (2016). Consistent evaluation of ACOS-GOSAT, BESD-SCIAMACHY, CarbonTracker, and MACC through comparisons to TCCON. *Atmospheric Measurement Techniques, 9*(2), 683–709. https://doi.org/10.5194/amt-9-683-2016.

Kuze, A., Suto, H., Nakajima, M., & Hamazaki, T. (2009). Thermal and near infrared sensor for carbon observation Fourier-transform spectrometer on the greenhouse gases observing satellite for greenhouse gases monitoring. *Applied Optics, 48*(35), 6716–6733. https://doi.org/10.1364/AO.48.006716.

Lauvaux, T., Gioli, B., Sarrat, C., Rayner, P. J., Ciais, P., Chevallier, F., et al. (2009). Bridging the gap between atmospheric concentrations and local ecosystem measurements. *Geophysical Research Letters, 36*(19). https://doi.org/10.1029/2009gl039574.

Lauvaux, T., Gurney, K. R., Miles, N. L., Davis, K. J., Richardson, S. J., Deng, A., et al. (2020). Policy-relevant assessment of urban CO_2 emissions. *Environmental Science & Technology, 54*(16), 10237–10245. https://doi.org/10.1021/acs.est.0c00343.

148 Section | B Methods

Lauvaux, T., Miles, N. L., Deng, A., Richardson, S. J., Cambaliza, M. O., Davis, K. J., et al. (2016). High-resolution atmospheric inversion of urban CO_2 emissions during the dormant season of the Indianapolis Flux Experiment (INFLUX). *Journal of Geophysical Research-Atmospheres*, *121*(10), 5213–5236. https://doi.org/10.1002/2015JD024473.

Lauvaux, T., Schuh, A. E., Uliasz, M., Richardson, S., Miles, N., Andrews, A. E., et al. (2012). Constraining the CO_2 budget of the corn belt: Exploring uncertainties from the assumptions in a mesoscale inverse system. *Atmospheric Chemistry and Physics*, *12*(1), 337–354. https://doi.org/10.5194/acp-12-337-2012.

Law, R. M., Rayner, P. J., Denning, A. S., Erickson, D., Fung, I. Y., Heimann, M., et al. (1996). Variations in modeled atmospheric transport of carbon dioxide and the consequences for CO_2 inversions. *Global Biogeochemical Cycles*, *10*(4), 783–796. https://doi.org/10.1029/96gb01892.

Le Quere, C., Andrew, R. M., Friedlingstein, P., Sitch, S., Hauck, J., Pongratz, J., et al. (2018). Global carbon budget 2018. *Earth System Science Data Discussions*, 1–3. https://doi.org/10.5194/essd-2018-120.

Levin, I., Hammer, S., Eichelmann, E., & Vogel, F. R. (2011). Verification of greenhouse gas emission reductions: the prospect of atmospheric monitoring in polluted areas. *Philosophical Transactions. Series A, Mathematical, Physical, and Engineering Sciences*, *369*(1943), 1906–1924. https://doi.org/10.1098/rsta.2010.0249.

Lian, J., Bréon, F.-M., Broquet, G., Lauvaux, T., Zheng, B., Ramonet, M., et al. (2021). Sensitivity to the sources of uncertainties in the modeling of atmospheric CO_2 concentration within and in the vicinity of Paris. *Atmospheric Chemistry and Physics*, *21*(13), 10707–10726. https://doi.org/10.5194/acp-21-10707-2021.

Liang, Q., Chipperfield, M. P., Fleming, E. L., Abraham, N. L., Braesicke, P., Burkholder, J. B., et al. (2017). Deriving global OH abundance and atmospheric lifetimes for long-lived gases: A search for CH_3CCl_3 alternatives. *Journal of Geophysical Research-Atmospheres*, *122*(21), 11914–11933. https://doi.org/10.1002/2017JD026926.

Liu, J., & Bowman, K. (2016). A method for independent validation of surface fluxes from atmospheric inversion: Application to CO_2. *Geophysical Research Letters*, *43*(7), 3502–3508. https://doi.org/10.1002/2016GL067828.

Liu, J., Bowman, K. W., & Lee, M. (2016). Comparison between the Local Ensemble Transform Kalman Filter (LETKF) and 4D-Var in atmospheric CO_2 flux inversion with the Goddard Earth Observing System-Chem model and the observation impact diagnostics from the LETKF. *Journal of Geophysical Research-Atmospheres*, *121*(21), 13,066–13,087. https://doi.org/10.1002/2016jd025100.

Liu, J., Bowman, K., Parazoo, N. C., Bloom, A. A., Wunch, D., Jiang, Z., et al. (2018). Detecting drought impact on terrestrial biosphere carbon fluxes over contiguous US with satellite observations. *Environmental Research Letters*, *13*(9). https://doi.org/10.1088/1748-9326/aad5ef, 095003.

Liu, J., Bowman, K. W., Schimel, D. S., Parazoo, N. C., Jiang, Z., Lee, M., et al. (2017). Contrasting carbon cycle responses of the tropical continents to the 2015–2016 El Niño. *Science*, *358*(6360). https://doi.org/10.1126/science.aam5690, eaam5690.

Marland, G., Hamal, K., & Jonas, M. (2009). How uncertain are estimates of CO_2 emissions? *Journal of Industrial Ecology*, *13*(1), 4–7. https://doi.org/10.1111/j.1530-9290.2009.00108.x.

Meesters, A. G. C. A., Tolk, L. F., Peters, W., Hutjes, R. W. A., Vellinga, O. S., Elbers, J. A., et al. (2012). Inverse carbon dioxide flux estimates for the Netherlands. *Journal of Geophysical Research*, *117*(D20), 807. https://doi.org/10.1029/2012JD017797.

Metropolis, N., Rosenbluth, A. W., Rosenbluth, M. N., Teller, A. H., & Teller, E. (1953). Equation of state calculations by fast computing machines. *Journal of Chemical Physics*, *21*(6), 1087–1092. https://doi.org/10.1063/1.1699114.

Michalak, A. M. (2008). Technical Note: Adapting a fixed-lag Kalman smoother to a geostatistical atmospheric inversion framework. *Atmospheric Chemistry and Physics*, *8*, 6789–6799.

Michalak, A. M., Hirsch, A., Bruhwiler, L., Gurney, K. R., Peters, W., & Tans, P. P. (2005). Maximum likelihood estimation of covariance parameters for Bayesian atmospheric trace gas surface flux inversions. *Journal of Geophysical Research-Atmospheres*, *110* (D24). https://doi.org/10.1029/2005jd005970.

Migliorini, S., Piccolo, C., & Rodgers, C. D. (2008). Use of the information content in satellite measurements for an efficient interface to data assimilation. *Monthly Weather Review*, *136*(7), 2633–2650. https://doi.org/10.1175/2007MWR2236.1.

Mikaloff Fletcher, S. E., Tans, P. P., Bruhwiler, L. M., Miller, J. B., & Heimann, M. (2004). CH_4 sources estimated from atmospheric observations of CH_4 and its $^{13}C/^{12}C$ isotopic ratios: 1. Inverse modeling of source processes. *Global Biogeochemical Cycles*, *18*(4). https://doi.org/ 10.1029/2004gb002223.

Miller, S. M., Miller, C. E., Commane, R., Chang, R. Y.-W., Dinardo, S. J., Henderson, J. M., et al. (2016). A multiyear estimate of methane fluxes in Alaska from CARVE atmospheric observations. *Global Biogeochemical Cycles*, *30*(10). https://doi.org/10.1002/2016GB005419, 2016GB005419.

Miller, C. B., & Wheeler, P. A. (2012). *Biological oceanography* (2nd ed.). Wiley-Blackwell.

Minschwaner, K., Salawitch, R. J., & McElroy, M. B. (1993). Absorption of solar radiation by O_2: Implications for O_3 and lifetimes of N_2O, $CFCl_3$, and CF_2Cl_2. *Journal of Geophysical Research*, *98*(D6). https://doi.org/10.1029/93JD00223, 10-543-10-561.

Monteil, G., Houweling, S., Dlugockenky, E. J., Maenhout, G., Vaughn, B. H., White, J. W. C., et al. (2011). Interpreting methane variations in the past two decades using measurements of CH_4 mixing ratio and isotopic composition. *Atmospheric Chemistry and Physics*, *11*(17), 9141–9153. https://doi.org/10.5194/acp-11-9141-2011.

Montzka, S. A., Krol, M., Dlugokencky, E., Hall, B., Jöckel, P., & Lelieveld, J. (2011). Small interannual variability of global atmospheric hydroxyl. *Science*, *331*(6013), 67–69. https://doi.org/ 10.1126/science.1197640.

Müller, J. F., & Brasseur, G. (1995). IMAGES: A three-dimensional chemical transport model of the global troposphere. *Journal of Geophysical Research-Atmospheres*, *100*(D8), 16445–16490. https://doi.org/10.1029/94JD03254.

Myhre, G., Shindell, D., Bréon, F. M., Collins, W., Fuglestvedt, J., Huang, J., et al. (2013). Chapter 8: Anthropogenic and natural radiative forcing. In D. Jacob, A. R. Ravishankara, & K. Shine (Eds.), *IPCC WG1 fifth assessment report*.

Nassar, R., Jones, D. B. A., Kulawik, S. S., Worden, J. R., Bowman, K. W., Andres, R. J., et al. (2011). Inverse modeling of CO_2 sources and sinks using satellite observations of CO_2 from TES and surface flask measurements. *Atmospheric Chemistry and Physics*, *11*(12), 6029–6047. https://doi.org/10.5194/acp-11-6029-2011.

Nevison, C., Andrews, A. E., Thoning, K., Dlugokencky, E., Sweeney, C., Miller, S., et al. (2018). Nitrous oxide emissions estimated with the CarbonTracker-Lagrange North American regional inversion framework. *Global Biogeochemical Cycles*. https://doi.org/10.1002/ 2017GB005759.

Nevison, C. D., Mahowald, N. M., Weiss, R. F., & Prinn, R. G. (2007). Interannual and seasonal variability in atmospheric N_2O. *Global Biogeochemical Cycles*, *21*(GB3017). https://doi.org/ 10.1029/2006gb002755.

Newman, S., Xu, X., Gurney, K. R., Hsu, Y.-K., Li, K. F., Jiang, X., et al. (2016). Toward consistency between trends in bottom-up CO_2 emissions and top-down atmospheric measurements in the Los Angeles megacity. *Atmospheric Chemistry and Physics*, *16*(6), 3843–3863. https://doi. org/10.5194/acp-16-3843-2016.

150 Section | B Methods

Nisbet, E. G., Dlugokencky, E. J., Manning, M. R., Lowry, D., Fisher, R. E., France, J. L., et al. (2016). Rising atmospheric methane: 2007–2014 growth and isotopic shift. *Global Biogeochemical Cycles*. https://doi.org/10.1002/2016GB005406, 2016GB005406.

Oda, T., Maksyutov, S., & Andres, R. J. (2018). The Open-source Data Inventory for Anthropogenic CO_2, version 2016 (ODIAC2016): A global monthly fossil fuel CO_2 gridded emissions data product for tracer transport simulations and surface flux inversions. *Earth System Science Data, 10*(1), 87–107. https://doi.org/10.5194/essd-10-87-2018.

O'Dell, C. W., Connor, B., Bösch, H., O'Brien, D., Frankenberg, C., Castano, R., et al. (2012). The ACOS CO_2 retrieval algorithm—Part 1: Description and validation against synthetic observations. *Atmospheric Measurement Techniques, 5*(1), 99–121. https://doi.org/10.5194/amt-5-99-2012.

O'Dell, C. W., Eldering, A., Wennberg, P. O., Crisp, D., Gunson, M. R., Fisher, B., et al. (2018). Improved retrievals of carbon dioxide from Orbiting Carbon Observatory-2 with the version 8 ACOS algorithm. *Atmospheric Measurement Techniques, 11*(12), 6539–6576. https://doi.org/10.5194/amt-11-6539-2018.

O'Keefe, A. (1998). Integrated cavity output analysis of ultra-weak absorption. *Chemical Physics Letters, 293*(5–6), 331–336. https://doi.org/10.1016/S0009-2614(98)00785-4.

Pacala, S., Breidenich, C., Brewer, P. G., Fung, I., Gunson, M. R., Heddle, G., et al. (2010). *Verifying greenhouse gas emissions: methods to support international climate agreements* (pp. 1–125). https://doi.org/10.17226/12883.

Paige, C. C., & Saunders, M. A. (1975). Solution of sparse indefinite systems of linear equations. *SIAM Journal on Numerical Analysis, 12*(4), 617–629. https://doi.org/10.1137/0712047.

Park, S., Croteau, P., Boering, K. A., Etheridge, D. M., Ferretti, D., Fraser, P. J., et al. (2012). Trends and seasonal cycles in the isotopic composition of nitrous oxide since 1940. *Nature Geoscience, 5*(4), 261–265. https://doi.org/10.1038/ngeo1421.

Patarasuk, R., Gurney, K. R., O'Keeffe, D., Song, Y., Huang, J., Rao, P., et al. (2016). Urban high-resolution fossil fuel CO_2 emissions quantification and exploration of emission drivers for potential policy applications. *Urban Ecosystem, 19*(3), 1013–1039. https://doi.org/10.1007/s11252-016-0553-1.

Patra, P. K., Houweling, S., Krol, M., Bousquet, P., Belikov, D., Bergmann, D., et al. (2011). TransCom model simulations of CH_4 and related species: linking transport, surface flux and chemical loss with CH_4 variability in the troposphere and lower stratosphere. *Atmospheric Chemistry and Physics, 11*(24), 12813–12837. https://doi.org/10.5194/acp-11-12813-2011.

Patra, P. K., Ishizawa, M., Maksyutov, S., Nakazawa, T., & Inoue, G. (2005). Role of biomass burning and climate anomalies for land-atmosphere carbon fluxes based on inverse modeling of atmospheric CO_2. *Global Biogeochemical Cycles, 19*(3). https://doi.org/10.1029/2004gb002258.

Patra, P. K., Krol, M. C., Montzka, S. A., Arnold, T., Atlas, E. L., Lintner, B. R., et al. (2014). Observational evidence for interhemispheric hydroxyl-radical parity. *Nature, 513*(7517), 219–223. https://doi.org/10.1038/nature13721.

Patra, P. K., Saeki, T., Dlugokencky, E. J., Ishijima, K., Umezawa, T., Ito, A., et al. (2016). Regional methane emission estimation based on observed atmospheric concentrations (2002–2012). *Journal of the Meteorological Society of Japan, 94*(1), 91–113. https://doi.org/10.2151/jmsj.2016-006.

Patra, P., Takigawa, M., Ishijima, K., Choi, B.-C., Cunnold, D. J., Dlugokencky, E., et al. (2009). Growth rate, seasonal, synoptic, diurnal variations and budget of methane in the lower atmosphere. *Journal of the Meteorological Society of Japan, 87*(4), 635–663. https://doi.org/10.2151/jmsj.87.635.

Peng, S. S., Piao, S. L., Bousquet, P., Ciais, P., Li, B. G., Lin, X., et al. (2016). Inventory of anthropogenic methane emissions in Mainland China from 1980 to 2010. *Atmospheric Chemistry and Physics Discussions, 2016*, 1–29. https://doi.org/10.5194/acp-2016-139.

Peng, S., Piao, S., Ciais, P., Myneni, R. B., Chen, A., Chevallier, F., et al. (2013). Asymmetric effects of daytime and night-time warming on Northern Hemisphere vegetation. *Nature*, *501*(7465), 88–92. https://doi.org/10.1038/nature12434.

Pérez, T., Trumbore, S. E., Tyler, S. C., Matson, P. A., Ortiz-Monasterio, I., Rahn, T., et al. (2001). Identifying the agricultural imprint on the global N_2O budget using stable isotopes. *Journal of Geophysical Research-Atmospheres*, *106*(D9), 9869–9878. https://doi.org/10.1029/2000JD900809.

Peters, W., Krol, M. C., Van der Werf, G. R., Houweling, S., Jones, C. D., Hughes, J., et al. (2010). Seven years of recent European net terrestrial carbon dioxide exchange constrained by atmospheric observations. *Global Change Biology*, *16*(4), 1317–1337. https://doi.org/10.1111/j.1365-2486.2009.02078.x.

Peters, W., Miller, J. B., Whitaker, J., Denning, A. S., Hirsch, A., Krol, M. C., et al. (2005). An ensemble data assimilation system to estimate CO_2 surface fluxes from atmospheric trace gas observations. *Journal of Geophysical Research*, *110*(D24), 419–421. https://doi.org/10.1029/2005JD006157.

Peylin, P., Law, R. M., Gurney, K. R., Chevallier, F., Jacobson, A. R., Maki, T., et al. (2013). Global atmospheric carbon budget: Results from an ensemble of atmospheric CO_2 inversions. *Biogeosciences*, *10*(10), 6699–6720. https://doi.org/10.5194/bg-10-6699-2013.

Pickett-Heaps, C. A., Rayner, P. J., Law, R. M., Ciais, P., Patra, P. K., Bousquet, P., et al. (2011). Atmospheric CO_2 inversion validation using vertical profile measurements: Analysis of four independent inversion models. *Journal of Geophysical Research*, *116*(D12). https://doi.org/10.1029/2010JD014887, GB1002.

Pison, I., Bousquet, P., Chevallier, F., Szopa, S., & Hauglustaine, D. (2009). Multi-species inverison of CH_4, CO and H_2 emissions from surface measurements. *Atmospheric Chemistry and Physics*, *9*, 5281–5297.

Polonsky, I. N., O'Brien, D. M., Kumer, J. B., O'Dell, C. W., & geoCARB Team. (2014). Performance of a geostationary mission, geoCARB, to measure CO_2, CH_4 and CO column-averaged concentrations. *Atmospheric Measurement Techniques*, *7*, 959–981.

Prather, M. J., Hsu, J., DeLuca, N. M., Jackman, C. H., Oman, L. D., Douglass, A. R., et al. (2015). Measuring and modeling the lifetime of nitrous oxide including its variability. *Journal of Geophysical Research-Atmospheres*, *120*, 5693–5705. https://doi.org/10.1002/2015JD023267.

Prather, M. J., Zhu, X., Strahan, S. E., Steenrod, S. D., & Rodriguez, J. M. (2008). Quantifying errors in trace species transport modeling. *Proceedings of the National Academy of Sciences of the United States of America*, *105*(50). https://doi.org/10.1073/pnas.0806541106. pnas.0806541106–19621.

Prinn, R. G., Cunnold, D., Rasmussen, R., Simmonds, P., Alyea, F., Crawford, A., et al. (1990). Atmospheric emissions and trends of nitrous oxide deduced from 10 years of ALE-GAGE data. *Journal of Geophysical Research*, *95*(D11), 18-369–18-385.

Prinn, R. G., Huang, J., Weiss, R. F., Cunnold, D. M., Fraser, P. J., Simmonds, P. G., et al. (2005). Evidence for variability of atmospheric hydroxyl radicals over the past quarter century. *Geophysical Research Letters*, *32*(7). https://doi.org/10.1029/2004GL022228.

Prinn, R. G., Weiss, R. F., Fraser, P. J., Simmonds, P. G., Cunnold, D. M., Alyea, F. N., et al. (2000). A history of chemically and radiatively important gases in air deduced from ALE/GAGE/AGAGE. *Journal of Geophysical Research*, *105*(D14), 17751–17792. https://doi.org/10.1029/2000jd900141.

Prinn, R. G., Weiss, R. F., Miller, B. R., Huang, J., Alyea, F. N., Cunnold, D. M., et al. (1995). Atmospheric trends and lifetime of CH_3CCl_3 and global OH concentrations. *Science*, *269*(5221), 187–192. https://doi.org/10.1126/science.269.5221.187.

152 Section | B Methods

Quay, P., Stutsman, J., Wilbur, D., Snover, A., Dlugokencky, E., & Brown, T. (1999). The isotopic composition of atmospheric methane. *Global Biogeochemical Cycles, 13*(2), 445–461. https://doi.org/10.1029/1998GB900006.

Rayner, P. J., Enting, I. G., Francey, R. J., & Langenfelds, R. (1999). Reconstructing the recent carbon cycle from atmospheric CO_2, $\delta^{13}C$ and O_2/N_2 observations. *Tellus B, 51*(2), 213–232. https://doi.org/10.3402/tellusb.v51i2.16273.

Rella, C. W., Hoffnagle, J., He, Y., & Tajima, S. (2015). Local- and regional-scale measurements of CH_4, d13CH_4, and C_2H_6 in the Uintah Basin using a mobile stable isotope analyzer. *Atmospheric Measurement Techniques, 8*(10), 4539–4559. https://doi.org/10.5194/amt-8-4539-2015.

Reuter, M., Buchwitz, M., Hilker, M., Heymann, J., Schneising, O., Pillai, D., et al. (2014). Satellite-inferred European carbon sink larger than expected. *Atmospheric Chemistry and Physics, 14* (24), 13739–13753. https://doi.org/10.5194/acp-14-13739-2014.

Rice, A. L., Butenhoff, C. L., Teama, D. G., Röger, F. H., Khalil, M. A. K., & Rasmussen, R. A. (2016). Atmospheric methane isotopic record favors fossil sources flat in 1980s and 1990s with recent increase. *Proceedings of the National Academy of Sciences of the United States of America.* https://doi.org/10.1073/pnas.1522923113.

Rigby, M., Manning, A. J., & Prinn, R. G. (2012). The value of high-frequency, high-precision methane isotopologue measurements for source and sink estimation. *Journal of Geophysical Research: Atmospheres (1984–2012), 117*(D12). https://doi.org/10.1029/2011JD017384.

Rigby, M., Montzka, S. A., Prinn, R. G., White, J. W. C., Young, D., O'Doherty, S., et al. (2017). Role of atmospheric oxidation in recent methane growth. *Proceedings of the National Academy of Sciences of the United States of America.* https://doi.org/10.1073/pnas.1616426114.

Rigby, M., Prinn, R. G., Fraser, P. J., Simmonds, P. G., Langenfelds, R. L., Huang, J., et al. (2008). Renewed growth of atmospheric methane. *Geophysical Research Letters, 35*(22). https://doi.org/10.1029/2008gl036037.

Rödenbeck, C., Gerbig, C., Trusilova, K., & Heimann, M. (2009). A two-step scheme for high-resolution regional atmospheric trace gas inversions based on independent models. *Atmospheric Chemistry and Physics, 9*(14), 5331–5342. https://doi.org/10.5194/acp-9-5331-2009.

Rödenbeck, C., Houweling, S., Gloor, M., & Heimann, M. (2003). CO_2 flux history 1982-2001 inferred from atmospheric data using a global inversion of atmospheric transport. *Atmospheric Chemistry and Physics, 3*, 1919–1964.

Rödenbeck, C., Zaehle, S., Keeling, R., & Heimann, M. (2018). How does the terrestrial carbon exchange respond to inter-annual climatic variations? A quantification based on atmospheric CO_2 data. *Biogeosciences, 15*(8), 2481–2498. https://doi.org/10.5194/bg-15-2481-2018.

Rodgers, C. D. (2000). *Inverse methods for atmospheric sounding: Theory and practice.* Singapore: World Scientific. https://doi.org/10.1142/3171.

Saeki, T., & Patra, P. K. (2017). Implications of overestimated anthropogenic CO_2 emissions on East Asian and global land CO_2 flux inversion. *Geoscience Letters, 4*(1), 9. https://doi.org/10.1186/s40562-017-0074-7.

Saunois, M., Bousquet, P., Poulter, B., Peregon, A., Ciais, P., Canadell, J. G., et al. (2016). The global methane budget: 2000-2012. *Earth System Science Data Discussions, 2016*, 1–79. https://doi.org/10.5194/essd-2016-25.

Saunois, M., Bousquet, P., Poulter, B., Peregon, A., Ciais, P., Canadell, J. G., et al. (2017). Variability and quasi-decadal changes in the methane budget over the period 2000-2012. *Atmospheric Chemistry and Physics Discussions, 2017*, 1–39. https://doi.org/10.5194/acp-2017-296.

Schwietzke, S., Sherwood, O. A., Bruhwiler, L. M. P., Miller, J. B., Etiope, G., Dlugokencky, E. J., et al. (2016). Upward revision of global fossil fuel methane emissions based on isotope database. *Nature, 538*(7623), 88–91. https://doi.org/10.1038/nature19797.

Seibert, P., & Frank, A. (2004). Source-receptor matrix calculation with a Lagrangian particle dispersion model in backward mode. *Atmospheric Chemistry and Physics, 4*(1), 51–63. https://doi.org/10.5194/acp-4-51-2004.

Spurr, R. J. D. (2006). VLIDORT: A linearized pseudo-spherical vector discrete ordinate radiative transfer code for forward model and retrieval studies in multilayer multiple scattering media. *Journal of Quantitative Spectroscopy and Radiative Transfer, 102*(2), 316–342. https://doi.org/10.1016/j.jqsrt.2006.05.005.

Spurr, R., Natraj, V., Lerot, C., Van Roozendael, M., & Loyola, D. (2013). Linearization of the Principal Component Analysis method for radiative transfer acceleration: Application to retrieval algorithms and sensitivity studies. *Journal of Quantitative Spectroscopy and Radiative Transfer, 125*, 1–17. https://doi.org/10.1016/j.jqsrt.2013.04.002.

Staufer, J., Broquet, G., Breon, F. M., Puygrenier, V., Chevallier, F., Xueref-Remy, I., et al. (2016). The first 1-year-long estimate of the Paris region fossil fuel CO_2 emissions based on atmospheric inversion. *Atmospheric Chemistry and Physics, 16*(22), 14703–14726. https://doi.org/10.5194/acp-16-14703-2016.

Stephens, B. B., Gurney, K. R., Tans, P. P., Sweeney, C., Peters, W., Bruhwiler, L., et al. (2007). Weak northern and strong tropical land carbon uptake from vertical profiles of atmospheric CO_2. *Science, 316*(5832), 1732–1735. https://doi.org/10.1126/science.1137004.

Stohl, A., Kim, J., Li, S., O'Doherty, S., Mühle, J., Salameh, P. K., et al. (2010). Hydrochlorofluorocarbon and hydrofluorocarbon emissions in East Asia determined by inverse modeling. *Atmospheric Chemistry and Physics, 10*(8), 3545–3560. https://doi.org/10.5194/acp-10-3545-2010.

Stohl, A., Seibert, P., Arduini, J., Eckhardt, S., Fraser, P., Greally, B. R., et al. (2009). An analytical inversion method for determining regional and global emissions of greenhouse gases: Sensitivity studies and application to halocarbons. *Atmospheric Chemistry and Physics, 9*(5), 1597–1620. https://doi.org/10.5194/acp-9-1597-2009.

Talagrand, O. (2014). Errors. A posteriori diagnostics. In *Advanced data assimilation for geosciences* (pp. 229–254). Oxford University Press. https://doi.org/10.1093/acprof:oso/9780198723844.003.0009.

Tamminen, J. (2004). Validation of nonlinear inverse algorithms with Markov chain Monte Carlo method. *Journal of Geophysical Research, 109*(D19), 237. https://doi.org/10.1029/2004JD004927.

Tans, P. P. (1997). The CO_2 lifetime concept should be banished; an editorial comment. *Climatic Change, 37*, 487–490.

Tans, P. P., Conway, T. J., & Nakazawa, T. (1989). Latitudinal distribution of the sources and sinks of atmospheric carbon dioxide derived from surface observations and an atmospheric transport model. *Journal of Geophysical Research: Atmospheres (1984–2012), 94*(D4), 5151–5172. https://doi.org/10.1029/JD094iD04p05151.

Tans, P. P., Fung, I. Y., & Takahashi, T. (1990). Observational constraints on the global atmospheric CO_2 budget. *Science, 247*, 1431–1438. https://doi.org/10.1126/science.247.4949.1431.

Tarantola, A. (2005). *Inverse problem theory and methods for model parameter estimation*. Philadelphia: Society for Industrial and Applied Mathematics.

Thompson, R. L., Chevallier, F., Crotwell, A. M., Dutton, G., Langenfelds, R. L., Prinn, R. G., et al. (2014). Nitrous oxide emissions 1999 to 2009 from a global atmospheric inversion. *Atmospheric Chemistry and Physics, 14*, 1–17. https://doi.org/10.5194/acp-14-1801-2014.

Thompson, R. L., Nisbet, E. G., Pisso, I., Stohl, A., Blake, D., Dlugokencky, E. J., et al. (2018). Variability in atmospheric methane from fossil fuel and microbial sources over the last three decades. *Geophysical Research Letters, 112*(7359), D04306–D04310. https://doi.org/10.1029/2018GL078127.

154 Section | B Methods

Thompson, R. L., Patra, P. K., Chevallier, F., Maksyutov, S., Law, R. M., Ziehn, T., et al. (2016). Top-down assessment of the Asian carbon budget since the mid 1990s. *Nature Communications*, 7. https://doi.org/10.1038/ncomms10724.

Thompson, R. L., Patra, P. K., Ishijima, K., Saikawa, E., Corazza, M., Karstens, U., et al. (2014). TransCom N_2O model inter-comparison-Part 1: Assessing the influence of transport and surface fluxes on tropospheric N_2O variability. *Atmospheric Chemistry and Physics*, *14*(8), 4349–4368. https://doi.org/10.5194/acp-14-4349-2014.

Thompson, R. L., & Stohl, A. (2014). FLEXINVERT: An atmospheric Bayesian inversion framework for determining surface fluxes of trace species using an optimized grid. *Geoscientific Model Development*, 7, 2223–2242. https://doi.org/10.5194/gmd-7-2223-2014.

Thompson, R. L., Stohl, A., Zhou, L. X., Dlugokencky, E., Fukuyama, Y., Tohjima, Y., et al. (2015). Methane emissions in East Asia for 2000-2011 estimated using an atmospheric Bayesian inversion. *Journal of Geophysical Research-Atmospheres*. https://doi.org/10.1002/2014JD022394.

Tippett, M. K., Anderson, J. L., Bishop, C. H., Hamill, T. M., & Whitaker, J. S. (2003). Ensemble square root filters. *American Meteorological Society*, *131*(7), 1485–1490.

Trampert, J., & Snieder, R. (1996). Model estimations biased by truncated expansions: Possible artefacts in seismic tomography. *Science*, *271*, 1257–1260.

Turnbull, J. C., Sweeney, C., Karion, A., Newberger, T., Lehman, S. J., Tans, P. P., et al. (2015). Toward quantification and source sector identification of fossil fuel CO_2 emissions from an urban area: Results from the INFLUX experiment. *Journal of Geophysical Research-Atmospheres*, *120*(1), 292–312. https://doi.org/10.1002/2014JD022555.

Turner, A. J., Frankenberg, C., Wennberg, P. O., & Jacob, D. J. (2017). Ambiguity in the causes for decadal trends in atmospheric methane and hydroxyl. *Proceedings of the National Academy of Sciences*. https://doi.org/10.1073/pnas.1616020114.

Uliasz, M. (1993). The atmospheric mesoscale dispersion modeling system. *Journal of Applied Meteorology*, *32*(1), 139–149. https://doi.org/10.1175/1520-0450(1993)032<0139:TAMDMS>2.0.CO;2.

Uliasz, M., & Pielke, R. A. (1991). Application of the receptor oriented approach in mesoscale dispersion modeling. In H. van Dop, & D. G. Steyn (Eds.), *Vol. 15. Air Pollution Modeling and Its Application VIII*. https://doi.org/10.1007/978-1-4615-3720-5_35. Billerica, Mass. Retrieved from.

van der Laan-Luijkx, I. T., van der Velde, I. R., Krol, M. C., Gatti, L. V., Domingues, L. G., Correia, C. S. C., et al. (2015). Response of the Amazon carbon balance to the 2010 drought derived with CarbonTracker South America. *Global Biogeochem Cycles*, *29*(7), 1092–1108. https://doi.org/10.1002/2014GB005082.

Warwick, N. J., Bekki, S., Law, K. S., Nisbet, E. G., & Pyle, J. A. (2002). The impact of meteorology on the interannual growth rate of atmospheric methane. *Geophysical Research Letters*, *29*(20), 8-1-8-4. https://doi.org/10.1029/2002GL015282.

Weiss, R. F. (1981). Determinations of carbon dioxide and methane by dual catalyst flame ionization chromatography and nitrous oxide by electron capture chromatography. *Journal of Chromatographic Science*, *19*(12), 611–616. https://doi.org/10.1093/chromsci/19.12.611.

Wells, K. C., Millet, D. B., Bousserez, N., Henze, D. K., Griffis, T. J., Chaliyakunnel, S., et al. (2018). Top-down constraints on global N_2O emissions at optimal resolution: Application of a new dimension reduction technique. *Atmospheric Chemistry and Physics*, *18*(2), 735–756.

Whiticar, M. (1993). Stable isotopes and global budgets. In M. A. K. Khalil (Ed.), *Vol. 13. Atmospheric methane: Sources, Sinks, and Role in Global Change*. Berlin: Springer.

Wolf, S., Keenan, T. F., Fisher, J. B., Baldocchi, D. D., Desai, A. R., Richardson, A. D., et al. (2016). Warm spring reduced carbon cycle impact of the 2012 US summer drought. *Proceedings of the National Academy of Sciences of the United States of America*, *113*(21), 5880–5885. https://doi.org/10.1073/pnas.1519620113.

Top-down approaches **Chapter | 4 155**

Worden, J. R., Bloom, A. A., Pandey, S., Jiang, Z., Worden, H. M., Walker, T. W., et al. (2017). Reduced biomass burning emissions reconcile conflicting estimates of the post-2006 atmospheric methane budget. *Nature Communications*, *8*(1), 2227. https://doi.org/10.1038/s41467-017-02246-0.

Worden, J. R., Doran, G., Kulawik, S., Eldering, A., Crisp, D., Frankenberg, C., et al. (2017). Evaluation and attribution of OCO-2 XCO_2 uncertainties. *Atmospheric Measurement Techniques*, *10*(7), 2759–2771. https://doi.org/10.5194/amt-10-2759-2017.

Worden, J., Kulawik, S. S., Shephard, M. W., Clough, S. A., Worden, H., Bowman, K., et al. (2004). Predicted errors of tropospheric emission spectrometer nadir retrievals from spectral window selection. *Journal of Geophysical Research-Atmospheres*, *109*(D9), 2356. https://doi.org/10.1029/2004JD004522.

Wunch, D., Toon, G. C., Blavier, J.-F. L., Washenfelder, R. A., Notholt, J., Connor, B. J., et al. (2011). The total carbon column observing network. *Philosophical Transactions of the Royal Society A: Mathematical, Physical and Engineering Sciences*, *369*(1943), 2087–2112. https://doi.org/10.1098/rsta.2010.0240.

Yadav, V., Michalak, A. M., Ray, J., & Shiga, Y. P. (2016). A statistical approach for isolating fossil fuel emissions in atmospheric inverse problems. *Journal of Geophysical Research-Atmospheres*, *121*(20). https://doi.org/10.1002/2016JD025642, 2016JD025642.

Zalicki, P., & Zare, R. N. (1995). Cavity ring-down spectroscopy for quantitative absorption measurements. *Journal of Chemical Physics*, *102*(7), 2708–2717. https://doi.org/10.1063/1.468647.

Zhao, C., Tans, P. P., & Thoning, K. W. (1997). A high precision manometric system for absolute calibrations of CO_2 in dry air. *Journal of Geophysical Research*, *102*(D5), 5885–5894.

Section C

Case Studies

Chapter 5

Current knowledge and uncertainties associated with the Arctic greenhouse gas budget

Eugénie S. Euskirchen[a], Lori M. Bruhwiler[b], Róisín Commane[c], Frans-Jan W. Parmentier[d,e], Christina Schädel[f], Edward A.G. Schuur[f], and Jennifer Watts[g]

[a]Institute of Arctic Biology, University of Alaska Fairbanks, Fairbanks, AK, United States, [b]National Oceanic and Atmospheric Administration, Boulder, CO, United States, [c]Department of Earth & Environmental Sciences, Lamont-Doherty Earth Observatory, Columbia University, Palisades, NY, United States, [d]Department of Physical Geography and Ecosystem Science, Lund University, Lund, Sweden, [e]Centre for Biogeochemistry in the Anthropocene, Department of Geosciences, University of Oslo, Oslo, Norway, [f]Center for Ecosystem Science and Society, Northern Arizona University, Flagstaff, AZ, United States, [g]Woodwell Climate Research Center, Falmouth, MA, United States

1 Introduction and background: Arctic ecosystems

In the Arctic region, understanding changes in carbon cycling and storage, including both carbon dioxide (CO_2) and methane (CH_4), is critical for understanding possible carbon cycle-climate feedbacks. This region contains the largest terrestrial carbon pool due to soil carbon storage, with an estimated 1035 Pg C in permafrost zone soils (ground that remains frozen for 2 or more years) up to 3 m depth (Schuur, McGuire, Romanovsky, Schädel, & Mack, 2018; Fig. 1). These permafrost carbon stores are vulnerable to warming, with a number of recent studies indicating greater vulnerability of ice-rich permafrost under warming (e.g., Hugelius et al., 2014; McGuire et al., 2018; Olefeldt et al., 2016; Schädel, Bader, Schuur, et al., 2016; Schädel, Schuur, Bracho, et al., 2014; Schuur, McGuire, Schädel, et al., 2015; Strauss et al., 2017).

The mean air temperature in the Arctic has warmed 2.7°C since 1971, with a more pronounced winter warming of 3.7°C (Box et al., 2019). This warming has caused reductions in snow and ice cover, which then decrease albedo, and

Balancing Greenhouse Gas Budgets. https://doi.org/10.1016/B978-0-12-814952-2.00007-1
Copyright © 2022 Elsevier Inc. All rights reserved.

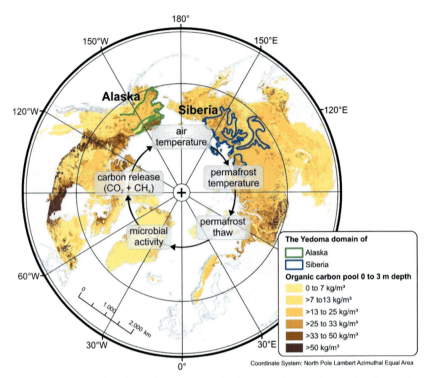

FIG. 1 Estimated 0–3 m deep carbon storage in the circum-Arctic permafrost region (inventories are summarized after Hugelius et al., 2014; Kanevskiy, Shur, Fortier, Jorgenson, & Stephani, 2011; Romanovskii, 1993), with a conceptual depiction of the permafrost-carbon climate feedback. The eastern boundary of Canadian Yedoma areas is uncertain, indicated as a dotted line at the Alaskan/Yukon state border. The Yedoma domain marks the potential maximum occurrence of widespread Yedoma deposits (organic-rich, Pleistocene-age permafrost with ice content of 50%–90% by volume; See also Fig. 2D) but also includes other deposits that formed after degradation of Yedoma, such as thermokarst deposits. In Western Siberia and the Russian Far East, Yedoma occurs sporadically, e.g., in river valleys. *(From Strauss, J., Schirrmeister, L., Grosse, G., Fortier, D., Hugelius, G., Knoblauch, C., et al. (2017). Deep Yedoma permafrost: A synthesis of depositional characteristics and carbon vulnerability. Earth-Science Reviews. https://doi.org/10.1016/j.earscirev.2017.07.007.)*

feedback to amplify the warming (Bokhorst et al., 2016; Chapin et al., 2005). Furthermore, warming is thickening the soil active layer (a soil layer above permafrost that freezes and thaws seasonally). Thickening of the active layer makes the large amounts of organic carbon (C) that are stored in these cold permafrost soils available for decomposition and release. This "permafrost carbon feedback" is estimated to transfer a large amount of carbon, 50–200 Pg C, to the atmosphere by the end of the 21st century (Schuur et al., 2015; Fig. 1). Warming coupled with increases in atmospheric CO_2 concentrations and increases in soil nutrients also stimulate arctic plant growth and vegetation C uptake. This potential increase in productivity may mitigate expected losses of C from permafrost soils (Keenan & Riley, 2018; Myers-Smith et al., 2011; Sturm, Racine, & Tape, 2001). Alternatively, other evidence, based primarily on satellite data, indicates that some regions are experiencing "browning," or decreases in plant growth, in response to warmer and/or wetter conditions in the Arctic (Bhatt et al., 2013; Lara, Nitze, Grosse, Martin, & McGuire, 2018; Phoenix & Bjerke, 2016). Consequently, these browning trends may exacerbate the carbon losses associated with permafrost thaw.

Cold and remote, the Arctic has traditionally been understudied compared to other regions further south. Field-based measurements of Arctic carbon began in the 1960s, although these were scarce and collected at only a few sites (Billings, Peterson, Shaver, & Trent, 1977; Bliss et al., 1973; Johnson & Kelley, 1970; Peterson & Billings, 1975). In recent years, numerous campaigns in the Arctic have been made through surface observations, aircraft studies, remote sensing, and modeling to better understand the Arctic greenhouse gas budget (Fisher et al., 2014; Forkel et al., 2016; Hayes et al., 2011; McGuire et al., 2012). Nevertheless, the scientific community has yet to reach an agreement on either the magnitude or the sign of the arctic terrestrial carbon budget. The Arctic Ocean is currently better constrained, with less variability across studies, and is estimated as a net carbon sink (MacGilchrist et al., 2014), although there is no consensus on the effect of sea ice melt on primary productivity and, therefore, the possible future status of the carbon sink strength of the Arctic Ocean (Cai et al., 2010; Jin et al., 2016; Sogaard et al., 2013). Recent data syntheses and modeling studies of arctic tundra net CO_2 balance have suggested that the tundra is either a CO_2 sink, a source, or approximately in balance (Belshe, Schuur, & Bolker, 2013; Fisher et al., 2014; McGuire et al., 2012; Virkkala, Aalto, Rogers, et al., 2021), with recent work attempting to reconcile these discrepancies (Schuur et al., 2018). More recent and mounting evidence suggests that the region is transitioning to a C source (e.g., Bradley-Cook, Petrenko, Friedland, & Virginia, 2016; Commane et al., 2017; Parazoo, Koven, Lawrence, Romanovsky, & Miller, 2018; Webb et al., 2016), especially when considering the offset of a summer C sink by winter respiration (Natali, Watts, Rogers, et al., 2019). Consequently, key uncertainties remain in the Arctic greenhouse gas budget, both at present and going forward under a warming climate.

In this chapter, we examine the main components of the greenhouse gas budgets from arctic terrestrial and aquatic environments, including how these components are measured and modeled at scales ranging from plots to regions. We discuss the uncertainties associated with these measurements and models and review the most current knowledge pertaining to estimates of these arctic greenhouse gas components. Finally, we introduce efforts currently underway to further advance our knowledge of the arctic greenhouse gas budget.

162 Section | C Case Studies

2 Methodologies

2.1 Components of the greenhouse gas budget of terrestrial arctic ecosystems

Greenhouse gas release from terrestrial arctic ecosystems is increasing due to enhanced microbial activity from warming and subsequent permafrost thaw. Carbon dioxide and methane are the major greenhouse gases that are released upon thaw, although significant amounts of nitrous oxide (N_2O) emissions have been measured in permafrost ecosystems (Abbott & Jones, 2015; Marushchak et al., 2011; Voigt, Lamprecht, Marushchak, et al., 2016). As much as one-quarter of the Arctic could be a source of N_2O emissions and potentially contribute to a noncarbon feedback to the climate (Voigt et al., 2017). Accounting for all three greenhouse gases together is crucial as they each exhibit a different forcing on the climate.

Warmer temperatures in the Arctic indirectly affect the carbon cycle by altering regional and local hydrology. An increase in permafrost thaw (Biskaborn et al., 2019; Romanovsky, Smith, & Christiansen, 2010; Romanovsky, Smith, Christiansen, et al., 2013) and near-surface degradation (Lawrence, Slater, Romanovsky, & Nicolsky, 2008; Slater & Lawrence, 2013) creates complex landscape patterns of wet and dry ecosystems due to the spatial patterning of ground ice (Schuur & Mack, 2018). In uplands, receding permafrost can dry soils as the water table overlying the permafrost surface descends (Lawrence, Koven, Swenson, Riley, & Slater, 2015). These dry soils tend to have less mineralization and release of CO_2 relative to wetter soils, although they also typically have minimal plant C uptake (Gulledge & Schimel, 1998). These uplands may also act as sinks of atmospheric CH_4 (Gulledge & Schimel, 1998; Jørgensen, Johansen, Westergaard-Nielsen, & Elberling, 2015). Alternatively, moistness of waterlogged soil can result from the thawing of ice-rich permafrost, which can also cause land surface collapse (Jorgenson, Shur, & Pullman, 2006; Liljedahl, Boike, Daanen, et al., 2016). This leads to anoxic soils where carbon is released by anaerobic microbial activity in the form of CO_2 and CH_4 (Schädel et al., 2016; Schuur et al., 2015).

Growing season carbon fluxes dominate greenhouse gas exchange in arctic ecosystems, and recent lengthening of the growing season in the Arctic by $\sim$2.5 days per decade may extend the period of time that terrestrial ecosystems take up CO_2 (Euskirchen et al., 2006) but may also increase CH_4 emissions (Pirk, Mastepanov, López-Blanco, et al., 2017). However, recent efforts have captured the importance of winter and shoulder season fluxes (e.g., Commane et al., 2017; Euskirchen, Bret-Harte, Edgar, Scott, & Shaver, 2012; Euskirchen, Bret-Harte, Shaver, Edgar, & Romanovsky, 2017; Oechel, Laskowski, Burba, Gioli, & Kalhori, 2014; Sweeney et al., 2016; Webb et al., 2016), including freeze-thaw events (e.g., Mastepanov et al., 2008; Raz-Yaseef, Torn, Wu, et al., 2016) on the annual C budget. Snow serves as an important insulator of the soil, acting as a porous medium in the diffusive exchange of gases. Microbial activity under the snow during below freezing temperatures contributes substantially to the annual greenhouse gas budget (Mikan, Schimel, & Doyle, 2002; Rivkina, Friedmann, McKay, & Gilichinsky, 2000; Schimel & Mikan, 2005). In addition, gases produced during fall and winter may get trapped by ice and snow cover and then later be released to the atmosphere in pulses during freeze-thaw events (e.g., Raz-Yaseef et al., 2016; Schimel & Clein, 1996). Large pulses during the autumn have also been attributed to reduced pore space during freeze-in, which leads to a physical squeezing of gases from the soil (Mastepanov et al., 2008; Pirk et al., 2015). Even though measurements of greenhouse gas exchange are difficult to conduct in the Arctic during the cold season, they are important in determining the annual greenhouse gas budget of the Arctic.

A unique attribute of Arctic ecosystems is a general absence of a large-scale disturbance regime (Walker & Walker, 1991), though paleo-fire records show that tundra fires have indeed occurred at varying frequencies and intervals over the past thousands of years during periods of warm summer temperatures and dry conditions (Hu et al., 2015). Infrequent tundra fires were recorded in northwestern Alaska in the 1970s (Wein, 1976). In recent years, the Arctic region has been marked by more frequent climatically driven disturbances. For example, the 2007 Anaktuvuk River Fire burned $1039 \, km^2$ of tussock tundra during an abnormally warm late summer in a region with no modern record of large wildfires (Mack et al., 2011), releasing a substantial 2.1 Tg C (primarily from the permafrost soils) to the atmosphere. Moreover, much of the region is vulnerable to thermokarst disturbance, or thawing of ice-rich permafrost followed by ground subsidence, with thermokarst vulnerable soils containing over half of the Arctic's belowground C pool (Grosse et al., 2011; Olefeldt et al., 2016). Land collapse and pattern formation following permafrost thaw and ice melt can greatly impact arctic C budgets by changing local hydrology and exposing stored C to warmer surface temperatures. These types of disturbances may have large impacts not only on the arctic carbon budget, but also on greenhouse gas release (Turetsky et al., 2019; Turetsky, Abbott, Jones, et al., 2020), and are particularly difficult to capture in coupled climate-carbon cycle models.

2.2 Methodologies for flux estimation in the Arctic

2.2.1 Chamber measurements and incubation studies in arctic terrestrial environments

Chamber measurements are commonly used in arctic field studies to study fine-scale ($100-10,000 \, cm^2$; Fig. 2A) gas exchange as they provide portable, low energy sampling of ecosystem and soil respiration and net ecosystem exchange (Virkkala, Virtanaen, Lehtonen, Rinne, & Luoto, 2018). Early versions of the closed chamber system were used to monitor soil respiration in Sweden in the early 1920s, where KOH and $Ba(OH)_2$ (alkali method) were used to absorb

FIG. 2 Arctic measurements and Yedoma permafrost. In (A) chamber measurements (foreground) and eddy covariance measurements (background) near Nuuk, Greenland, (B) winter eddy covariance measurements at the Imnavait Creek Watershed in Northern Alaska, (C) eddy covariance measurements on a thermokarst lake near Fairbanks, Alaska, and (D) ice-rich Yedoma permafrost along the banks of the Kolyma River in Eastern Siberia. *(Photo credits: E. Euskirchen.)*

CO_2 from air samples collected by the chamber (Pumpanen, Longdoz, & Kutsch, 2010). Open, dynamic steady-state flow chambers were introduced in the 1930s in an attempt to reduce underestimation of gradients driving diffusive flux (Pumpanen et al., 2010). The alkali absorption methods were increasingly replaced in the 1970s by infrared gas analyzers (IRGA) which allowed rapid and more efficient sampling of CO_2 concentrations and rates of gas exchange (Fang & Moncrieff, 1996). Present-day designs typically use a closed static chamber, a closed dynamic chamber, or an open dynamic chamber with CO_2 concentrations estimated with an IRGA (Pumpanen et al., 2004, 2010). Automatic chamber measurements are still relatively rare in the Arctic (Virkkala et al., 2018) because of their complex installation combined with larger power and maintenance requirements, but they have been successfully conducted over the longer term at some field sites (e.g., Mauritz et al., 2017).

Given the absence of power and system maintenance requirements for sample collection, soda lime adsorption techniques have remained a reliable method for measuring soil CO_2 efflux beneath the snowpack during winter seasons (Grogan, 2012; Rogers, Sullivan, & Welker, 2011). More recently, the "forced diffusion" method has become available for use with portable chambers by employing a membrane of a known flow rate, or diffusivity, which, when coupled with measurements of CO_2 concentration within the chamber, allows for the calculation of CO_2 emission rates from the soil (Risk, Nickerson, Creelman, McArthur, & Ownens, 2011). This method is permitting advances in shoulder and cold season chamber measurements, as well as use in even more

Arctic greenhouse gas budgets Chapter | 5   **165**

remote areas, due to its low power requirement and absence of moving parts (Kim, Park, Lee, & Risk, 2016). However, in general, chamber measurements are still primarily conducted at well-characterized field sites in Alaska and Fennoscandia, with relatively few measurements in Arctic Russia, Canada, and Greenland (Virkkala et al., 2018). Measurements of CH_4 and N_2O using forced diffusion methodology are not yet available.

Soil incubation studies are a commonly used method with soils from arctic terrestrial permafrost environments (Brach et al., 2016; Dutta, Schuur, Neff, & Zimov, 2006; Knoblauch, Beer, Sosnin, Wagner, & Pfeiffer, 2013). This approach is used to estimate the decomposability of soil organic matter by measuring greenhouse gas release as carbon is mineralized from soils under controlled conditions. Incubation studies benefit from their abstraction from the field as abiotic factors can be controlled or manipulated, allowing microbial and soil organic matter responses to experimental treatments (e.g., temperature, water, nutrients) to be isolated. Furthermore, laboratory incubation studies are useful to compare the carbon mineralization potential of different regions, ecosystems, or soil types and they allow for comparisons between wide ranges of soil organic matter quality. Incubation durations may vary from less than 1 day to up to many years. Short-term incubations (a few days to a few months) provide information on how much carbon is readily decomposable and may be closer to the initial conditions experienced within the soil profile. Long-term incubations (months to years) may diverge further from the conditions found within the profile, but can give insights into the potential decomposability of slower turning over carbon, which is needed given that the largest fraction of carbon turns over slowly, and is particularly important in permafrost environments (decades to centuries, Schädel et al., 2014).

2.2.2 Arctic micrometeorological eddy covariance measurements in terrestrial ecosystems

The spatial and temporal resolution of eddy covariance instruments mounted on towers in arctic terrestrial ecosystems (Fig. 2B) has increased substantially since initial flux tower measurements began 25 years ago, as illustrated in a recently designed map of arctic flux towers and their dates of operation (http://cosima.nceas.ucsb.edu/carbon-flux-sites/). In the Alaskan Low Arctic, flux tower systems were first deployed in the mid-1990s in the coastal wet tundra (Vourlitis & Oechel, 1997). In the Russian Low Arctic, flux tower measurements began in the early 2000s in the wet sedge and low shrub tundra (Zomolodchikov, Karelin, Ivaschnenko, Oechel, & Hastings, 2003), with measurements beginning in the mid-2000s in the Canadian Low Arctic heath and shrub tundra (LaFleur & Humphreys, 2007). In the colder High Arctic, measurements began in Greenland in 1996 (Soegaard & Nordstroem, 1999). While these initial arctic tundra eddy covariance studies only took place during the growing season when the ecosystems act as a net sink of CO_2, more recently,

166 Section | C Case Studies

measurements at some sites are being conducted over the full annual cycle and documenting cold season release of CO_2 (Celis et al., 2017; Euskirchen et al., 2012; Luers, Westermann, Piel, & Boike, 2014; Oechel et al., 2014). Eddy covariance measurements of CH_4 were initiated in the Arctic as early as 1998 in Lapland, Finland (Hargreves, Fowler, Pitcairn, & Aurela, 2001), although sites with both CO_2 and CH_4 are still relatively rare. Few of these flux tower CH_4 sites operate year-round (Zona et al., 2016), largely due to power requirements.

A small number of short-term aircraft campaigns have measured eddy covariance trace gas fluxes in the Arctic, providing a larger regional context for the tower fluxes, but with some methodological differences, than tower-based fluxes (Gioli et al., 2004; Oechel, Vourlitis, Brooks, Crawford, & Dumas, 1998). The first airborne flux measurements were made over the Alaskan tundra during Atmospheric Boundary Layer Experiment (ABLE) 3A in 1988 (Fan et al., 1992; Harriss, Sachse, et al., 1992; Harriss, Wofsy, et al., 1992). Later studies focused on carbon fluxes associated with the tundra on the North Slope of Alaska (Sayres et al., 2017; Zulueta, Oechel, Loescher, Lawrence, & Paw U, 2011) and the Mackenzie River delta in Canada (Kohnert, Serafimovich, Metzger, Hartmann, & Sachs, 2017). One study also measured fluxes of N_2O emissions from the North Slope tundra over a short period (Wilkerson et al., 2019). A few airborne campaigns have also occurred in Scandinavia (Gioli et al., 2004).

Sources of instrumental error from arctic eddy covariance measurements are often related to malfunction due to ice and snow buildup on the instrumentation as well as the self-heating effect of open-path gas analyzers (Burba, McDermitt, Grelle, Anderson, & Xu, 2008; Kittler et al., 2017). Furthermore, given the remote location of the sites, often with an absence of line power, data loss due to power failures from alternative sources (wind, solar), particularly during the cold season, may be more substantial than at warmer, more accessible sites. In addition, eddy covariance measurements from aircraft are expensive and typically restricted to the growing season and favorable flying weather.

2.2.3 Concentration measurements from aircraft and tall towers

Atmospheric observations of CO_2, CH_4, and the gas concentrations of many other atmospheric trace species have been collected in the Arctic for over 40 years (Fig. 3). These long-term observations are essential for monitoring changes in arctic carbon fluxes. The earliest measurement sites are Utqiagvik (formerly Barrow), Alaska and Alert, Canada. The Barrow Observatory is operated by NOAA (the National Oceanic and Atmospheric Administration), and data records start in 1971. The Alert station is operated by Environment and Climate Change Canada, and data records start in 1985.

Several approaches are used to monitor atmospheric carbon in the Arctic and globally. At many sites, discrete measurements, often at weekly intervals, are

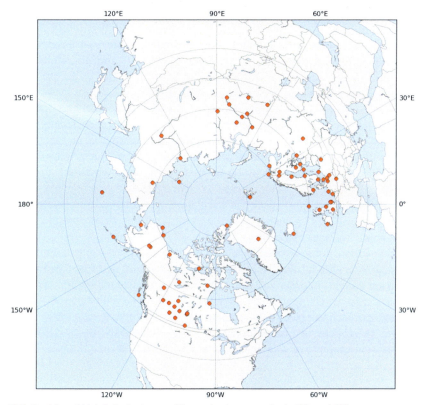

FIG. 3 Map of high latitude sites used in monitoring atmospheric CO_2 and CH_4.

collected and sent to a lab for analysis (such as the NOAA Earth System Research Laboratory in Boulder, CO). The use of a common analysis system has advantages for data quality, since analyses can be subject to strict quality control and calibrated to international scales. This rigorous approach provides the best opportunity to use long data records to detect small trends and changes in gradients that can reveal changes in greenhouse gas fluxes (e.g., Dlugokencky et al., 2005).

Understanding changes in the atmospheric carbon budgets for a particular region, such as the Arctic, cannot come from just a regional network: global observations are also required. This is because changes in horizontal gradients can be important indicators of changes in carbon fluxes (e.g., Dlugokencky et al., 2009). Furthermore, regional emissions may be a small part of the global carbon budget, and therefore transport from outside the region of interest may be the leading term in the regional budget. Through atmospheric circulation, emissions from lower latitudes are transported north to the Arctic. For example, CH_4 emissions from arctic wetlands are significant, but small

168 Section | C Case Studies

compared to tropical wetland emissions and anthropogenic emissions occurring at lower latitudes.

Long-term atmospheric monitoring is also carried out from aircraft and towers at multiple altitudes (Fig. 3). Strategies for analysis from aircraft include collecting discrete samples at multiple altitudes/intake heights or the use of an in situ analyzer, usually a spectroscopic instrument (see Chapter 4). Vertical profiles are useful for constraining models of the long-range transport of emissions. Networks of monitoring towers operating in the Arctic include those associated with Environment Canada, the JR-Station in Siberia, and the Japanese National Institute of Environmental Studies, along with some Arctic observations through the European Union International Carbon Observing System. Observations from towers can be used to characterize local-to-regional emissions as determined by atmospheric circulation. In addition, if measurements are collected at high frequency, short-term variability such as diurnal cycles and synoptic changes can be resolved.

Observations using aircraft have also been collected in "campaign mode." This approach differs from typical monitoring systems in that frequent aircraft flights are used to characterize carbon fluxes over a limited region for a specific period of time. By flying transects up- and downwind of a source region, a mass balance approach can be used to estimate fluxes, assuming the wind velocity is known. The Carbon in Arctic Reservoirs Vulnerability Experiment (CARVE) project is an example of a campaign approach (https://carve.ornl.gov).

Global air sampling networks, such as the NOAA Greenhouse Gas Reference Network (https://www.esrl.noaa.gov/gmd/ccgg/), were originally established to monitor "background" atmospheric trace species far from strong local sources. But sampling well-mixed air by definition means that information about the spatial and temporal distribution of fluxes is lost. Observation sites have been established near sources of interest, for example near wetlands. These sites often use continuous analyzers situated on towers. Examples are recent sites set up by the Integrated Carbon Observing System, EU, and the Max Planck Institute for Biogeochemistry in Jena, which recently established a site at Ambarchik, Siberia (Reum et al., 2018).

Long-term observations of atmospheric CO_2 and CH_4 have been used to determine whether high latitude carbon fluxes are changing over time due to climate change. Graven et al. (2013) and Barlow, Palmer, Bruhwiler, and Tans (2015) found increases in the amplitude of the CO_2 seasonal cycle at high latitude observation sites (e.g., Utqiagvik) relative to lower latitude sites (Mauna Loa) and concluded that productivity (carbon uptake) in northern ecosystems is increasing. Increases in late growing season CO_2 emissions, possibly linked to increased respiration in warming soils, were found by Commane et al. (2017) using the long-term Utqiagvik Observatory record. For CH_4, no significant longer-term trends in annual Arctic emissions have yet been found, although an increase in early winter CH_4 emissions has been documented since 2010 (Sweeney et al., 2016) and shorter-term trends have been identified (e.g., Thompson et al., 2017). Current understanding of arctic CH_4 emissions predicts increases due to warming peatland soils and thawing permafrost, but it is possible that changes are currently too small to be detected by global network observations or that the less sensitive coastal and remote Arctic sites may miss emissions. It is also possible that our understanding of methane emissions from arctic ecosystems is not complete, particularly as to how they are controlled by microbial dynamics in soil. For example, Oh et al. (2016) suggested that methanotrophy may be more temperature-dependent than methanogenesis such that microbial consumption of CH_4 can exceed warming-induced increases. Also incomplete is our understanding of arctic hydrology and how it may evolve over time. If arctic soils are becoming drier, then this may account for the lack of an observed increase in methane emissions.

Although tall towers can provide regional sampling, air sampling networks and in situ observation techniques for measuring atmospheric greenhouse gas concentrations still have limited spatial and temporal coverage. Large regions may be logistically difficult to sample, and data sharing among researchers at various international institutions at required temporal scales is often limited. Data collected from space-based platforms offer potentially greater spatial and temporal coverage, but there are currently some important limitations of these systems. Space-based observing systems remotely sense atmospheric column average CO_2 and CH_4 either in thermal infrared (TIR) or shortwave infrared wavelengths (SWIR). TIR retrievals (such as from the CrIS or IIASI instruments) are sensitive mainly to the upper troposphere, and this limits information about near-surface processes. SWIR retrievals (GOSAT, OCO-2, TROPOMI) have greater sensitivity to lower levels in the atmosphere, but these are still column averages rather than being representative of purely near-surface levels. For SWIR instruments, there are no data collected during winter, and few good retrievals at high latitudes during other seasons due to long atmospheric paths and persistent aerosols and clouds. It is possible that the use of active sensors will improve overall coverage in the Arctic.

2.2.4 Arctic freshwater/fluvial flux (DOC/DIC/POC) measurements

The exchange of carbon in freshwater environments, including lakes, rivers, and streams, is often more challenging to quantify than the exchange of greenhouse gases in terrestrial environments. This is due to the dynamic nature of freshwater systems, and also because they include a larger diversity of fluxes. Freshwater exchange not only includes vertical fluxes of CO_2 and CH_4 into the atmosphere, but carbon may also be transported laterally as dissolved organic carbon (DOC), dissolved inorganic carbon (DIC), and particulate organic carbon (POC).

Water-atmosphere fluxes can, in principle, be measured analogous to terrestrial installations, with floating flux chambers. However, this removes the influence of wind, which is important for the transfer of gases from the water surface

170 Section | C Case Studies

to the atmosphere. Floating chambers are also difficult to operate in stormy weather, even though these events are important for lake mixing. Moreover, fluxes can be highly variable spatially, but this variability is not captured at the lake surface when there is wave action. This is especially true for CH_4 fluxes. Ebullition of CH_4 is highly localized in hotspots, which means that a large number of chambers are needed to achieve a statistically reliable estimate of the total CH_4 flux of a lake (Wik, Thornton, Bastviken, Uhlbäck, & Crill, 2016). Fortunately, these hotspots become apparent in winter, when lakes freeze over and bubbles are incorporated into the ice. Winter is therefore the best time of year to locate hotspots and to install bubble traps that capture methane gas floating up to the lake surface (Walter, Zimov, Chanton, Verbyla, & Chapin, 2006). However, this may be impractical for a majority of arctic sites that are difficult to access in winter.

In recent years, greenhouse gas exchange from lakes has been increasingly measured with eddy covariance, either from the shore (Jammet, Crill, Dengel, & Friborg, 2015) or from floating platforms (Erkkilä et al., 2018; Fig. 2C). The former has the disadvantage that it only captures lake fluxes when the wind is coming from the right direction, while the latter is more challenging since compensation is needed for movement of the platform with the waves (Mammarella et al., 2015). Disturbance of the wind flow by the platform on which the tower is installed also needs to be considered. In the Arctic, lake ice formed during the winter may also upset the platform upon thaw in the spring. Nonetheless, lake-based eddy covariance systems have the advantage that they measure continuously and integrate all fluxes from a lake, including hotspots.

While lake-atmosphere fluxes are challenging to measure, it is even more difficult to track the lateral movement of C through streams and rivers toward the ocean (Vonk & Gustafsson, 2013). Estimates of how much carbon flows through river systems can be made by measuring DOC, DIC, and POC in water samples over time. These are then scaled with the total discharge of a river. When these measurements are performed at the mouths of rivers, they are a reliable estimate of how much carbon is ultimately delivered to the ocean (e.g., Tank, Striegl, Mcclelland, & Kokelj, 2016), but it can be difficult to trace the source sites contributing to this flux.

It remains uncertain how much C is lost to the atmosphere on the way to the mouth of a river. Regional C loss from inland waters is estimated as the sum of river lateral export to the coast, C outgassing from rivers occurring during transport, C outgassing from lakes, and C removal from the transport system through lake burial (Stackpoole et al., 2017). Sampling campaigns along a transect upstream can provide valuable insights in the variation of carbon flowing through a river system, but spatial differences may not necessarily be due to carbon loss to the atmosphere since some may be buried in river sediment over short or long timescales. The determination of the partitioning between burial and release is critical to assess how much carbon that enters freshwater systems may end up in the atmosphere as CO_2 or CH_4.

2.2.5 Arctic marine/oceanic fluxes

Ocean-atmosphere flux measurements of GHGs have been performed from ships for many decades by measuring the difference in the partial pressure or concentration of a greenhouse gas within the surface ocean and the atmosphere. The concentration in water is typically measured through headspace equilibration: a bottle is filled with seawater, and a neutral gas (i.e., nitrogen or helium) is inserted to create a headspace. After vigorous shaking, the dissolved gas equilibrates with the headspace, and the concentration of the gas in the headspace can be measured using spectroscopic or gas chromatography techniques. The amount of greenhouse gas in the original seawater is then back-calculated through known solubility values. The concentration in the atmosphere can be determined by either a continuous gas analyzer or by analyzing discrete air samples.

Once the concentration in the atmosphere and the seawater is known, the diffusive flux to the atmosphere is typically determined through the following equation (Wanninkhof, 1992):

$$flux = k\left(C_w - C_{eq}\right) \tag{1}$$

C_w is the measured concentration of the greenhouse gas in the seawater, while C_{eq} is the concentration that would be expected in seawater if it was in equilibrium with the atmosphere. k is the gas transfer coefficient, which depends on wind speed measured at a height of 10 m.

Since its introduction, this diffusive flux calculation method has been applied across the world's oceans (as described further in Chapter 11), as well as on lakes, but it requires careful parameterization. Together with progressing insight, this has led to numerous adjustments of this method to suit specific situations (e.g., Cole & Caraco, 1998; MacIntyre et al., 2010; Wanninkhof, 2014). However, common parameterizations may not work well in the Arctic due to the presence of sea ice, lower winter temperatures, and more frequent high intensity storms. Moreover, the method ignores emissions hotspots that may be important contributors of CH_4 to the atmosphere. Therefore, in recent years, eddy covariance is often used in conjunction with the method by Wanninkhof (1992). Once air flow disturbances from the ship, and its movement, are compensated for, this method can provide an independent estimate of the sea-to-air flux (see, e.g., Blomquist, Huebert, Fairall, et al., 2014; Thornton et al., 2020).

In addition to flux measurements of GHGs as described above, measurements of primary productivity from both ice algae and phytoplankton in the water column are important to consider in the Arctic Ocean carbon budget. These organisms transform dissolved inorganic carbon into organic material through photosynthesis, and their activity is therefore a strong driver of ocean CO_2 uptake. Primary productivity in the Arctic Ocean is highly seasonal as it is particularly dependent on light as well as nutrient availability. The spring melt and retreat of sea ice improves light availability, in addition to the increase in

172 Section | C Case Studies

daylight. Remote sensing techniques are used to provide an estimate of chlorophyll-*a* concentrations, which are a proxy for the amount of algal biomass. Primary production estimates are then determined based on combining the chlorophyll-*a* concentrations with drivers of primary production, including incoming solar radiation, sea surface temperature, and mixed layer depths (Babin et al., 2015; Lee et al., 2015). Inherent to remote sensing techniques, these estimates can be confounded by the presence of sea ice and clouds.

2.3 Top-down and bottom-up methods for estimating carbon fluxes in the Arctic

Methods for estimating arctic carbon budgets fall into two categories, top-down and bottom-up (McGuire et al., 2012). Bottom-up approaches generally use process-based information or models and scale these up to pan-Arctic scales. Examples of this approach include upscaling of eddy flux observations to Arctic scales (e.g., Ueyama et al., 2013; Virkkala et al., 2021), or using a process model of emissions informed by flux observations to predict Arctic-wide fluxes (e.g., Qiu et al., 2018). In contrast, top-down approaches use atmospheric observations to infer fluxes. Atmospheric observations contain integrated signals from fluxes encountered by air parcels on their way to the measurement site, and these signals must be deconvoluted spatially and temporally to estimate fluxes at the desired resolution. This necessarily requires information about atmospheric transport, which is supplied by a model that can range in complexity from a simple box model to a full atmospheric transport model. Mass balance approaches that use upwind and downwind observations from aircraft transects, and wind velocities from observations or analyses, to estimate regional fluxes can be considered as box models, while inverse techniques combined with fully resolved atmospheric transport models are examples of a more detailed and complex approach (see Chapter 4).

Atmospheric inversions employ statistical optimization methods to determine the fluxes (of, e.g., CO_2, CH_4, and N_2O) that are in optimal agreement with the observations and prior information. The fluxes are necessarily discretized in space and time (e.g., Enting, 2018; Tarantola, 1987, see also Chapter 4). Atmospheric inversions can be thought of as consisting of two steps. The first step involves modeling the state of the atmosphere, i.e., the concentrations of the trace gas species of interest, e.g., CO_2. This is done using an atmospheric transport model and a first guess of the fluxes based on prior information. This is the so-called "forward problem." Second, to find the optimal fluxes, the atmospheric transport is effectively "inverted" to relate the concentrations to the fluxes. This is the so-called "inverse problem." While the forward problem has a unique solution, the inversion problem does not: it is mathematically underdetermined. Prior information usually comes from emission inventories and process-based models of the natural fluxes. The sparseness of observations limits the accuracy at which the fluxes can be optimized, as well as the degree of spatial and temporal information about the fluxes. In regions where data are sparse or nonexistent, the optimized fluxes remain close to the prior estimate, and thus nothing new is learned compared to the prior information.

A potential source of error in atmospheric inversions comes from the uncertainty in the modeled atmospheric transport. This can arise if processes, such as mixing in the planetary boundary layer, or deep convection, are not well described by the model. In addition, transport in models is often only resolved on spatial scales from 10s to 100s of kilometers, which cannot capture local air circulation, such as in mountains and valleys and coastal locations. Furthermore, coarse discretization of the fluxes may mean that the influence of local or point sources on observations may not be well represented. Increasing horizontal and vertical resolution of a transport model may improve the representation of transport in models, and this has led some to use higher resolution (10 km) regional models for atmospheric inversions (e.g., Miller et al., 2016). However, the use of regional transport models requires specification of boundary conditions; these are difficult to determine, and biased boundary conditions can lead to biased flux estimates (refer to Chapter 4 for a complete discussion of inverse model uncertainty and error).

Both the bottom-up and top-down approaches suffer from limited observations. In the case of process models and scaled-up flux tower observations, flux data may be too sparse to adequately capture the true ecological diversity. For CH_4, as an example, many Arctic flux observations may be made near wetlands and a few near drier upland environments, leading to a bias toward larger emissions in scaled-up flux products. For top-down approaches, sparse observations lead to larger posterior uncertainties and solutions that remain close to prior estimates.

Estimated CO_2 fluxes from four atmospheric inversion models show small but statistically significant trends for regions between 50° and 60° N, and further north, 60°–90° N (Fig. 4). The trend for both regions is $\sim$0.01 Pg C/yr. This is in contrast to the study of Welp et al. (2016), who did not find significant trends north of 60° N from flux inversions. On the other hand, 11 global inversions using surface observations collected by the Global Carbon Project CH_4 synthesis (Saunois, 2020) show little change over the past two decades. The regional inverse model of Thompson et al. (2017) found that CH_4 emissions are increasing north of 50° N, but much of this increase occurred outside of the Arctic.

2.4 Terrestrial ecosystem and land surface models in the Arctic

Large-scale, process-based biogeochemical and Earth system models with simulations over arctic regions exhibit substantial carbon cycle uncertainty, ranging from estimates of carbon sources, sinks, or neutrality (Fisher et al., 2014; McGuire et al., 2012; Melton et al., 2013). The models show variation in the timing of peak summer carbon uptake as well as large differences in the amount

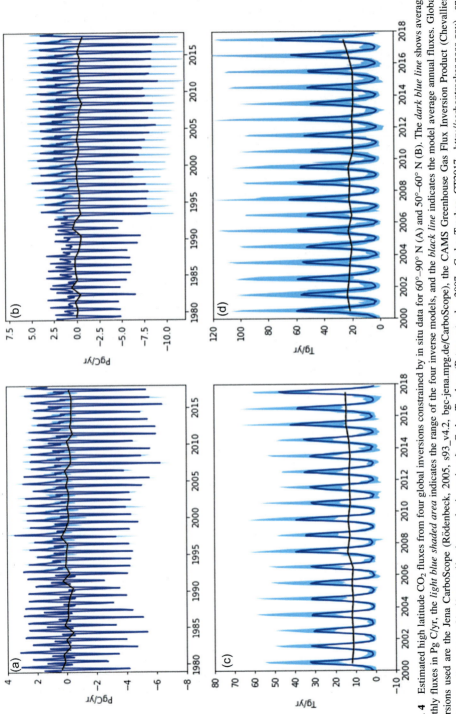

FIG. 4 Estimated high latitude CO_2 fluxes from four global inversions constrained by in situ data for 60°–90° N (A) and 50°–60° N (B). The *dark blue line* shows average monthly fluxes in Pg C/yr, the *light blue shaded area* indicates the range of the four inverse models, and the *black line* indicates the model average annual fluxes. Global inversions used are the Jena CarboScope (Rödenbeck, 2005, s93_v4.2, bgc-jena.mpg.de/CarboScope), the CAMS Greenhouse Gas Flux Inversion Product (Chevallier, 2018, https://apps.ecmwf.int/datasets/data/cams-ghg-inversions/), CarbonTracker (Peters et al., 2007, CarbonTracker CT2017, http://carbontracker.noaa.gov), and CarbonTracker-EU ((van der Laan-Luijkx et al., 2017), http://www.carbontracker.eu). Estimated natural high latitude CH_4 emissions constrained by data for 11 inversions submitted to the Global Carbon Project CH_4 Budget Update (Saunois, 2020) for 60°–90° N (C) and 50°–60° N (D). The *dark and light blue lines* indicate the model mean and spread, and the *black line* indicates the model average annual emissions.

of winter respiration (Commane et al., 2017). Furthermore, many large-scale model simulations over the Arctic lack an adequate representation of the highly heterogeneous tundra vegetation, which is often simulated simply as a C3 grass (Wullschleger et al., 2014). Similarly, only in recent years have more models begun to include a detailed representation of the permafrost regime, including representation of soil C at depth (Koven et al., 2011; Lawrence & Slater, 2005; Schaefer, Zhang, Bruhwiler, & Barrett, 2011; Zhuang, Romanovsky, & McGuire, 2001). Despite representing permafrost, nearly all land surface models do not consider thermokarst disturbance, a complex subgrid cell process that is difficult to represent (Aas et al., 2019), but may result in release of C through both soil organic carbon (SOC) erosion and mineralization of SOC into both CO_2 and CH_4 (Turetsky et al., 2020). Model representation of feedbacks between the carbon and nitrogen cycles (Koven, Lawrence, & Riley, 2015; McGuire et al., 1997a, 1997b) is also critical given that nitrogen is a key limiting nutrient to plant productivity in the Arctic.

Models simulated over the Arctic often lack observational data for model parameterization and verification, with large sampling biases in field studies (Metcalfe et al., 2018). In particular, 31% of the arctic field studies published in the peer-reviewed literature come from sites located within 50 km of two study areas, Toolik Field Station in Alaska and Abisko Field Station in Sweden (Metcalfe et al., 2018). Vulnerable warming areas such as the Canadian Arctic archipelago and the Arctic coastline of Russia are poorly sampled. Even long-term monitoring programs, such as the Circumpolar Active Layer Monitoring (CALM) Program, are subject to biases, particularly in complex terrain where finer-scale measurements are needed to resolve the spatial structure of the active layer thickness (Fagan & Nelson, 2016), and ground subsidence is not considered. Modelers have identified key data gaps, with primary needs including soil carbon, net primary productivity, plant biomass, soil moisture, plant functional types, and gross primary productivity. Secondary needs include soil respiration, litter biomass, active layer thickness, freeze/thaw dynamics, net ecosystem exchange, soil temperature, evapotranspiration, water table, permafrost, soil vertical profile, and leaf area index (Fisher et al., 2018). Many of these variables are required both temporally and spatially.

A key carbon dynamic that these models have examined in recent years is the trade-off between the loss of soil carbon from emissions due to permafrost thaw versus gains in ecosystem carbon due to increasing plant productivity increasing C storage (Harden et al., 2019; McGuire et al., 2018; Mekonnen, Riley, & Grant, 2018). Initially, these models were simulated and evaluated with respect to soil thermal dynamics and the amount of area that would be impacted by permafrost thaw through the 21st century. The range of permafrost loss simulated across these models is wide, with estimates ranging between 2% and 99% loss of the present-day continuous permafrost, depending on the model, some of which include little permafrost initially, and the warming scenario used (Euskirchen et al., 2006; Lawrence & Slater, 2005; Koven,

176 Section | C Case Studies

Riley, & Stern, 2013; McGuire et al., 2016). Despite these losses of permafrost, models generally estimate increases in both soil and vegetation C storage through the 21st century, attributed to the effect of atmospheric [CO_2] fertilization on plant productivity (McGuire et al., 2016, 2018), and also because not all models incorporate nutrient limitations on plant growth (Xia et al., 2017) or adequately model shoulder and winter season respiration (Commane et al., 2017). However, in the current suite of models, this increase in C storage is expected to diminish, with substantial losses occurring by 2300 under the RCP8.5 high emissions scenario (208 Pg C loss, across models; McGuire et al., 2018). These losses can act as a feedback to climate, influencing 0.5%–17% of the change in the global mean temperature by 2300, with this range representing differences in land surface models (Burke et al., 2017).

2.5 Review of Arctic GHG estimates by sector and associated key uncertainties

The ocean is consistently estimated as a CO_2 sink, but the terrestrial CO_2 budget is less well understood (Table 1; Fig. 5). Terrestrial estimates vary from a CO_2 source to sink to neutral. Evidence is more conclusive with CH_4, indicating a terrestrial CH_4 source, while the Arctic Ocean is also estimated as a source, but with a large uncertainty of the magnitude at present and in the future. In this section, we review the Arctic GHG estimates on the land (including lakes, streams, and rivers) and marine sectors, as well as the uncertainties associated with these sectors.

Terrestrial CO$_2$: A recent estimate of the Arctic tundra net CO_2 uptake, taking into account eddy covariance and chamber measurements, indicates that the tundra biome is a slight sink of annual average NEE from 1990 to 2015 at $-13 \, \text{Tg C yr}^{-1}$ (with annual ranges extending from -81 to $62 \, \text{Tg C yr}^{-1}$ Virkkala et al., 2021). Statistical models applied in this same study show a slight source of annual average NEE from 1990 to 2015 at $+10 \, \text{g C m}^{-2} \, \text{yr}^{-1}$. These estimates extend the analysis of McGuire et al. (2012) that indicated that, when taken together, wet and dry Arctic tundra were a sink of $-202 \, \text{Tg C yr}^{-1}$ in the early 2000s and that of Belshe et al. (2013) that found an average annual NEE of $462 \pm 378 \, \text{Tg C yr}^{-1}$ from the mid-1908s until the early 2000s. As permafrost continues to thaw under a warming climate, the rate at which previously frozen soil carbon can be transferred to the atmosphere depends on the SOC quality and decomposability (lability), microbial properties, and environmental drivers, all factors which can be difficult to measure and predict (Schädel et al., 2014; Schuur et al., 2015). Thus, it is important to continue to assess the annual tundra biome carbon budget as more data become available. As mentioned above, Arctic greening and increases in plant productivity may offset losses from increases in heterotrophic respiration. Less well documented is evidence of Arctic browning, or a decline in plant productivity and biomass associated with extreme weather events, such as extreme heat, frost droughts, or rain on snow

Arctic greenhouse gas budgets **Chapter | 5** **177**

TABLE 1 Terrestrial and aquatic sources and sinks of greenhouse gases in the Arctic, level of uncertainty, and expected impacts of continued warming.

Type of flux	Source or sink	Level of certainty	Uncertainty and possible trajectory
CO_2 fluxes			
Net terrestrial flux	Sink, neutral or source	Low to medium	• Increases in plant productivity, including tundra shrubs, increase CO_2 uptake. • Some uncertainty associated with warmer, drier summers, which may inhibit productivity. • Increases in microbial respiration associated with warming permafrost, particularly in the cold seasons and winter may increase CO_2 losses. • Also influenced by disturbance, including C losses from tundra fire and thermokarst
Tundra lakes	Source	Medium	• Dependent on the trajectory of geomorphological change and the formation of thermokarst lakes versus infilling of lakes by sediment and fen or bog vegetation
Net marine flux	Sink	Medium to high	• Impacts on plant productivity related to sea ice decline are uncertain • Changes in water temperature and water temperature-dependent CO_2 uptake.
CH_4 fluxes			
Tundra wetlands	Source	Medium	• Uncertainty in the magnitude of winter and shoulder season emissions • Changes in plant community composition and plant-mediated CH_4 flux may increase emissions • Changes in wetland extent due to permafrost thaw
Tundra lakes	Source	Medium to high	• Dependent on the trajectory of geomorphological change and the formation of thermokarst lakes versus loss or infilling of lakes by sediment and fen or bog vegetation
Ocean flux	Source	Low	• Dependent on vulnerability of subsea permafrost and methane hydrates to warming and oxidation

Continued

178 Section | C Case Studies

TABLE 1 Terrestrial and aquatic sources and sinks of greenhouse gases in the Arctic, level of uncertainty, and expected impacts of continued warming—cont'd

Type of flux	Source or sink	Level of certainty	Uncertainty and possible trajectory
Lateral carbon flow			
Dissolved and particulate OC/IC fluxes	Source	Low	• Dependent on terrestrial productivity and groundwater flow in thawing permafrost environments
Coastal erosion	Source	Medium	• Related to precipitation amounts, exposure of coastline in the absence of sea ice, and carbon content (yedoma soils) of affected areas
Ice sheet and glacier OC flux	Source	Medium	• Related to melt rates
N_2O fluxes			
Terrestrial flux	Source	Low	• Dependent on the trajectory of permafrost warming and thaw
Marine flux	Source	Low	• Dependent on sea ice extent and meltwater

Modified from Parmentier, F. J. W., Christensen, T. R., Rysgaard, S., Bendtsen, J., Glud, R. N., Else, B., et al. (2017). A synthesis of the arctic terrestrial and marine carbon cycles under pressure from a dwindling cryosphere. *Ambio, 46*, 53–69. https://doi.org/10.1007/s13280-016-0872-8.

(Parmentier et al., 2018; Treharne, Bjerke, Tømmervik, Stendardi, & Phoenix, 2018). This arctic browning may thus counterbalance, or even outweigh, increases in productivity from Arctic greening, although these studies of CO_2 exchange in conjunction with browning have not been scaled up to cover the entire Arctic.

Terrestrial CH_4: While arctic wetlands have long been regarded as emitters of CH_4 in the absence of warming, additional warming may promote increases in microbial CH_4 production and permafrost thaw may influence hydrology, either increasing or decreasing wetland area and extent, which controls CH_4 production. Current estimates of natural CH_4 emissions from the Arctic wetland tundra fall in the range of 11–39 Tg CH_4 yr^{-1} (Parmentier et al., 2017). Currently, most measurements of CH_4 fluxes take place in the snow-free

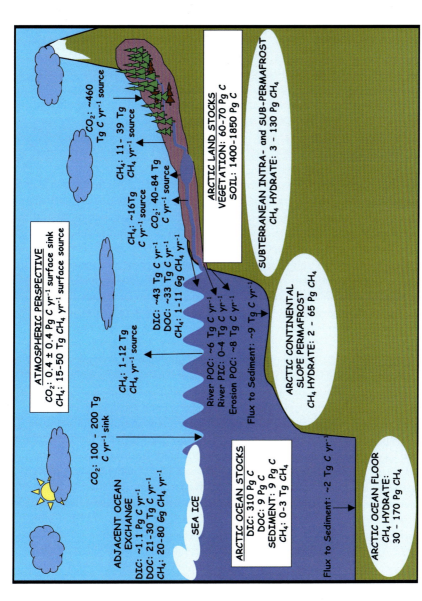

FIG. 5 The current state of the Arctic carbon cycle. Values shown are the ranges of uncertainty. *(Based on McGuire et al. (2009). Reprinted with permission from Wiley.)*

season using chambers and eddy covariance towers, although some observational efforts have taken place in winter (Mastepanov et al., 2008; Pirk et al., 2016; Taylor, Celis, Ledman, Bracho, & Schuur, 2018; Whalen & Reeburgh, 1992; Zona et al., 2016). Winter CH_4 emissions have been shown to act as a substantial contribution to the annual terrestrial CH_4 budget in some areas (Mastepanov et al., 2008; Zona et al., 2016), but other studies, highlighting the heterogeneity of the tundra landscape, show minimal winter emissions (Castro-Morales et al., 2018). Plant-mediated emissions of CH_4 are a primary mechanism for transport of CH_4 to the atmosphere in wetland tundra, with CH_4 flux rates differing among plant species (Andersen, Lara, Tweedie, & Lougheed, 2017). Thus, Arctic CH_4 exchange will be influenced by any long-term shifts in wetland tundra plant communities under environmental change, including increases in surface temperature and hydrological changes associated with thawing permafrost.

Lake CO_2: Lakes are not only a pronounced feature of the Arctic landscape, but also an important component of GHG exchange across this region. More than 10% of the net ecosystem production of terrestrial ecosystems can be exported as SOC to lakes, streams, and rivers across the Arctic (Weyhenmeyer et al., 2012). Arctic lakes have generally been considered sources of CO_2, although many of these studies focus on thermokarst systems in the yedoma (organic-rich Pleistocene-age permafrost with high ice content) of Alaska and Siberia (Elder et al., 2018; Walter-Anthony, Zimov, Chanton, Verbyla, & Chapin, 2014). Other evidence indicates that lakes in arid, topographically flat, low elevation permafrost regions may have weak connectivity to groundwater or streamflow, thereby reducing their ability to process SOC exports and emit CO_2 (Bogard et al., 2019). Other work indicates that lakes may evade carbon to the atmosphere (Rocher-Ros et al., 2017). Consequently, considerable uncertainty still exists in the overall emissions of CO_2 from Arctic lakes.

Lake CH_4: Based on a data synthesis across boreal and arctic lakes, approximately two-thirds of the total CH_4 emissions from landscapes north of latitude 50° N originate from freshwater systems ($16.5 \pm 9.2 \, Tg \, CH_4 \, yr^{-1}$), with thermokarst lakes contributing about 25% of this total from freshwater ecosystems (Wik, Varner, Anthony, MacIntyre, & Bastviken, 2016). Nonthermokarst, postglacial lakes emit less CH_4 per unit area (Sepulveda-Jauregui, Walter Anthony, Martinez-Cruz, Greene, & Thalasso, 2015; Wik, Varner, et al., 2016). However, since they cover a larger area, they may emit half of the total CH_4 emissions attributed to northern lakes and ponds (Wik, Varner, et al., 2016).

Arctic lakes are underrepresented in observational studies on the lake-atmosphere GHG exchange. Studies are generally short-term (<3 years) and do not include the full annual cycle, including ice breakup when spring emissions peak due in part to the release of seasonally ice-trapped bubbles (Sepulveda-Jauregui et al., 2015). Consequently, arctic lakes are also not well represented in modeling approaches, although recent work has begun to address this issue (Bayer, Gustafsson, Brakebusch, & Beer, 2019; Tan et al., 2017).

The distribution of lakes across the Arctic landscape is also shifting, with some regions reporting a trend toward lake drying and loss due to greater lake evapotranspiration (Bring et al., 2016; Finger Higgens et al., 2019) or loss due to permafrost thaw in the previously frozen and impermeable layer below the lake (Smith, Sheng, MacDonald, & Hinzmann, 2005). Other studies report lake formation as a result of permafrost thaw (Walter-Anthony et al., 2018). This change in distribution will likely influence arctic lake GHG emissions in the future. However, these changes are difficult to discern because doing so requires analysis of high-resolution, long-term remotely sensed datasets over a given region, while also considering other factors that influence lake levels, such as periodic flooding.

2.5.1 Lateral flow: Rivers and streams

Carbon may flow laterally from streams and rivers into the Arctic Ocean, or may release CO_2 through evasion. Combined, arctic rivers are estimated to deliver $82 \, \text{Tg C yr}^{-1}$ to the Arctic Ocean, with $43 \, \text{Tg C}$ as DIC, $33 \, \text{Tg}$ as DOC, and $6 \, \text{Tg C}$ as POC (McGuire et al., 2009). Thawing permafrost may release additional carbon to streams and rivers as both DOC and POC (Vonk & Gustafsson, 2013; Wild et al., 2019). POC is also released through coastal erosion, particularly along coastlines with high yedoma (Couture, Irrgang, Pollard, Lantuit, & Fritz, 2018; Günther, Overduin, Sandakov, Grosse, & Grigoriev, 2013). The occurrence of evasion in rivers and streams in the Arctic was identified decades ago (Kling, Kipphut, & Miller, 1991), but our understanding of the temporal and spatial variation of evasion across the Arctic is still low (Denfeld, Frey, Sobczak, Mann, & Holmes, 2013). Glaciers and ice sheets store and release organic carbon following microbial processes, entering the Arctic Ocean as both DOC and POC (Hood, Battin, Fellman, O'Neel, & Spencer, 2015). The Greenland Ice Sheet is estimated to release 0.13–$0.17 \, \text{Tg C yr}^{-1}$ as DOC and 0.36–$1.52 \, \text{Tg C yr}^{-1}$ as POC, both of which may increase under a warming climate (Lawson et al., 2014).

Recent work highlights the uncertainty in the Arctic lateral flow of carbon. Li Yung Lung et al. (2018) emphasize data needs in Canada, where more than two-thirds of the land area drain in the Arctic Ocean and Hudson Bay. Nevertheless, long-term, continuous datasets only exist for two of the Canadian Arctic watersheds (the Mackenzie and the Peel). Stackpoole et al. (2017) assess Alaskan inland waters and also highlight the lack of long-term datasets from both rivers and streams. These data needs inhibit accurate assessment of the freshwater carbon fluxes in this region. Furthermore, there is a need to better integrate inland water C fluxes with terrestrial C fluxes since the total lateral C export and CO_2 fluxes from inland waters can represent a significant proportion of terrestrial C uptake (Stackpoole et al., 2017). Finally, other work highlights the need to advance our understanding of how groundwater contributions in seasonally thawed permafrost environments contribute to C export to rivers and streams (Neilson et al., 2018).

182 Section | C Case Studies

Ocean CO_2: The Arctic Ocean is consistently estimated as a sink of CO_2 of ~100–200 Tg C yr^{-1} (Schuster et al., 2013; Yasunaka et al., 2018) based on a combination of measurements and modeling, although other more regional estimates show considerable variation (Parmentier et al., 2017). This sink is largely due to primary production during the summer and the high solubility of CO_2 in low temperature water. Since photosynthesis and plant productivity depend on light and nutrients, the Arctic Ocean productivity onset is generally related to sea ice melt in the spring, which releases nutrients into the water and permits light penetration. As Arctic sea ice declines (Perovich et al., 2018), there is an expected response in terms of a shift in productivity, although the mechanisms driving this response are uncertain. Some research suggests that the decrease in the summer ice extent and increased light availability may increase primary productivity in some assemblages of phytoplankton (Arrigo & van Dijken, 2011, 2015; Schuback, Hope, Tremblay, Maldonado, & Tortell, 2017), while other assemblages may remain constrained by low temperatures and nitrate depletion (Schuback et al., 2017). On the whole, the primary production response to sea ice loss is both seasonally and spatially variable (e.g., Hill, Ardyna, Lee, & Varela, 2018; Tremblay et al., 2015). Furthermore, since cold water can dissolve and absorb more CO_2 than warm water, some additional carbon may be released into the atmosphere with warming summer waters in the future (Cai et al., 2010). In the winter, cold water again takes up more carbon, although this too could be less than in previous years if the winter water temperatures begin to warm. Consequently, there is uncertainty in future Arctic Ocean CO_2 sink in both primary production and water temperature-dependent CO_2 uptake.

Ocean CH_4: Emissions of CH_4 from the Arctic Ocean may be related to thawing subsea permafrost and destabilized CH_4 hydrates in ocean seabed sediments. While some earlier studies warned of a large release of CH_4 from these sources (e.g., Shakhova et al., 2010b), more recent work suggests that oxidation and dispersion within the ocean can limit atmospheric emissions and that emission estimates need to be adjusted downward significantly (Berchet et al., 2016; Sparrow et al., 2018). There is also uncertainty regarding the size of the methane hydrate stores that are vulnerable to warming (Boswell & Collett, 2011; Shakhova et al., 2010a). Furthermore, other sources of CH_4 emissions from the Arctic Ocean are related to biological productivity in the surface waters, which are also subject to uncertainty (Kort et al., 2012). Overall, estimates of CH_4 release from the Arctic Ocean range from 1 to 17 Tg CH_4 yr^{-1} (Christensen, Arora, Gauss, Höglund, & Parmentier, 2019), with some recent studies finding emissions on the lower end of this range of 2–3 Tg CH_4 yr^{-1} (Thornton et al., 2020; Tohjima et al., 2020).

2.5.2 Nitrous oxide

While N_2O concentrations have been rising in the atmosphere (Syakila & Kroeze, 2011), studies of N_2O emissions from the Arctic have been scarce.

This is in part because permafrost environments have generally been considered negligible emitters of N_2O since nitrogen supply is limited (Rodionow, Flessa, Kazansky, & Guggenberger, 2006). Arctic N_2O emissions occur naturally when nitrogen is released from litter and soil during mineralization, though the denitrification process could increase considerably as permafrost soils thaw and decompose more rapidly. In situ plot-scale evidence is accumulating to suggest that N_2O emissions from permafrost soils may be high, particularly in conjunction with warming and thaw (Voigt et al., 2016, 2017), and equivalent to daily emissions from tropical forests. Landscape-scale studies with aircraft eddy covariance measurements indicate large variability in N_2O fluxes over the North Slope of Alaska (Wilkerson et al., 2019). In the Arctic marine environment, N_2O concentrations show dependencies associated with melt-water (lower N_2O) and under multiyear sea ice (higher N_2O; Kitidis, Upstill-Goddard, & Anderson, 2010). Thus, sea ice melt and formation may control fluxes of N_2O from the sea-air surface (Randall et al., 2012). Overall, since measurements of N_2O in the arctic are scarce, it is unclear if elevated N_2O measurements from some terrestrial studies represent an increasing trend because previous N_2O data have been so limited. Eddy covariance measurements of N_2O, similar to those ongoing for CO_2 and CH_4, would be ideal, but these N_2O measurement systems are challenging to operate due to their high power requirements while the regions of interest are often off-grid.

3 Uncertainty and reducing uncertainty

As highlighted above, a key source of uncertainty in GHG budgets in the Arctic is a lack of data from this remote region with few long-term monitoring programs and experiments. This geographical gap includes much of Siberia and the Arctic Archipelago. Alaska, parts of Canada, and Scandinavia are better represented, although studies tend to concentrate near roads, population areas, or field stations (e.g., Metcalfe et al., 2018, as described above). This lack of data has been long recognized, and initiatives to develop coordinated programs and networks to address data gaps have been put forth over the past several decades, several of which we describe below.

One initiative, Sustaining Arctic Observing Networks (SAON; https://www. arcticobserving.org/), has been developed to promote and sustain an open and well-integrated Arctic Observing System, coordinating across stakeholders and knowledge systems. This includes in situ, remote sensing, and community-based observations. The Arctic Great Rivers Observatory (ArcticGRO), a component of the US National Science Foundation's Arctic Observing Network, is an international effort to collect and synthesize time-series data from the six largest arctic rivers, Yukon, MacKenzie, Kolyma, Lena, Yenisey, and Ob'. The long-running International Tundra Experiment (ITEX), begun in 1990, examines the response of cold-adapted tundra vegetation to environmental change, specifically an increase in summer temperature, using long- and short-term experiments at over two dozen sites (Arft et al., 1999). More recently, ShrubHub has been developed to document changes in woody vegetation in tundra ecosystems (Bjorkman et al., 2018). The International Network for Terrestrial Research and Monitoring (INTERACT) is a circum-Arctic infrastructure network of 86 terrestrial field stations developed to build capacity for research and monitoring in the Arctic. Founded in 1983, the International Permafrost Association (IPA) fosters and disseminates knowledge concerning permafrost and, in particular, developed the widely used "Circum-Arctic Map of Permafrost and Ground-Ice Conditions" (Brown, Ferrians, Heginbottom, & Melnikov, 1997). The Permafrost Carbon Network (PCN) was formed to link biological C cycle research with well-developed networks in the physical sciences focused on the thermal state of permafrost. The PCN produces new knowledge through research synthesis to quantify the role of permafrost C in driving future climate change. Collectively, these initiatives, and others, have advanced and continue to advance the suite of arctic observations that can be incorporated to reach a better understanding of greenhouse gas exchange in the region.

Recent advances in remote sensing observations, including optical satellite imagery and airborne and spaceborne light detection and ranging data (LiDAR), are also filling critical data needs. Remotely sensed maps of the normalized difference vegetation index (NDVI) are available since 1982, documenting changes in arctic tundra vegetation greenness (Bhatt et al., 2013, 2017). A high resolution (0.25 m) LiDAR digital elevation model, combined with multispectral imagery and ground-based plant community measurements, has also recently been employed to develop an approach to upscale sparse field measurements of tundra vegetation (Langford et al., 2016). While permafrost distribution and integrity cannot be monitored directly from space, and is typically measured through boreholes, satellite data of topographic changes, thermal conditions, snow cover, and air temperature, combined with in situ measurements, have recently been applied to create a higher resolution permafrost map (Obu et al., 2019) than that of Brown et al. (1997). Analyses of the historical satellite image time series can be used to identify and map landscape changes resulting from permafrost thaw (Nitze, Grosse, Jones, et al., 2018). Additional advancements with space-based observations are still needed. For example, process-based modeling of CH_4 requires detailed information on the ground-surface conditions, which are not readily available from remote-sensing techniques.

4 Perspective and future opportunities

4.1 The current status of the GHG budget of the arctic terrestrial and marine environments

The Arctic is experiencing amplified climate change, with possibly profound impacts on the C budget of the region. We cannot count on arctic ecosystems to store as much C as they have in the past, and evidence indicates that they will likely store much less. The permafrost carbon pool is particularly vulnerable to thaw and carbon release, creating a positive feedback to climate warming (Schuur et al., 2015). Carbon release through glacier and ice sheet mass loss, coastal erosion, browning of tundra vegetation, and warming ocean waters all represent additional positive GHG feedbacks to warming. Negative GHG feedbacks to climate warming include increased CO_2 uptake from increases in tundra plant productivity and infilling of lakes by sediment or fen/bog vegetation (Table 1). It is critical to, at a minimum, maintain coverage and preferably increase spatial and temporal coverage of key in situ measurements, such as carbon fluxes, plant greenness and productivity, permafrost temperature and integrity, coastal erosion, and river exports. Long-term data are constantly threatened, particularly in the Arctic, where the costs of data collection in remote areas can be prohibitive.

4.2 Future perspectives: Improving the Arctic GHG budget

Models suggest that simulated plant growth may offset permafrost C losses until 2300 (e.g., McGuire et al., 2018). However, other research indicates that the remaining permissible carbon emissions under a maximum warming of 1.5°C (IPCC, 2018) may be 20% smaller when taking into account permafrost and wetland emissions by 2100 (Comyn-Platt et al., 2018). These inconsistencies across studies need further evaluation, incorporating newly available data streams, of key model processes that drive GHG budgets, including plant productivity under [CO_2] fertilization and CO_2 emissions during fall and winter. And, perhaps most essential, the models are not structured to capture "abrupt thaw," leading to even greater uncertainty.

Many studies focus on understanding changes in carbon cycling under warming arctic temperatures, but it is also important to consider changes in precipitation in the Arctic, where strong increases in rainfall are expected (Bintanja & Andry, 2017). This will mobilize additional organic matter and sediments into coastal nearshore zones (Coch et al., 2018). Moreover, recent evidence shows that rainfall may also interact with deep soil temperatures to control CH_4 emissions in thawing permafrost (Neumann et al., 2019), representing a modeling gap in permafrost environments. Changes in winter precipitation, including both increases and reductions in snowfall, may also play an important role in determining the sink or source strength of tundra ecosystems, but these dynamics have been recognized only recently (Blanc-Betes, Welker, Sturchio, Chanton, & Gonzalez-Meler, 2016; Lupascu et al., 2018). Snow cover depth on sea ice may also influence carbon cycling in arctic marine environments, but again these processes have also been investigated only recently (Søgaard, Deming, Meire, & Rysgaard, 2019).

The human footprint on the GHG budget is smaller in the Arctic compared to other regions. However, developmental impacts stemming from oil, gas, and mineral extraction have been escalating, thereby increasing this footprint

186 Section | C Case Studies

(Sidorstov, 2016) and shifting the Arctic to become a critical geopolitical issue (Brutschin & Schubert, 2016). Changes in shipping routes associated with the opening of the Northwest Passage due to declining sea ice may also impact the GHG budget of the region (Schröder, Reimer, & Jochmann, 2017). "Last chance tourism" in the Arctic promotes additional industrial activity (Veijola & Strauss-Mazzullo, 2019). In the future, it will become increasingly important to incorporate these developmental impacts into the Arctic GHG budget. Non-GHG feedbacks to warming in the Arctic are also critical to consider, including the albedo feedback from changes in outgoing radiation with reductions in sea ice and snow cover, which can have a strong influence over global warming (Cvijanovic & Caldeira, 2015; Duan, Cao, & Caldeira, 2019; Vavrus, 2007). Both the permafrost carbon and snow and ice albedo feedbacks are expected to result in a large economic impact of climate change, with recent estimates ranging from $24.8 trillion to $66.9 trillion by 2300 (Yumashev, Hope, Schaefer, et al., 2019). Working toward a better understanding of Arctic GHG budgets, with increased spatial and temporal coverage of key measurements, not only provides key knowledge in assessing climate change impacts, but also provides a record of the staggering pace of arctic change.

Acknowledgments

Support for this work was through the National Science Foundation Arctic Observatory Network and the National Science Foundation Research, Synthesis, and Knowledge Transfer in a Changing Arctic: Science Support for the Study of Environmental Arctic Change (SEARCH). Additional support was provided from the Department of Energy through the Next-Generation Ecosystem Experiment (NGEE-Arctic).

References

Aas, K. S., Martin, L., Nitzbon, J., Langer, M., Boike, J., Lee, H., et al. (2019). Thaw processes in ice-rich permafrost landscapes represented with laterally coupled tiles in a land surface model. *The Cryosphere*, *13*(2), 591–609. https://doi.org/10.5194/tc-13-591-2019.

Abbott, B. W., & Jones, J. B. (2015). Permafrost collapse alters soil carbon stocks, respiration, CH4, and N2O in upland tundra. *Global Change Biology*, *21*, 4570–4587.

Andersen, C. G., Lara, M. J., Tweedie, C. E., & Lougheed, V. L. (2017). Rising plant-mediated methane emissions from arctic wetlands. *Global Change Biology*, *23*, 1128–1139. https://doi.org/10.1111/gcb.13469.

Arft, A. M., Walker, M. D., Gurevitch, J., Alatalo, J. M., Bret-Harte, M. S., Dale, M., et al. (1999). Response patterns of tundra plant species to experimental warming: A meta-analysis of the International Tundra Experiment. *Ecological Monographs*, *69*, 491–511.

Arrigo, K. R., & van Dijken, G. L. (2011). Secular trends in Arctic Ocean net primary production. *Journal of Geophysical Research*, *116*, C09011. https://doi.org/10.1029/2011JC007151.

Arrigo, K. R., & van Dijken, G. L. (2015). Continued increases in Arctic Ocean primary production. *Progress in Oceanography*, *136*, 60–70. https://doi.org/10.1016/j.pocean.2015.05.002.

Babin, M., Bélanger, S., Ellinsten, I., Forest, A., Le Fouest, V., Lacour, T., et al. (2015). Estimation of primary production in the Arctic Ocean using ocean colour remote sensing and coupled physical-biological models: Strengths, limitations and how they compare. *Progress in Oceanography*, *139*, 197–220. https://doi.org/10.1016/j.pocean.2015.08.008.

Barlow, J. M., Palmer, P. I., Bruhwiler, L. M., & Tans, P. (2015). Analysis of CO_2 mole fraction data: First evidence of large-scale changes in CO_2 uptake at high northern latitudes. *Atmospheric Chemistry and Physics*, *15*, 13739–13758. https://doi.org/10.5194/acp-15-13739-2015.

Bayer, T. K., Gustafsson, E., Brakebusch, M., & Beer, C. (2019). Future carbon emission from boreal and permafrost lakes are sensitive to atmospheric carbon concentrations and catchment organic carbon loads. *Journal of Geophysical Research: Biogeosciences*. https://doi.org/10.1029/2018JG004978.

Belshe, E. F., Schuur, E. A. G., & Bolker, B. M. (2013). Tundra ecosystems observed to be CO_2 sources due to differential amplification of the carbon cycle. *Ecology Letters*, *16*, 1307–1315.

Berchet, A., Bousquet, P., Pison, I., Locatelli, R., Chevallier, F., Paris, J.-D., et al. (2016). Atmospheric constraints on the methane emissions from the East Siberian Shelf. *Atmospheric Chemistry and Physics*, *16*, 4147–4157. https://doi.org/10.5194/acp-16-4147-2016.

Bhatt, U. S., Walker, D. A., Raynolds, M. K., Bienick, P. A., Epstein, H. E., Comiso, J. C., et al. (2013). Recent declines in warming and vegetation greening trends over pan-Arctic tundra. *Remote Sensing*, *5*, 4229–4254.

Bhatt, U. S., Walker, D., Raynolds, R., Bieniek, P., Epstein, H., Comiso, J., et al. (2017). Changing seasonality of panarctic tundra vegetation in relationship to climatic variables. *Environmental Research Letters*, *12*, 055003.

Billings, W. D., Peterson, K. M., Shaver, G. R., & Trent, A. W. (1977). Root growth, respiration, and carbon dioxide evolution in an arctic tundra soil. *Arctic and Alpine Research*, *9*, 129–137. https://doi.org/10.1080/00040851.1977.12003908.

Bintanja, R., & Andry, O. (2017). Towards a rain-dominated Arctic. *Nature Climate Change*, *7*, 263–268.

Biskaborn, B. K., Smith, S. L., Noetzli, J., Matthes, H., Vieira, G., Streletskiy, D. A., et al. (2019). Permafrost is warming at a global scale. *Nature Communications*, *10*, 264. https://doi.org/10.1038/s41467-018-08240-4.

Bjorkman, A. D., Myers-Smith, I. H., Elmendorf, S. C., Normand, S., Rüger, N., et al. (2018). Plant functional trait change across a warming tundra biome. *Nature*, *562*, 57–62.

Blanc-Betes, E., Welker, J. M., Sturchio, N. C., Chanton, J. P., & Gonzalez-Meler, M. A. (2016). Winter precipitation and snow accumulation drive the methane sink or source strength of Arctic tussock tundra. *Global Change Biology*, *2016*. https://doi.org/10.1111/gcb.13242.

Bliss, L. C., Courtin, G. M., Pattie, D. L., Riewe, R. R., Whitfield, D. W. A., & Widden, P. (1973). Arctic tundra ecosystems. *Annual Review of Ecology and Systematics*, *4*, 359–399.

Blomquist, B. W., Huebert, B. J., Fairall, C. W., et al. (2014). Advances in air–sea CO_2 flux measurement by Eddy correlation. *Boundary-Layer Meteorology*, *152*, 245. https://doi.org/10.1007/s10546-014-9926-2.

Bogard, M. J., Kuhn, C. D., Johnston, S. E., Striegl, R. G., Holtgrieve, G. W., Dornblaser, M. M., et al. (2019). Negligible cycling of terrestrial carbon in many lakes of the arid circumpolar landscape. *Nature Geoscience*, *12*, 180–185.

Bokhorst, S., Pedersen, S. H., Brucker, L., Anisimov, O., Bjerke, J. W., Brown, R. D., et al. (2016). Changing arctic snow cover: A review of recent developments and assessment of future needs for observations, modelling, and impacts. *Ambio*, *45*, 516–537. https://doi.org/10.1007/s13280-016-0770-0.

Boswell, R., & Collett, T. S. (2011). Current perspectives on gas hydrate resources. *Energy and Environmental Science*, *4*, 1206–1215.

188 Section | C Case Studies

Box, J. E., Colgan, W. T., Christensen, T. R., Schmidt, N. M., Lund, M., Parmentier, F.-J. W., et al. (2019). Key indicators of Arctic climate change: 1971–2017. *Environmental Research Letters, 14*. https://doi.org/10.1088/1748-9326/aafc1b.

Brach, R., Natali, S., Pegoraro, E., Crummer, K. G., Schadel, C., Celis, G., et al. (2016). Temperature sensitivity of organic matter decomposition of permafrost-region soils during laboratory incubations. *Soil Biology and Biochemistry, 97*, 1–14.

Bradley-Cook, J. I., Petrenko, C. L., Friedland, A. J., & Virginia, R. A. (2016). Temperature sensitivity of mineral soil carbon decomposition in shrub and graminoid tundra, west Greenland. *Climate Change Responses*. https://doi.org/10.1186/s40665-016-0016-1.

Bring, A., Fedorova, I., Dibike, Y., Hinzman, L., Mård, J., Mernild, S. H., et al. (2016). Arctic terrestrial hydrology: A synthesis of processes, regional effects, and research challenges. *Journal of Geophysical Research: Biogeosciences, 121*, 621–649. https://doi.org/10.1002/2015JG003131.

Brown, J., Ferrians, O. J., Jr., Heginbottom, J. A., & Melnikov, E. S. (1997). *Circum-Arctic map of permafrost and ground-ice conditions*. Reston: US Geological Survey.

Brutschin, E., & Schubert, S. R. (2016). Icy waters, hot tempers, and high stakes: Geopolitics and geoeconomics of the Arctic. *Energy Research and Social Science, 16*, 147–159.

Burba, G. G., McDermitt, D. K., Grelle, A., Anderson, D. J., & Xu, L. (2008). Addressing the influence of instrument surface heat exchange on the measurements of CO2 flux from open-path gas analyzers. *Global Change Biology, 14*, 1–23.

Burke, E. J., Ekici, A., Huang, Y., Chadburn, S. E., Huntingford, C., Ciais, P., et al. (2017). Quantifying uncertainties of permafrost carbon-climate feedbacks. *Biogeosciences, 14*, 3051–3066.

Cai, W. J., Chen, L. Q., Chen, B. S., Gao, Z. Y., Lee, S. H., Chen, J. F., et al. (2010). Decrease in the CO2 uptake capacity in an ice-free Arctic Ocean Basin. *Science, 329*, 556–559. https://doi.org/10.1126/science.1189338.

Castro-Morales, K., Kleinen, T., Kaiser, S., Zaehle, S., Kittler, F., Kwon, M. J., et al. (2018). Year-round simulated methane emissions from a permafrost ecosystem in Northeast Siberia. *Biogeosciences, 15*, 2691–2722. https://doi.org/10.5194/bg-15-2691-2018.

Celis, G., Mauritz, M., Bracho, R., Salmon, V. G., Webb, E. E., Hutchings, J., et al. (2017). Tundra is a consistent source of CO2 at a site with progressive permafrost thaw during six years of chamber and eddy covariance measurements. *Journal of Geophysical Research: Biogeosciences, 122*, 1471–1485. https://doi.org/10.1002/2016JG003671.

Chapin, F. S., III, Strum, M., Serreze, M. C., McFadden, J. P., Key, J. R., Lloyd, A. H., et al. (2005). Role of land-surface changes in arctic summer warming. *Science, 310*, 657–660.

Chevallier, F. (2018). *Validation report for the inverted CO2 fluxes*. v18r1-version 1.0, ECMWF Copernicus Report Copernicus Atmosphere Monitoring Service. https://atmosphere. copernicus.eu/sites/default/files/2019-01/CAMS73_2018SC1_D73.1.4.1-2017-v0_201812_ v1_final.pdf. https://apps.ecmwf.int/datasets/data/cams-ghg-inversions/.

Christensen, T. R., Arora, V. K., Gauss, M., Höglund, I. L., & Parmentier, F.-J. W. (2019). Tracing the climate signal: Mitigation of anthropogenic methane emissions can outweigh a large Arctic natural emission increase. *Scientific Reports, 9*, e1146. https://doi.org/10.1038/s41598-018-37719-9.

Coch, C., Lamoureux, S., Knoblauch, C., Eischeid, I., Fritz, M., Obu, J., et al. (2018). Summer rainfall dissolved organic carbon, solute, and sediment fluxes in a small Arctic coastal catchment on Herschel Island (Yukon Territory, Canada). *Arctic Science, 4*, 750–780. https://doi.org/10.1139/as-2018-0010.

Cole, J. J., & Caraco, N. F. (1998). Atmospheric exchange of carbon dioxide in a low-wind oligotrophic lake measured by the addition of SF_6. *Limnology and Oceanography, 43*, 647–656.

Commane, R., Lindaas, J., Benmergui, J., Luus, K., Chang, R., Daube, B., et al. (2017). Carbon dioxide sources from Alaska driven by increasing early winter respiration from Arctic tundra. *PNAS, 114*, 5361–5366. https://doi.org/10.1073/pnas.1618567114.

Comyn-Platt, E., Hayman, G., Huntingford, C., Chadburn, S. E., Burke, E. J., Harper, A. B., et al. (2018). Carbon budgets for 1.5 and 2 °C targets lowered by natural wetland and permafrost feedbacks. *Nature Geoscience, 11*, 568–573.

Couture, N. J., Irrgang, A., Pollard, W., Lantuit, H., & Fritz, M. (2018). Coastal erosion of permafrost soils along the Yukon Coastal Plain and fluxes of organic carbon to the Canadian Beaufort Sea. *Journal of Geophysical Research: Biogeosciences, 123*, 406–422. https://doi.org/10.1002/2017JG004166.

Cvijanovic, I., & Caldeira, K. (2015). Atmospheric impacts of sea ice decline in CO_2 induced global warming. *Climate Dynamics, 44*, 1173–1186.

Denfeld, B. A., Frey, K. E., Sobczak, W. V., Mann, P. J., & Holmes, R. M. (2013). Summer CO_2 evasion from streams and rivers in the Kolyma River basin, northeast Siberia. *Polar Research, 32*, 1–15. https://doi.org/10.3402/polar.v32i0.19704.

Dlugokencky, E. J., Bruhwiler, L., White, J. W. C., Emmonds, L. K., Novelli, P. C., Montzka, S. A., et al. (2009). Observational constraints on recent increases in the atmospheric CH4 burden. *Geophysical Research Letters, 36*, L18803. https://doi.org/10.1029/2009GL039780.

Dlugokencky, E. J., Myers, R. C., Lang, P. M., Masarie, K. A., Crotwell, A. M., Thoning, K. W., et al. (2005). Conversion of NOAA atmospheric dry air CH4 mole fractions to a gravimetrically prepared standard scale. *Journal of Geophysical Research, 110*. https://doi.org/10.1029/2005JD006035.

Duan, L., Cao, L., & Caldeira, K. (2019). Estimating contributions of sea ice and land snow to climate feedback. *Journal of Geophysical Research: Atmospheres, 124*, 199–208. https://doi.org/10.1029/2018JD029093.

Dutta, K., Schuur, E. A. G., Neff, J. C., & Zimov, S. A. (2006). Potential carbon release from permafrost soils of Northeastern Siberia. *Global Change Biology, 12*, 2336–2351. https://doi.org/10.1111/j.1365-2486.2006.01259.x.

Elder, C. D., Schweiger, M., Lam, B., Crook, E. D., Xu, X., Walker, J., et al. (2018). Greenhouse gas emissions from diverse Arctic Alaskan lakes are dominated by young carbon. *Nature Climate Change, 8*, 166–171.

Enting, I. (2018). Estimation and inversion across the spectrum of carbon cycle modelling. *AIMS Geosciences, 4*, 126–143.

Erkkilä, K.-M., Ojala, A., Bastviken, D., Biermann, T., Heiskanen, J. J., Lindroth, A., et al. (2018). Methane and carbon dioxide fluxes over a lake: Comparison between eddy covariance, floating chambers and boundary layer method. *Biogeosciences, 15*, 429–445. https://doi.org/10.5194/bg-15-429-2018.

Euskirchen, E. S., Bret-Harte, M. S., Edgar, C., Scott, G. J., & Shaver, G. R. (2012). Seasonal patterns of carbon and water fluxes in three representative ecosystems in the northern foothills of the Brooks Range, Alaska. *Ecosphere, 3*, 1–19.

Euskirchen, E. S., Bret-Harte, M. S., Shaver, G. R., Edgar, C. W., & Romanovsky, V. E. (2017). Long-term release of carbon dioxide from arctic tundra ecosystems in northern Alaska. *Ecosystems, 20*, 960–974. https://doi.org/10.1007/s10021-016-0085-9.

Euskirchen, E. S., McGuire, A. D., Kickligher, D. W., Zhuang, Q., Clein, J. S., Dargaville, R. J., et al. (2006). Importance of recent shifts in soil thermal dynamics on growing season length, productivity, and carbon sequestration in terrestrial high-latitude ecosystems. *Global Change Biology, 12*, 731–750.

190 Section | C Case Studies

Fagan, J. D., & Nelson, F. E. (2016). Spatial sampling design in the circumpolar active layer monitoring programme. *Permafrost and Periglacial Processes, 28*, 42–51. https://doi.org/10.1002/ppp.1904.

Fan, S. M., Wofsy, S. C., Bakwin, P. S., Jacob, D. J., Anderson, S. M., Kebabian, P. L., et al. (1992). Micrometeorological measurements of CH_4 and CO_2 exchange between the atmosphere and subarctic tundra. *Journal of Geophysical Research, 97*, 16627. https://doi.org/10.1029/91JD02531.

Fang, C., & Moncrieff, J. B. (1996). An open-top chamber for measuring soil respiration and the influence of pressure difference on CO_2 efflux measurement. *Functional Ecology, 12*, 319–325.

Finger Higgens, R. A., Chipman, J. W., Lutz, D. A., Culler, L. E., Virginia, R. A., & Ogden, L. A. (2019). Changing lake dynamics indicate a drier Arctic in western Greenland. *Journal of Geophysical Research: Biogeosciences, 124*, 870–883. https://doi.org/10.1029/2018JG004879.

Fisher, J., Hayes, D., Schwalm, C. R., Huntzinger, D., Stofferahn, E., Schaefer, K., et al. (2018). Missing pieces to modeling the Arctic-Boreal puzzle. *Environmental Research Letters, 13*, 020202. https://doi.org/10.1088/1748-9326/aa9d9a.

Fisher, J. B., Sikka, M., Oechel, W. C., Huntzinger, D. N., Melton, J. R., Koven, C. D., et al. (2014). Carbon cycle uncertainty in the Alaskan Arctic. *Biogeosciences, 11*, 4271–4288. https://doi.org/10.5194/bg-11-4271-2014.

Forkel, M., Carvalhais, N., Rödenbeck, C., Keeling, R. F., Heimann, M., Thonicke, K., et al. (2016). Enhanced seasonal CO_2 exchange caused by amplified plant productivity in northern ecosystems. *Science, 6274*, 696–699.

Gioli, B., Miglietta, F., De Martino, B., Hutjes, R. W. A., Dolman, H. A. J., Lindroth, A., et al. (2004). Comparison between tower and aircraft-based eddy covariance fluxes in five European regions. *Agricultural and Forest Meteorology, 127*, 1–16.

Graven, H. D., Keeling, R. F., Piper, S. C., Patra, P. K., Stephens, B. B., et al. (2013). Enhanced seasonal exchange of CO2 by northern ecosystems since 1960. *Science, 341*, 1085–1089.

Grogan, P. (2012). Cold season respiration across a low arctic landscape: The influence of vegetation type, snow depth, and interannual climatic variation. *Arctic, Antarctic, and Alpine Research, 44*, 446–456.

Grosse, G., et al. (2011). Vulnerability of high-latitude soil organic carbon in North America to disturbance. *Journal of Geophysical Research, 116*, G00K06. https://doi.org/10.1029/2010JG001507.

Gulledge, J., & Schimel, J. P. (1998). Moisture control over atmospheric CH_4 consumption and CO_2 production in diverse Alaskan soils. *Soil Biology and Biochemistry, 30*, 1127–1132.

Günther, F., Overduin, P. P., Sandakov, A. V., Grosse, G., & Grigoriev, M. N. (2013). Short- and long-term thermo-erosion of ice-rich permafrost coasts in the Laptev Sea region. *Biogeosciences, 10*, 4297–4318. https://doi.org/10.5194/bg-10-4297-2013.

Harden, J. W., O'Donnell, J. A., Heckman, K. A., Sulman, B. J., Koven, C. D., Ping, C.-L., et al. (2019). Beneath the arctic greening: Will soils loose or gain carbon or perhaps a little of both? *Soil Discussions.* https://doi.org/10.5194/soil-2018-41.

Hargreves, K. J., Fowler, D., Pitcairn, C. E. R., & Aurela, M. (2001). Annual methane emission from Finnish mires estimated from eddy covariance campaign measurements. *Theoretical and Applied Climatology, 70*, 203–213.

Harriss, R. C., Sachse, G. W., Hill, G. F., Wade, L., Bartlett, K. B., Collins, J. E., et al. (1992). Carbon monoxide and methane in the North American arctic and subarctic troposphere: July–August 1988. *Journal of Geophysical Research – Atmospheres, 97*, 16589–16599.

Harriss, R. C., Wofsy, S., Bartlett, D. S., Shipham, M. C., Jacob, D. J., Hoell, J. M., et al. (1992). The Arctic Boundary Layer Expedition (ABLE 3A): July–August 1988. *Journal of Geophysical Research: Atmospheres, 97*(D15), 16383–16394. https://doi.org/10.1029/91JD02109.

Arctic greenhouse gas budgets **Chapter | 5** **191**

Hayes, D. J., McGuire, A. D., Kicklighter, D. W., Gurney, K. R., Burnside, T. J., & Melillo, J. M. (2011). Is the northern high-latitude land-based CO2 sink weakening? *Global Biogeochemical Cycles, 25*, GB3018. https://doi.org/10.1029/2010GB003813.

Hill, V., Ardyna, M., Lee, S. H., & Varela, D. E. (2018). Decadal trends in phytoplankton production in the Pacific Arctic Region from 1950 to 2012. *Deep-Sea Research Part II, 152*, 82–94. https://doi.org/10.1016/j.dsr2.2016.12.015.

Hood, E., Battin, T. J., Fellman, J., O'Neel, S., & Spencer, R. G. M. (2015). Storage and release of organic carbon from glaciers and ice sheets. *Nature Geoscience, 8*, 91–96. https://doi.org/10.1038/ngeo2331.

Hu, F. S., Higuera, P. E., Duffy, P., Chipman, M. L., Rocha, A. V., Young, A. M., et al. (2015). Arctic tundra fires: Natural variability and responses to climate change. *Frontiers in Ecology and the Environment, 13*, 369–377. https://doi.org/10.1890/150063.

Hugelius, G., Strauss, J., Zubrzycki, S., Harden, J. W., Schuur, E. A. G., Ping, C.-L., et al. (2014). Estimated stocks of circumpolar permafrost carbon with quantified uncertainty ranges and identified data gaps. *Biogeosciences, 11*, 6573–6593. https://doi.org/10.5194/bg-11-6573-2014.

IPCC. (2018). *Global warming of 1.5 °C: An IPCC special report on the impacts of global warming of 1.5 °C above pre-industrial levels and related global greenhouse gas emission pathways, in the context of strengthening the global response to the threat of climate change, sustainable development, and efforts to eradicate poverty.* Summary for Policymakers (IPCC SR1.5). Released October 6, 2018.

Jammet, M., Crill, P., Dengel, S., & Friborg, T. (2015). Large methane emissions from a subarctic lake during spring thaw: Mechanisms and landscape significance. *Journal of Geophysical Research: Biogeosciences, 120*, 2289–2305. https://doi.org/10.1002/2015JG003137.

Jin, M., Popova, E. E., Zhang, J., Ji, R., Pendleton, D., Varpe, Ø., et al. (2016). Ecosystem model intercomparison of under-ice and total primary production in the Arctic Ocean. *Journal of Geophysical Research: Oceans, 121*, 934–948. https://doi.org/10.1002/2015JC011183.

Johnson, P. L., & Kelley, J. J., Jr. (1970). Dynamics of carbon dioxide and productivity in an Arctic biosphere. *Ecology, 51*, 73–80.

Jørgensen, C. J., Johansen, K. M. L., Westergaard-Nielsen, A., & Elberling, B. (2015). Net regional methane sink in High Arctic soils of northeast Greenland. *Nature Geoscience, 8*, 20–23.

Jorgenson, M. T., Shur, Y. L., & Pullman, E. R. (2006). Abrupt increase in permafrost degradation in Arctic Alaska. *Geophysical Research Letters, 33*, L02503.

Kanevskiy, M., Shur, Y., Fortier, D., Jorgenson, M. T., & Stephani, E. (2011). Cryostratigraphy of late Pleistocene syngenetic permafrost (yedoma) in northern Alaska, Itkillik River exposure. *Quaternary Research, 75*, 584–596. https://doi.org/10.1016/j.yqres.2010.12.003.

Keenan, T. F., & Riley, W. J. (2018). Greening of the land surface in the world's cold regions consistent with recent warming. *Nature Climate Change, 8*, 825–828.

Kim, Y., Park, S.-J., Lee, B.-Y., & Risk, D. (2016). Continuous measurement of soil carbon efflux with Forced Diffusion (FD) chambers in a tundra ecosystem of Alaska. *Science of the Total Environment, 566–567*, 175–184. https://doi.org/10.1016/j.scitotenv.2016.05.052.

Kitidis, V., Upstill-Goddard, R. C., & Anderson, L. G. (2010). Methane and nitrous oxide in surface water along the North-West Passage, Arctic Ocean. *Marine Chemistry, 121*, 80–86. https://doi.org/10.1016/j.marchem.2010.03.006.

Kittler, F., Eugster, W., Foken, T., Heimann, M., Kolle, O., & Gockede, M. (2017). High quality eddy-covariance CO_2 budgets under cold climate conditions. *Journal of Geophysical Research – Biogeosciences, 122*, 2064–2084. https://doi.org/10.1002/2017JG003830.

Kling, G. W., Kipphut, G. W., & Miller, M. C. (1991). Arctic lakes and streams as gas conduits to the atmosphere: Implications for tundra carbon budgets. *Science, 251*, 298–301.

192 Section | C Case Studies

Knoblauch, C., Beer, C., Sosnin, A., Wagner, D., & Pfeiffer, E.-M. (2013). Predicting long-term carbon mineralization and trace gas production from thawing permafrost of Northeast Siberia. *Global Change Biology, 19*, 1160–1172. https://doi.org/10.1111/gcb.12116.

Kohnert, K., Serafimovich, A., Metzger, S., Hartmann, J., & Sachs, T. (2017). Strong geologic methane emissions from discontinuous terrestrial permafrost in the Mackenzie Delta, Canada. *Scientific Reports, 7*. https://doi.org/10.1038/s41598-017-05783-2.

Kort, E. A., Wofsy, S. C., Daube, B. C., Diao, M., Elkins, J. W., Gao, R. S., et al. (2012). Atmospheric observations of Arctic Ocean methane emissions up to 82° north. *Nature Geoscience, 5*, 318–321.

Koven, C. D., Lawrence, D. M., & Riley, W. J. (2015). Permafrost carbon–climate feedback is sensitive to deep soil carbon decomposability but not deep soil nitrogen dynamics. *Proceedings of the National Academy of Sciences, 112*, 3752–3757.

Koven, C. D., Riley, W. J., & Stern, A. (2013). Analysis of permafrost thermal dynamics and response to climate change in the CMIP5 earth system models. *Journal of Climate, 26*, 1877–1900.

Koven, C. D., Ringeval, B., Friedlingstein, P., Ciais, P., Cadule, P., Khvorostyanov, D., et al. (2011). Permafrost carbon-climate feedbacks accelerate global warming. *Proceedings of the National Academy of Sciences, 108*, 14769–14774.

LaFleur, P. M., & Humpheys, E. R. (2007). Spring warming and carbon dioxide exchange over low Arctic tundra in central Canada. *Global Change Biology, 14*, 740–756. https://doi.org/10.1111/j.1365-2486.2007.01529.x.

Langford, Z., Kumar, J., Hoffman, F. M., Norby, R. J., Wullschleger, S. D., Sloan, V. L., et al. (2016). Mapping Arctic plant functional type distributions in the Barrow Environmental Observatory using WorldView-2 and LiDAR Datasets. *Remote Sensing, 8*, 733. https://doi.org/10.3390/rs8090733.

Lara, M. J., Nitze, I., Grosse, G., Martin, P., & McGuire, A. D. (2018). Reduced arctic tundra productivity linked with landform and climate change interactions. *Scientific Reports, 8*, 2345. https://doi.org/10.1038/s41598-018-20692-8.

Lawrence, D. M., Koven, C. D., Swenson, S. C., Riley, W. J., & Slater, A. G. (2015). Permafrost thaw and resulting soil moisture changes regulate projected high-latitude CO2 and CH4 emissions. *Environmental Research Letters, 10*, 94011.

Lawrence, D. M., & Slater, A. G. (2005). A projection of severe near-surface permafrost degradation during the 21st century. *Geophysical Research Letters, 32*, L24401. https://doi.org/10.1029/2005GL025080.

Lawrence, D. M., Slater, A. G., Romanovsky, V. E., & Nicolsky, D. J. (2008). Sensitivity of a model projection of near-surface permafrost degradation to soil column depth and representation of soil organic matter. *Journal of Geophysical Research - Earth Surface, 113*, F02011.

Lawson, E. C., Wadham, J. L., Tranter, M., Stibal, M., Lis, G. P., Butler, C. E. H., et al. (2014). Greenland Ice Sheet exports labile organic carbon to the Arctic oceans. *Biogeosciences, 11*, 4015–4028. https://doi.org/10.5194/bg-11-4015-2014.

Lee, Y. J., Matrai, P. A., Friedrichs, M. A. M., Saba, V. S., Antoine, D., Ardyna, M., et al. (2015). An assessment of phytoplankton primary productivity in the Arctic Ocean from satellite ocean color/in situ chlorophyll-a based models. *Journal of Geophysical Research, Oceans, 120*, 6508–6541. https://doi.org/10.1002/2015JC011018.

Li Yung Lung, J. Y. S., Tank, S. E., Spence, C., Yang, D., Bonsal, B., McClelland, J. W., et al. (2018). Seasonal and geographic variation in dissolved carbon biogeochemistry of rivers draining to the Canadian Arctic Ocean and Hudson Bay. *Journal of Geophysical Research: Biogeosciences, 123*, 3371–3386.

Liljedahl, A. K., Boike, J., Daanen, R. P., et al. (2016). Pan-Arctic ice-wedge degradation in warming permafrost and its influence on tundra hydrology. *Nature Geoscience*, *9*, 312–318.

Luers, J., Westermann, S., Piel, K., & Boike, J. (2014). Annual CO_2 budget and seasonal CO_2 exchange signals at a high Arctic permafrost site on Spitsbergen, Svalbard archipelago. *Biogeosciences*, *11*, 6307–6322.

Lupascu, M., Czimczik, C. I., Welker, M. C., Ziolkowski, L. A., Cooper, E. J., & Welker, J. M. (2018). Winter ecosystem respiration and sources of CO2 from the High Arctic tundra of Svalbard: Response to a deeper snow experiment. *Journal of Geophysical Research: Biogeosciences*, *123*, 2627–2642. https://doi.org/10.1029/2018JG004396.

MacGilchrist, G. A., Naveira Garabato, A. C., Tsubouchi, T., Bacon, S., Torres-Valdés, S., & Azetsu-Scott, K. (2014). The Arctic Ocean carbon sink. *Deep-Sea Research I*, *86*, 39–55.

MacIntyre, S., Jonsson, A., Jansson, M., Aberg, J., Turney, D. E., & Miller, S. D. (2010). Buoyancy flux, turbulence, and the gas transfer coefficient in a stratified lake. *Geophysical Research Letters*, *37*, L24604. https://doi.org/10.1029/2010GL044164.

Mack, M. C., Bret-Harte, M. S., Hollingsworth, T. N., Jandt, R. R., Schuur, E. A. G., Shaver, G. R., et al. (2011). Carbon loss from an unprecedented Arctic tundra wildfire. *Nature*, *475*, 489–492.

Mammarella, I., Nordbo, A., Rannik, Ü., Haapanala, S., Levula, J., Laakso, H., et al. (2015). Carbon dioxide and energy fluxes over a small boreal lake in Southern Finland. *Journal of Geophysical Research – Biogeosciences*, *120*, 1296–1314. https://doi.org/10.1002/2014JG002873.

Marushchak, M. E., Pitkamaki, A., Koponen, H., Biasi, C., Seppala, M., & Martikainen, P. J. (2011). Hot spots for nitrous oxide emissions found in different types of permafrost peatlands. *Global Change Biology*, *17*, 2601–2614.

Mastepanov, M., Sigsgaard, C., Dlugokencky, E. J., Houweling, S., Ström, L., Tamstorf, M. P., et al. (2008). Large tundra methane burst during onset of freezing. *Nature*, *456*, 628–631. https://doi.org/10.1038/nature07464.

Mauritz, M., Bracho, R., Celis, G., Hutchings, J., Natali, S. M., Pegoraro, E., et al. (2017). Nonlinear CO2 flux response to seven years of experimentally induced permafrost thaw. *Global Change Biology*, *23*, 3646–3666. https://doi.org/10.1111/gcb.13661.

McGuire, A. D., Anderson, L. G., Christensen, T. R., Dallimore, S., Guo, L., Hayes, D. J., et al. (2009). Sensitivity of the carbon cycle in the Arctic to climate change. *Ecological Monographs*, *79*, 523–555.

McGuire, A. D., Christensen, T. R., Hayes, D. J., Heroult, A., Euskirchen, E., Yi, Y., et al. (2012). An assessment of the carbon balance of the Arctic tundra: Comparisons among observations, process models, and atmospheric inversions. *Biogeosciences*, *9*, 3185–3204.

McGuire, A. D., Koven, C., Lawrence, D. M., Clein, J. S., Xia, J., Beer, C., et al. (2016). Variability in the sensitivity among model simulations of permafrost and carbon dynamics in the permafrost region between 1960 and 2009. *Global Biogeochemical Cycles*, *30*(7), 1015–1037. https://doi.org/10.1002/2016GB005405.

McGuire, A. D., Lawrence, D. M., Koven, C., Clein, J. S., Burke, E., Chen, G., et al. (2018). Dependence of the evolution of carbon dynamics in the northern permafrost region on the trajectory of climate change. *Proceedings of the National Academy of Sciences*, *115*(15). https://doi.org/10.1073/pnas.1719903115.

McGuire, A. D., Melillo, J. M., Kicklighter, D. W., Pan, Y., Xiao, X., Helfrich, J., et al. (1997a). Equilibrium responses of global net primary production and carbon storage to doubled atmospheric carbon dioxide: Sensitivity to changes in vegetation nitrogen concentration. *Global Biogeochemical Cycle*, *11*, 173–189. https://doi.org/10.1029/97GB00059.

McGuire, A. D., Melillo, J. M., Kicklighter, D. W., Pan, Y., Xiao, X., Helfrich, J., et al. (1997b). Sensitivity of the carbon cycle in the Arctic to climate change. *Ecological Monographs*, *79*(4), 523–555.

194 Section | C Case Studies

Mekonnen, Z. A., Riley, W. J., & Grant, R. F. (2018). 21st century tundra shrubification could enhance net carbon uptake of North America Arctic tundra under an RCP8.5 climate trajectory. *Environmental Research Letters, 13*(5), 054029. https://doi.org/10.1088/1748-9326/aabf28.

Melton, J. R., Wania, R., Hodson, E., Poulter, B., Ringeval, B., Spahni, R., et al. (2013). Present state of global wetland extent and wetland methane modelling: Conclusions from a model inter-comparison project (WETCHIMP). *Biogeosciences, 10*, 753–788.

Metcalfe, D. B., Hermans, T. D. G., Ahlstrand, J., Becker, M., Berggren, M., Björk, R. G., et al. (2018). Patchy field sampling biases understanding of climate change impacts across the Arctic. *Nature Ecology and Evolution, 2*(9), 1443–1448. https://doi.org/10.1038/s41559-018-0612-5.

Mikan, C. J., Schimel, J. P., & Doyle, A. P. (2002). Temperature controls of microbial respiration in arctic tundra soils above and below freezing. *Soil Biology & Biochemistry, 34*, 1785–1795.

Miller, S. M., Miller, C. E., Commane, R., Chang, R. Y.-W., Dinardo, S. J., Henderson, J. M., et al. (2016). A multiyear estimate of methane fluxes in Alaska from CARVE atmospheric observations. *Global Biogeochemical Cycles, 30*, 1441–1453. https://doi.org/10.1002/2016GB005419.

Myers-Smith, I. H., Forbes, B. C., Wilmking, M., Hallinger, M., Lanztz, T., Blok, D., et al. (2011). Shrub expansion in tundra ecosystems: Dynamics, impacts and research priorities. *Environment Research Letters, 6*, 45509.

Natali, S. M., Watts, J. D., Rogers, B. M., et al. (2019). Large loss of CO_2 in winter observed across the northern permafrost region. *Nature Climate Change, 9*, 852–857. https://doi.org/10.1038/s41558-019-0592-8.

Neilson, B. T., Cardenas, M. B., O'Connor, M. T., Rasmussen, M. T., King, T. V., & Kling, G. W. (2018). Groundwater flow and exchange across the land surface explain carbon export patterns in continuous permafrost watersheds. *Geophysical Research Letters, 45*, 7596–7605. https://doi.org/10.1029/2018GL078140.

Neumann, R. B., Moorberg, C. J., Lundquist, J. D., Turner, J. C., Waldrop, M. P., McFarland, J. W., et al. (2019). Warming effects of spring rainfall increase methane emissions from thawing permafrost. *Geophysical Research Letters, 46*, 1393–1401. https://doi.org/10.1029/2018GL081274.

Nitze, I., Grosse, G., Jones, B. M., et al. (2018). Remote sensing quantifies widespread abundance of permafrost region disturbances across the Arctic and Subarctic. *Nature Communications, 9*, 5423. https://doi-org.wv-o-ursus-proxy02.ursus.maine.edu/10.1038/s41467-018-07663-3.

Obu, J., Westermann, S., Bartsch, A., Berdnikov, N., Christiansen, H. H., Dashtseren, A., et al. (2019). Northern Hemisphere permafrost map based on TTOP modelling for 2000-2016 at 1 km^2 scale. *Earth-Science Reviews, 193*, 299–316.

Oechel, W. C., Laskowski, C. A., Burba, G., Gioli, B., & Kalhori, A. A. M. (2014). Annual patterns and budget of CO_2 flux in an Arctic tussock tundra ecosystem. *Journal Geophysical Research – Biogeosciences, 119*, 323–339. https://doi.org/10.1002/2013JG002431.

Oechel, W. C., Vourlitis, G. L., Brooks, S. B., Crawford, T. L., & Dumas, E. J. (1998). Intercomparison between chamber, tower, and aircraft net CO2 exchange and energy fluxes measured during the Arctic system sciences land–atmosphere–ice interaction (ARCSS-LAII) flux study. *Journal of Geophysical Research, 103*, 28993–29003.

Oh, Y., Stackhouse, B., Lau, M. C. Y., Xu, X., Trugman, A. T., et al. (2016). A scalable model for methane consumption in arctic mineral soils. *Geophysical Research Letters, 43*, 5143–5150. https://doi.org/10.1002/2016GL069049.

Olefeldt, D., Goswami, S., Grosse, G., Hayes, D., Hugelius, G., Kuhry, P., et al. (2016). Circumpolar distribution and carbon storage of thermokarst landscapes. *Natute Communications, 7*, 13043. https://doi.org/10.1038/ncomms13043.

Parazoo, N. C., Koven, C. D., Lawrence, D. M., Romanovsky, V., & Miller, C. E. (2018). Detecting the permafrost carbon feedback: Talik formation and increased cold-season respiration as precursors to sink-to-source transitions. *The Cryosphere, 12*, 123–144. https://doi.org/10.5194/tc-12-123-2018.

Parmentier, F. J. W., Christensen, T. R., Rysgaard, S., Bendtsen, J., Glud, R. N., Else, B., et al. (2017). A synthesis of the arctic terrestrial and marine carbon cycles under pressure from a dwindling cryosphere. *Ambio, 46*, 53–69. https://doi.org/10.1007/s13280-016-0872-8.

Parmentier, F. J. W., Rasse, D. P., Lund, M., Bjerke, J. W., Drake, B. G., Weldon, S., et al. (2018). Vulnerability and resilience of the carbon exchange of a subarctic peatland to an extreme winter event. *Environmental Research Letters, 13*(6), 065009. https://doi.org/10.1088/1748-9326/aabff3.

Perovich, D., et al. (2018). *Sea ice* (in Arctic Report Card 2018) https://www.arctic.noaa.gov/Report-Card.

Peters, W., Jacobson, A. R., Sweeney, C., Andrews, A. E., Conway, T. J., Masarie, K., et al. (2007). An atmospheric perspective on North American carbon dioxide exchange: CarbonTracker. *Proceedings of the National Academy of Sciences, 104*, 18925–18930. https://doi.org/10.1073/pnas.0708986104.

Peterson, K., & Billings, W. (1975). Carbon dioxide flux from tundra soils and vegetation as related to temperature at Barrow, Alaska. *The American Midland Naturalist, 94*, 88–98. https://doi.org/10.2307/2424540.

Phoenix, G. K., & Bjerke, J. W. (2016). Arctic browning: Extreme events and trends reversing arctic greening. *Global Change Biology, 22*, 2960–2962.

Pirk, N., Mastepanov, M., López-Blanco, E., et al. (2017). Toward a statistical description of methane emissions from arctic wetlands. *Ambio, 46*(Suppl. 1), 70. https://doi.org/10.1007/s13280-016-0893-3.

Pirk, N., Santos, T., Gustafson, C., Johansson, A. J., Tufvesson, F., Parmentier, F. J. W., et al. (2015). Methane emission bursts from permafrost environments during autumn freeze-in: New insights from ground penetrating radar. *Geophysical Research Letters, 42*(6732), 6738. https://doi.org/10.1002/2015gl065034.

Pirk, N., Tamstorf, M. P., Lund, M., Mastepanov, M., Pedersen, S. H., Mylius, M. R., et al. (2016). Snowpack fluxes of methane and carbon dioxide from high Arctic tundra. *Journal Geophysical Research – Biogeosciences, 121*. https://doi.org/10.1002/2016JG003486.

Pumpanen, J., Kolari, P., Ilvesniemi, H., Minkkinen, K., Vesala, T., Niinistö, S., et al. (2004). Comparison of different chamber technique for measuring soil CO_2 flux. *Agricultural and Forest Meteorology, 123*, 159–176.

Pumpanen, J., Longdoz, B., & Kutsch, W. L. (2010). Field measurements of soil respiration: Principles and constraints, potentials and limitations of different methods. In *Soil carbon dynamics: An integrated methodology* (pp. 16–33). Cambridge University Press. https://doi.org/10.1017/CB09780511794.003.

Qiu, C., Zhu, D., Ciais, P., Guenet, B., Krinner, G., Peng, S., et al. (2018). ORCHIDEE-PEAT, a model for northern peatland CO2, water and energy fluxes on daily to annual scales. *Geoscientific Model Development, 11*, 497–519.

Randall, K., Scarratt, M., Levasseur, M., Michaud, S., Xie, H., & Gosselin, M. (2012). First measurements of nitrous oxide in Arctic sea ice. *Journal of Geophysical Research, 117*, C00G15. https://doi.org/10.1029/2011JC007340.

Raz-Yaseef, N., Torn, M. S., Wu, Y., et al. (2016). Large CO_2 and CH_4 emissions from polygonal tundra during spring thaw in northern Alaska. *Geophysical Research Letters, 44*, 504–513.

196 Section | C Case Studies

Reum, F., Göckede, M., Lavrič, J. V., Kolle, O., Zimov, S., Zimov, N., et al. (2018). Accurate measurements of atmospheric carbon dioxide and methane mole fractions at the Siberian coastal site Ambarchik. *Atmospheric Measurement Techniques Discussions*. https://doi.org/10.5194/amt-2018-325.

Risk, R., Nickerson, N., Creelman, C., McArthur, G., & Ownens, J. (2011). Forced Diffusion soil flux: A new technique for continuous monitoring of soil gas efflux. *Agricultural and Forest Meteorology, 151*, 1622–1631.

Rivkina, E. M., Friedmann, E. I., McKay, C. P., & Gilichinsky, D. A. (2000). Metabolic activity of permafrost bacteria below the freezing point. *Applied and Environmental Microbiology, 66*, 3230–3233.

Rocher-Ros, G., Giesler, R., Lundin, E., Salimi, S., Jonsson, A., & Karlsson, J. (2017). Large lakes dominate CO_2 evasion from lakes in an arctic catchment. *Geophysical Research Letters, 44*, 12,254–12,261. https://doi.org/10.1002/2017GL076146.

Rödenbeck, C. (2005). *Estimating CO_2 sources and sinks from atmospheric mixing ratio measurements using a global inversion of atmospheric transport*. Technical Report 6, Version v2.0 Jena: Max Planck Institute for Biogeochemistry. bgc-jena.mpg.de/CarboScope.

Rodionow, A., Flessa, H., Kazansky, O., & Guggenberger, G. (2006). Organic matter composition and potential trace gas production of permafrost soils in the forest tundra in northern Siberia. *Geoderma, 135*, 49–62.

Rogers, M. C., Sullivan, P. F., & Welker, J. M. (2011). Evidence of nonlinearity in the response of net ecosystem CO2 exchange to increasing levels of winter snow depth in the high arctic of northwest Greenland. *Arctic, Antarctic, and Alpine Research, 43*(1), 95–106. https://doi.org/10.1657/1938-4246-43.1.95.

Romanovskii, N. N. (1993). *Fundamentals of cryogenesis of lithosphere*. Moscow: Moscow University Press.

Romanovsky, V. E., Smith, S. L., & Christiansen, H. H. (2010). Permafrost thermal state in the polar Northern Hemisphere during the international polar year 2007–2009: A synthesis. *Permafrost and Periglacial Processes, 21*, 106–116.

Romanovsky, V. E., Smith, S. L., Christiansen, H. H., et al. (2013). *Permafrost [Arctic Report Card 2011]*. http://www.arctic.noaa.gov/report11/.

Saunois, M. (2020). Global methane budget 2000–2017. *Earth System Science Data, 12*, 1561–1623. https://doi.org/10.5194/essd-12-1561-2020.

Sayres, D. S., Dobosy, R., Healy, C., Dumas, E., Kochendorfer, J., Munster, J., et al. (2017). Arctic regional methane fluxes by ecotype as derived using eddy covariance from a low-flying aircraft. *Atmospheric Chemistry and Physics, 17*, 8619–8633. https://doi.org/10.5194/acp-17-8619-2017.

Schädel, C., Bader, M. K. F., Schuur, E. A. G., et al. (2016). Potential carbon emissions dominated by carbon dioxide from thawed permafrost soils. *Nature Climate Change, 6*, 950–953.

Schädel, C., Schuur, E. A. G., Bracho, R., et al. (2014). Circumpolar assessment of permafrost C quality and its vulnerability over time using long-term incubation data. *Global Change Biology, 20*, 641–652.

Schaefer, K., Zhang, T., Bruhwiler, L., & Barrett, A. P. (2011). Amount and timing of permafrost carbon release in response to climate warming. *Tellus B, 63*, 165–180.

Schimel, J. P., & Clein, J. S. (1996). Microbial response to freeze-thaw cycles in tundra and taiga soils. *Soil Biology and Biochemistry, 28*, 1061–1066.

Schimel, J. P., & Mikan, C. (2005). Changing microbial substrate use in Arctic tundra soils through a freeze-thaw cycle. *Soil Biology & Biochemistry, 37*, 1411–1418.

Schröder, C., Reimer, N., & Jochmann, P. (2017). Environmental impact of exhaust emissions by Arctic shipping. *Ambio, 46*, 400–409.

Schuback, N., Hope, C. J. M., Tremblay, J.-E., Maldonado, M. T., & Tortell, P. D. (2017). Primary productivity 2017. Primary productivity and the coupling of photosynthetic electron transport and carbon fixation in the Arctic Ocean. *Limnology and Oceanography, 62*, 898–921.

Schuster, U., McKinley, G. A., Bates, N., Chevallier, F., Doney, S. C., Fay, A. R., et al. (2013). An assessment of the Atlantic and Arctic sea–air CO2 fluxes, 1990–2009. *Biogeosciences, 10*, 607–627. https://doi.org/10.5194/bg-10-607-2013.

Schuur, E. A. G., & Mack, M. C. (2018). Ecological response to permafrost thaw and consequences for local and global ecosystem services. *Annual Review of Ecology, Evolution, and Systematics, 49*, 279–301.

Schuur, E. A. G., McGuire, A. D., Romanovsky, V., Schädel, C., & Mack, M. (2018). Arctic and boreal carbon. In N. Cavallaro, G. Shrestha, R. Birdsey, M. A. Mayes, R. G. Najjar, S. C. Reed, P. Romero-Lankao, & Z. Zhu (Eds.), *Second state of the carbon cycle report (SOCCR2): A sustained assessment report* (pp. 428–468). Washington, DC: U.S. Global Change Research Program. https://doi.org/10.7930/SOCCR2.2018.Ch11 (chapter 11).

Schuur, E. A. G., McGuire, A. D., Schädel, C., et al. (2015). Climate change and the permafrost carbon feedback. *Nature, 520*, 171–179.

Sepulveda-Jauregui, A., Walter Anthony, K. M., Martinez-Cruz, K., Greene, S., & Thalasso, F. (2015). Methane and carbon dioxide emissions from 40 lakes along a north–south latitudinal transect in Alaska. *Biogeosciences, 12*, 3197–3223. https://doi.org/10.5194/bg-12-3197-2015.

Shakhova, N., et al. (2010a). Geochemical and geophysical evidence of methane release over the East Siberian Arctic Shelf. *Journal of Geophysical Research, 115*, C08007. https://doi.org/10.1029/2009JC005602.

Shakhova, N., et al. (2010b). Extensive methane venting to the atmosphere from sediments of the East Siberian Arctic Shelf. *Science, 327*, 1246–1250. https://doi.org/10.1126/science.1182221.

Sidorstov, R. (2016). A perfect moment during imperfect times: Arctic energy research in a low-carbon era. *Energy Research & Social Science, 16*, 1–7.

Slater, A. G., & Lawrence, D. M. (2013). Diagnosing present and future permafrost from climate models. *Journal of Climate, 26*, 5608–5623.

Smith, L. C., Sheng, Y., MacDonald, G. M., & Hinzmann, L. D. (2005). Disappearing Arctic Lakes. *Science, 308*, 1429. https://doi.org/10.1126/science.1108142.

Soegaard, H., & Nordstroem, C. (1999). Carbon dioxide exchange in a high-arctic fen estimated by eddy covariance measurements and modelling. *Global Change Biology, 5*, 547–562.

Søgaard, D. H., Deming, J. W., Meire, L., & Rysgaard, S. (2019). Effects of microbial processes and $CaCO_3$ dynamics on inorganic carbon cycling in snow-covered Arctic winter sea ice. *Marine Ecology Progress Series, 61*, 31–44.

Sogaard, D. H., Thomas, D. N., Rysgaard, S., Glud, R. N., Norman, L., Kaartakallio, H., et al. (2013). The relative contributions of biological and abiotic processes to car-bon dynamics in subarctic sea-ice. *Polar Biology, 36*, 1761–1777.

Sparrow, K. J., Kessler, J. D., Southon, J. R., Garcia-Tigreros, F., Schreiner, K. M., Ruppel, C. D., et al. (2018). Limited contribution of ancient methane to surface waters of the U.S. Beaufort sea shelf. *Science Advances, 4*(1), eaao4842. https://doi.org/10.1126/sciadv.aao4842.

Stackpoole, S. M., Butman, D. E., Clow, D. W., Verdin, K. L., Gagliotti, B. V., Genet, H., et al. (2017). Inland waters and their role in the carbon cycle of Alaska. *Ecological Applications, 27*, 1403–1420.

Strauss, J., Schirrmeister, L., Grosse, G., Fortier, D., Hugelius, G., Knoblauch, C., et al. (2017). Deep Yedoma permafrost: A synthesis of depositional characteristics and carbon vulnerability. *Earth-Science Reviews*. https://doi.org/10.1016/j.earscirev.2017.07.007.

Sturm, M., Racine, C., & Tape, K. (2001). Climate change. Increasing shrub abundance in the Arctic. *Nature, 411*, 546–547.

Sweeney, C., Dlugokencky, E., Miller, C. E., Wofsy, S., Karion, A., Dinardo, S., et al. (2016). No significant increase in long-term CH4 emissions on North Slope of Alaska despite significant increase in air temperature. *Geophysical Research Letters, 43*(12), 6604–6611. https://doi.org/10.1002/2016GL069292.

Syakila, A., & Kroeze, C. (2011). The global nitrous oxide budget revisited. *Greenhouse Gas Measurement and Management, 1*, 17–26. https://doi.org/10.3763/ghgmm.2010.0007.

Tan, Z., Zhuang, Q., Shurpali, N. J., Marushchak, M. E., Biasi, C., Eugster, W., et al. (2017). Modeling CO_2 emissions from Arctic lakes: Model development and site-level study. *Journal of Advances in Modeling Earth Systems, 9*, 2190–2213. https://doi.org/10.1002/2017MS001028.

Tank, S. E., Striegl, R. G., Mcclelland, J. W., & Kokelj, S. V. (2016). Multi-decadal increases in dissolved organic carbon and alkalinity flux from the Mackenzie drainage basin to the Arctic Ocean. *Environmental Research Letters, 11*, 1–10. https://doi.org/10.1088/1748-9326/11/5/054015.

Tarantola, A. (1987). *Inverse problem theory methods for data fitting and model parameter estimation*. New York: Elsevier Sci.

Taylor, M. A., Celis, G., Ledman, J. D., Bracho, R., & Schuur, E. A. G. (2018). Methane efflux measured by Eddy covariance in Alaskan upland tundra undergoing permafrost degradation. *Journal of Geophysical Research: Biogeosciences, 123*, 2695–2710. https://doi.org/10.1029/2018JG004444.

Thompson, R. L., Sasakawa, M., Machida, T., Aalto, T., Worthy, D., Lavric, J. V., et al. (2017). Methane fluxes in the high northern latitudes for 2005–2013 estimated using a Bayesian atmospheric inversion. *Atmospheric Chemistry and Physics, 17*(5), P3553–P3572. https://doi.org/10.5194/acp-17-3553-2017.

Thornton, B. F., Prytherch, J., Andersson, K., Brooks, I. M., Salisbury, D., Tjernström, M., et al. (2020). Shipborne eddy covariance observations of methane fluxes constrain Arctic sea emissions. *Science Advances, 6*, eaay7934. https://doi.org/10.1126/sciadv.aay7934.

Tohjima, Y., Zeng, J., Shirai, T., Niwa, Y., Ishidoya, S., Taketani, F., et al. (2020). Estimation of CH4 emissions from the East Siberian Arctic Shelf based on atmospheric observations aboard the R/V Mirai during fall cruises from 2012 to 2017. *Polar Science*, 100571. https://doi.org/10.1016/j.polar.2020.100571.

Treharne, R., Bjerke, J. W., Tømmervik, H., Stendardi, L., & Phoenix, G. K. (2018). Arctic browning: Impacts of extreme climatic events on heathland ecosystem CO2 fluxes. *Global Change Biology*, 1–15. https://doi.org/10.1111/gcb.14500.

Tremblay, J.-É., Anderson, L. G., Matrai, P., Bélanger, S., Michel, C., Coupel, P., et al. (2015). Global and regional drivers of nutrient supply, primary production and CO_2 drawdown in the changing Arctic Ocean. *Progress in Oceanography, 139*, 171–196. https://doi.org/10.1016/j.pocean.2015.08.009.

Turetsky, M. R., Abbott, B. W., Jones, M. C., Anthony, K. W., Olefeldt, D., Schuur, E. A. G., et al. (2019). Permafrost collapse is accelerating carbon release. *Nature, 569*, 32.

Turetsky, M. R., Abbott, B. W., Jones, M. C., et al. (2020). Carbon release through abrupt permafrost thaw. *Nature Geoscience, 13*, 138–143. https://doi.org/10.1038/s41561-019-0526-0.

Ueyama, M., Ichii, K., Iwata, H., Euskirchen, E. S., Zona, D., Rocha, A. V., et al. (2013). Upscaling terrestrial carbon dioxide fluxes in Alaska with satellite remote sensing and support vector regression. *Journal of Geophysical Research – Biogeosciences, 118*, 1–16. https://doi.org/10.1002/jgrg.20095.

van der Laan-Luijkx, I. T., van der Velde, I. R., van der Veen, E., Tsuruta, A., Stanislawska, K., Babenhauserheide, A., et al. (2017). The Carbon Tracker Data Assimilation Shell (CTDAS) v1.0: Implementation and global carbon balance 2001–2015. *Geoscientific Model Development, 10*(7), 2785–2800. https://doi.org/10.5194/gmd-10-2785-2017.

Vavrus, S. (2007). The role of terrestrial snow cover in the climate system. *Climate Dynamics, 29*(1), 73–88. https://doi.org/10.1007/s00382-007-0226-0.

Veijola, S., & Strauss-Mazzullo, H. (2019). Tourism at the crossroads of contesting paradigms of Arctic development. In M. Finger, & L. Heininen (Eds.), *The globalArctic handbook*. Cham: Springer.

Virkkala, A.-M., Aalto, J., Rogers, B. M., et al. (2021). Statistical upscaling of ecosystem CO_2 fluxes across the terrestrial tundra and boreal domain: Regional patterns and uncertainties. *Global Change Biology, 2021*. https://doi.org/10.1111/gcb.15659.

Virkkala, A.-M., Virtanaen, T., Lehtonen, A., Rinne, J., & Luoto, M. (2018). The current state of CO_2 flux chamber studies in the Arctic tundra: A review. *Progress in Physical Geography, 42*, 162–184.

Voigt, C., Lamprecht, R. E., Marushchak, M. E., et al. (2016). Warming of subarctic tundra increases emissions of all three important greenhouse gases—Carbon dioxide, methane, and nitrous oxide. *Global Change Biology, 28*, 1365–2486.

Voigt, C., Marushchak, M. E., Lamprecht, R. E., Jackowicz-Korczyński, M., Lindgren, A., Mastepanov, M., et al. (2017). Increased nitrous oxide emissions from Arctic peatlands after permafrost thaw. *Proceedings of the National Academy of Sciences of the United States of America, 114*, 6238–6243. https://doi.org/10.1073/pnas.1702902114.

Vonk, J. E., & Gustafsson, Ö. (2013). Permafrost-carbon complexities. *Nature Geoscience, 6*, 675–676.

Vourlitis, G. L., & Oechel, W. C. (1997). Landscape-scale CO_2, H_2O vapour and energy flux of moist-wet coastal tundra ecosystems over two growing seasons. *Journal of Ecology, 85*, 575–590.

Walker, D. A., & Walker, M. D. (1991). History and pattern of disturbance in Alaskan Arctic terrestrial ecosystems: A hierarchical approach to analysing landscape change. *Journal of Applied Ecology, 28*, 244–276.

Walter, K. M., Zimov, S. A., Chanton, J. P., Verbyla, D., & Chapin, F. S. (2006). Methane bubbling from Siberian thaw lakes as a positive feedback to climate warming. *Nature, 443*, 71–75.

Walter-Anthony, K., Schneider Von Deimling, T., Nitze, I., Frolking, S., Emond, A., Daanen, R., et al. (2018). 21st-century modeled permafrost carbon emissions accelerated by abrupt thaw beneath lakes. *Nature Communications, 9*, 3262.

Walter-Anthony, K. M., Zimov, S. A., Chanton, J. P., Verbyla, D., & Chapin, F. S., III. (2014). A shift of thermokarst lakes from carbon sources to sinks during the Holocene epoch. *Nature, 511*, 452–456.

Wanninkhof, R. (1992). Relationship between gas exchange and wind speed over the ocean. *Journal of Geophysical Research, 97*, 7373–7381. https://doi.org/10.1029/92JC00188.

Wanninkhof, R. (2014). Relationship between wind speed and gas exchange over the ocean revisited. *Limnology and Oceanography: Methods, 12*, 351–362.

Webb, E. E., Schuur, E. A. G., Natali, S. M., Oken, K. L., Bracho, R., Krapek, J. P., et al. (2016). Increased wintertime CO_2 loss as a result of sustained tundra warming. *Journal of Geophysical Research – Biogeosciences, 121,* 249–265. https://doi.org/10.1002/2014JG002795.

Wein, R. W. (1976). Frequency and characteristics of Arctic tundra fires. *Arctic, 29,* 213–222.

Welp, L. R., Patra, P. K., Rödenbeck, C., Nemani, R., Bi, J., Piper, S. C., et al. (2016). Increasing summer net CO_2 uptake in high northern ecosystems inferred from atmospheric inversions and comparisons to remote-sensing NDVI. *Atmospheric Chemistry and Physics, 16,* 9047–9066. https://doi.org/10.5194/acp-16-9047-2016.

Weyhenmeyer, G. A., Fröberg, M., Karltun, E., Khalili, M., Kothawala, D., Temnerud, J., et al. (2012). Selective decay of terrestrial organic carbon during transport from land to sea. *Global Change Biology, 18,* 349–355. https://doi.org/10.1111/j.1365-2486.2011.02544.x.

Whalen, S. C., & Reeburgh, W. S. (1992). Interannual variations in tundra methane emission: A 4-year time series at fixed sites. *Global Biogeochemical Cycles, 6,* 139–159. https://doi.org/10.1029/92GB00430.

Wik, M., Thornton, B. F., Bastviken, D., Uhlbäck, J., & Crill, P. M. (2016). Biased sampling of methane release from northern lakes: A problem for extrapolation. *Geophysical Research Letters, 43,* 1256–1262. https://doi.org/10.1002/2015GL066501.

Wik, M., Varner, R. K., Anthony, K. M. W., MacIntyre, S., & Bastviken, D. (2016). Climate-sensitive northern lakes and ponds are critical components of methane release. *Nature Geoscience, 9,* 99–105.

Wild, B., Andersson, A., Bröder, L., Vonk, J., Hugelius, G., McClelland, J. W., et al. (2019). Rivers across the Siberian Arctic unearth the patterns of carbon release from thawing permafrost. *Proceedings of the National Academy of Sciences, 116,* 10280–10285. https://doi.org/10.1073/pnas.1811797116.

Wilkerson, J., Dobosy, R., Sayres, D. S., Healy, C., Dumas, E., Baker, B., et al. (2019). Permafrost nitrous oxide emissions observed on a landscape scale using the airborne Eddy-covariance method. *Atmospheric Chemistry and Physics, 19*(7), 4257–4268. https://doi.org/10.5194/acp-19-4257-2019.

Wullschleger, S. D., Epstein, H. E., Box, E. O., Euskirchen, E. S., Goswami, S., Iversen, C. M., et al. (2014). Plant functional types in Earth System Models: Past experiences and future directions for application of dynamic vegetation models in high-latitude ecosystems. *Annals of Botany, 114,* 1–16.

Xia, J., McGuire, A. D., Lawrence, D., Burke, E., Chen, X., Delire, C., et al. (2017). Terrestrial ecosystem model performance in simulating productivity and its vulnerability to climate change in the northern permafrost region. *Journal of Geophysical Research – Biogeosciences, 122,* 430–446.

Yasunaka, S., Siswanto, O. A., Hoppema, M., Watanabe, E., Fransson, A. I., Chierici, M., et al. (2018). Arctic Ocean CO_2 uptake: An improved multiyear estimate of the air-sea CO2 flux incorporating chlorophyll a concentrations. *Biogeosciences, 15,* 1643–1661.

Yumashev, D., Hope, C., Schaefer, K., et al. (2019). Climate policy implications of nonlinear decline of Arctic land permafrost and other cryosphere elements. *Nature Communications, 10,* 1900. https://doi.org/10.1038/s41467-019-09863-x.

Zhuang, Q., Romanovsky, V. E., & McGuire, A. D. (2001). Incorporation of a permafrost model into a large-scale ecosystem model: evaluation of temporal and spatial scaling issues in simulating soil thermal dynamics. *Journal of Geophysical Research – Biogeosciences, 106,* 649–670.

Zomolodchikov, D. G., Karelin, D. V., Ivaschnenko, A. I. V., Oechel, W. C., & Hastings, J. S. (2003). CO_2 flux measurements in Russian Far East tundra using eddy covariance and closed chamber techniques. *Tellus B, 4,* 879–892.

Zona, D., Gioli, B., Commane, R., Lindaas, J., Wofsy, S. C., Miller, C. E., et al. (2016). Cold season emissions dominate the Arctic tundra methane budget. *Proceedings of the National Academy of Sciences of the United States of America, 113,* 40–45. https://doi.org/10.1073/pnas.1516017113.

Zulueta, R. C., Oechel, W. C., Loescher, H. W., Lawrence, W. T., & Paw U, K. T. (2011). Aircraft-derived regional scale CO_2 fluxes from vegetated drained thaw-lake basins and interstitial tundra on the Arctic coastal plain of Alaska. *Global Change Biology, 17,* 2781–2802. https://doi.org/10.1111/j.1365-2486.2011.02433.x.

Chapter 6

Boreal forests

Daniel J. Hayes[a], David E. Butman[b], Grant M. Domke[c], Joshua B. Fisher[d], Christopher S.R. Neigh[e], and Lisa R. Welp[f]

[a]*School of Forest Resources, University of Maine, Orono, ME, United States,* [b]*Department of Civil and Environmental Engineering, University of Washington, Seattle, WA, United States,* [c]*US Department of Agriculture, Forest Service Northern Research Station, St. Paul, MN, United States,* [d]*Joint Institute for Regional Earth System Science and Engineering, University of California, Los Angeles, CA, United States,* [e]*Code 618, Biospheric Sciences Laboratory, NASA Goddard Space Flight Center, Greenbelt, MD, United States,* [f]*Earth, Atmospheric, and Planetary Sciences, Purdue University, West Lafayette, IN, United States*

1 Carbon in boreal forests

The boreal forest is one of the largest biomes on Earth, occurring in the northern high latitude regions mostly between about 50°N and 65°N (Fig. 1). The biome is dominated by coniferous trees, often occurring as extensive areas of pure-species, even-aged stands regenerated from large, stand-replacing disturbances. Its northern extent lies along the tundra-taiga ecotone; on its southern extent, boreal forest types begin to transition to more temperate conifer species and a greater mix of hardwoods (Goldblum & Rigg, 2010; Montesano, Neigh, Macander, Feng, & Noojipady, 2020). Natural disturbances are a prominent feature of the boreal forest, with high severity fires and insect outbreaks occurring on relatively frequent cycles (Angelstam & Kuuluvainen, 2004; Cogbill, 2007) over a range of spatial scales (Gromtsev, 2002; Hunter, 1993; Kasischke et al., 2010a, 2010b, 2010c; Taylor, Carroll, Alfaro, & Safranyik, 2006; Weed, Ayres, & Hicke, 2013). Changes in these natural disturbance regimes—including fire, insect outbreaks, storms, and drought—will play a large role in the future carbon budget of the boreal forest (Kasischke et al., 2010a, 2010b, 2010c; Navarro, Morin, Bergeron, & Girona, 2018).

A variety of terrain and ecosystem types create a diversity of hydrologic systems, from fast-flowing rivers in steep terrain to flat, poorly drained landscapes with numerous lakes and wetlands. The boreal forest region gives rise to several very large rivers that drain extensive watersheds and often include substantial delta systems at their confluence. Soils vary across the region according to climate, substrate, forest type, landscape position, and drainage pattern (Kuhry et al., 2013). Permafrost-affected landscapes are an important feature of the

Balancing Greenhouse Gas Budgets. https://doi.org/10.1016/B978-0-12-814952-2.00025-3
Copyright © 2022 Elsevier Inc. All rights reserved.

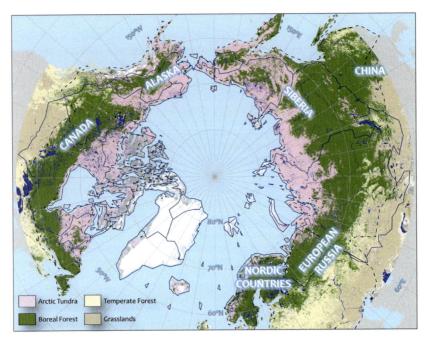

FIG. 1 The geographic scope of this chapter follows the distribution of the boreal forest biome across the circumpolar northern high latitudes. The boreal forest is mostly located between 50°N and 65°N latitude, bordered by grasslands or temperate forest to the south and arctic tundra to the north. This domain encompasses several major regions discussed in this chapter—including the Nordic countries (Norway, Sweden, and Finland), European and Siberian Russia, China, Alaska, and Canada—where national forest inventories and other "bottom-up" budgets are typically conducted. The boreal forest is contained within three "Transcom" regions that have typically served as the boundaries *(dotted lines)* for large-scale atmospheric inversion modeling and "top-down" constraints in model-data comparison studies of land-atmosphere carbon exchange (McGuire et al., 2012). The solid *blue lines* represent the boundaries of the major watersheds of the Arctic Basin, within which most of the boreal forest is contained and where the lateral export of carbon through the river systems can be monitored as the "sideways" constraints on regional budgets. *(Modified from Hayes, D. J., McGuire, A. D., Kicklighter, D. W., Gurney, K. R., Burnside, T. J., & Melillo, J. M. (2011). Is the northern high-latitude land-based CO_2 sink weakening? Global Biogeochemical Cycles https://doi.org/10.1029/2010GB003813.)*

boreal forest, covering about 80% of the region (Helbig et al., 2016; Zhang, Barry, Knowles, Heginbottom, & Brown, 2008). The current and future trajectories in ecosystem composition, structure, and function across the boreal forest are determined by complex interactions and feedbacks among climate warming, wildfire, hydrology, and permafrost thaw (Baltzer, Veness, Chasmer, Sniderhan, & Quinton, 2014; Carpino, Berg, Quinton, & Adams, 2018).

The boreal forest region mainly spans eight nations. Canada, the Nordic countries (Norway, Sweden, and Finland), and Russia make up the core of this region, and the vast majority of the forest of these nations is considered to be of the boreal type. The United States, in Alaska, and China, in its northeastern region, both contain significant areas of boreal forest. Boreal forests are also found in smaller areas of Southern Greenland, Iceland, and the Faroe Islands. Compared to other global regions, the boreal forest shows a relatively small footprint of direct human development, limited to a few centers of urban, agricultural, and resource extraction land uses. The area of boreal forest has remained relatively stable in recent decades, compared to the greater amounts of forest land cover and land-use change found in the temperate and tropical biomes (Pan et al., 2011). Despite the low population density across the boreal region, about two-thirds of its area is under some form of management, primarily for the harvesting of wood (Gauthier, Bernier, Kuuluvainen, Shvidenko, & Schepaschenko, 2015; Saucier, Baldwin, Krestov, & Jorgenson, 2015). The Nordic countries have a highly mechanized and efficient forest industry. Large portions of the Canadian and Western Russian boreal forest are also under active management, whereas expansive regions of Alaska, Northwestern Canada, and Siberia remain largely unmanaged with limited commercial forestry or wood harvesting.

1.1 The major components of the boreal forest carbon budget

The boreal forest region acts as a net source of greenhouse gases to the atmosphere through emissions primarily from fossil fuel combustion, biomass burning in wildfire, and outgassing from inland waters and wetland ecosystems (Le Quéré et al., 2017; Pastor et al., 2003; Price et al., 2013; Saunois, Jackson, Bousquet, Poulter, & Canadell, 2016). Fossil fuel emissions are tracked and reported at national levels and so do not necessarily follow the borders of the boreal region itself. Excluding the United States and China, the core boreal countries have accounted for around 10% of global emissions, and this proportion has been declining over recent decades. Since 2008, the Russian Federation is ranked fourth globally in total fossil fuel emissions and Canada is ranked ninth, while Finland, Sweden, and Norway ranked 59th, 62nd, and 65th among world nations, respectively (Le Quéré et al., 2016). While the statistics are difficult to track, it is likely that a substantial portion of the carbon embedded in fossil fuels and harvested wood products originating in the boreal forest region is exported to, consumed in, and emitted from other more populated regions of the world.

Boreal forests contain substantial stocks of carbon in their ecosystems, about one-third of the global terrestrial storage total (Bradshaw & Warkentin, 2015; McGuire et al., 2009; Pan et al., 2011). Boreal forests contain about 25%–30% of the live biomass carbon of global forests, and some studies have estimated up to 60% of the soil carbon (Dixon et al., 1994; Fyles et al., 2000; Kasischke, 2000). Despite the limited vegetation productivity, carbon accumulates in boreal forest soils as the rates of organic matter decomposition are reduced by the cold and water-saturated conditions (Hobbie, Schimel, Trumbore, & Randerson, 2000). A significant portion of the total boreal forest carbon stock is found in the region's peatlands (Hugelius et al., 2014; Kasischke, 2000).

Excluding the anthropogenic source from fossil fuel combustion, the ecosystems of the boreal forest biome have long been considered to be acting as a net sink in terms of the global terrestrial carbon cycle (Ciais et al., 2010, 2010; Goodale et al., 2002; Myneni et al., 2001). A global-scale analysis of forest inventory information shows that boreal forests as a whole had a consistent net rate of carbon accumulation in the 1990s and 2000s, accounting for 20%–22% of the global carbon sink in established forests during those decades (Pan et al., 2011). Historically, carbon accumulation in biomass has been limited by low vegetation productivity, but the relatively high rates of disturbance across the region have resulted in tree mortality and the transfer of large amounts of carbon from the live to dead organic matter pools (Kurz et al., 2008; Kurz et al., 2013; Kurz, Stinson, & Rampley, 2008; Kurz, Stinson, Rampley, Dymond, & Neilson, 2008; Stinson et al., 2011). The combustion of biomass and organic matter from Boreal Forest vegetation and soils in wildfires makes a globally significant contribution of greenhouse gases (GHGs) to the atmosphere on an annual basis (Van Der Werf et al., 2017). The boreal region is also a major contributor to biogenic methane (CH_4) emissions at the global scale due to its extensive peatland and wetland areas and the wet, organic-rich soils commonly found in boreal forest ecosystems (McGuire et al., 2009; Roulet, Ash, & Moore, 1992). Increases in such climate- and disturbance-driven carbon losses could lead to a weakening of the boreal land sink.

The high latitudes are warming faster than anywhere else on Earth (AMAP, 2017; Gauthier et al., 2015; Serreze & Barry, 2011; Walsh, 2014), and this is driving an increase in "natural" disturbances (e.g., drought, wildfire, insect outbreaks) throughout the boreal forest region (Kasischke et al., 2010a, 2010b, 2010c; Kasischke & Turetsky, 2006; Kurz, Dymond, Stinson, Rampley, et al., 2008; Kurz, Stinson, & Rampley, 2008; Kurz, Stinson, Rampley, Dymond, & Neilson, 2008; Turetsky et al., 2011). These factors cause ecosystem carbon losses from thawing permafrost (Hayes et al., 2014; Schuur et al., 2015) and disturbance emissions (Amiro et al., 2011; Balshi et al., 2009a, 2009b; Chen, Hayes, & David McGuire, 2017), thus leading to a weakening of the Boreal Forest sink for global atmospheric CO_2 (Hayes et al., 2011; Hayes, McGuire, Kicklighter, Burnside, & Melillo, 2011; Kurz, Dymond, Stinson, Rampley, et al., 2008; Kurz, Stinson, & Rampley, 2008; Kurz, Stinson, Rampley, Dymond, & Neilson, 2008; Ma et al., 2012). Model projections suggest that high-latitude ecosystems could gain carbon over the next century under lesser global warming scenarios, but otherwise are likely to lose substantial amounts of carbon after the year 2100 in the absence of an aggressive climate change mitigation pathway (McGuire et al., 2018).

2 Estimating carbon stocks and fluxes in boreal forests

The overall impact of the boreal forest on the atmospheric GHG budget is determined by the imbalance between the anthropogenic and natural sources of carbon dioxide (CO_2) and CH_4 versus the carbon taken up by its natural and managed ecosystems (Fig. 2). The major, continental-scale sources of GHGs from the boreal region originate by: (1) fossil fuel combustion, (2) wildfire

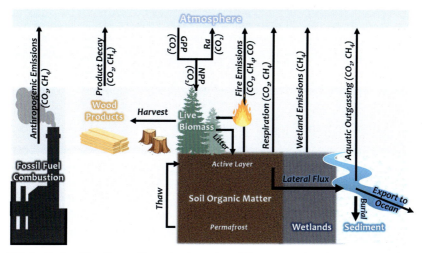

FIG. 2 Pool and flux diagram illustrating the stocks and flows among the major components of the boreal forest carbon budget. The net exchange of carbon-containing greenhouse gases (GHGs: CO_2, CH_4, CO) between the atmosphere and the land surface over a regional-scale domain is estimated directly using top-down approaches such as atmospheric inversion modeling. The net ecosystem exchange (NEE) of CO_2 is also estimated at local scales with tower-based, eddy-covariance measurements that are partitioned into ecosystem uptake (gross primary productivity, GPP) and ecosystem respiration (plant respiration, Ra, plus the heterotrophic respiration of dead organic matter in litter and soils). These fluxes can then be upscaled from the tower footprints ($\sim 1\,km^2$) to broader regions of the boreal forest with spatial modeling based on remote sensing. Both methods rely on observations to estimate NEE at shorter time periods (subdaily to monthly) and then are aggregated to interannual time scales over the length of the observational record. Regional-scale land-atmosphere exchange is estimated indirectly by summing across the bottom-up inventory or modeling of the major GHG source and sink components—both natural and anthropogenic—on land. Fossil fuel emissions are inventoried and reported with relatively small uncertainty at state/province to national levels. Point sources of anthropogenic GHG emissions are more difficult to track at local scales. National forest inventory (NFI) programs are typically cited as the best available information on carbon stocks and stock changes across the boreal forest domain. As a key part of these inventories in boreal forests, GHG emissions from wildfire are estimated based on aerial survey and satellite remote sensing of burned areas combined with models of fuel loads and fire behavior. While NFIs provide reliable estimates of the change in tree biomass and harvested wood product pools between two points in time from the periodic measurement of the plot networks, other pools—especially soils—are undersampled and thus more uncertain. Furthermore, large areas of the boreal forest are designated "unmanaged" and not subject to GHG reporting for international agreements and so are not included in NFIs. Other research-driven methods such as terrestrial biosphere modeling are used to fill in these gaps in undersampled pools and noninventoried geographies. There are no formal carbon inventories for nonforest ecosystems, such as peatlands and other wetlands, that cover large areas of the boreal region and emit large amounts of CH_4 to the atmosphere. These estimates can be included in the budgets based on various scaling methods from "measure-and-multiply" extrapolation of field studies to more detailed, process-based modeling. Finally, the regional budget is closed by estimating the lateral flux of terrestrial carbon to the aquatic system, the amount buried in sediments, and the outgassing of GHGs from the water column back to the atmosphere. These fluxes are typically derived from empirical estimates constrained by measurements of dissolved carbon concentrations in the water leaving stream and river systems. *(Modified from McGuire, A.D., Hayes, D. J., Kicklighter, D. W., Manizza, M., Zhuang, Q., Chen, M., et al. (2010). An analysis of the carbon balance of the Arctic Basin from 1997 to 2006. Tellus, Series B: Chemical and Physical Meteorology, 62(5). https://doi.org/10.1111/j.1600-0889.2010.00497.x.)*

208 Section | C Case Studies

and other disturbances, and (3) emissions from wetlands and aquatic ecosystems. These sources are partially offset by sinks in natural and managed ecosystems driven by plant photosynthesis that converts CO_2 into biomass. Carbon is then stored for a longer term in boreal forest ecosystems as live biomass and dead organic matter both above- and below-ground after losses via emissions to the atmosphere and lateral export through the aquatic system.

2.1 Sampling boreal forest carbon stocks

Basic biometric measurements, field surveys, and ground-based plot networks form the basis for "bottom-up" estimates of terrestrial ecosystem carbon stocks and stock changes (see Chapter 3). Total carbon estimation in boreal forest systems requires sampled, field-based measures of the major carbon pools in the overall budget, i.e., above- and below-ground biomass, litter and woody debris, and soil organic carbon. These measures are then summed to estimates of growing stock (i.e., volume of wood) that are then converted to live biomass using wood density and allometric equations parameterized by species group, age, site, and/or geography. Such measurement campaigns are not always consistent in methodologies or comprehensive in their sampling of the major pools, i.e., the components that are more difficult to measure such as forest floor litter, downed woody debris, and soil organic matter can be significantly undersampled. The lack of in situ data characterizing these pools leads to large uncertainties in bottom-up budgets—particularly in the boreal region where soil carbon is the largest pool but has the most uncertain estimate.

While there has been a wealth of these data collected over the decades across representative boreal forest sites and research areas (Gower et al., 1997; Pattison et al., 2018; Schulze et al., 1999), they are still limited in spatial coverage of this large and mostly remote region. These research data often come from "one-off" studies that are out of date and limited in their temporal scope (Botkin & Simpson, 1990; Fisher et al., 2018a, 2018b; Gower et al., 2001). The amount and spatial coverage of available, field-based measurement data for boreal forest carbon stocks tend to be limited by accessibility, regional extent, and institutional investment in inventory and research. More formal forest inventories are driven by the economic value of management for wood fiber, and are not necessarily designed to produce a biome-scale GHG budget. The spatial density of data collection is highest in the Nordic countries, given their long and active history of boreal forest research combined with the smaller geographical area to cover (Næsset et al., 2004). Similarly, the interior boreal forest of Alaska is relatively well studied, whereas data become sparse in more remote and inaccessible regions. Both Canada and Russia have important research areas and inventory networks, but the sheer size of these areas results in less data coverage especially in the remote and largely "unmanaged" extents of their boreal forest (Schimel et al., 2015). These field surveys form the basis of the national forest inventories (NFIs) that vary by country in terms of their spatial coverage, sampling intensity, measured components, repeat frequency, and scaling methods.

When linked to ground-based measurements, remote sensing data collected from satellites or aircraft can expand existing plot networks and field studies over time and space. Remote sensing has been a critical tool in estimating and tracking biomass carbon over large and remote boreal forests with few and scattered field sites (Margolis et al., 2015; Neigh et al., 2013). Boreal forests have historically served as important test areas demonstrating the use of high-resolution optical sensors from low-altitude aircraft, including traditional aerial photography (Leckie & Gillis, 1995; Maclean & Martin, 1984; Magnusson, Fransson, & Olsson, 2007), digital photogrammetry (Bohlin, Wallerman, & Fransson, 2012; Næsset, 2002; White et al., 2013), and hyperspectral imaging (Halme, Pellikka, & Mõttus, 2019). Active remote sensing systems, i.e., airborne LiDAR (light detection and ranging) and synthetic aperture radar (SAR), are capable of 3D characterization of forest structure and commonly used to estimate above-ground tree biomass. The sensitivity of SAR backscatter retrievals to above-ground biomass (AGB) has been demonstrated across a range of boreal forest conditions, structures, and geographies (Neumann, Saatchi, Ulander, & Fransson, 2012; Rignot, Williams, & Viereck, 1994; Sandberg, Ulander, Fransson, Holmgren, & Le Toan, 2011).

Airborne laser scanning (ALS), or LiDAR, is considered the most promising approach for accurate, high-resolution biomass mapping (Boudreau et al., 2008; Montesano et al., 2014) in large part because of its extensive application in boreal forests, particularly in Canada (Lim, Treitz, Wulder, St-Ongé, & Flood, 2003; Treitz et al., 2012) and the Nordic countries (Gobakken et al., 2012; Hyyppä et al., 2008; Næsset, 2004). ALS acquisitions can be linked to plot-based inventories to map forest carbon and other management-relevant forest attributes over larger areas (White et al., 2013, 2017; Wulder et al., 2012). These small-footprint LiDAR data show accurate and consistent model results in estimating forest biomass, but with the consequence of reduced sampling area. Having few or no publicly available ALS data is a major limitation on reducing uncertainties of biomass estimates in many areas needed to capture the range of forest types over the boreal region. This is particularly pertinent in places such as Siberia and its large, remote areas of deciduous needleleaf Larch forests. Current spaceborne LiDAR assets are rapidly increasing the sampling of vegetation structure in forests worldwide. NASA's Ice, Cloud, and Land Elevation Satellite-2 (ICESat-2) is a photon-counting laser on board a polar orbiting satellite that can be used to measure forest canopy height and estimate biomass across the boreal forest biome (Narine et al., 2019; Queinnec, White, & Coops, 2021). NASA's Global Ecosystem Dynamics Investigation (GEDI) instrument is based on the International Space Station (ISS) and collects LiDAR waveforms in snapshots along its orbital track (Dubayah et al., 2020). While GEDI data are being used in spatial models to map above-ground biomass in temperate and tropical forests, the latitudinal maxima of the ISS transits at 52° north and south limits its useability for boreal forests. Near-future missions bring the promise of integrating with expanding capabilities, notably the European Space Agency's BIOMASS mission that will use satellite-based SAR to map global carbon stocks (Quegan et al., 2019).

2.2 Sampling boreal ecosystem carbon fluxes

The eddy covariance flux technique measures ecosystem-scale carbon exchange using tower-based instrumentation with footprints on the order of one or more square kilometers (Baldocchi, 2003). This technique has been applied extensively in studies of boreal forest carbon balance and variability to climate and disturbance (e.g., Barr et al., 2002; Chen et al., 1999; Grant et al., 2009; Kurbatova, Li, Varlagin, Xiao, & Vygodskaya, 2008; Lagergren et al., 2008). These net ecosystem exchange measurements are most appropriate for characterizing patterns in the biological carbon fluxes (i.e., gross primary productivity and total ecosystem respiration) over short time periods (daily to interannual) and fine spatial scales ($\sim$1 km). Scaling these flux measurements for regional accounting of net carbon change is a challenge, however, where a limited number of tower sites do not capture the variability known to be important in determining carbon budgets across the boreal forest domain (Chen et al., 2011; Goulden et al., 2011; Zha et al., 2013), i.e., climate and abiotic conditions, forest and other ecosystem types, and forest age and disturbance. Yet, eddy covariance flux studies have provided a wealth of understanding on the processes and controls of boreal carbon—including CO_2 and CH_4 exchange in other ecosystems such as wetlands (Rinne et al., 2007; Wang et al., 2018) and lakes (Huotari et al., 2011; Podgrajsek et al., 2016) in addition to upland forests —and thus are used as the basis for upscaling flux estimates and calibrating and evaluating ecosystem models (Clein et al., 2002; Ueyama et al., 2016; Virkkala et al., 2021).

2.3 Carbon emissions from wildfire

Disturbance, particularly fire, is a primary driver of forest carbon dynamics across the boreal region. Fire in the boreal forest is characterized by large, stand-replacing wildfires; prescribed and managed fires used for forest management or land clearing are not a significant component of the regional carbon budget. North American and Eurasian boreal fire regimes are known to be different in many ways (De Groot et al., 2013; Rogers, Soja, Goulden, & Randerson, 2015), but extensive areas of interior Alaska, Western Canada, and Siberia have been impacted by frequent, large, and severe wildfires in recent decades (Balshi et al., 2009a, 2009b; Kharuk et al., 2021; Krylov et al., 2014). Carbon losses from wildfire are often already, albeit implicitly, accounted for in the calculation of stock change between two inventory dates (Harris et al., 2016). Alternatively, fire emissions can be explicitly quantified in inventory- or process-based models by mapping burned area and estimating emissions (Chen et al., 2017; Domke et al., 2021; French et al., 2011; Van Der Werf et al., 2017).

There are various programs that actively map burned areas across the circumpolar Boreal Forest, primarily based on remote sensing. Burned area perimeters have been mapped historically from field, aerial, and satellite observations by the Alaska Fire Service (Kasischke, Williams, & Barry, 2002). The Canadian National Fire Database maintains a collection of fire locations and burned area perimeters available since 1986 (Stocks et al., 2003) from various sources by fire management agencies among the provinces, territories, and Parks Canada. Maps of detected burn areas across the vast forests of Siberia since the 1980s have been developed by coarse resolution satellite imagery (Sukhinin et al., 2004). Burned area is currently mapped globally using Moderate Resolution Imaging Spectrometer (MODIS) data (Van Der Werf et al., 2017), which closely agrees with these other, regional data sets (Hayes, McGuire, Kicklighter, Burnside, & Melillo, 2011; Hayes, McGuire, Kicklighter, Gurney, et al., 2011). However, models using similar burned area estimates will show larger ranges in their emissions estimates due to differences in fuel loads and combustion factors. Fuel loads are determined by simulations of biomass and soil organic matter pools with a process-based ecosystem model, moderated by the proportional carbon combustion both in the above-ground vegetation as well as to some depth of the soil (Kasischke & Bruhwiler, 2002). Boreal forest fires on average tend to have a larger portion of soil carbon consumed compared to other forests (Van Der Werf et al., 2017), which results in a higher amount of smoldering relative to flaming emissions and thus releases higher CH_4 and CO emissions than other forests (French, Kasischke, & Williams, 2003).

2.4 Carbon in the aquatic system

The riverine transport of carbon from terrestrial watersheds to the world's oceans is a major component of the global carbon cycle (Li et al., 2017). While biomass stocks and land-atmosphere fluxes often receive more attention in carbon accounting, including the lateral export of carbon through the aquatic system directly impacts the estimates of carbon uptake in terrestrial ecosystems. Inland waters of the boreal forest biome in particular have some of the highest concentrations, quantities, and fluxes of dissolved organic and inorganic, and particulate, carbon (DOC, DIC, POC) (Holmes et al., 2012; Morison et al., 2012; Vonk et al., 2015). The total aquatic flux from boreal forest ecosystems is related to temperature and precipitation, river morphology, wetland, and permafrost area, among other complex factors (Laudon et al., 2011).

These land-water fluxes are critical to account for where they are otherwise assumed to be lost to the atmosphere by inventory-based assessments or stored in the ecosystem by land-atmosphere flux studies (Butman et al., 2018; Hayes & Turner, 2012). The quantities and rates of carbon transfers from land to aquatic systems remain poorly constrained spatially and are critical to properly balance inventory-based estimates of land-atmosphere fluxes. Aquatic carbon fluxes integrate landscapes within whole watersheds, from headwaters to the coast. Furthermore, carbon is produced and consumed along the length of the hydrologic network thus making direct comparisons between terrestrial and aquatic carbon fluxes difficult across biomes or ecosystem types. These differences in

212 Section | C Case Studies

scale and carbon processes have created methodological challenges for integrating these two flux pathways in carbon budgets. Aquatic fluxes also integrate across ecosystem types where measured and reported within large basins that span biomes, and thus are not easily partitioned between boreal forests and arctic tundra, for example.

The export of carbon from boreal forest ecosystems occurs largely from the drainage basins of the Arctic Ocean and its marginal seas, thus representing the key connection between the terrestrial and marine carbon cycle of the Circumpolar North (McGuire et al., 2009). Estimates of riverine DOC export to the Arctic Ocean can be derived and scaled for several large watershed domains (Lammers, Shiklomanov, Vörösmarty, Fekete, & Peterson, 2001) based on empirical relationships between concentration and water discharge data collected at gauging stations at the mouths of the major arctic-boreal river systems (Manizza et al., 2009; McClelland et al., 2008). While these estimates provide a constraint on the lateral export to the ocean, it may represent only half of the total aquatic flux that also includes quantities of carbon that are outgassed from (as both CO_2 and CH_4), and buried in the sediments of, inland water bodies (Cole et al., 2007). These other flux components are often estimated based on geospatial maps to upscale sample measurements to full watersheds based on empirical relationships with the controlling factors (Stackpoole et al., 2017). Although it is highly variable by ecosystem type, DOC production can also be considered as a proportion of terrestrial inputs from primary production, which can be simulated with an ecosystem model (Genet et al., 2013).

Biogeochemical process models can incorporate DOC flux into the aquatic system as a function of decomposition in the soil pool, but they otherwise do not simulate the fate of this carbon through the inland water network (Kicklighter et al., 2013). Process model estimates can be connected with other models to simulate the atmosphere-land-ocean as a large-scale, integrated carbon budget (McGuire et al., 2010). Less attention has been given to the lateral transport of inorganic carbon, inclusive of dissolved CO_2 (Tank et al., 2012). The sources and fate of inorganic carbon integrate both the long- and short-term carbon cycles through weathering and ecosystem respiration. Chemical tracers can be used to untangle the sources of DIC, but challenges still remain in estimating the atmospheric emissions of CO_2 along riverine networks.

3 Carbon accounting in boreal forests

For the purposes of scientific study and policy analysis, carbon budget accounting is often performed at regional scales and summarized over annual to decadal time periods. Estimation of carbon stock change for accounting efforts requires repeated measures of field plots and continuous forest inventory programs at the national level. These inventories estimate the standing carbon stocks at each successive time period, tracking the change in the major above- and belowground pools. Inventories also typically track the harvest, removal, and fate of wood products and their emissions as part of the overall budget. National-scale forest inventories, however, are challenging to conduct and require significant resources. Indeed, there are large areas of remote and "unmanaged" boreal forest that are not included in formal inventories, and the remeasurement cycle can be too coarse to attribute fine-scale process. Here, remote sensing and other scaling approaches can be used to help fill in these spatial and temporal gaps in carbon budget accounting.

3.1 National forest inventories

Most countries in the temperate and boreal biomes have established national forest inventory (NFI) programs with repeated measurement of permanent sample plots (Pan et al., 2011). In the 1920s, the first sample-based NFIs were established in the boreal forest—in Norway, Finland, and Sweden (Tomppo, Gschwantner, Lawrence, & McRoberts, 2010). Most modern-day NFIs in the boreal forest and elsewhere are based on statistical sampling methods where plots are randomly or systematically located across all forested areas of the country, or at least the managed portions (McRoberts, Tomppo, & Næsset, 2010). A census of the trees are made at each plot along with various measurements of each including species, diameter, height, and condition that can then be used in allometric equations to estimate tree biomass (Xing et al., 2019). Additional measurements are taken at each or a subset of the plots to estimate plot-level carbon by including other important pools such as understory vegetation, woody debris, litter, and soils (Banfield, Bhatti, Jiang, & Apps, 2002; Shaw et al., 2014). The plot-based carbon estimates are then scaled up to the national level by some type of modeling approach, which differs across the NFIs of different countries (Kurz et al., 2013; Woodall, Heath, Domke, & Nichols, 2011). Estimates of total forest carbon for the full inventoried domain can be imputed over the plot network (Wilson, Woodall, & Griffith, 2013), often making use of remote sensing and spatial modeling (Beaudoin et al., 2014; Kangas et al., 2018).

Changes in the stocks of live and dead organic matter pools in the forest are determined from NFIs by direct plot remeasurement and/or with some combination of spatial modeling and remote sensing approaches. The "stock-change" approach used in forests of the continental US, for example, is based on the difference between complete inventories at two points in time, thus capturing the total change in ecosystem carbon (Hou et al., 2021). However, the US Forest Inventory and Analysis (FIA) program does not have the density of plots in the boreal forest of interior Alaska as it does elsewhere in the country, and therefore relies on remote sensing data to fill in the gaps (Babcock et al., 2018). In the Nordic countries, carbon stock change is estimated from the NFIs based on remeasurement of a subset of each nation's permanent plot network on a 5-year-cycle (Kangas et al., 2018). Area-based modeling using airborne laser scanning data has become a major component of the inventories over the last decade in Norway, Finland, and Sweden (Maltamo & Packalen, 2014; Næsset, 2014).

The Russian Federation has an NFI system based on a sample of ground-based inventory plots across the country's extensive forest areas, but uncertainty arises from how current the measurements are, as well as differences in assessment methods (Shvidenko & Nilsson, 2002). Inventory-based data have been compiled for Russia in broader-scale carbon budget assessments (Pan et al., 2011) and compared with other, model-driven estimates (Dolman et al., 2012).

Alternatively, Canada's national forest carbon inventory is based on the "gain-loss" method, which starts with a complete inventory that then is updated by modeling forward the components of change, including growth, mortality, decomposition, and disturbance (Kurz et al., 2009; Stinson et al., 2011). The accounting relies heavily on empirical and observational data, including for both forest growth and yield as well as disturbance characterization and mapping. The carbon budget model ingests this information and then simulates annual carbon stock changes in forest biomass. Carbon stock changes in the dead organic matter pools are directly linked to the better-known biomass dynamics, or net primary productivity. Dead wood, litter, and soil are calculated as the mass balances from inputs (through litterfall, biomass turnover, and disturbance inputs) and losses (through decomposition, transfers by harvesting, and losses to the atmosphere during disturbances such as fire). The "gain-loss" approach has been adopted by the FIA for estimates of the managed forest land in Alaska that lack remeasurements (Domke et al., 2021).

3.2 Carbon in harvested wood products

Harvested wood products (HWPs) represent an important pool of carbon, particularly in intensively managed forests like those found in the Nordic countries and other areas of the boreal region (Triviño et al., 2015). The HWP pool must be accounted for in NFIs and other carbon budget accounting efforts as a key component due to its potential for long-term carbon sequestration as well as by replacing GHG emissions from other, nonrenewable sources (Chen, Ter-Mikaelian, Yang, & Colombo, 2018; Johnston & Radeloff, 2019; Zhang, Chen, Dias, & Yang, 2020). Forest management practices and the fate of HWP play a large role in determining whether forests overall will act as net sources or sinks of carbon (Birdsey, Pregitzer, & Lucier, 2006; Paradis, Thiffault, & Achim, 2019). Most countries, including in the Boreal region, account for carbon in HWP using simple spreadsheet models, which have default assumptions for inputs to short- and long-term product pools (Bergman, Puettmann, Taylor, & Skog, 2014; Jasinevičius, Lindner, Pingoud, & Tykkylainen, 2015; Skog, Pingoud, & Smith, 2004). Some portion of the carbon removed in harvest, about 20%–40% (Hayes et al., 2012; Smith, Heath, Skog, & Birdsey, 2006), is emitted during processing into wood products. This processing, or "primary consumption," is assumed to occur largely at the mill. The remainder is assigned to an "in use" product pool of various half-life (e.g., pulp and paper vs sawlogs), solid-waste disposal (i.e., landfills), or exported out of the country or reporting zone. Carbon is emitted as 'secondary consumption' from each of these pools as it decays at a certain percent over some time frame, typically calculated as a constant proportion per year over 10–100 years. For the purposes of reporting to the United Nations Framework Convention on Climate Change (UNFCCC), the International Panel on Climate Change (IPCC) recommends calculations based on the "production approach" where the carbon emissions from the decay of HWP stocks are accounted for in the country or reporting zone where the wood was originally grown and harvested, regardless of the locations of eventual primary and secondary consumption of the products (Buendia et al., 2019; Penman, Gytarsky, Hiraishi, Irving, & Krug, 2006).

3.3 Managed vs unmanaged forest lands

There are large areas of the boreal Forest where resources for NFIs are not cost-effective, prioritized, or practical. These areas are typically remote and largely unaffected in a significant way by direct anthropogenic activities such as land-use conversion, harvest, or fire suppression (Ogle et al., 2018). Approximately 51 and 118 million hectares are designated as "Unmanaged Forest Lands" in Alaska and Canada, respectively (Kurz et al., 2018; Pan et al., 2011; Fig. 3). While the US FIA program covers the coastal temperate forests of Southeast Alaska, the boreal forests of the interior are not currently included in the

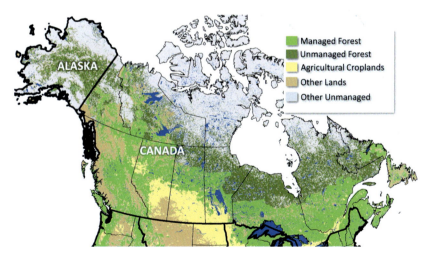

FIG. 3 Map of the managed versus unmanaged forest areas as designated in Canada and Alaska. *(Modified from Hayes, D. J., Vargas, R, Alin, S.R., Conant, R. T., Hutyra, L. R., Jacobson, A. R., et al. (2018). Chapter 2: The North American carbon budget. In N. Cavallaro, G. Shrestha, R. Birdsey, M. A. Mayes, R. G. Najjar, S. C. Reed, P. Romero-Lankao, & Z. Zhu (Eds.), Second state of the carbon cycle report (SOCCR2): A sustained assessment report (pp. 71–108). Washington, DC: U.S. Global Change Research Program. https://doi.org/10.7930/SOCCR2.2018.Ch2.)*

inventory. The criteria for, and proportion of, managed forest across Canada varies by province but are generally determined by those areas designated for timber harvesting and/or fire suppression (Kurz et al., 2009). In Alaska, new inventory plots are being installed in interior boreal forests over a 15-year period, and supplemented by advanced remote sensing data and spatial modeling techniques (Andersen et al., 2015; Babcock et al., 2018). While not reported in its national GHG inventory, unmanaged forests are included in Canada's deforestation monitoring program to track resource extraction and land-use change in order to update its land designations (Dyk, Leckie, Tinis, & Ortlepp, 2015).

Carbon sources and sinks in these unmanaged forests are dominated by the cycles of growth, succession, and natural disturbances. For example, studies show declines in the growth and high fire losses in unmanaged larch forests across the permafrost zone in Siberia (Kharuk et al., 2019; Kharuk et al., 2021). In lieu of inventories, process-based modeling can be calibrated to available ground data and used to simulate these ecosystem dynamics. Modeling studies suggest that these areas have operated as a small carbon sink in recent decades (McGuire et al., 2009) although that near-balance may be tipping toward increasing sources from wildfire (Hayes, McGuire, Kicklighter, Burnside, & Melillo, 2011; Hayes, McGuire, Kicklighter, Gurney, et al., 2011; Walker et al., 2019) and permafrost thaw (Hayes et al., 2014). Compared to the large carbon sink estimated in forests of European Russia, the boreal forests of North America are thought to be only small sinks or sources (Pan et al., 2011). As such, it is generally assumed in carbon budget assessments would not significantly change the estimates currently reported by the national-level inventories of the boreal countries (Pan et al., 2011).

3.4 The role of remote sensing in boreal forest inventories

Spatial statistical models take in situ measurements from representative locations and fill in spatial gaps by connecting them to wall-to-wall maps of environmental variables derived from remote sensing (Jung et al., 2020; Tramontana et al., 2016). Multitemporal remote sensing plays an important role in Boreal Forest inventory updates. The availability of global satellite optical, SAR, and LiDAR data—linked to ground measurements—has led to the development of wall-to-wall, regional, and circumboreal maps of AGB (Matasci et al., 2018; Neigh et al., 2013; Santoro et al., 2015). Of course, remote sensing approaches by nature are most effective at estimating above-ground carbon pools, i.e., AGB. Below-ground and other nonliving carbon pools are more difficult to estimate with remote sensing as well as in the field, and thus are often undersampled in inventories compared to live tree biomass. As such, field and remotely sensed data need to be integrated with other flux measurements and modeling frameworks (Hopkinson et al., 2016; Kimball, Keyser, Running, & Saatchi, 2000; Liu, Chen, Cihlar, & Park, 1997; Sitch et al., 2007) in order to account for a more comprehensive budget of regional-scale boreal forest carbon stock changes over time.

4 Regional-scale modeling

In order to expand the sampling of fine-scale measurements and observations to the broader scope of time and space required for regional budgets, some type of scaling approach is required. Scaling involves extrapolation of both the sample data and the understanding of fine-scale mechanisms across a hierarchy of coarser levels. Various modeling frameworks have been designed to temporally extrapolate sample flux measurements using diagnostic (Rödenbeck, Zaehle, Keeling, & Heimann, 2018) and prognostic (McGuire et al., 2016) approaches. The surface-atmosphere exchange of trace gases can be estimated essentially by two complementary approaches, using what are generally categorized as either "top-down" or "bottom-up" methodologies (McGuire et al., 2012). Top-down estimates of CO_2 or CH_4 flux are based on measurements of these trace gas concentrations from networks of atmospheric monitoring stations. Atmospheric inversion models (AIMs) use initial estimates of the land flux combined with an atmospheric transport model, then adjust the pattern of land fluxes until inferred gas concentrations closely match observations to achieve an optimized posterior flux estimate (Chen, Chen, & Worthy, 2005; Ciais, Canadell, et al., 2010; Ciais, Rayner, et al., 2010). Bottom-up models rely on ground-based, in situ measurements of carbon stocks and fluxes at representative locations to calibrate terrestrial biosphere models (TBMs). These ecosystem process simulations can be extrapolated to regional carbon budget estimates using maps of plant functional types and analyzed or predicted over time using climate, disturbance, and other forcing data (Fisher et al., 2014; Fisher, Huntzinger, Schwalm, & Sitch, 2014).

Atmosphere-based approaches are useful as a top-down constraint on regional carbon budgets because they estimate the total net surface-atmosphere exchange of CO_2 or CH_4 directly as one integrated flux. Furthermore, AIM is currently the only methodology that fully integrates aquatic contributions to atmospheric carbon, albeit with limited ability to differentiate between terrestrial and aquatic sources. However, there are large uncertainties in AIM estimates over the high latitudes that arise from the sparse observation network, transport model error, and incorrect boundary conditions (Dargaville, Baker, Rödenbeck, Rayner, & Ciais, 2006). AIMs have tended to estimate much larger carbon sinks over the boreal regions than the inventories or TBMs (Dargaville et al., 2002; Hayes, McGuire, Kicklighter, Burnside, & Melillo, 2011; Hayes, McGuire, Kicklighter, Gurney, et al., 2011). Analyses that have incorporated improvements on these data and modeling methods show that carbon uptake by boreal forests is not as strong as suggested by previous studies (Stephens et al., 2007). Regional inversions over the boreal biome do show increases in both the seasonal cycle of CO_2 exchange (Forkel et al., 2016) and the interannual net uptake of carbon in these ecosystems (Welp et al., 2016). There are uncertainties in these assessments that are associated with the role of increasing active layer depths and its impact on soil carbon respiration (Carvalhais et al., 2014). AIM results are shown to be consistent with satellite remote sensing

218 Section | C Case Studies

observations of "greening" (Beck & Goetz, 2011), despite increasing fire over many areas and browning trends in some (Verbyla, 2011). An atmospheric inversion study of CH_4 fluxes in the northern high latitudes shows positive trends in emissions from both natural (wetlands) and anthropogenic (oil and gas development) sources across the boreal region (Thompson et al., 2017). Overall, top-down modeling frameworks continue to improve their capabilities for reducing uncertainties in increasingly regional-scale estimates of land-atmosphere carbon exchange (Jacobson et al., 2018). These improvements are being driven by advances in statistical upscaling and machine-learning approaches based on rapidly increasing atmospheric GHG observations from aircraft and satellites.

Simulation experiments with TBMs allow scientists to explore hypotheses related to driver sensitivity and attribution of carbon dynamics in boreal forests and other ecosystems (Amthor et al., 2001; McGuire et al., 2001). TBMs have been applied to carbon budget estimation in boreal forests, including the impacts of climate, CO_2 fertilization, nutrient dynamics, disturbance, vegetation shifts, and land use (Euskirchen, McGuire, Chapin, Yi, & Thompson, 2009; Hayes, McGuire, Kicklighter, Burnside, & Melillo, 2011; Hayes, McGuire, Kicklighter, Gurney, et al., 2011; Kalliokoski, Mäkelä, Fronzek, Minunno, & Peltoniemi, 2018; Mekonnen, Riley, Randerson, Grant, & Rogers, 2019). Comparisons among TBM ensembles can show considerable disagreement in carbon budget estimates for both current (Fisher, Huntzinger, et al., 2014; Fisher, Sikka, et al., 2014) and future conditions (McGuire et al., 2018). These results highlight the need for TBMs to improve the representation of key, boreal-specific processes such as wildfire and permafrost dynamics (Horvath et al., 2021; Koven, Riley, & Stern, 2013; Wang, Baccini, Farina, Randerson, & Friedl, 2021) along with the high-quality in situ and remotely sensed data sets needed for benchmarking (Fisher et al., 2018a, 2018b; Stofferahn et al., 2019). Importantly, TBMs need to tackle challenging issues of scale, particularly in capturing the high subgrid heterogeneity of structure and processes in boreal forest ecosystems. TBM frameworks are increasingly working toward the development and application of individual-based models that incorporate tree and plant demography and dynamics in boreal ecosystems. Furthermore, grid-designed TBMs are not based along hydrologic networks, so additional model development is needed to properly integrate land-water transfers of carbon.

5 Synthesis

The circumboreal biome is large, remote, and relatively unmanaged compared to other, more densely populated regions of the world. However, the region's carbon budget is highly dynamic as driven by variability in climate, disturbance, and land use (Schuur, McGuire, Romanovsky, Schädel, & Mack, 2018). Despite the lower population density of this large region, the boreal nations together account for a significant portion of global fossil fuel emissions. Deforestation is not as prevalent in the boreal region as it is in temperate and tropical forests, but there are large sources of carbon from wildfire and other disturbances along with emissions from lakes and wetlands, including as CH_4. Some portion of these carbon emissions are offset by natural sinks in boreal forest ecosystems, which can store large quantities in its vegetation and, especially, soils. Rates of carbon uptake by these ecosystems are slow, however, due to short growing seasons, cold and wet conditions, and unproductive tree species. Frequent disturbances directly and immediately emit GHGs to the atmosphere and transfer carbon to dead organic matter pools (Kurz et al., 2013) while also resulting in the suppression of the long-term biomass carbon sink (Wang et al., 2021). Forest management in the boreal region results in a substantial portion of carbon removed from the forest and stored in long-term wood product pools (Stinson et al., 2011). Considering all of these dynamics, the boreal carbon sink is relatively stable overall as a result of a loss of uptake in biomass of Canada's forests offset by an increased carbon sink in the other boreal regions (Pan et al., 2011).

There is large uncertainty around how the boreal carbon cycle will respond to future changes (Kurz, Dymond, Stinson, Rampley, et al., 2008; Kurz, Stinson, & Rampley, 2008; Kurz, Stinson, Rampley, Dymond, & Neilson, 2008), and so it is critical that all of these various dynamics are carefully and consistently accounted for in both scientific assessments and GHG inventory reporting. Toward this end, there are many field studies and research networks across the boreal region that provide in situ data on forest carbon stocks. These data have been used to extrapolate carbon budget information over larger regions by their use in calibrating and validating remote sensing and process-based models. Relative to other large regions, boreal forest carbon resources are well inventoried, with Canada, Russia, and the Nordic countries, all having formal NFIs that form the basis for their GHG reporting from the land use sector. Despite this existing information and accounting tools, there remain challenges in representing all of the geographies and carbon pools of such a large domain. There are pools and fluxes that are undersampled in these inventories, and the global land surface models are missing many of the boreal-specific mechanisms altogether. In particular, model comparison studies have shown large uncertainty in the simulation of soil carbon stocks in high-latitude ecosystems, largely a function of differing initial conditions (Huntzinger et al., 2020). There are many opportunities to improve carbon data and accounting in boreal forests coming online, including programs being developed that link inventory with remote sensing and modeling for noninventoried forests such as in Alaska and Northern Canada.

The calculation of the contribution of forest carbon to net-zero emissions targets is complicated by the managed land proxy used for national GHG reporting (Grassi et al., 2021). Countries only get "credit" for these offsets in their managed lands, but in many nations substantial areas of boreal forest are considered "unmanaged" because there is no evidence that direct human intervention has influenced its condition (Ogle & Kurz, 2021). Defining the land base for what is "managed forest" and thus included in GHG reporting will have implications for policy actions to mitigate GHG emissions (Grassi et al., 2017). The IPCC provides guidelines for how countries may define what is "managed" versus "unmanaged" lands for the purposes of UNFCCC reporting (IPCC, 2010), a distinction adopted by the United States and Canada. With respect to scientific assessments more generally, the distinction is made more about whether or not forest areas are regularly and comprehensively inventoried. Indeed, large areas of unmanaged forests in the boreal region lack sufficient ground data and reporting information for the level of carbon accounting that is consistent with the other forestland areas that are included in the NFIs. These unmanaged, or noninventoried, forest areas are not included in national GHG reporting, which results in a significant discrepancy with estimates of forest-based sinks from global models.

While carbon sources and sinks in forests and other lands are directly impacted by anthropogenic management, they are also a function of "natural" processes of the ecosystem (Ogle et al., 2018). National GHG inventories focus on anthropogenic sources because policy interventions and active management can directly influence emissions and offsets. However, other nonanthropogenic carbon sources are significant in the boreal region as a result of prevalent disturbances such as wildfire and insect outbreaks. There are also important, but uncertain, emissions from other wetland and permafrost ecosystems that are not likely to be impacted by direct management actions. Solutions to this discrepancy in attributing anthropogenic versus natural impacts on GHG emissions and offsets involve alternative approaches that can isolate and quantify the impacts of management on the forest carbon budget from those sources and sinks controlled by disturbances and natural processes (Grassi et al., 2018; Kurz et al., 2018).

From a scientific perspective, confidence in the estimates of regional-scale carbon budgets is expected to increase in the near future with more observations, improved data, and better understanding of the processes. Atmospheric models such as NOAA's CarbonTracker system are increasingly being constrained by regional atmospheric inversions with greater numbers of observations. Aided by focused field and airborne data campaigns like NASA's Arctic-Boreal Vulnerability Experiment, TBMs are being developed and tested to incorporate key ecosystem processes needed to better simulate boreal forest carbon dynamics (Fisher et al., 2018a, 2018b). Although there is value in comparing the various top-down and bottom-up approaches for estimating and carbon fluxes, the greatest progress can be made by integrating them more formally in accounting frameworks (Hayes et al., 2018). Meanwhile, NFI programs of the boreal countries will continue to provide high-quality, ground-based, and updated information on carbon stocks across the region. Much of the leading edge of research on remote sensing for forest inventory is being demonstrated in boreal forests (White, Chen, Woods, Low, & Nasonova, 2019; Wulder, Bater,

Coops, Hilker, & White, 2008), and new data and tools such as spaceborne LiDAR are becoming increasingly available for this work (Popescu et al., 2018). Future space-based remote sensing observations hold promise for the development of a continuous monitoring system comprehensive of the key components of the boreal forest carbon cycle (Duncan et al., 2020).

All of these regional-scale inventory, modeling, and accounting approaches discussed in this chapter will be needed to fill in reporting gaps in noninventoried lands and undersampled components of the boreal forest carbon budget, with careful attention to the consistency and compatibility of these scientific assessments with the policy requirements of GHG reporting. As the scientific community continues to refine methodologies to quantify, monitor, and predict changes in carbon storage and flux across the boreal region, new insights and approaches are needed that acknowledge natural systems as continuums, without discrete boundaries, and where the flows of carbon are highly connected. Forest inventory, remote sensing, and modeling approaches have traditionally discretized landscapes as our understanding of individual ecosystems has evolved. This partitioning of landscapes is used as the framework for GHG accounting and reporting as it provides pathways for management decisions to be implemented. However, as changes in climate disproportionately impact large regions of the boreal forest, many of our traditional definitions of ecosystems may change as well. Carbon budgets and GHG accounting systems will thus require methodology and modeling frameworks to be flexible into the future, including the incorporation of growing observation networks and new data streams.

Acknowledgments

DJH was supported in part by a National Aeronautics and Space Administration (NASA) grant from the Carbon Monitoring System (CMS) program (Grant Number 80NSSC21K0966). JBF contributed to this work at the Jet Propulsion Laboratory, California Institute of Technology, under a contract with the National Aeronautics and Space Administration. California Institute of Technology. Government sponsorship acknowledged. JBF was supported by the NASA Arctic-Boreal Vulnerability Experiment (ABoVE). Copyright 2021. All rights reserved.

References

AMAP. (2017). Snow, water, ice and permafrost in the Arctic. In *AMAP report to the Arctic council chapter 4*. https://doi.org/10.1029/2002WR001512.

Amiro, B. D., Todd, J. B., Wotton, B. M., Logan, K. A., Flannigan, M. D., Stocks, B. J., et al. (2011). Direct carbon emissions from Canadian forest fires, 1959-1999. *Canadian Journal of Forest Research*. https://doi.org/10.1139/x00-197.

Amthor, J. S., Chen, J. M., Clein, J. S., Frolking, S. E., Goulden, M. L., Grant, R. F., et al. (2001). Boreal forest CO_2 exchange and evapotranspiration predicted by nine ecosystem process models: Intermodel comparisons and relationships to field measurements. *Journal of Geophysical Research Atmospheres*, *106*(D24), 33623–33648. Blackwell Publishing Ltd https://doi.org/10.1029/2000JD900850.

Andersen, H.-E., Babcock, C., Pattison, R., Cook, B., Morton, D., & Finley, A. (2015). *The 2014 tanana inventory pilot: A USFS-NASA partnership to leverage advanced remote sensing technologies for forest inventory.*

Angelstam, P., & Kuuluvainen, T. (2004). Boreal forest disturbance regimes, successional dynamics and landscape structures—A European perspective. *Ecological Bulletins, 51,* 117–136.

Babcock, C., Finley, A. O., Andersen, H. E., Pattison, R., Cook, B. D., Morton, D. C., et al. (2018). Geostatistical estimation of forest biomass in interior Alaska combining Landsat-derived tree cover, sampled airborne lidar and field observations. *Remote Sensing of Environment, 212,* 212–230. https://doi.org/10.1016/j.rse.2018.04.044.

Baldocchi, D. D. (2003). Assessing the eddy covariance technique for evaluating carbon dioxide exchange rates of ecosystems: Past, present and future. *Global Change Biology.* https://doi.org/10.1046/j.1365-2486.2003.00629.x.

Balshi, M. S., McGuire, A. D., Duffy, P., Flannigan, M., Kicklighter, D. W., & Melillo, J. (2009a). Vulnerability of carbon storage in North American boreal forests to wildfires during the 21st century. *Global Change Biology.* https://doi.org/10.1111/j.1365-2486.2009.01877.x.

Balshi, M. S., McGuire, A. D., Duffy, P., Flannigan, M., Walsh, J., & Melillo, J. (2009b). Assessing the response of area burned to changing climate in western boreal North America using a Multivariate Adaptive Regression Splines (MARS) approach. *Global Change Biology.* https://doi.org/10.1111/j.1365-2486.2008.01679.x.

Baltzer, J. L., Veness, T., Chasmer, L. E., Sniderhan, A. E., & Quinton, W. L. (2014). Forests on thawing permafrost: Fragmentation, edge effects, and net forest loss. *Global Change Biology.* https://doi.org/10.1111/gcb.12349.

Banfield, G. E., Bhatti, J. S., Jiang, H., & Apps, M. J. (2002). Variability in regional scale estimates of carbon stocks in boreal forest ecosystems: Results from West-Central Alberta. *Forest Ecology and Management, 169*(1–2), 15–27. https://doi.org/10.1016/S0378-1127(02)00292-X.

Barr, A. G., Griffis, T. J., Black, T. A., Lee, X., Staebler, R. M., Fuentes, J. D., et al. (2002). Comparing the carbon budgets of boreal and temperate deciduous forest stands. *Canadian Journal of Forest Research.* https://doi.org/10.1139/x01-131.

Beaudoin, A., Bernier, P. Y., Guindon, L., Villemaire, P., Guo, X. J., Stinson, G., et al. (2014). Mapping attributes of Canada's forests at moderate resolution through kNN and MODIS imagery. *Canadian Journal of Forest Research, 44*(5), 521–532. https://doi.org/10.1139/cjfr-2013-0401.

Beck, P. S. A., & Goetz, S. J. (2011). Satellite observations of high northern latitude vegetation productivity changes between 1982 and 2008: Ecological variability and regional differences. *Environmental Research Letters, 6*(4). https://doi.org/10.1088/1748-9326/6/4/045501, 045501.

Bergman, R., Puettmann, M., Taylor, A., & Skog, K. E. (2014). The carbon impacts of wood products. *Forest Products Journal, 64*(7–8), 220–231. Forest Products Society 10.13073/FPJ-D-14-00047.

Birdsey, R., Pregitzer, K., & Lucier, A. (2006). Forest carbon management in the United States. *Journal of Environment Quality.* https://doi.org/10.2134/jeq2005.0162.

Bohlin, J., Wallerman, J., & Fransson, J. E. S. (2012). Forest variable estimation using photogrammetric matching of digital aerial images in combination with a high-resolution DEM. *Scandinavian Journal of Forest Research.* https://doi.org/10.1080/02827581.2012.686625.

Botkin, D. B., & Simpson, L. G. (1990). Biomass of the North American boreal forest: A step toward accurate global measures. *Biogeochemistry, 9,* 161–174.

Boudreau, J., Nelson, R. F., Margolis, H. A., Beaudoin, A., Guindon, L., & Kimes, D. S. (2008). Regional aboveground forest biomass using airborne and spaceborne LiDAR in Québec. *Remote Sensing of Environment.* https://doi.org/10.1016/j.rse.2008.06.003.

Bradshaw, C. J. A., & Warkentin, I. G. (2015). Global estimates of boreal forest carbon stocks and flux. *Global and Planetary Change*. https://doi.org/10.1016/j.gloplacha.2015.02.004.

Buendia, C., Tanabe, E., Kranjc, K., Baasansuren, A., Fukuda, J., Ngarize, M., et al. (2019). *2019 refinement to the 2006 IPCC guidelines for National Greenhouse Gas Inventories Task Force on National Greenhouse Gas Inventories*. www.ipcc-nggip.iges.or.jp.

Butman, D., Striegl, R., Stackpoole, S., del Giorgio, P., Prairie, Y., Pilcher, D., et al. (2018). Chapter 14: Inland waters. In N. Cavallaro, & G. Shrestha (Eds.), *Second state of the carbon cycle report*. https://doi.org/10.7930/SOCCR2.2018.Ch14.

Carpino, O. A., Berg, A. A., Quinton, W. L., & Adams, J. R. (2018). Climate change and permafrost thaw-induced boreal forest loss in northwestern Canada. *Environmental Research Letters*. https://doi.org/10.1088/1748-9326/aad74e.

Carvalhais, N., Forkel, M., Khomik, M., Bellarby, J., Jung, M., Migliavacca, M., et al. (2014). Global covariation of carbon turnover times with climate in terrestrial ecosystems. *Nature*, *514*(7521), 213–217. https://doi.org/10.1038/nature13731.

Chen, W. J., Black, T. A., Yang, P. C., Barr, A. G., Neumann, H. H., Nesic, Z., et al. (1999). Effects of climatic variability on the annual carbon sequestration by a boreal aspen forest. *Global Change Biology*. https://doi.org/10.1046/j.1365-2486.1998.00201.x.

Chen, B., Chen, J. M., & Worthy, D. E. J. (2005). Interannual variability in the atmospheric CO_2 rectification over a boreal forest region. *Journal of Geophysical Research D: Atmospheres*, *110*(16), 1–12. https://doi.org/10.1029/2004JD005546.

Chen, B., Coops, N. C., Fu, D., Margolis, H. A., Amiro, B. D., Barr, A. G., et al. (2011). Assessing eddy-covariance flux tower location bias across the Fluxnet-Canada research network based on remote sensing and footprint modelling. *Agricultural and Forest Meteorology*. https://doi.org/10.1016/j.agrformet.2010.09.005.

Chen, G., Hayes, D. J., & David McGuire, A. (2017). Contributions of wildland fire to terrestrial ecosystem carbon dynamics in North America from 1990 to 2012. *Global Biogeochemical Cycles*, *31*(5). https://doi.org/10.1002/2016GB005548.

Chen, J., Ter-Mikaelian, M. T., Yang, H., & Colombo, S. J. (2018). Assessing the greenhouse gas effects of harvested wood products manufactured from managed forests in Canada. *Forestry*, *91*(2), 193–205. https://doi.org/10.1093/forestry/cpx056.

Ciais, P., Canadell, J. G., Luyssaert, S., Chevallier, F., Shvidenko, A., Poussi, Z., et al. (2010). Can we reconcile atmospheric estimates of the Northern terrestrial carbon sink with land-based accounting? *Current Opinion in Environmental Sustainability*. https://doi.org/10.1016/j.cosust.2010.06.008.

Ciais, P., Rayner, P., Chevallier, F., Bousquet, P., Logan, M., Peylin, P., et al. (2010). Atmospheric inversions for estimating CO_2 fluxes: Methods and perspectives. *Climatic Change*, *103*(1–2), 69–92. https://doi.org/10.1007/s10584-010-9909-3.

Clein, J. S., McGuire, A. D., Zhang, X., Kicklighter, D. W., Melillo, J. M., Wofsy, S. C., et al. (2002). Historical and projected carbon balance of mature black spruce ecosystems across north america: The role of carbon-nitrogen interactions. *Plant and Soil*. https://doi.org/10.1023/A:1019673420225.

Cogbill, C. V. (2007). Dynamics of the boreal forests of the Laurentian highlands, Canada. *Canadian Journal of Forest Research*. https://doi.org/10.1139/x85-043.

Cole, J. J., Prairie, Y. T., Caraco, N. F., McDowell, W. H., Tranvik, L. J., Striegl, R. G., et al. (2007). Plumbing the global carbon cycle: Integrating inland waters into the terrestrial carbon budget. *Ecosystems*. https://doi.org/10.1007/s10021-006-9013-8.

Dargaville, R., Baker, D., Rödenbeck, C., Rayner, P., & Ciais, P. (2006). Estimating high latitude carbon fluxes with inversions of atmospheric CO_2. *Mitigation and Adaptation Strategies for Global Change*, *11*(4), 769–782. https://doi.org/10.1007/s11027-005-9018-1.

224 Section | C Case Studies

Dargaville, R. J., Heimann, M., McGuire, A. D., Prentice, I. C., Kicklighter, D. W., Joos, F., et al. (2002). Evaluation of terrestrial carbon cycle models with atmospheric CO_2 measurements: Results from transient simulations considering increasing CO_2, climate, and land-use effects. *Global Biogeochemical Cycles, 16*(4), 1092. https://doi.org/10.1029/2001gb001426.

De Groot, W. J., Cantin, A. S., Flannigan, M. D., Soja, A. J., Gowman, L. M., & Newbery, A. (2013). A comparison of Canadian and Russian boreal forest fire regimes. *Forest Ecology and Management.* https://doi.org/10.1016/j.foreco.2012.07.033.

Dixon, R. K., Brown, S., Houghton, R. A., Solomon, A. M., Trexler, M. C., & Wisniewski, J. (1994). Carbon pools and flux of global forest ecosystems. *Science.* https://doi.org/10.1126/science.263.5144.185.

Dolman, A. J., Shvidenko, A., Schepaschenko, D., Ciais, P., Tchebakova, N., Chen, T., et al. (2012). An estimate of the terrestrial carbon budget of Russia using inventory-based, eddy covariance and inversion methods. *Biogeosciences, 9*(12), 5323–5340. https://doi.org/10.5194/bg-9-5323-2012.

Domke, G. M., Walters, B. F., Nowak, D. J., Smith, J. E., Nichols, M. C., Ogle, S. M., et al. (2021). *Greenhouse gas emissions and removals from forest land, woodlands, and urban trees in the United States, 1990–2019. Resource update FS–307* (p. 5). Madison, WI: U.S. Department of Agriculture, Forest Service, Northern Research Station (plus 2 appendixes) https://doi.org/10.2737/FS-RU-307.

Dubayah, R., Blair, J. B., Goetz, S., Fatoyinbo, L., Hansen, M., Healey, S., et al. (2020). The global ecosystem dynamics investigation: High-resolution laser ranging of the Earth's forests and topography. *Science of Remote Sensing, 1*, 100002.

Duncan, B. N., Ott, L. E., Abshire, J. B., Brucker, L., Carroll, M. L., Carton, J., et al. (2020). Space-based observations for understanding changes in the Arctic-Boreal Zone. *Reviews of Geophysics, 58*(1), e2019RG000652.

Dyk, A., Leckie, D. G., Tinis, S., & Ortlepp, S. M. (2015). *Canada's National Deforestation Monitoring System: System description.*

Euskirchen, E. S., McGuire, A. D., Chapin, F. S., Yi, S., & Thompson, C. C. (2009). Changes in vegetation in northern Alaska under scenarios of climate change, 2003-2100: Implications for climate feedbacks. *Ecological Applications, 19*(4), 1022–1043. https://doi.org/10.1890/08-0806.1.

Fisher, J. B., Hayes, D. J., Schwalm, C. R., Huntzinger, D. N., Stofferahn, E., Schaefer, K., et al. (2018b). Missing pieces to modeling the Arctic-Boreal puzzle. *Environmental Research Letters, 13*(2). https://doi.org/10.1088/1748-9326/aa9d9a, 020202.

Fisher, J. B., Hayes, D. J., Schwalm, C. R., Huntzinger, D. N., Stofferahn, E., Schaefer, K., et al. (2018a). Missing pieces to modeling the Arctic-Boreal puzzle. *Environmental Research Letters.* https://doi.org/10.1088/1748-9326/aa9d9a.

Fisher, J. B., Huntzinger, D. N., Schwalm, C. R., & Sitch, S. (2014). Modeling the terrestrial biosphere. *Annual Review of Environment and Resources.* https://doi.org/10.1146/annurev-environ-012913-093456.

Fisher, J. B., Sikka, M., Oechel, W. C., Huntzinger, D. N., Melton, J. R., Koven, C. D., et al. (2014). Carbon cycle uncertainty in the Alaskan Arctic. *Biogeosciences, 11*(15). https://doi.org/10.5194/bg-11-4271-2014.

Forkel, M., Carvalhais, N., Rödenbeck, C., Keeling, R., Heimann, M., Thonicke, K., et al. (2016). Enhanced seasonal CO_2 exchange caused by amplified plant productivity in northern ecosystems. *Science, 351*(6274), 696–699. https://doi.org/10.1126/science.aac4971.

French, N. H. F., De Groot, W. J., Jenkins, L. K., Rogers, B. M., Alvarado, E., Amiro, B., et al. (2011). Model comparisons for estimating carbon emissions from North American wildland fire. *Journal of Geophysical Research: Biogeosciences*, *116*(2). https://doi.org/10.1029/2010JG001469. 0–05.

French, N. H. F., Kasischke, E. S., & Williams, D. G. (2003). Variability in the emission of carbon-based trace gases from wildfire in the Alaskan boreal forest. *Journal of Geophysical Research: Atmospheres*, *108*(1), 8151. https://doi.org/10.1029/2001jd000480.

Fyles, I. H., Shaw, C. H., Apps, M. J., Karjalainen, T., Stocks, B. J., Running, S. W., et al. (2000). The role of boreal forests and forestry in the global carbon budget: A synthesis. In *Proceedings of IBFRA 2000 conference May 8-12, 2000.*

Gauthier, S., Bernier, P., Kuuluvainen, T., Shvidenko, A. Z., & Schepaschenko, D. G. (2015). Boreal forest health and global change. *Science.* https://doi.org/10.1126/science.aaa9092.

Genet, H., McGuire, A. D., Barrett, K., Breen, A., Euskirchen, E. S., Johnstone, J. F., et al. (2013). Modeling the effects of fire severity and climate warming on active layer thickness and soil carbon storage of black spruce forests across the landscape in interior Alaska. *Environmental Research Letters*, *8*(4). https://doi.org/10.1088/1748-9326/8/4/045016, 045016.

Gobakken, T., Næsset, E., Nelson, R., Bollandsås, O. M., Gregoire, T. G., Ståhl, G., et al. (2012). Estimating biomass in Hedmark County, Norway using national forest inventory field plots and airborne laser scanning. *Remote Sensing of Environment.* https://doi.org/10.1016/j.rse.2012.01.025.

Goldblum, D., & Rigg, L. S. (2010). The deciduous forest—Boreal forest ecotone. *Geography Compass*, *4*(7), 701–717.

Goodale, C. L., Apps, M. J., Birdsey, R. A., Field, C. B., Heath, L. S., Houghton, R. A., et al. (2002). Forest carbon sinks in the Northern hemisphere. *Ecological Applications.* https://doi.org/10.1890/1051-0761(2002)012[0891:FCSITN]2.0.CO;2.

Goulden, M. L., Mcmillan, A. M. S., Winston, G. C., Rocha, A. V., Manies, K. L., Harden, J. W., et al. (2011). Patterns of NPP, GPP, respiration, and NEP during boreal forest succession. *Global Change Biology.* https://doi.org/10.1111/j.1365-2486.2010.02274.x.

Gower, S. T., Krankina, O., Olson, R. J., Apps, M., Linder, S., & Wang, C. (2001). Net primary production and carbon allocation patterns of boreal forest ecosystems. *Ecological Applications.* https://doi.org/10.1890/1051-0761(2001)011[1395:NPPACA]2.0.CO;2.

Gower, S. T., Vogel, J. G., Norman, J. M., Kucharik, C. J., Steele, S. J., & Stow, T. K. (1997). Carbon distribution and aboveground net primary production in aspen, jack pine, and black spruce stands in Saskatchewan and Manitoba, Canada. *Journal of Geophysical Research Atmospheres.* https://doi.org/10.1029/97jd02317.

Grant, R. F., Barr, A. G., Black, T. A., Margolis, H. A., Dunn, A. L., Metsaranta, J., et al. (2009). Interannual variation in net ecosystem productivity of Canadian forests as affected by regional weather patterns—A Fluxnet-Canada synthesis. *Agricultural and Forest Meteorology.* https://doi.org/10.1016/j.agrformet.2009.07.010.

Grassi, G., House, J., Dentener, F., Federici, S., Den Elzen, M., & Penman, J. (2017). The key role of forests in meeting climate targets requires science for credible mitigation. *Nature Climate Change.* https://doi.org/10.1038/nclimate3227.

Grassi, G., House, J., Kurz, W. A., Cescatti, A., Houghton, R. A., Peters, G. P., et al. (2018). Reconciling global-model estimates and country reporting of anthropogenic forest CO_2 sinks. *Nature Climate Change*, *8*(10), 914–920. https://doi.org/10.1038/s41558-018-0283-x.

Grassi, G., Stehfest, E., Rogelj, J., van Vuuren, D., Cescatti, A., House, J., et al. (2021). Critical adjustment of land mitigation pathways for assessing countries' climate progress. *Nature Climate Change*, *11*(5), 425–434.

Gromtsev, A. (2002). Natural disturbance dynamics in the boreal forests of European Russia: A review. *Silva Fennica*, *36*, 41–55.

Halme, E., Pellikka, P., & Mõttus, M. (2019). Utility of hyperspectral compared to multispectral remote sensing data in estimating forest biomass and structure variables in Finnish boreal forest. *International Journal of Applied Earth Observation and Geoinformation.* https://doi.org/10.1016/j.jag.2019.101942.

Harris, N. L., Hagen, S. C., Saatchi, S. S., Pearson, T. R. H., Woodall, C. W., Domke, G. M., et al. (2016). Attribution of net carbon change by disturbance type across forest lands of the conterminous United States. *Carbon Balance and Management, 11*(1), 24. https://doi.org/10.1186/s13021-016-0066-5.

Hayes, D. J., Kicklighter, D. W., McGuire, A. D., Chen, M., Zhuang, Q., Yuan, F., et al. (2014). The impacts of recent permafrost thaw on land-atmosphere greenhouse gas exchange. *Environmental Research Letters, 9*(4). https://doi.org/10.1088/1748-9326/9/4/045005.

Hayes, D. J., McGuire, A. D., Kicklighter, D. W., Burnside, T. J., & Melillo, J. M. (2011). The effects of land cover and land use change on the contemporary carbon balance of the arctic and boreal terrestrial ecosystems of Northern Eurasia. In *Eurasian Arctic land cover and land use in a changing climate.* https://doi.org/10.1007/978-90-481-9118-5_6.

Hayes, D. J., McGuire, A. D., Kicklighter, D. W., Gurney, K. R., Burnside, T. J., & Melillo, J. M. (2011). Is the northern high-latitude land-based CO_2 sink weakening? *Global Biogeochemical Cycles.* https://doi.org/10.1029/2010GB003813.

Hayes, D., & Turner, D. (2012). The need for "apples-to-apples" comparisons of carbon dioxide source and sink estimates. *Eos.* https://doi.org/10.1029/2012EO410007.

Hayes, D. J., Turner, D. P., Stinson, G., McGuire, A. D., Wei, Y., West, T. O., et al. (2012). Reconciling estimates of the contemporary North American carbon balance among terrestrial biosphere models, atmospheric inversions, and a new approach for estimating net ecosystem exchange from inventory-based data. *Global Change Biology, 18*(4). https://doi.org/10.1111/j.1365-2486.2011.02627.x.

Hayes, D. J., Vargas, R., Alin, S. R., Conant, R. T., Hutyra, L. R., Jacobson, A. R., et al. (2018). Chapter 2: The North American carbon budget. In N. Cavallaro, G. Shrestha, R. Birdsey, M. A. Mayes, R. G. Najjar, S. C. Reed, P. Romero-Lankao, & Z. Zhu (Eds.), *Second state of the carbon cycle report (SOCCR2): A sustained assessment report* (pp. 71–108). Washington, DC: U.S. Global Change Research Program. https://doi.org/10.7930/SOCCR2.2018.Ch2.

Helbig, M., Wischnewski, K., Kljun, N., Chasmer, L. E., Quinton, W. L., Detto, M., et al. (2016). Regional atmospheric cooling and wetting effect of permafrost thaw-induced boreal forest loss. *Global Change Biology.* https://doi.org/10.1111/gcb.13348.

Hobbie, S. E., Schimel, J. P., Trumbore, S. E., & Randerson, J. R. (2000). Controls over carbon storage and turnover in high-latitude soils. *Global Change Biology.* https://doi.org/10.1046/j.1365-2486.2000.06021.x.

Holmes, R. M., McClelland, J. W., Peterson, B. J., Tank, S. E., Bulygina, E., Eglinton, T. I., et al. (2012). Seasonal and annual fluxes of nutrients and organic matter from large rivers to the Arctic Ocean and surrounding seas. *Estuaries and Coasts, 35*(2), 369–382. https://doi.org/10.1007/s12237-011-9386-6.

Hopkinson, C., Chasmer, L., Barr, A. G., Kljun, N., Black, T. A., & McCaughey, J. H. (2016). Monitoring boreal forest biomass and carbon storage change by integrating airborne laser scanning, Biometry and eddy covariance data. *Remote Sensing of Environment.* https://doi.org/10.1016/j.rse.2016.04.010.

Horvath, P., Tang, H., Halvorsen, R., Stordal, F., Merete Tallaksen, L., Koren Berntsen, T., et al. (2021). Improving the representation of high-latitude vegetation distribution in dynamic global vegetation models. *Biogeosciences, 18*(1), 95–112. https://doi.org/10.5194/bg-18-95-2021.

Hou, Z., Domke, G. M., Russell, M. B., Coulston, J. W., Nelson, M. D., Xu, Q., et al. (2021). Updating annual state- and county-level forest inventory estimates with data assimilation and FIA data. *Forest Ecology and Management, 483*, 118777. https://doi.org/10.1016/j.foreco.2020.118777.

Hugelius, G., Strauss, J., Zubrzycki, S., Harden, J. W., Schuur, E. A. G., Ping, C. L., et al. (2014). Estimated stocks of circumpolar permafrost carbon with quantified uncertainty ranges and identified data gaps. *Biogeosciences*. https://doi.org/10.5194/bg-11-6573-2014.

Hunter, M. L. (1993). Natural fire regimes as spatial models for managing boreal forests. *Biological Conservation*. https://doi.org/10.1016/0006-3207(93)90440-C.

Huntzinger, D. N., Schaefer, K., Schwalm, C., Fisher, J. B., Hayes, D. J., Stofferahn, E., … Tian, H. (2020). Evaluation of simulated soil carbon dynamics in Arctic-Boreal ecosystems. *Environmental Research Letters, 15*(2), 025005.

Huotari, J., Ojala, A., Peltomaa, E., Nordbo, A., Launiainen, S., Pumpanen, J., et al. (2011). Long-term direct CO_2 flux measurements over a boreal lake: Five years of eddy covariance data. *Geophysical Research Letters*. https://doi.org/10.1029/2011GL048753.

Hyyppä, J., Hyyppä, H., Leckie, D., Gougeon, F., Yu, X., & Maltamo, M. (2008). Review of methods of small-footprint airborne laser scanning for extracting forest inventory data in boreal forests. *International Journal of Remote Sensing*. https://doi.org/10.1080/01431160701736489.

Intergovernmental Panel on Climate Change (IPCC). (2010). Revisiting the use of managed land as a proxy for estimating national anthropogenic emissions and removals [online]. In H. S. Eggleston, N. Srivastava, K. Tanabe, & J. Baasansuren (Eds.), *Intergovernmental panel on climate change expert meeting report, São José dos Campos, Brazil, 5–7 May 2009*. Hayama, Japan: Institute for Global Environmental Strategies. Available from https://www.ipcc-nggip.iges.or.jp/public/mtdocs/pdfiles/0905_MLP_Report.pdf.

Jacobson, A. R., Miller, J. B., Ballantyne, A., Basu, S., Bruhwiler, L., Chatterjee, A., et al. (2018). Chapter 8: Observations of atmospheric carbon dioxide and methane. In N. Cavallaro, G. Shrestha, R. Birdsey, M. A. Mayes, R. G. Najjar, S. C. Reed, P. Romero-Lankao, & Z. Zhu (Eds.), *Second state of the carbon cycle report (SOCCR2): A sustained assessment report* (pp. 337–364). Washington, DC: U.S. Global Change Research Program. https://doi.org/10.7930/SOCCR2.2018.Ch8.

Jasinevičius, G., Lindner, M., Pingoud, K., & Tykkylainen, M. (2015). Review of models for carbon accounting in harvested wood products. *International Wood Products Journal, 6*(4), 198–212. https://doi.org/10.1080/20426445.2015.1104078.

Johnston, C. M. T., & Radeloff, V. C. (2019). Global mitigation potential of carbon stored in harvested wood products. *Proceedings of the National Academy of Sciences of the United States of America, 116*(29), 14526–14531. https://doi.org/10.1073/pnas.1904231116.

Jung, M., Schwalm, C., Migliavacca, M., Walther, S., Camps-Valls, G., Koirala, S., et al. (2020). Scaling carbon fluxes from eddy covariance sites to globe: Synthesis and evaluation of the FLUXCOM approach. *Biogeosciences*. https://doi.org/10.5194/bg-17-1343-2020.

Kalliokoski, T., Mäkelä, A., Fronzek, S., Minunno, F., & Peltoniemi, M. (2018). Decomposing sources of uncertainty in climate change projections of boreal forest primary production. *Agricultural and Forest Meteorology, 262*, 192–205. https://doi.org/10.1016/j.agrformet.2018.06.030.

Kangas, A., Astrup, R., Breidenbach, J., Fridman, J., Gobakken, T., Korhonen, K. T., et al. (2018). Remote sensing and forest inventories in Nordic countries–roadmap for the future. *Scandinavian Journal of Forest Research, 33*(4), 397–412. Taylor and Francis AS https://doi.org/10.1080/02827581.2017.1416666.

228 Section | C Case Studies

Kasischke, E. S. (2000). *Boreal ecosystems in the global carbon cycle.* https://doi.org/10.1007/978-0-387-21629-4_2.

Kasischke, E. S., & Bruhwiler, L. P. (2002). Emissions of carbon dioxide, carbon monoxide, and methane from boreal forest fires in 1998. *Journal of Geophysical Research.* https://doi.org/10.1029/2001jd000461.

Kasischke, E. S., & Turetsky, M. R. (2006). Recent changes in the fire regime across the North American boreal region—Spatial and temporal patterns of burning across Canada and Alaska. *Geophysical Research Letters.* https://doi.org/10.1029/2006GL025677.

Kasischke, E. S., Verbyla, D. L., Rupp, S., McGuire, D., Murphy, K. A., Jandt, R., et al. (2010a). Alaska's changing fire regime—Implications for the vulnerability of its boreal forests. *Canadian Journal of Forest Research.* https://doi.org/10.1139/X10-061.

Kasischke, E. S., Verbyla, D., Rupp, T. S., McGuire, A. D., Murphy, K. A., Jandt, R., et al. (2010b). Alaska's changing fire regime—Implications for the vulnerability of its boreal forests. *Canadian Journal of Forest Research, 40*, 1313–1324.

Kasischke, E. S., Verbyla, D. L., Rupp, T. S., McGuire, A. D., Murphy, K. A., Jandt, R., et al. (2010c). Alaska's changing fire regime—Implications for the vulnerability of its boreal forests. This article is one of a selection of papers from The Dynamics of Change in Alaska's Boreal Forests: Resilience and Vulnerability in Response to Climate Warming *Canadian Journal of Forest Research.* https://doi.org/10.1139/X10-098.

Kasischke, E. S., Williams, D., & Barry, D. (2002). Analysis of the patterns of large fires in the boreal forest region of Alaska. *International Journal of Wildland Fire, 11*(2), 131–144. https://doi.org/10.1071/WF02023.

Kharuk, V. I., Ponomarev, E. I., Ivanova, G. A., Dvinskaya, M. L., Coogan, S. C., & Flannigan, M. D. (2021). Wildfires in the Siberian taiga. *Ambio,* 1–22. https://doi.org/10.1007/s13280-020-01490-x.

Kharuk, V. I., Ranson, K. J., Petrov, I. A., Dvinskaya, M. L., Im, S. T., & Golyukov, A. S. (2019). Larch (Larix Dahurica Turcz) growth response to climate change in the Siberian permafrost zone. *Regional Environmental Change, 19*(1), 233–243. https://doi.org/10.1007/s10113-018-1401-z.

Kicklighter, D. W., Hayes, D. J., Mcclelland, J. W., Peterson, B. J., McGuire, A. D., & Melillo, J. M. (2013). Insights and issues with simulating terrestrial DOC loading of Arctic river networks. *Ecological Applications, 23*(8). https://doi.org/10.1890/11-1050.1.

Kimball, J. S., Keyser, A. R., Running, S. W., & Saatchi, S. S. (2000). Regional assessment of boreal forest productivity using an ecological process model and remote sensing parameter maps. *Tree Physiology.* https://doi.org/10.1093/treephys/20.11.761.

Koven, C. D., Riley, W. J., & Stern, A. (2013). Analysis of permafrost thermal dynamics and response to climate change in the CMIP5 earth system models. *Journal of Climate, 26*(6), 1877–1900. https://doi.org/10.1175/JCLI-D-12-00228.1.

Krylov, A., McCarty, J. L., Potapov, P., Loboda, T., Tyukavina, A., Turubanova, S., et al. (2014). Remote sensing estimates of stand-replacement fires in Russia, 2002-2011. *Environmental Research Letters, 9*(10), 105007. https://doi.org/10.1088/1748-9326/9/10/105007.

Kuhry, P., Grosse, G., Harden, J. W., Hugelius, G., Koven, C. D., Ping, C. L., et al. (2013). Characterisation of the permafrost carbon pool. *Permafrost and Periglacial Processes.* https://doi.org/10.1002/ppp.1782.

Kurbatova, J., Li, C., Varlagin, A., Xiao, X., & Vygodskaya, N. (2008). Modeling carbon dynamics in two adjacent spruce forests with different soil conditions in Russia. *Biogeosciences.* https://doi.org/10.5194/bg-5-969-2008.

Kurz, W. A., Dymond, C. C., Stinson, G., Rampley, G. J., Neilson, E. T., Carroll, A. L., et al. (2008). Mountain pine beetle and forest carbon feedback to climate change. *Nature*. https://doi.org/10.1038/nature06777.

Kurz, W. A., Dymond, C. C., White, T. M., Stinson, G., Shaw, C. H., Rampley, G. J., et al. (2009). CBM-CFS3: A model of carbon-dynamics in forestry and land-use change implementing IPCC standards. *Ecological Modelling*. https://doi.org/10.1016/j.ecolmodel.2008.10.018.

Kurz, W. A., Hayne, S., Fellows, M., Macdonald, J. D., Metsaranta, J. M., Hafer, M., et al. (2018). Quantifying the impacts of human activities on reported greenhouse gas emissions and removals in Canada's managed forest: Conceptual framework and implementation. *Canadian Journal of Forest Research*, 48(10), 1227–1240. https://doi.org/10.1139/cjfr-2018-0176.

Kurz, W. A., Shaw, C. H., Boisvenue, C., Stinson, G., Metsaranta, J., Leckie, D., et al. (2013). Carbon in Canada's boreal forest—A synthesis. *Environmental Reviews*. https://doi.org/10.1139/er-2013-0041.

Kurz, W. A., Stinson, G., & Rampley, G. (2008). Could increased boreal forest ecosystem productivity offset carbon losses from increased disturbances? *Philosophical Transactions of the Royal Society B: Biological Sciences*. https://doi.org/10.1098/rstb.2007.2198.

Kurz, W. A., Stinson, G., Rampley, G. J., Dymond, C. C., & Neilson, E. T. (2008). Risk of natural disturbances makes future contribution of Canada's forests to the global carbon cycle highly uncertain. *Proceedings of the National Academy of Sciences of the United States of America*. https://doi.org/10.1073/pnas.0708133105.

Lagergren, F., Lindroth, A., Dellwik, E., Ibrom, A., Lankreijer, H., Launiainen, S., et al. (2008). Biophysical controls on CO2 fluxes of three Northern forests based on long-term eddy covariance data. *Tellus, Series B: Chemical and Physical Meteorology*. https://doi.org/10.1111/j.1600-0889.2006.00324.x.

Lammers, R. B., Shiklomanov, A. I., Vörösmarty, C. J., Fekete, B. M., & Peterson, B. J. (2001). Assessment of contemporary Arctic river runoff based on observational discharge records. *Journal of Geophysical Research Atmospheres*, 106(D4), 3321–3334. https://doi.org/10.1029/2000JD900444.

Laudon, H., Berggren, M., Ågren, A., Buffam, I., Bishop, K., Grabs, T., et al. (2011). Patterns and dynamics of dissolved organic carbon (DOC) in boreal streams: The role of processes, connectivity, and scaling. *Ecosystems*, 14(6), 880–893. https://doi.org/10.1007/s10021-011-9452-8.

Le Quéré, C., Andrew, R. M., Canadell, J. G., Sitch, S., Ivar Korsbakken, J., Peters, G. P., et al. (2016). Global carbon budget 2016. *Earth System Science Data*. https://doi.org/10.5194/essd-8-605-2016.

Le Quéré, C., Andrew, R. M., Friedlingstein, P., Sitch, S., Pongratz, J., Manning, A. C., et al. (2017). Global carbon budget 2017. *Earth System Science Data Discussions*. https://doi.org/10.5194/essd-2017-123.

Leckie, D. G., & Gillis, M. D. (1995). Forest inventory in Canada with emphasis on map production. *Forestry Chronicle*. https://doi.org/10.5558/tfc71074-1.

Li, M., Peng, C., Wang, M., Xue, W., Zhang, K., Wang, K., et al. (2017). The carbon flux of global rivers: A re-evaluation of amount and spatial patterns. *Ecological Indicators*, 80, 40–51. https://doi.org/10.1016/j.ecolind.2017.04.049.

Lim, K., Treitz, P., Wulder, M., St-Ongé, B., & Flood, M. (2003). LiDAR remote sensing of forest structure. *Progress in Physical Geography*. https://doi.org/10.1191/0309133303pp360ra.

Liu, J., Chen, J. M., Cihlar, J., & Park, W. M. (1997). A process-based boreal ecosystem productivity simulator using remote sensing inputs. *Remote Sensing of Environment*. https://doi.org/10.1016/S0034-4257(97)00089-8.

230 Section | C Case Studies

Ma, Z., Peng, C., Zhu, Q., Chen, H., Yu, G., Li, W., et al. (2012). Regional drought-induced reduction in the biomass carbon sink of Canada's boreal forests. *Proceedings of the National Academy of Sciences of the United States of America*. https://doi.org/10.1073/pnas.1111576109.

Maclean, G. A., & Martin, G. L. (1984). Merchantable timber volume estimation using cross-sectional photogrammetric and densitometric methods. *Canadian Journal of Forest Research*. https://doi.org/10.1139/x84-142.

Magnusson, M., Fransson, J. E. S., & Olsson, H. (2007). Aerial photo-interpretation using Z/I DMC images for estimation of forest variables. *Scandinavian Journal of Forest Research*, *22*(3), 254–266. https://doi.org/10.1080/02827580701262964.

Maltamo, M., & Packalen, P. (2014). *Species-specific management inventory in Finland* (pp. 241–252). https://doi.org/10.1007/978-94-017-8663-8_12.

Manizza, M., Follows, M. J., Dutkiewicz, S., McClelland, J. W., Menemenlis, D., Hill, C. N., et al. (2009). Modeling transport and fate of riverine dissolved organic carbon in the Arctic Ocean. *Global Biogeochemical Cycles*, *23*(4). https://doi.org/10.1029/2008GB003396.

Margolis, H. A., Nelson, R. F., Montesano, P. M., Beaudoin, A., Sun, G., Andersen, H. E., et al. (2015). Combining satellite lidar, airborne lidar, and ground plots to estimate the amount and distribution of aboveground biomass in the boreal forest of North America. *Canadian Journal of Forest Research*, *45*(7), 838–855.

Matasci, G., Hermosilla, T., Wulder, M. A., White, J. C., Coops, N. C., Hobart, G. W., et al. (2018). Three decades of forest structural dynamics over Canada's forested ecosystems using Landsat time-series and lidar plots. *Remote Sensing of Environment*. https://doi.org/10.1016/j.rse.2018.07.024.

McClelland, J. W., Holmes, R. M., Peterson, B. J., Amon, R., Brabets, T., Cooper, L., et al. (2008). Development of Pan-Arctic database for river chemistry. *Eos*, *89*(24), 217–218. https://doi.org/10.1029/2008EO240001.

McGuire, A. D., Anderson, L. G., Christensen, T. R., Scott, D., Laodong, G., Hayes, D. J., et al. (2009). Sensitivity of the carbon cycle in the Arctic to climate change. *Ecological Monographs*, *79*(4), 523–555. https://doi.org/10.1890/08-2025.1.

McGuire, A. D., Christensen, T. R., Hayes, D., Heroult, A., Euskirchen, E., Kimball, J. S., et al. (2012). An assessment of the carbon balance of Arctic tundra: Comparisons among observations, process models, and atmospheric inversions. *Biogeosciences*, *9*(8). https://doi.org/10.5194/bg-9-3185-2012.

McGuire, A. D., Hayes, D. J., Kicklighter, D. W., Manizza, M., Zhuang, Q., Chen, M., et al. (2010). An analysis of the carbon balance of the Arctic Basin from 1997 to 2006. *Tellus, Series B: Chemical and Physical Meteorology*, *62*(5). https://doi.org/10.1111/j.1600-0889.2010.00497.x.

McGuire, A. D., Koven, C., Lawrence, D. M., Clein, J. S., Xia, J., Beer, C., et al. (2016). Variability in the sensitivity among model simulations of permafrost and carbon dynamics in the permafrost region between 1960 and 2009. *Global Biogeochemical Cycles*, *30*(7). https://doi.org/10.1002/2016GB005405.

McGuire, A. D., Lawrence, D. M., Koven, C., Clein, J. S., Burke, E., Chen, G., et al. (2018). Dependence of the evolution of carbon dynamics in the northern permafrost region on the trajectory of climate change. *Proceedings of the National Academy of Sciences of the United States of America*. https://doi.org/10.1073/pnas.1719903115.

McGuire, A. D., Sitch, S., Clein, J. S., Dargaville, R., Esser, G., Foley, J., et al. (2001). Carbon balance of the terrestrial biosphere in the twentieth century: Analyses of CO2, climate and land use effects with four process-based ecosytem models. *Global Biogeochemical Cycles*. https://doi.org/10.1029/2000GB001298.

Boreal forests **Chapter | 6 231**

McRoberts, R. E., Tomppo, E. O., & Næsset, E. (2010). Advances and emerging issues in national forest inventories. *Scandinavian Journal of Forest Research*, *25*(4), 368–381. Taylor & Francis Group https://doi.org/10.1080/02827581.2010.496739.

Mekonnen, Z. A., Riley, W. J., Randerson, J. T., Grant, R. F., & Rogers, B. M. (2019). Expansion of high-latitude deciduous forests driven by interactions between climate warming and fire. *Nature Plants*, *5*(9), 952–958. Palgrave Macmillan Ltd https://doi.org/10.1038/s41477-019-0495-8.

Montesano, P. M., Neigh, C. S. R., Macander, M., Feng, M., & Noojipady, P. (2020). The bioclimatic extent and pattern of the cold edge of the Boreal forest: The circumpolar taiga-tundra ecotone. *Environmental Research Letters*, *15*(10), 105019. https://doi.org/10.1088/1748-9326/abb2c7.

Montesano, P. M., Nelson, R. F., Dubayah, R. O., Sun, G., Cook, B. D., Ranson, K. J. R., et al. (2014). The uncertainty of biomass estimates from LiDAR and SAR across a boreal forest structure gradient. *Remote Sensing of Environment*. https://doi.org/10.1016/j.rse.2014.01.027.

Morison, J., Kwok, R., Peralta-Ferriz, C., Alkire, M., Rigor, I., Andersen, R., et al. (2012). Changing Arctic Ocean freshwater pathways. *Nature*, *481*(7379), 66–70. https://doi.org/10.1038/nature10705.

Myneni, R. B., Dong, J., Tucker, C. J., Kaufmann, R. K., Kauppi, P. E., Liski, J., et al. (2001). A large carbon sink in the woody biomass of Northern forests. *Proceedings of the National Academy of Sciences of the United States of America*. https://doi.org/10.1073/pnas.261555198.

Næsset, E. (2002). Determination of mean tree height of forest stands by digital photogrammetry. *Scandinavian Journal of Forest Research*. https://doi.org/10.1080/028275802320435469.

Næsset, E. (2004). Practical large-scale forest stand inventory using a small-footprint airborne scanning laser. *Scandinavian Journal of Forest Research*. https://doi.org/10.1080/02827580310019257.

Næsset, E. (2014). *Area-based inventory in Norway—From innovation to an operational reality* (pp. 215–240). https://doi.org/10.1007/978-94-017-8663-8_11.

Næsset, E., Gobakken, T., Holmgren, J., Hyyppä, H., Hyyppä, J., Maltamo, M., et al. (2004). Laser scanning of forest resources: The Nordic experience. *Scandinavian Journal of Forest Research*, *19*(6), 482–499. https://doi.org/10.1080/02827580410019553.

Narine, L. L., Popescu, S., Neuenschwander, A., Zhou, T., Srinivasan, S., & Harbeck, K. (2019). Estimating aboveground biomass and forest canopy cover with simulated ICESat-2 data. *Remote Sensing of Environment*, *224*, 1–11.

Navarro, L., Morin, H., Bergeron, Y., & Girona, M. M. (2018). Changes in spatiotemporal patterns of 20th century spruce budworm outbreaks in eastern Canadian boreal forests. *Frontiers in Plant Science*. https://doi.org/10.3389/fpls.2018.01905.

Neigh, C. S. R., Nelson, R. F., Ranson, K. J., Margolis, H. A., Montesano, P. M., Sun, G., et al. (2013). Taking stock of circumboreal forest carbon with ground measurements, airborne and spaceborne LiDAR. *Remote Sensing of Environment*. https://doi.org/10.1016/j.rse.2013.06.019.

Neumann, M., Saatchi, S. S., Ulander, L. M. H., & Fransson, J. E. S. (2012). Assessing performance of L- and P-band polarimetric interferometric SAR data in estimating boreal forest aboveground biomass. *IEEE Transactions on Geoscience and Remote Sensing*. https://doi.org/10.1109/TGRS.2011.2176133.

Ogle, S. M., Domke, G., Kurz, W. A., Rocha, M. T., Huffman, T., Swan, A., et al. (2018). Delineating managed land for reporting national greenhouse gas emissions and removals to the United Nations framework convention on climate change. *Carbon Balance and Management*, *13*(1), 1–13. Springer https://doi.org/10.1186/s13021-018-0095-3.

Ogle, S. M., & Kurz, W. A. (2021). Land-based emissions. *Nature Climate Change*, *11*(5), 382–383.

232 Section | C Case Studies

Pan, Y., Birdsey, R. A., Fang, J., Houghton, R., Kauppi, P. E., Kurz, W. A., et al. (2011). A large and persistent carbon sink in the world's forests. *Science, 333*(6045), 988–993. https://doi.org/10.1126/science.1201609.

Paradis, L., Thiffault, E., & Achim, A. (2019). Comparison of carbon balance and climate change mitigation potential of forest management strategies in the boreal forest of Quebec (Canada). *Forestry, 92*(3), 264–277. https://doi.org/10.1093/forestry/cpz004.

Pastor, J., Solin, J., Bridgham, S. D., Updegraff, K., Harth, C., Weishampel, P., et al. (2003). Global warming and the export of dissolved organic carbon from boreal peatlands. *Oikos.* https://doi.org/10.1034/j.1600-0706.2003.11774.x.

Pattison, R., Andersen, H. E., Gray, A., Schulz, B., Smith, R. J., & Jovan, S. (2018). *Forests of the Tanana Valley state forest and Tetlin National Wildlife Refuge, Alaska: Results of the 2014 pilot inventory.* Gen. Tech. Rep. PNW-GTR-967 (p. 80). Portland, OR: US Department of Agriculture, Forest Service, Pacific Northwest Research Station. 967.

Penman, J., Gytarsky, M., Hiraishi, T., Irving, W., & Krug, T. (2006). 2006 IPCC—Guidelines for National Greenhouse Gas Inventories. In *Directrices para los inventarios nacionales GEI.* https://www.ipcc.ch/report/2006-ipcc-guidelines-for-national-greenhouse-gas-inventories/.

Podgrajsek, E., Sahlée, E., Bastviken, D., Natchimuthu, S., Kljun, N., Chmiel, H. E., et al. (2016). Methane fluxes from a small boreal lake measured with the eddy covariance method. *Limnology and Oceanography.* https://doi.org/10.1002/lno.10245.

Popescu, S. C., Zhou, T., Nelson, R., Neuenschwander, A., Sheridan, R., Narine, L., et al. (2018). Photon counting LiDAR: An adaptive ground and canopy height retrieval algorithm for ICESat-2 data. *Remote Sensing of Environment, 208*, 154–170. https://doi.org/10.1016/j.rse.2018.02.019.

Price, D. T., Alfaro, R. I., Brown, K. J., Flannigan, M. D., Fleming, R. A., Hogg, E. H., et al. (2013). Anticipating the consequences of climate change for Canada's boreal forest ecosystems. *Environmental Reviews.* https://doi.org/10.1139/er-2013-0042.

Quegan, S., Le Toan, T., Chave, J., Dall, J., Exbrayat, J. F., Minh, D. H. T., et al. (2019). The European Space Agency BIOMASS mission: Measuring forest above-ground biomass from space. *Remote Sensing of Environment, 227*, 44–60.

Queinnec, M., White, J. C., & Coops, N. C. (2021). Comparing airborne and spaceborne photon-counting LiDAR canopy structural estimates across different boreal forest types. *Remote Sensing of Environment, 262*, 112510.

Rignot, E., Williams, C., & Viereck, L. (1994). Radar estimates of aboveground biomass in boreal forests of interior Alaska. *IEEE Transactions on Geoscience and Remote Sensing.* https://doi.org/10.1109/36.312903.

Rinne, J., Riutta, T., Pihlatie, M., Aurela, M., Haapanala, S., Tuovinen, J. P., et al. (2007). Annual cycle of methane emission from a boreal fen measured by the eddy covariance technique. *Tellus, Series B: Chemical and Physical Meteorology.* https://doi.org/10.1111/j.1600-0889.2007.00261.x.

Rödenbeck, C., Zaehle, S., Keeling, R., & Heimann, M. (2018). How does the terrestrial carbon exchange respond to inter-annual climatic variations? A quantification based on atmospheric CO2 data. *Biogeosciences.* https://doi.org/10.5194/bg-15-2481-2018.

Rogers, B. M., Soja, A. J., Goulden, M. L., & Randerson, J. T. (2015). Influence of tree species on continental differences in boreal fires and climate feedbacks. *Nature Geoscience.* https://doi.org/10.1038/ngeo2352.

Roulet, N. T., Ash, R., & Moore, T. R. (1992). Low boreal wetlands as a source of atmospheric methane. *Journal of Geophysical Research.* https://doi.org/10.1029/91JD03109.

Sandberg, G., Ulander, L. M. H., Fransson, J. E. S., Holmgren, J., & Le Toan, T. (2011). L- and P-band backscatter intensity for biomass retrieval in hemiboreal forest. *Remote Sensing of Environment.* https://doi.org/10.1016/j.rse.2010.03.018.

Santoro, M., Beaudoin, A., Beer, C., Cartus, O., Fransson, J. E. S., Hall, R. J., et al. (2015). Forest growing stock volume of the northern hemisphere: Spatially explicit estimates for 2010 derived from Envisat ASAR. *Remote Sensing of Environment.* https://doi.org/10.1016/j.rse.2015.07.005.

Saucier, J. P., Baldwin, K., Krestov, P., & Jorgenson, T. (2015). Boreal forests. In *Routledge handbook of forest ecology.* https://doi.org/10.4324/9781315818290.

Saunois, M., Jackson, R. B., Bousquet, P., Poulter, B., & Canadell, J. G. (2016). The growing role of methane in anthropogenic climate change. *Environmental Research Letters.* https://doi.org/10.1088/1748-9326/11/12/120207.

Schimel, D., Pavlick, R., Fisher, J. B., Asner, G. P., Saatchi, S., Townsend, P., et al. (2015). Observing terrestrial ecosystems and the carbon cycle from space. *Global Change Biology, 21*(5), 1762–1776.

Schulze, E. D., Lloyd, J., Kelliher, F. M., Wirth, C., Rebmann, C., Luhker, B., et al. (1999). Productivity of forests in the eurosiberian boreal region and their potential to act as a carbon sink—A synthesis. *Global Change Biology.* https://doi.org/10.1046/j.1365-2486.1999.00266.x.

Schuur, E. A. G., McGuire, A. D., Romanovsky, V., Schädel, C., & Mack, M. (2018). Chapter 11: Arctic and boreal carbon. In N. Cavallaro, G. Shrestha, R. Birdsey, M. A. Mayes, R. G. Najjar, S. C. Reed, P. Romero-Lankao, & Z. Zhu (Eds.), *Second state of the carbon cycle report (SOCCR2): A sustained assessment report* (pp. 428–468). Washington, DC: U.S. Global Change Research Program. https://doi.org/10.7930/SOCCR2.2018.Ch11.

Schuur, E. A. G., McGuire, A. D., Schädel, C., Grosse, G., Harden, J. W., Hayes, D. J., et al. (2015). Climate change and the permafrost carbon feedback. *Nature.* https://doi.org/10.1038/nature14338.

Serreze, M. C., & Barry, R. G. (2011). Processes and impacts of Arctic amplification: A research synthesis. *Global and Planetary Change.* https://doi.org/10.1016/j.gloplacha.2011.03.004.

Shaw, C. H., Hilger, A. B., Metsaranta, J., Kurz, W. A., Russo, G., Eichel, F., et al. (2014). Evaluation of simulated estimates of forest ecosystem carbon stocks using ground plot data from Canada's National Forest Inventory. *Ecological Modelling, 272,* 323–347. https://doi.org/10.1016/j.ecolmodel.2013.10.005.

Shvidenko, A., & Nilsson, S. (2002). Dynamics of Russian forests and the carbon budget in 1961-1998: An assessment based on long-term forest inventory data. *Climatic Change, 55*(1–2), 5–37. https://doi.org/10.1023/A:1020243304744.

Sitch, S., David McGuire, A., Kimball, J., Gedney, N., Gamon, J., Engstrom, R., et al. (2007). Assessing the carbon balance of circumpolar Arctic tundra using remote sensing and process modeling. *Ecological Applications.* https://doi.org/10.1890/1051-0761(2007)017[0213:ATCBOC]2.0.CO;2.

Skog, K. E., Pingoud, K., & Smith, J. E. (2004). A method countries can use to estimate changes in carbon stored in harvested wood products and the uncertainty of such estimates. *Environmental Management.* https://doi.org/10.1007/s00267-003-9118-1.

Smith, J. E., Heath, L. S., Skog, K. E., & Birdsey, R. A. (2006). *Methods for calculating forest ecosystem and harvested carbon with standard estimates for forest types of the United States.* Gen. Tech. Rep. NE-343 *Vol. 343* (p. 216). Newtown Square, PA: U.S. Department of Agriculture, Forest Service, Northeastern Research Station. https://doi.org/10.2737/NE-GTR-343.

234 Section | C Case Studies

Stackpoole, S. M., Butman, D. E., Clow, D. W., Verdin, K. L., Gaglioti, B. V., Genet, H., et al. (2017). Inland waters and their role in the carbon cycle of Alaska. *Ecological Applications, 27*(5), 1403–1420. https://doi.org/10.1002/eap.1552.

Stephens, B. B., Gurney, K. R., Tans, P. P., Sweeney, C., Peters, W., Bruhwiler, L., et al. (2007). Weak northern and strong tropical land carbon uptake from vertical profiles of atmospheric CO2. *Science, 316*(5832), 1732–1735. https://doi.org/10.1126/science.1137004.

Stinson, G., Kurz, W. A., Smyth, C. E., Neilson, E. T., Dymond, C. C., Metsaranta, J. M., et al. (2011). An inventory-based analysis of Canada's managed forest carbon dynamics, 1990 to 2008. *Global Change Biology*. https://doi.org/10.1111/j.1365-2486.2010.02369.x.

Stocks, B. J., Mason, J. A., Todd, J. B., Bosch, E. M., Wotton, B. M., Amiro, B. D., et al. (2003). Large forest fires in Canada, 1959-1997. *Journal of Geophysical Research: Atmospheres, 108* (1), 1959–1997. https://doi.org/10.1029/2001jd000484.

Stofferahn, E., Fisher, J. B., Hayes, D. J., Schwalm, C. R., Huntzinger, D. N., Hantson, W., et al. (2019). The Arctic-Boreal vulnerability experiment model benchmarking system. *Environmental Research Letters, 14*(5). https://doi.org/10.1088/1748-9326/ab10fa.

Sukhinin, A. I., French, N. H. F., Kasischke, E. S., Hewson, J. H., Soja, A. J., Csiszar, I. A., et al. (2004). AVHRR-based mapping of fires in Russia: New products for fire management and carbon cycle studies. *Remote Sensing of Environment, 93*(4), 546–564. https://doi.org/10.1016/j.rse.2004.08.011.

Tank, S. E., Raymond, P. A., Striegl, R. G., McClelland, J. W., Holmes, R. M., Fiske, G. J., et al. (2012). A land-to-ocean perspective on the magnitude, source and implication of DIC flux from major Arctic rivers to the Arctic Ocean. *Global Biogeochemical Cycles, 26*(4). https://doi.org/10.1029/2011GB004192, 2012, GB4018.

Taylor, S. W., Carroll, A. L., Alfaro, R. I., & Safranyik, L. (2006). Forest, climate and mountain pine beetle outbreak dynamics in western Canada. In L. Safranyik, & W. Wilson (Eds.), *The mountain pine beetle: A synthesis of biology, management, and impacts on lodgepole pine* (p. 304). Victoria, British Columbia: Natural Resources Canada, Canadian Forest Service, Pacific Forestry Centre. https://www.for.gov.bc.ca/hfd/library/documents/bib96122.pdf.

Thompson, R. L., Sasakawa, M., Machida, T., Aalto, T., Worthy, D., Lavric, J. V., et al. (2017). Methane fluxes in the high northern latitudes for 2005-2013 estimated using a Bayesian atmospheric inversion. *Atmospheric Chemistry and Physics, 17*(5), 3553–3572. https://doi.org/10.5194/acp-17-3553-2017.

Tomppo, E., Gschwantner, T., Lawrence, M., & McRoberts, R. E. (2010). *National forest inventories: Pathways for common reporting*. https://doi.org/10.1007/978-90-481-3233-1.

Tramontana, G., Jung, M., Schwalm, C. R., Ichii, K., Camps-Valls, G., Ráduly, B., et al. (2016). Predicting carbon dioxide and energy fluxes across global FLUXNET sites with regression algorithms. *Biogeosciences*. https://doi.org/10.5194/bg-13-4291-2016.

Treitz, P., Lim, K., Woods, M., Pitt, D., Nesbitt, D., & Etheridge, D. (2012). LiDAR sampling density for forest resource inventories in Ontario, Canada. *Remote Sensing*. https://doi.org/10.3390/rs4040830.

Triviño, M., Juutinen, A., Mazziotta, A., Miettinen, K., Podkopaev, D., Reunanen, P., et al. (2015). Managing a boreal forest landscape for providing timber, storing and sequestering carbon. *Ecosystem Services, 14*, 179–189. https://doi.org/10.1016/j.ecoser.2015.02.003.

Turetsky, M. R., Kane, E. S., Harden, J. W., Ottmar, R. D., Manies, K. L., Hoy, E., et al. (2011). Recent acceleration of biomass burning and carbon losses in Alaskan forests and peatlands. *Nature Geoscience*. https://doi.org/10.1038/ngeo1027.

Ueyama, M., Tahara, N., Iwata, H., Euskirchen, E. S., Ikawa, H., Kobayashi, H., et al. (2016). Optimization of a biochemical model with eddy covariance measurements in black spruce forests of

Alaska for estimating CO2 fertilization effects. *Agricultural and Forest Meteorology.* https://doi.org/10.1016/j.agrformet.2016.03.007.

Van Der Werf, G. R., Randerson, J. T., Giglio, L., Van Leeuwen, T. T., Chen, Y., Rogers, B. M., et al. (2017). Global fire emissions estimates during 1997-2016. *Earth System Science Data.* https://doi.org/10.5194/essd-9-697-2017.

Verbyla, D. (2011). Browning boreal forests of western North America. *Environmental Research Letters.* https://doi.org/10.1088/1748-9326/6/4/041003.

Virkkala, A., Aalto, J., Rogers, B. M., Tagesson, T., Treat, C. C., Natali, S. M., et al. (2021). Statistical upscaling of ecosystem CO_2 fluxes across the terrestrial tundra and boreal domain: Regional patterns and uncertainties. *Global Change Biology.* https://doi.org/10.1111/gcb.15659. gcb.15659.

Vonk, J. E., Tank, S. E., Bowden, W. B., Laurion, I., Vincent, W. F., Alekseychik, P., et al. (2015). Reviews and syntheses: Effects of permafrost thaw on Arctic aquatic ecosystems. *Biogeosciences, 12*(23), 7129–7167. Copernicus GmbH https://doi.org/10.5194/bg-12-7129-2015.

Walker, X. J., Baltzer, J. L., Cumming, S. G., Day, N. J., Ebert, C., Goetz, S., et al. (2019). Increasing wildfires threaten historic carbon sink of boreal forest soils. *Nature, 572*(7770), 520–523. https://doi.org/10.1038/s41586-019-1474-y.

Walsh, J. E. (2014). Intensified warming of the Arctic: Causes and impacts on middle latitudes. *Global and Planetary Change.* https://doi.org/10.1016/j.gloplacha.2014.03.003.

Wang, J. A., Baccini, A., Farina, M., Randerson, J. T., & Friedl, M. A. (2021). Disturbance suppresses the aboveground carbon sink in North American boreal forests. *Nature Climate Change, 11*(5), 435–441. https://doi.org/10.1038/s41558-021-01027-4.

Wang, M., Wu, J., Lafleur, P. M., Luan, J., Chen, H., & Zhu, X. (2018). Temporal shifts in controls over methane emissions from a boreal bog. *Agricultural and Forest Meteorology.* https://doi.org/10.1016/j.agrformet.2018.07.002.

Weed, A. S., Ayres, M. P., & Hicke, J. A. (2013). Consequences of climate change for biotic disturbances in North American forests. *Ecological Monographs.* https://doi.org/10.1890/13-0160.1.

Welp, L. R., Patra, P. K., Rödenbeck, C., Nemani, R., Bi, J., Piper, S. C., et al. (2016). Increasing summer net CO2 uptake in high northern ecosystems inferred from atmospheric inversions and comparisons to remote-sensing NDVI. *Atmospheric Chemistry and Physics.* https://doi.org/10.5194/acp-16-9047-2016.

White, J. C., Chen, H., Woods, M. E., Low, B., & Nasonova, S. (2019). The Petawawa research forest: Establishment of a remote sensing supersite. *The Forestry Chronicle, 95*(3), 149–156.

White, J. C., Tompalski, P., Vastaranta, M. A., Wulder, M. A., Saarinen, N. P., Stepper, C., et al. (2017). *A model development and application guide for generating an enhanced forest inventory using airborne laser scanning data and an area-based approach.* Natural Resources Canada.

White, J. C., Wulder, M. A., Varhola, A., Vastaranta, M., Coops, N. C., Cook, B. D., et al. (2013). A best practices guide for generating forest inventory attributes from airborne laser scanning data using an area-based approach. *The Forestry Chronicle, 89*(6), 722–723.

White, J. C., Wulder, M. A., Vastaranta, M., Coops, N. C., Pitt, D., & Woods, M. (2013). The utility of image-based point clouds for forest inventory: A comparison with airborne laser scanning. *Forests.* https://doi.org/10.3390/f4030518.

Wilson, B. T., Woodall, C. W., & Griffith, D. M. (2013). Imputing forest carbon stock estimates from inventory plots to a nationally continuous coverage. *Carbon Balance and Management, 8*(1), 1–15. https://doi.org/10.1186/1750-0680-8-1.

Woodall, C. W., Heath, L. S., Domke, G. M., & Nichols, M. C. (2011). *Methods and equations for estimating aboveground volume, biomass, and carbon for trees in the U. S. forest inventory, 2010*. Forest Service.

Wulder, M. A., Bater, C. W., Coops, N. C., Hilker, T., & White, J. C. (2008). The role of LiDAR in sustainable forest management. *The Forestry Chronicle, 84*(6), 807–826.

Wulder, M. A., White, J. C., Nelson, R. F., Næsset, E., Ørka, H. O., Coops, N. C., et al. (2012). Lidar sampling for large-area forest characterization: A review. *Remote Sensing of Environment, 121,* 196–209.

Xing, D., Bergeron, J. A. C., Solarik, K. A., Tomm, B., Macdonald, S. E., Spence, J. R., et al. (2019). Challenges in estimating forest biomass: Use of allometric equations for three boreal tree species. *Canadian Journal of Forest Research, 49*(12), 1613–1622. https://doi.org/10.1139/cjfr-2019-0258.

Zha, T. S., Barr, A. G., Bernier, P. Y., Lavigne, M. B., Trofymow, J. A., Amiro, B. D., et al. (2013). Gross and aboveground net primary production at Canadian forest carbon flux sites. *Agricultural and Forest Meteorology*. https://doi.org/10.1016/j.agrformet.2013.02.004.

Zhang, T., Barry, R. G., Knowles, K., Heginbottom, J. A., & Brown, J. (2008). Statistics and characteristics of permafrost and ground-ice distribution in the Northern hemisphere. *Polar Geography*. https://doi.org/10.1080/10889370802175895.

Zhang, X., Chen, J., Dias, A. C., & Yang, H. (2020). Improving carbon stock estimates for in-use harvested wood products by linking production and consumption—A global case study. *Environmental Science and Technology, 54*(5), 2565–2574. https://doi.org/10.1021/acs.est.9b05721.

Chapter 7

State of science in carbon budget assessments for temperate forests and grasslands

Masayuki Kondo[a,b], Richard Birdsey[c], Thomas A.M. Pugh[d,e], Ronny Lauerwald[f], Peter A. Raymond[g], Shuli Niu[h], and Kim Naudts[i]

[a]Institute for Space-Earth Environmental Research, Nagoya University, Nagoya, Japan, [b]Center for Global Environmental Research, National Institute for Environmental Studies, Tsukuba, Japan, [c]Woodwell Climate Research Center, Falmouth, MA, United States, [d]Department of Physical Geography and Ecosystem Science, Lund University, Lund, Sweden, [e]School of Geography Earth & Environmental Sciences and Birmingham Institute of Forest Research, University of Birmingham, Birmingham, United Kingdom, [f]UMR Ecosys, Université Paris-Saclay, Paris, France, [g]Yale School of the Environment, Yale University, New Haven, CT, United States, [h]Key Laboratory of Ecosystem Network Observation and Modeling, Institute of Geographic Sciences and Natural Resources Research, Chinese Academy of Sciences, Beijing, People's Republic of China, [i]Department of Earth Sciences, Vrije Universiteit Amsterdam, Amsterdam, The Netherlands

1 Introduction and background

Temperate ecosystems (e.g., forests and grasslands) are one of the major terrestrial biomes on Earth and play a major role in absorbing atmospheric greenhouse gases. Typically, this type of ecosystem is found in regions with moderate temperatures and precipitations, such as the Northern Hemisphere (e.g., North America, Europe, and East Asia) and a part of the Southern Hemisphere (small parts of South America, Africa, and Oceania). Throughout the history of ecological studies, many in situ observations including forest inventories (Goodale et al., 2002; Pan et al., 2011) and micrometeorological flux measurements (Baldocchi, Chu, & Reichstein, 2017) have been conducted in temperate ecosystems because they are present in heavily populated regions in the Northern Hemisphere. Similarly, the foundation of current process-based models, such as the theoretical and empirical relations between carbon fluxes (e.g., photosynthesis and ecosystem respiration) and meteorological variables, has been established based on studies of temperate ecosystems. Therefore, a large part of

Balancing Greenhouse Gas Budgets. https://doi.org/10.1016/B978-0-12-814952-2.00011-3
Copyright © 2022 Elsevier Inc. All rights reserved.

238 Section | C Case Studies

today's understanding of the terrestrial carbon cycle is founded on studies of temperate ecosystems.

With the abundance of observations and well-evaluated theoretical and empirical relations, temperate regions have an advantage in carbon budget assessment. Carbon budget assessment is an integration of the sinks and sources of carbon fluxes of the terrestrial biosphere, such as photosynthesis, the respiration of vegetation, decomposition of soil organic matters, fire emissions, carbon fluxes resulting from the human impact of land-use and land-cover changes, riverine carbon exports, shallow coastal carbon fluxes (known as blue carbon), lateral carbon transport via wood and crop product trade, and emissions of biogenic volatile organic compounds. The net carbon flux (a net exchange of carbon uptake and release) and associated component fluxes can be estimated by several independent methods, including process-based models, ecosystem inventories, micrometeorological flux measurements, satellite observations from microwave sensors, and national statistics, with each having its own range of application. The availability of diverse data allows for carbon budget assessment in temperate regions to be more comprehensive and include component fluxes and process understanding that cannot be realized in other regions (Hayes et al., 2012; King et al., 2015; Luyssaert et al., 2012; Piao et al., 2012; Williams, Gu, MacLean, Masek, & Collatz, 2016).

The contributions of component fluxes to the carbon budget largely vary depending on the characteristics of the region of interest and the accuracy of each flux estimation. In temperate regions, land-use changes are one of the key components of the carbon budget. When discussing the impact of land use today, the greatest concern has always been for tropical forests, which are facing ongoing active deforestation. However, land-use changes in temperate ecosystems cannot be ignored because of their legacy effect. Being present in highly populated regions, temperate ecosystems have been subjected to highly active land-use changes, mainly from forest to agriculture, until the 1970s in North America and Europe, and until the 1980s in East Asia before country-level management suppressed excessive land-use changes. Today, temperate ecosystems are in the midst of regrowth following reforestation, potentially playing a key role in the enhancement of land carbon uptake in recent decades although clearing of land for settlements and associated land uses driven by population increase is still a significant factor (Birdsey, Pregitzer, & Lucier, 2006; Kondo et al., 2018; Pugh et al., 2019). As an obvious consequence of today's active industry, carbon emissions from fossil fuel burning and cement production are substantially larger in temperate regions than in other global regions, but the research community has yet to establish confidence in how much of the carbon released from industry has been taken up by natural ecosystems and regrowth.

This chapter presents the current state of the science in carbon budget assessment for temperate ecosystems, focusing on three temperate regions in the Northern Hemisphere that are characterized by a high density of observations and abundant model experiments (i.e., North America, Europe, and East Asia). We review several methods of net carbon flux and component flux estimation and illustrate roles of the key components in the temperate carbon budget, such as land-use changes and forest regrowth, and discuss uncertainties in the current carbon budget of temperate ecosystems that the research community needs to resolve. Finally, we summarize progress in carbon budget estimation and discuss how it can be improved further and become more useful in policy making.

2 Methodologies for flux estimations in temperate regions

2.1 Net carbon flux estimations

Four fundamental methods form the basis of carbon budget assessment in temperate regions: (i) inventory- and satellite-based carbon stock changes, (ii) micrometeorological flux upscaling, (iii) process-based terrestrial biosphere models, and (iv) atmospheric inversions. These methods are founded on different principles and use different base data. Thus, they are completely independent. Comprehensive comparison of these methods leads to a better understanding of the current state of carbon budget estimation and is a step toward compiling a more reliable carbon budget in temperate regions.

2.1.1 Carbon stock changes

The carbon stock of terrestrial ecosystems (e.g., biomasses of leaves, stems, roots, and soils) changes in response to the sinks and sources of terrestrial carbon fluxes. That is, the change in carbon stock at two or more points in time represents a quantity equivalent to the terrestrial carbon budget for the same time period. Currently, there are two approaches to estimate carbon stock changes for temperate ecosystems that are based on different data sources: forest inventory and satellite data.

Forest inventory is a collection of ecological and physiological information about forests, including the species, diameter at breast height (DBH), forest area, and carbon stock of leaves, stems, stumps, branches, bark, seeds, roots, foliage, litters, and soil (Goodale et al., 2002; Hayes et al., 2012). In general, countries in temperate regions have established forest inventories that provide a sound basis for estimating carbon stocks and stock changes (e.g., the US Forest Inventory and Analysis (FIA), Europe EFISCEN inventory, and Chinese and Japanese national forest inventories). As demonstrated in early studies and standardized in the IPCC guidelines, regional-scale carbon stock changes can be estimated by integrating national inventories with information about the forest area and density obtained from Global Forest Resources Assessment (FRA) reports or other sources (Pan et al., 2011). However, there are two major limitations to this method. First, measurements of carbon stocks in woody debris, forest floor, understory biomass, and soil are often not performed; they are typically supplemented using simple empirical models parameterized from ecosystem studies that relate these variables to observed forest characteristics from the inventory. In addition, when soil carbon stocks are measured, the measurements do not usually exceed a soil depth of 1 m, which may be insufficient to account

240 Section | C Case Studies

for changes in the total soil carbon stock. Second, some of today's inventories do not account for fluxes from land-use changes, as they have been performed in forests that remain forests over time. To provide a more comprehensive carbon budget, a land-use change flux estimate from a bookkeeping model (as described in Section 2.2.1) or from measurements at sample sites that have changed land use as in the US FIA need to be added to regionally upscaled forest inventories that only cover forests remaining forests.

In addition to forest inventories, spatially explicit carbon stock changes can be estimated via the vegetation optical depth (VOD), which represents the degree of microwave attenuation within vegetation. Spatial coverage of the VOD can be derived from passive microwave sensors such as the Special Sensor Microwave Imager (SSM/I), Advanced Microwave Scanning Radiometer for Earth Observation System (AMSR-E), FengYun-3B Microwave Radiometer Imager (MWRI), Soil Moisture and Ocean Salinity (SMOS), Tropical Rainfall Measuring Mission Microwave Imager (TMI), and WindSat. The sensitivity of the VOD to the vegetation water content varies by wavelength; long wavelengths are more sensitive to deeper vegetation layers whereas short wavelengths are more sensitive to leaf moisture. Currently, there are VOD estimates based on the Ku-band, X-band, C-band (Moesinger et al., 2020), L-band (Fan et al., 2019), and a combination of those (Liu et al., 2015). The VOD is known to be relatively well correlated with aboveground biomass. Thus, multiple years of vegetation biomass can be estimated from an empirical relationship between satellite VOD data and a reference map of aboveground biomass (e.g., Saatchi et al., 2011), which enables the calculation of spatially explicit carbon stock changes. However, as in forest inventories, VOD-based carbon stock changes cannot explicitly account for belowground biomass because of the limited applicability of microwave sensors. The currently available options to supplement belowground stock changes include setting belowground biomass to vary as a fraction of aboveground biomass and using estimates from forest inventories.

2.1.2 Upscaling of eddy covariance flux measurements

The micrometeorology method is used to obtain the gas exchange flux by measuring the turbulence near the formation and the change in trace gas concentration. The main methods of measuring the gas exchange flux between vegetation and the atmosphere by micrometeorology are the eddy covariance technique, relaxed eddy accumulation, aerodynamic profile method, and Bowen-ratio energy balance. The eddy covariance technique produces a direct measure of the net CO_2 exchange across the canopy-atmosphere interface, and it is the global standard method for measuring CO_2 and H_2O fluxes (Baldocchi, Valentini, Running, Oechel, & Dahlman, 1996; Papale et al., 2006). This method is accomplished by using micrometeorological theory to interpret measurements of the covariance between the vertical wind velocity and scalar concentration fluctuations and can be used to measure ecosystem CO_2 exchange across a spectrum of time scales, from hours to years (Baldocchi, 2003). The eddy covariance technique samples atmospheric turbulent motions to determine the net difference in material moving across the canopy-atmosphere interface. In practice, this task is accomplished by statistical analysis of the instantaneous vertical mass flux density. Until today, advanced networks of micrometeorological observations have been established in the continental United States and Canada (AmeriFlux), Europe (CarboEurope), and Asia (AsiaFlux), leading to the establishment of the global network FLUXNET.

The prosperity of eddy flux observation networks allows for upscaling of eddy flux observations at the regional and global scales by empirical machine learning algorithms (Jung et al., 2011, 2017; Kondo, Ichii, Takagi, & Sasakawa, 2015). Currently, two ensembles of multiple empirical upscaling of eddy flux data are available from the collaborative FLUXCOM initiative: FLUXCOM_RS and FLUXCOM_MET (Jung et al., 2017; Zscheischler et al., 2017). The FLUXCOM_RS and FLUXCOM_MET datasets are based on an ensemble of multiple machine learning algorithms, but the former is primarily trained with satellite remote sensing data and the latter is trained with meteorological data. Although FLUXCOM provides datasets that make the best use of current eddy covariance observations, it still has room for improvement. The carbon budget of forests and grassland can vary significantly with different gap-filling methods on a time series (Moffat et al., 2007; Soloway, Amiro, Dunn, & Wofsy, 2017) and with the use of open or closed path analyzers for eddy flux measurement (Haslwanter, Hammerle, & Wohlfahrt, 2009). Potentially large uncertainty coming from these aspects has not been addressed yet, as the FLUXCOM datasets are based on the FLUXNET data that are gap-filled by a dedicated method and include measurements obtained by both open and closed path analyzers. In addition, a weakness in the FLUXCOM data is that they lack an explicit treatment of the CO_2 fertilization effect (Jung et al., 2020), which is the major effect inducing an increasing trend of carbon uptake in the land since around the 1960s (Keenan et al., 2016; Kondo et al., 2018). Thus, the FLUXCOM data are not ideal for trend estimation of carbon fluxes.

2.1.3 Terrestrial biosphere models

Process-based terrestrial biosphere models (TBMs) are a widely used tool to calculate the exchange of carbon between the land surface and the atmosphere and are a fundamental part of the methodology underlying recent global and regional carbon budget assessments (Kondo et al., 2020; Le Quéré et al., 2018). TBMs are used to calculate the exchange of carbon and in some cases N_2O and CH_4, between ecosystems and the atmosphere starting from the fundamental ecological processes of photosynthesis and respiration. The core of these models is pools of carbon representing vegetation, litter, and soil components of the ecosystem, between which fluxes are passed according to a range of processes. It is also becoming common for TBMs to simulate cooccurring pools of nitrogen to account for nutrient cycling (e.g., Smith et al., 2014; Zaehle & Friend, 2010). TBMs typically group the world's plants into a handful of plant functional types (PFTs), which are explicitly simulated. In many TBMs, known as dynamic global vegetation models (DGVMs), rather than being prescribed, these PFTs are allowed to compete with each other for light and water (and in some TBMs, nitrogen or phosphorus) to determine a "natural" vegetation composition at a given location. Rules governing tissue turnover and plant mortality determine how long carbon remains in vegetation before being passed to a litter pool. Soil and litter carbon stocks are contained in a number of pools, each subject to a characteristic respiration rate to which temperature and water content modifiers are typically applied. Respired carbon is lost to the atmosphere, and carbon can move between pools according to defined rates, such as in the transformation from litter to soil organic matter. Combining photosynthesis and respiration fluxes together yields the net ecosystem exchange of carbon with the atmosphere (NEE). More details about the general structure of TBMs are given in Prentice et al. (2007) and in the individual model description papers (see Le Quéré et al., 2018 for a list).

As the lag between an event or environmental change and full realization of the resulting changes in carbon emissions or uptake can be on the order of centuries, simulations of carbon budgets have to be initialized more than a century prior to the period of interest. This poses a problem, as sufficient observations of ecosystem states are not available to initialize models outside of the recent historical period. Thus, TBMs rely on the equilibrium assumption, undergoing a spin-up period of hundreds or thousands of years under constant environmental forcing to bring carbon stocks and fluxes into equilibrium with these conditions. For recent historical carbon budgets, this spin-up is typically applied to a pre-industrial period from 1700 to 1900. Following spin-up, the models are run under transient forcing of the climate, atmospheric CO_2 mixing ratio, land-use and management and, where relevant, nitrogen deposition for the period of interest. Typically, such simulations are conducted at spatial resolutions of 0.5 degrees $\times 0.5$ degrees or coarser, but higher resolution simulations have also been employed where driving data are available (Kim, Kerns, Drapek, Pitts, & Halofsky, 2018; Shafer, Bartlein, Gray, & Pelltier, 2015). In this way, an estimate of the net carbon flux over the vegetated land surface is calculated and presented as a total number, subsuming both the natural ecosystem dynamics and land-use changes, or as the residual uptake by terrestrial ecosystems once emissions due to land-use changes have been subtracted.

2.1.4 Atmospheric inversions

Atmospheric inversions are a numerical method that reflects the characteristics of CO_2 observations measured at tall towers in atmosphere-land and atmosphere-ocean carbon flux estimation (Chevallier et al., 2010; Maki et al., 2010; Niwa et al., 2017; Peters et al., 2007; Rödenbeck, Zaehle, Keeling, & Heimann, 2018; Saeki & Patra, 2017). This model is based on the principle that CO_2 is a stable and long-lived compound, so carbon sinks and sources on land and

Temperate carbon budget assessments Chapter | 7 **243**

ocean surfaces are directly reflected in atmospheric CO_2 concentrations. As the name suggests, the method inversely calculates a net carbon flux (a sum of all components of carbon uptake and release) at land and ocean surfaces through processes of transport and accumulation of CO_2 in the atmosphere. Atmospheric inversions use a CO_2 transport model and prior information about carbon sinks and sources as a basis (e.g., land and ocean carbon fluxes, and fossil fuel emissions) and seek a solution that minimizes the difference between the observed and simulated atmospheric CO_2 concentrations using data assimilation techniques (Bayes statistics, ensemble Kalman filter, etc.).

The development of today's atmospheric inversions would not have been possible without the global expansion of observation networks. Since the start of continuous in situ measurements at the Mauna Loa Observatory in 1957, a number of tower measurement sites have been established in various regions around the world, forming global networks (e.g., World Data Center for Greenhouse Gases, WDCGG; and the observation package—ObsPack—from the National Oceanic and Atmospheric Administration Earth System Research Laboratory, NOAA/ESRL). Today, there are more than 200 in situ observation sites, and CO_2 observations from ships (e.g., Ship of Opportunity: SOOP) and aircraft (e.g., Comprehensive Observation Network for TRace gases by AIr-Liner: CONTRAIL, and HIAPER Pole-to-Pole Observations: HIPPO) are also available. In the 1980s and 1990s, when observational data were relatively few, the net carbon flux of multiple inversions showed variations of more than $3 \, \text{Pg}$ (petagram $= 10^{15} \, \text{g}$) $C \, \text{year}^{-1}$ at the global and hemispherical scales (Gurney et al., 2002; Jacobson, Fletcher, Gruber, Sarmiento, & Gloor, 2007; Peylin, Baker, Sarmiento, Ciais, & Bousquet, 2002; Rödenbeck, Houweling, Gloor, & Heimann, 2003), but it was reduced to $1 \, \text{Pg} \, C \, \text{year}^{-1}$ because the observation network and atmospheric transport model made great strides (Gaubert et al., 2019; Kondo et al., 2020; Stephens et al., 2007). These developments led to today's understanding of the global carbon budget in which tropical regions are carbon neutral, and the mid-high latitudes of the Northern Hemisphere account for most of the global net sink.

In the carbon budget assessment using multiple inversions, it is necessary to adjust the difference between the respective fossil fuel emissions prescribed in each inversion and a reference emission estimate. This "fossil fuel correction" is necessary to reduce variability in posterior net carbon fluxes, as differences in prescribed fossil fuel emissions largely affect posterior fluxes (Kondo et al., 2020; Peylin et al., 2013), especially for recent periods for which the uncertainty in fossil fuel emissions remains large (Ballantyne et al., 2015).

2.2 Components of the carbon budget in temperate regions

The terrestrial carbon cycle consists of multiple fluxes that enter and leave through the terrestrial biosphere (Fig. 1). The level of complexity is largely different among carbon fluxes, as are the availability of observational constraints

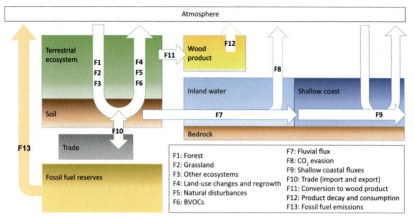

FIG. 1 Schematic representation of carbon fluxes through the atmosphere, biosphere, hydrosphere, and other carbon reservoirs.

and the accuracy of flux estimation. This section describes methodologies for carbon flux estimation that are relevant to the carbon budget of temperate forests and grasslands.

2.2.1 Land-use changes and regrowth

Land-use activities such as deforestation, forest degradation, and shifting cultivation, significantly contribute to global carbon emissions and have the potential for large additional emissions in future decades (Fearnside, 2000; Miles & Kapos, 2008). On the other hand, regrowth of terrestrial ecosystems after large-scale land-use changes is one of the key factors accounting for the recent carbon uptake enhancement, of which the eastern part of North America, Europe, and the southern part of East Asia have been identified as hot spots (Kondo et al., 2018; Pugh et al., 2019). Net carbon fluxes resulting from land-use changes and regrowth can be estimated by quantifying the changes in carbon stocks or the changes in carbon fluxes. This net land-use change flux is typically estimated by either bookkeeping approaches, using TBMs, or inventory data that include monitoring sample plots for land-use change (i.e., the US FIA). This section describes the available sources of land-use change data and covers how land-use change fluxes are calculated using bookkeeping approaches and TBMs.

  (i) **Land-use datasets**

   The most commonly used global land-use dataset is the HistorY Database of the global Environment (HYDE) (Klein Goldewijk, Beusen, Doelman, & Stehfest, 2017), which underlies calculations of the Global Carbon Project (Le Quéré et al., 2018). HYDE provides the fractions of cropland, intensive pastureland, and rangeland over the last 12,000 years, with annual updates. Using a recent satellite land-cover map as the baseline, it uses the UN Food and Agricultural Organization (FAO) national statistics and subnational statistics to map changes in land use back to 1960, with estimates further back in time using a population-based modeling approach (Klein Goldewijk et al., 2017). Land-Use Harmonization (LUH) is the extended version of HYDE (Hurtt et al., 2020). It combines the HYDE cropland and pasture status with the wood harvest status based on the FAO national wood harvest statistics to extend global land-use change patterns, including transitions of cropland, pasture, and primary and secondary forest and nonforest including grassland and shrub.

(ii) **Bookkeeping calculations**

The bookkeeping approach uses observation-based soil and vegetation carbon stock densities for each land-use type. In principle, the emissions associated with a land-use transition are the difference between the carbon stock densities for the two land-use types. However, since an instantaneous emission equal to this difference would be unrealistic to spread the emissions appropriately across time, empirical response curves are used to define the rate of transition between these stocks following land-use change. These response curves vary with land-use type and region.

There are two main bookkeeping models in use: the model of Houghton, which pioneered the bookkeeping methodology (Houghton et al., 1983; Houghton & Nassikas, 2017), and the BLUE model of Hansis, Davis, and Pongratz (2015). The bookkeeping approaches are not necessarily gridded. The Houghton approach operates at the country level and is forced directly by FAO estimates of land-use change and wood harvest. Rather than being forced with a transition matrix, it uses a rule-based approach for land-use transition choices, e.g., preferentially expanding pastures into nonforested ecosystems (Houghton & Nassikas, 2017). The BLUE model uses gridded land-use information, simulating land use, land-use change, and wood harvest at a $0.5\,degrees \times 0.5\,degrees$ resolution (Hansis et al., 2015). The latest estimates from both bookkeeping models also include carbon emissions from peat burning and drainage, based on additional datasets (Le Quéré et al., 2018). Because bookkeeping models use observation-based carbon stock densities, they do not account for the effect of environmental change. This means that they do not include the change in sink capacity due to environmental change.

(iii) **Process-based calculations**

TBM estimates of land-use change and management emissions generally rely on factorial simulations in which fluxes from a simulation with all forcings are compared with simulations where one or more forcings are held constant (i.e., land-use change forcing). However, carbon fluxes due to land-use changes have several facets (Fig. 2), which require different sets of simulations to quantify. The myriad of different ways in which the flux of carbon can be conceptualized and simulated has led to large inconsistencies in the results reported by different studies, making cross-comparison

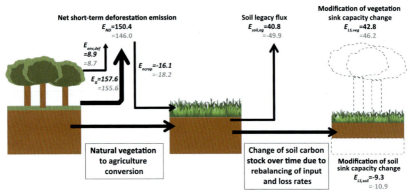

FIG. 2 Decomposition of the land-use change flux, E_{LUC}, into component parts representing the emissions from net short-term conversion (E_{ND}, composed of emissions from vegetation carbon lost that would have been there in the absence of environmental change, $E_{veg,old}$, and that biomass which is only there because of environmental change to date, $E_{veg,env}$), legacy fluxes due to vegetation regrowth ($E_{veg,leg}$) and changes in soil carbon stock resulting from modified input and loss fluxes ($E_{soil,leg}$), and from modified sink capacity in the vegetation ($E_{SC,veg}$) and soil ($E_{SC,soil}$). Fluxes are accumulated over 1850–2012 and given in Pg C, as described in Pugh et al. (2015).

challenging. For instance, only direct emissions from the action of land-use changes (e.g., deforestation) may be included, or legacy fluxes from soil can be incorporated. The differences in how environmental changes affect the new ecosystem compared with the previous one may be included (the change in sink capacity).

Although many options are available, the most widely applied method is that used in the Global Carbon Budget (Le Quéré et al., 2018), which calculates emissions due to land-use changes through differencing two TBM simulations. Both simulations consider fully varying environmental forcing (e.g., climate, atmospheric CO_2 mixing ratio, nitrogen deposition) since 1860, but only one of which includes varying land-use and land management throughout the simulation, and the other simulation uses fixed 1860 land-use and management throughout. The emissions calculation based on taking a difference between two TBM simulations has become a common approach, but the choice and use of the land-use datasets differ among TBMs; for example, whether to use HYDE or the additional variables added to HYDE by LUH (e.g., forest and nonforest distinction, and primary and secondary forest distinction), as a forcing data set. Furthermore, TBM simulations account for the net effect of land-use changes on the terrestrial carbon cycle, including instantaneous and legacy emissions, and regrowth flux, but specific schemes for land-use change model were left to the discretion of the model developers, which means that fundamental assumptions and levels of complexity in model largely vary among TBMs.

Alongside advances in land-use representation in TBMs, the consideration of forest dynamics has also advanced greatly, which led to a better realization of regrowth flux. The standard TBM approach simulates each PFT as an average individual, whose size depends on the amount of carbon it contains relative to its areal coverage. This can lead to strange situations through which conversion of cropland to forest leads to a merging of mature and newly established vegetation carbon pools, such that the average amount of carbon per area for the forest PFT actually shrinks. It also fails to take into account differences in tree growth and mortality rates that are dependent upon their size or age, which can substantially affect carbon exchange (Haverd, Smith, Nieradzik, & Briggs, 2014; Pugh et al., 2019); a recent study suggested that the carbon balances of temperate forests are substantially perturbed by their relatively young age distribution (Pugh et al., 2019). A minority of TBMs can explicitly simulate tree demography, adopting a "cohort"-based approach, in which cohorts of trees of similar age or size are modeled together, allowing a multilayer canopy to be simulated and for trees to be tracked throughout their life span (Fisher et al., 2018; Haverd et al., 2014; Koven et al., 2020; Naudts et al., 2016; Smith et al., 2014). This allows for these TBMs to make use of emerging datasets on forest age alongside emerging representations of forest disturbances beyond fire (Chen et al., 2018; Kautz, Anthoni, Meddens, Pugh, & Arneth, 2018). As the method of net land-use change, flux estimation is based on factorial simulations with and without varying land-use and management changes, estimation of legacy flux is based on the same principle. The difference is whether the method involves differencing simulations of net carbon flux or NEE (Kondo et al., 2018); the former extracts the net effect of land-use changes on net carbon fluxes and the latter extracts that on NEE. Another approach is to explicitly extract regrowth flux from the carbon cycle simulation.

The net land-use change flux estimated by TBMs is still highly uncertain. According to an ensemble of TBMs (Kondo et al., 2018), a net effect of land-use changes on temperate ecosystems results in carbon emissions of $0.79 \, \mathrm{Pg\,C\,year}^{-1}$ for the 2000s, whereas the bookkeeping model of Houghton and Nassikas (2017) suggests a carbon sink of $0.21 \, \mathrm{Pg\,C\,year}^{-1}$ for the same period (Table 1). TBMs indicate that the largest carbon uptake induced by regrowth is found in North America, which neutralizes gross land-use change emissions from that region, whereas weaker effects of regrowth are overwhelmed by gross land-use change emissions in Europe and East Asia. Contrary to those TBM simulations, Houghton and Nassikas (2017) indicate that the net land-use change flux induces a net sink in all the three temperate regions, implying regrowth flux overwhelms gross emissions. This suggests the possibilities of current TBMs underestimating carbon sink induced by regrowth, overestimating gross emissions, or both.

248 Section | C Case Studies

TABLE 1 Estimates of land-use change fluxes by an ensemble of TBMs from Kondo et al. (2018) and a bookkeeping model from Houghton and Nassikas (2017). All units are in Pg C year^{-1}.

		North America	Europe	East Asia
TBMs (Kondo et al., 2018)	Gross emissions	0.22 ± 0.17	0.49 ± 0.28	0.37 ± 0.34
TBMs (Kondo et al., 2018)	Regrowth flux	-0.22 ± 0.16	-0.06 ± 0.33	-0.01 ± 0.33
TBMs (Kondo et al., 2018)	Net land-use change flux	0.0 ± 0.05	0.43 ± 0.07	0.36 ± 0.18
H&N (Houghton & Nassikas, 2017)	Net land-use change flux	-0.062	-0.106	-0.062

2.2.2 Fluvial flux

Rivers are important transport routes of terrestrially derived organic carbon to the oceans. Organic carbon is mobilized from soil and litter in dissolved form (DOC) by leaching and in particulate form (POC) by erosion or by litter falling directly into streams and rivers. Dissolved inorganic carbon (DIC) is produced in the process of chemical rock weathering, where carbonic acid, which is formed when CO_2 dissolves in water, reacts with silicate and carbonate minerals. The DIC in rivers mainly consists of dissolved carbonates, which were formed as weathering products. As the larger part of this DIC originates from dissolved CO_2, chemical rock weathering represents a CO_2 sink. Over Earth's history, chemical rock weathering has been the most important sink for atmospheric CO_2 (Berner, Lasaga, & Garrels, 1983). When carbonate minerals are involved in the weathering process, a portion of the produced DIC originates from lithogenic sources.

The exports of carbon from single rivers to the coast can be obtained from the observed carbon concentrations at the river's mouth and from gauged streamflow data. Observation of the DOC, POC, and DIC concentrations requires discrete sampling of river water and laboratory measurements. In addition, automated optical sensors can be used for high-frequency observations of DOC (Ruhala & Zarnetske, 2017). However, the use of this methodology is still limited. Regional and national water quality surveys in Europe, North America, and Australia often conduct sampling campaigns at biweekly to monthly intervals for the most important rivers (Hartmann, Lauerwald, & Moosdorf, 2014). For other parts of the world, observational time series are limited and often restricted to a few sampling campaigns related to scientific field trips.

As all river systems are not monitored with regard to fluvial carbon exports, different extrapolation techniques have been used to estimate land-ocean carbon exports at the global scale and for different climate zones. A first assessment of the land-ocean carbon exports per climate zone was achieved by Meybeck (1993). Meybeck assembled the observed carbon exports from 40 large river basins covering 38%–45% of the world's exorheic area, i.e., the continental area draining to the sea. Based on these data, he calculated the average fluvial carbon export rates normalized by basin area ($g\,Cm^{-2}\,year^{-1}$) for the cold, temperate, humid, tropical, and arid areas (including savannahs). Multiplying these average export rates by the total exorheic area of each zone, he estimated the total land-ocean carbon exports from each of the climate zones and for the globe.

A different approach to obtain regional to global scale estimates of land-ocean carbon exports is the use of empirical models. To set up the empirical model, a database is created that combines the observed fluvial carbon exports and the major environmental properties of the corresponding river basins derived from geo-data, including climate, land cover, lithology (i.e., the chemical properties of the underlying bedrock) and soil properties. Based on this database and the use of multivariate statistics, an empirical model is designed to predict fluvial carbon exports from a river basin based on a set of river basin properties. The empirical model is used with the corresponding geophysical data to obtain a global, spatially explicit estimate of land-ocean carbon transfers.

The best known global, spatially explicit estimates of land-ocean carbon exports are those by Ludwig, Amiotte-Suchet, Munhoven, and Probst (1998) and GlobalNEWS v2 (Mayorga et al., 2010). Ludwig et al. (1998) predicted fluvial DOC export from the average runoff, terrain slope, and soil organic carbon concentration (SOC) within the river basin. The GlobalNEWS v2 dataset uses the DOC model established by Harrison et al. (2005), which predicts fluvial DOC export from runoff and the areal proportion of wetlands in the river basin. Wetlands are important sources of DOC to the river network, and according to the empirical model by Harrison et al. (2005), wetland DOC yields are more than three times higher than those from dry soils.

To predict fluvial POC exports, both Ludwig et al. (1998) and the Global-NEWS v2 group first predict the fluvial exports of total suspended sediments (F_{TSS}) and then estimate the percentage of POC in the suspended sediments. Ludwig et al. (1998) estimated F_{TSS} based on runoff, terrain slope, and the Fournier Index of precipitation intensity, which are the most important predictors used in most erosion and sediment transport models (Renard, Foster, Weesies, & Porter, 1991; Syvitski, Morehead, & Nicholson, 1998). Soil erosion and fluvial sediment transport increase with the runoff and slope gradients of hillslopes. The Fournier Index of precipitation intensity, which has a positive effect on erosion and sediment transport, accentuates the impact of single months with high

250 Section | C Case Studies

precipitation, which can be more important for annual sediment transport than the average annual precipitation because erosion and sediment transport tend to increase overproportionally with precipitation and runoff.

For a spatially explicit estimation of fluvial DIC fluxes and the amount of CO_2 consumed in the weathering process, Ludwig et al. (1998) used the GEM-CO_2 model developed by Amiotte-Suchet and Probst (1995), whereas GlobalNEWS v2 used the estimates by Hartmann, Jansen, Dürr, Kempe, and Köhler (2009). GEM-CO_2 and the Hartmann et al. (2009) model have basically the same structure, with a fixed HCO_3^- concentration for each lithological class, which is multiplied by the runoff to estimate the fluvial export flux of DIC. For GEM-CO_2, a coarse lithological map was used to distinguish seven lithological classes (Amiotte-Suchet & Probst, 1995), whereas Hartmann et al. (2009) used a more detailed map distinguishing 15 lithological classes (Dürr, Meybeck, & Dürr, 2005). In contrast to the models used to estimate the fluvial export flux of DOC and POC, both models to estimate the flux of DIC globally were calibrated on small headwater catchments restricted to a relatively small region: GEM-CO_2 was calibrated on 232 French headwater catchments, and Hartmann et al. (2009) calibrated their model on 382 Japanese catchments. Ludwig et al. (1998) evaluated GEM-CO_2 against observed DIC flux from large world rivers and corrected the estimates accordingly for different climate zones.

Despite the different methodologies and datasets used, the estimates of land-ocean carbon exports listed in Table 2 are comparable. According to those

TABLE 2 Estimates of land-ocean C exports at global scale and from the temperate zone. All units are in Tg C year^{-1}.

	DOC	POC	DIC$_{lith}$	DIC$_{atm}$	TAC
Meybeck (1993) – Global	199	99		244	542
– Temperate	32.2	33.7		100	166
Ludwig et al. (1998) – Global	205	187	96.2	233	625
– Temperate	38.3	44.5	28.4	52.8	136
GlobalNEWS v2 – Global	168	112	88	237	517
– Temperate	32.1	29.0	24.4	49.1	110

Results are listed for dissolved organic carbon (DOC), particulate organic carbon (POC) and dissolved inorganic carbon (DIC). DIC is further subdivides into DIC being contributed by the CO_2 taken up in the weathering process (DIC$_{atm}$) and that being contributed by weathered carbonate minerals (DIC$_{lith}$). The total atmospheric carbon (TAC) exported by the rivers is equal to the sum of DOC, POC and DIC$_{atm}$.

estimates, the temperate zone contributes 16%–19% of the global DOC exports and 24%–34% to the global POC exports. However, Meybeck's estimate of the weathering CO_2 sink in the temperate zone is about twice as high as the estimates derived from empirical models. As Ludwig et al. (1998) pointed out, this may be related to the fact that Meybeck did not account for the spatial distribution of lithology and runoff as major controls on rock weathering in his extrapolation. The estimates derived from the two empirical models are very similar, despite the fact that very different calibration data sets were used. According to these estimates, the temperate zone contributes 21%–23% of the global weathering CO_2 sink.

2.2.3 CO_2 evasion from river and lake

Carbon emissions from inland waters such as rivers and lakes (referred to as CO_2 evasion) are an often poorly quantified component of the carbon cycle. However, their contributions to the carbon budget are potentially large as recent global estimates suggest $0.35–1.8\,Pg\,C\,year^{-1}$ from rivers and streams (Cole et al., 2007; Raymond et al., 2013), and $0.32–0.8\,Pg\,C\,year^{-1}$ from lakes (Barros et al., 2011; Raymond et al., 2013). Although tropical regions are major emission sources from inland waters, emissions from temperate regions may not be negligible because of major lakes and rivers (e.g., Mississippi River, Elbe River, and Yellow River) in these regions.

Several studies quantified CO_2 evasions in temperate regions, but they often focused on particular rivers or lakes. An available option to obtain temperate CO_2 evasions is to cut off temperate regions from spatially explicit global estimates. Global river CO_2 evasion can be derived from the empirical river water pCO_2 (the partial pressure of carbon dioxide) model and global maps of stream surface area and gas exchange velocities (Lauerwald, Laruelle, Hartmann, Ciais, & Regnier, 2015). Likewise, global lake CO_2 evasion can be estimated based on lake pCO_2, total lake/reservoir surface area, and total CO_2 evasions for 231 coastal regions (Raymond et al., 2013), subsequently downscaled to a continuous grid scale via the Global Lakes and Wetland Database (Zscheischler et al., 2017). These methods estimate $76\,Tg$ (teragram $= 10^{12}\,g$) $C\,year^{-1}$ of CO_2 evasion from rivers and $156\,Tg\,C\,year^{-1}$ from lakes in temperate regions, of which North America contributes 59% and 77%, respectively.

2.2.4 Fire emissions

Global carbon emissions from fires are difficult to quantify and can potentially influence the interannual variability and long-term trends in atmospheric CO_2 concentrations. Previous studies extrapolated regional carbon losses from fires by developing inventories of aggregated fuel, combustion factors, and fire return times for different biome types and using global vegetation maps for estimation (Crutzen & Andreae, 1990; Seiler & Crutzen, 1980). The spatial and temporal variability of carbon emissions from forest and grassland fires can

252 Section | C Case Studies

be assessed by an empirical model that combines satellite-derived estimates of burned area, fire activity, and a process-based plant productivity simulation (the Global Fire Emissions Database (GFED), van der Werf et al., 2010), or a linear relationship between fuel consumption and total emitted fire radiative energy (the Global Fire Assimilation System (GFAS), Kaiser et al., 2012). Despite notable differences found in tropical savanna and cropland, the latest GFED and GFAS agree in having about $100 \, \mathrm{Tg} \, C \, \mathrm{year}^{-1}$ of fire emissions in temperate regions (Shi, Matsunaga, Saito, Yamaguchi, & Chen, 2015). Fire disturbance and emissions have been represented in some of the early TBMs (Sitch et al., 2003) and since then a wide range of models of varying complexity is available (Hantson et al., 2012), but their representation is still not ubiquitous in models used for global carbon budgets (Le Quéré et al., 2018).

2.2.5 Biogenic volatile organic compounds

In addition to CO_2 and CH_4, terrestrial ecosystems emit carbon in the form of volatile organic compounds, referred to as biogenic volatile organic compounds (BVOCs). Isoprene and monoterpenes are the most abundant BVOC species, accounting for 70% and 11% of the whole species, respectively (Guenther et al., 2012). The rest of the species occupies smaller fractions: methanol (6%), acetone (3%), sesquiterpenes (2.5%), and other species each contributing less than 2% (Sindelarova et al., 2014). As much as BVOCs influence air quality, their radiative forcing is an issue of climate change today. BVOCs are short-lived compounds stable from a few hours to months, but eventually turn into CO_2 via several reaction pathways.

Estimation of BVOC emissions is pioneered by a process-based model of Guenther et al. (1995), whose underlying theories, forcings, and model designs are similar to TBMs, but more focused on BVOCs. Alongside the advancement of TBMs, BVOC models have advanced in recent years. Current estimates by these models suggest that BVOCs are a large carbon emission source, which amounts to $0.76–1.15 \, \mathrm{Pg} \, C \, \mathrm{year}^{-1}$ at the global scale (Guenther et al., 1995: Sindelarova et al., 2014). However, the contribution of temperate regions to the global BVOC emissions is relatively low ($\sim$15%), while that of tropical regions is substantially higher (Sindelarova et al., 2014).

3 Review of the carbon budget of temperate forests and grasslands

When assessing the carbon budget, it is important to evaluate the results of the multiple methods mentioned above under a consistent definition of the net carbon flux (Bastos et al., 2020; Kondo et al., 2020). Since this chapter focuses on a carbon budget comprised mainly of forests and grasslands in temperate regions, here we define the net carbon flux as the atmosphere-land CO_2 exchange. This corresponds to components F1 through F6 of the carbon budget illustrated in Fig. 1.

3.1 Adjustments for the carbon budget

The four fundamental methods for carbon budget estimation differ in components accounting for the net carbon flux (Table 3). Therefore, adjustments to the results of those methods are needed to estimate the carbon budget of temperate ecosystems, which include fluxes, such as photosynthesis, autotrophic and heterotrophic respiration, fire emissions, land-use change emissions, and BVOC emissions. Carbon stock changes represent the carbon budget of terrestrial ecosystems as they are the observed residual of carbon in and out of ecosystems. However, carbon stock changes from the compilation of forest inventories only account for the carbon budget of intact forests (except the inventories in the United States), missing the contribution from grasslands and other ecosystems, and fluxes from land-use changes, whereas those based on the VOD data account for both forests and grasslands, and the effect of land-use changes on those ecosystems, as they are spatially explicit. Currently, the inventory-based method cannot account for carbon stock changes in grasslands, as a global set of grassland inventories is not available, but the contribution of land-use change fluxes can be supplemented by a bookkeeping model if not included in the inventory, as such that the H&N model supplemented land-use change fluxes in forest carbon budget assessment of Pan et al. (2011).

The micrometeorology method measures the net CO_2 exchange across the canopy-atmosphere interface. Therefore, upscaling of eddy flux observations captures regional photosynthesis, ecosystem respiration, and their net exchange (i.e., NEE), but does not include other components, such as fire emissions, BVOC emissions, and fluxes associated with land-use changes. Those missing components can be supplemented by independent estimates of component fluxes (Zscheischler et al., 2017).

TBMs and atmospheric inversions also differ in a definition of the net carbon flux. TBMs can provide the process-based carbon budget for forests and other ecosystems, as they consider numerous processes of atmosphere-land biogeochemistry, including advanced land-use change submodels in the latest development (Pugh et al., 2019). Being based on atmospheric CO_2 measurements, the carbon budget estimated by atmospheric inversions inevitably includes all components of atmosphere-land carbon exchange. However, carbon budget estimates can be different between the two methods largely due to the missing representation of carbon releases in TBMs. By design, TBMs assume an equilibrium state of carbon balance in the preindustrial period i.e., the magnitude of carbon uptake changes by processes of carbon release to the atmosphere and to the hydrosphere represented in TBMs (Fig. 3). Ito (2019) demonstrated that including the representation of carbon flows such as wood harvest, fluvial fluxes, and BVOCs emissions in a TBM induces a shift of the global net land-atmosphere flux from carbon sink of 3.21 to $6.85 \, PgC \, year^{-1}$. Having TBMs not fully representing components of carbon release, a currently available option for estimating the net atmosphere-land flux is to add the amount of misrepresented carbon release to the carbon sink in

TABLE 3 Components of the carbon budget included in the carbon accounting methods.

	CO$_2$ fertilization	Climate variability	Land-use change fluxes	Fluvial flux	CO$_2$ evasions	Fire emissions	BVOCs
Inventory-based carbon stock changes (forests only)	✓	✓	(only in a few inventories)				✓
VOD-based carbon stock changes	✓	✓	✓			✓	✓
Eddy flux upscaling		✓					
Terrestrial biosphere model	✓	✓	✓			✓ (only in a few TBMs)	✓ (only in a few TBMs)
Atmospheric inversion	✓	✓	✓		✓	✓	✓

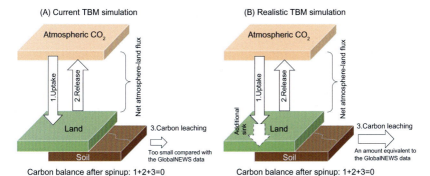

FIG. 3 Schematic representation that illustrates a difference in the magnitude of carbon sink estimated by TBMs. Carbon uptake after the spin-up run can largely differ between (A) TBMs that simulate a smaller amount of fluvial flux and (B) that simulate an amount of fluvial flux equivalent to the GlobalNEWS data.

TBMs (Fig. 3) or to exclude fluvial flux and CO_2 evasions from atmospheric inversions for having a consistent definition between the two methods (Kondo et al., 2020).

3.2 Carbon budget assessment

This section compares the four carbon accounting methods, carbon stock changes based on forest inventories and VOD data, eddy flux upscaling, TBMs, and atmospheric inversions, each adjusted to yield the carbon budget of forests and grasslands.

(i) Carbon stock changes

According to the compilation of global forest inventories of Pan et al. (2011), temperate forests acted as a net sink of $0.72\,\text{Pg}\,\text{C}\,\text{year}^{-1}$ for 2000–07, and each of major temperate regions of the Northern Hemisphere (temperate North America, Europe, and East Asia) accounts for a nearly equivalent net sink of 0.22–$0.25\,\text{Pg}\,\text{C}\,\text{year}^{-1}$. Compared with forest inventories, carbon stock changes based on the harmonized multiband VOD data (Liu et al., 2015) estimated a lower net sink of $0.48\,\text{Pg}\,\text{C}\,\text{year}^{-1}$ for temperate regions in the 2000s (Kondo et al., 2020). Nevertheless, as in forest inventories, this data shows an equivalent net sink of 0.14–$0.19\,\text{Pg}\,\text{C}\,\text{year}^{-1}$ in North America, Europe, and East Asia.

(ii) Upscaling of eddy flux observations

Despite the abovementioned adjustment for the carbon budget, the two datasets of eddy flux upscaling, FLUXCOM_RS and FLUXCOM_MET, show a substantially large carbon sink of 4.23 and $6.55\,\text{Pg}\,\text{C}\,\text{year}^{-1}$ in temperate regions, respectively. These are exceptionally larger net sinks than the other methods that are commonly found in North America, Europe, and East Asia (Fig. 4). Even before the FLUXCOM datasets, it has been known

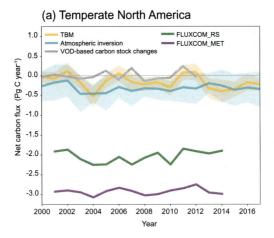
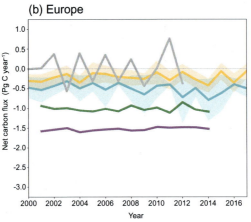
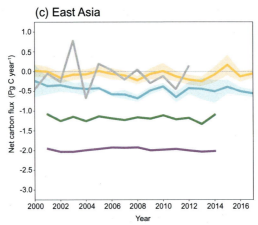

FIG. 4 Interannual variability in the net carbon flux for (A) temperate North America, (B) Europe, (C) East Asia, by the fundamental carbon accounting methods including two carbon stock changes (based on forest inventories and VOD data, respectively), two upscalings of eddy flux observations (FLUXCOM_RS and FLUXCOM_MET), TBMs, and atmospheric inversions.

that the empirical upscaling of eddy flux tends to estimate a large net sink in global regions (Jung et al., 2011; Kondo et al., 2015). The three well-known datasets of carbon emissions from fossil fuel burning and cement production, Carbon Dioxide Information Analysis Center (CDIAC), the Emissions Database for Global Atmospheric Research (EDGAR), Fossil Fuel Data Assimilation System (FFDAS), suggest that $4.98 \pm 0.32\,Pg\,C\,year^{-1}$ of CO_2 was emitted from temperate regions of the Northern Hemisphere in the 2000s. If the carbon budgets estimated by FLUXCOM_RS and FLUX-COM_MET are realistic, then it is suggesting that carbon uptake by temperate forests and grasslands are large enough to offset all fossil fuel emissions from those regions. That is hardly possible because tower observations of atmospheric CO_2 in temperate regions (https://gaw.kishou.go.jp/) show a continuous increase in atmospheric CO_2 concentrations. Even considering croplands, this large carbon sink can only be reduced by less than $1.0\,Pg\,C\,year^{-1}$ (Wolf et al., 2015), which makes it difficult to remain consistent with tower observations of atmospheric CO_2.

(iii) TBMs and atmospheric inversions

With necessary adjustments implemented, ensembles of TBMs and atmospheric inversions estimated a net sink of 0.42 ± 0.26 and $0.70 \pm 0.86\,Pg\,C\,year^{-1}$ in temperate regions for the 2000s, respectively (Kondo et al., 2020). A mean carbon sink is approximately 60% greater in atmospheric inversions than in TBMs, whereas a model-by-model variation is larger in atmospheric inversions than in TBMs. These results pose questions on the reliability of both methods. Considering the difference in net land-use change flux between TBMs and a bookkeeping model (Table 1), the potentially underestimated carbon uptake by regrowth may be responsible for the difference between TBMs and atmospheric inversions. Meanwhile, a large model-by-model variation in atmospheric inversions indicates that constraints by tower observations of atmospheric CO_2 do not necessarily lead to the convergence of carbon budget estimates. Nevertheless, these two model estimates commonly show that East Asia is the region that accounts for the largest temperate carbon sink, followed by North America and Europe (Kondo et al., 2020). Excluding the empirical upscaling of eddy flux observations, TBMs and atmospheric inversions correspond to the upper and lower ranges of the temperate carbon budget estimates (Fig. 5), suggesting that temperate forests and grasslands are compensating for 8%–14% of fossil fuel emissions in temperate regions.

4 Uncertainties in carbon fluxes

Among the global regions, our understanding of the carbon cycle is advanced in the temperate regions, which have a high density of in situ observation data that can be used to infer regional carbon budgets and constrain the processes of model simulations. Nevertheless, even in such temperate regions, there are still

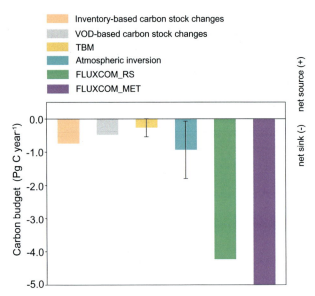

FIG. 5 Comparison of the temperate carbon budget for the 2000s estimated by two carbon stock changes (based on forest inventories and VOD data, respectively), two upscalings of eddy flux observations (FLUXCOM_RS and FLUXCOM_MET), TBMs, and atmospheric inversions.

large uncertainties in carbon budget estimation (Table 4). As highlighted above, the key uncertainties for the temperate regions are that (i) the carbon budgets based on different observation data of terrestrial ecosystems (e.g., in situ and remote sensing observations) are not necessarily consistent (ii), the upscaling of eddy flux observations is largely biased toward a net sink, and (iii) the models constrained by atmospheric observations and process-based models still differ in the magnitude of a carbon budget and estimation range.

4.1 Reliability and uncertainty in observational methods

Carbon stock change estimates based on VOD data underestimate a net sink compared with those of forest inventories. Most countries in the temperate zone of the Northern Hemisphere have well-established and funded forest inventories covering all forestlands and are repeated at regular intervals, but inventories of grasslands are less comprehensive, and some large areas are not inventoried. It is expected that spatially explicit carbon stock changes based on VOD data infer a greater net carbon sink than those based on forest inventory with the amount of an additional sink by grasslands, but the results are contrary to the expectation. Further comparison is required to detail various aspects of carbon stock changes, including the reliability of empirical methods for upscaling forest inventories, and the degree of representativeness of the VOD-biomass relation depending on the microwave wavelength.

TABLE 4 Uncertainties in the four carbon accounting methods.

Method	Uncertainty
Carbon stock changes	• A lack of belowground inventories • Less comprehensive inventories of grasslands • Reliability of empirical methods for upscaling forest inventories • Degree of representativeness of the VOD-biomass relation
Eddy flux upscaling	• Unrealistically large net carbon sink, which is enough to offset fossil fuel emissions in temperate regions • "Black-boxed" calculation processes
Terrestrial biosphere model	• Disagreement with bookkeeping models • Processes of CO_2 uptake during recovery or regrowth after large-scale disturbances • A lack of forcing data on large-scale forest management practices
Atmospheric inversion	• Large model-by-model variations • Differences in model resolution, the period during which data assimilation is conducted, the rates at which CO_2 is transported from a source region to neighboring regions through the model atmosphere, and transport model errors among inversions • Variability in the temperate carbon budgets influences budget estimates of other regions via "dipole effect"

The regional upscaling of eddy flux observations estimates the largest carbon sink in temperate regions among the methods that we have investigated. However, a net sink indicated by this method is too large for temperate ecosystems and counterintuitive because this method does not account for the CO_2 fertilization effect that is largely responsible for recent carbon uptake by terrestrial ecosystems (Keenan et al., 2016; Kondo et al., 2018). Jung et al. (2020) suggest that the too strong global carbon sink estimated by FLUX-COM is originated in tropical forests due to a lack of site history effects on NEE, but that unlikely justifies the cause. Compared with NEE estimates from a set of TBMs, the FLUXCOM data show a substantially larger carbon sink in both the Northern extratropical and Southern+tropical regions (Fig. 6). We need to address potential issues of this method further, such as whether the method is properly designed to reflect the eddy flux observations and whether a large carbon sink is originated from the upscaling method rather than from eddy flux observations. However, it is currently difficult to resolve these issues, as this method is based on machine learning algorithms, whose calculation processes remain "black-boxed." A realistic step toward improving this method may be to revisit fundamental issues in eddy flux observations, such as the difference in

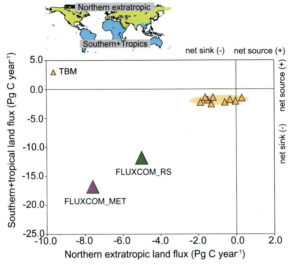

FIG. 6 Net ecosystem exchange (NEE) estimates for northern extratropical and southern+tropical regions by nine TBMs (*orange shade* represents 1σ error ellipse), FLUXCOM_MET and FLUXCOM_RS. Negative values represent a net sink of CO_2 and positive values a net source of CO_2.

site-level carbon budget due to gap-filling methods and the use of open and closed path analyzers.

The model-by-model variations are nearly fourfold larger among atmospheric inversions than among TBMs. This result is problematic because variations between atmospheric inversions in one region affect other regions connected via atmospheric transport, such as temperate and boreal North America, Europe and boreal Asia, and temperate Asia and South Asia (Kondo et al., 2020). In atmospheric inversions, the sink-source compensation between two or more regions is known as the dipole effect (Niwa & Fujii, 2020; Peylin et al., 2002). This effect is intrinsic to the design of inversion systems, where the carbon budgets of neighboring regions connected via wind paths are tightly anticorrelated because the sum of the regions is better constrained from the large-scale atmospheric signals than the individual regions. While additional CO_2 observations could provide better constraints on the inversion system, this alone is unlikely to resolve the large variability among inversions. As notable model-by-model variability was found in Europe, one of the regions characterized by a high density of in situ CO_2 observations, we need to acknowledge the possibility that modeling issues are responsible for this variability. They include differences in model resolution, the period during which data assimilation is conducted, the rates at which CO_2 is transported from a source region to neighboring regions through the model atmosphere, and transport model errors among inversions. For example, the degree to which a regional budget reflects localized fossil fuel signals or CO_2 measurement signals varies with the resolution of the transport models. These differences might have caused the large variability in the European CO_2 budgets, which could propagate into the budget estimates for boreal Asia via the dipole effect.

These results highlight that despite the abundant observations in temperate regions, we cannot be confident in whether the science community is using them properly on carbon budget estimation. This issue is not constrained only to temperate regions, but also to the global regions. The availability of observations (inventory, satellite, and tower observations) does not necessarily lead to a better estimate of the carbon budget. Rather, it requires us to revisit how to make the best use of them in empirical and process modeling.

4.2 Uncertainty in components of the temperate carbon budget

Just as we need to address the uncertainty in observational methods, it is necessary to acknowledge the uncertainty in processes of the carbon cycle in temperate regions. One of the key uncertainties concerns the processes of CO_2 uptake during recovery or regrowth after large-scale disturbances (deforestation, wood harvest activities, and natural fires). Just as agricultural lands differ greatly from natural grasslands in their carbon dynamics, managed forests differ greatly from natural forests, particularly in terms of the species mixtures and the crucial additional element of wood harvest. Representations of detailed forest management practices such as planting, thinning, and increment-related harvesting have only been published for one TBM to date (Luyssaert et al., 2018; Naudts et al., 2016), although they are in development for several others. One clear problem here is a lack of forcing data on large-scale forest management practices, which limits our understanding of how forests regrow in the transition from large-scale land-use changes to management. As a consequence, future projections of regrowth flux remain unconstrained, limiting understanding of its future role in mitigating anthropogenic carbon emissions.

There are also components of the temperate carbon budget that need to be addressed further. Blue carbon is the carbon stored in shallow coastal ecosystems, such as mangrove forests, saltmarsh, Sargassum forests, and kelp forests. Exchange of carbon between the atmosphere and those ecosystems could be nonnegligible in temperate regions. For instance, a comparison of observational and process-based methods suggests a net carbon uptake in North American shallow coasts to be $94\,Tg\,C\,year^{-1}$ at most (USGCRP, 2018). However, a question remains whether the shallow coastal carbon exchange should be added to the carbon budget of the land or ocean or treated independently. Atmospheric inversions account for all the components of the net carbon exchange between the atmosphere and land-ocean surfaces, but it is difficult to determine whether the contribution of blue carbon is included in the land or ocean because a spatial resolution is too coarse to realize the shallow coastal land cover in this method. Wood products resulting from land-use changes or imported from other regions are also an uncertain component of the carbon budget that needs to be addressed in future. In some TBMs, a proportion of harvested wood is converted to wood products, but rates at which carbon stored in them returns to the atmosphere via decay are based on assumptions without a sound basis. To the best of our knowledge, the decay of imported wood products has not been addressed in carbon budget assessment though it may be counted in the exporting region.

5 Perspective and future opportunities for policy decision-making

The research of carbon budgets in temperate ecosystems is more advanced than in other regions, both in terms of monitoring and modeling. Hence, it is critical to understand whether current observations and models are sophisticated enough for policies. This section describes the key progress made in the carbon budget assessment in past decades, and how it should be further advanced to be useful for policy decision-making.

5.1 Progress over past decades

The main areas of progress in temperate regions have involved: (1) the establishment of sound and continuous forest inventories; (2) advances in remote sensing technology and increasing use in operational monitoring; (3) extensive deployment of eddy flux towers and direct measurements of CO_2 concentrations and exchange between the land and atmosphere; and (4) improved analysis methods, as represented by improved atmospheric (top-down) and process-based (bottom-up) estimates of the terrestrial carbon sink. Pioneering the development of observational networks and modeling methods, temperate regions have served as a good test bed for carbon budget estimation, and the region is expected to serve such a role in the future.

Among the various major advancements in the temperate carbon cycle in past decades, the most important one may be the advancements in land-use modeling, which add the capability for TBMs to simulate the anthropogenic impacts on ecosystem carbon fluxes. Early assessments of carbon fluxes made the assumption that all land was covered by natural lands, i.e., unmanaged, vegetation, or simulated agricultural lands as grasslands. This relatively simplistic approach has resulted in large biases in the carbon budget (Arneth et al., 2017; Kondo et al., 2020; Pugh et al., 2015) and is particularly relevant in a temperate zone where the vast majority of the land is strongly influenced by human activities (Ellis & Ramankutty, 2008). To represent the variety of real-world land-uses, TBMs have undergone somewhat of a revolution in recent years, and explicitly represent a suite of managed vegetation types. It is now common to represent major crop types using dedicated PFTs that better reflect their growth, allocation, and life cycle behavior than grasses and to simulate explicit management actions such as sowing, harvest, tillage, irrigation, and fertilization

(Bondeau et al., 2007; Drewniak, Song, Prell, Kotamarthi, & Jacob, 2013; Levis, Hartman, & Bonan, 2014; Olin et al., 2015). Despite uncertainty in modeling land-use changes remains, TBMs are certainly advancing as a useful tool for forest management policies.

5.2 Future perspective

Policy makers will take action about climate change even in the absence of perfect information. In temperate regions, if available data and methods converge to the common characterization of the carbon cycle of temperate forests and grasslands, they would provide a solid foundation for decision-making at large geographic scales (e.g., countries or regions) and aid in monitoring the effects of policies on greenhouse gases. Exceptions may involve geographic and content gaps and the ability to deploy cost-effective monitoring at the smaller scales at which land-use and management decisions are made. However, advanced remote sensing technologies such as LiDAR and improvements in modeling could be deployed if policy makers can agree on international accounting rules and reporting guidelines, and if funding is available to support a continuous and credible monitoring program. The details of selected policies and programs will influence the exact nature of required monitoring, and these decisions are not independent. That is, future policies and programs need to be informed by the availability of monitoring to verify the actions and potential consequences, and the monitoring must be tailored to support policies.

5.3 Toward policy-driven carbon budgets

Monitoring needs to support reporting requirements such that activities to manage greenhouse gases are explicitly tracked, and effects on greenhouse gases that are not a direct consequence of the activity are "factored out." For example, the impact of modifying forest management practices to increase CO_2 uptake and store more carbon in harvested wood may be reduced if the area of the activity is subject to a natural disturbance such as wildfire or drought. Separating the effects of different factors may be essential to evaluate policies and crediting actions. Research over the last two decades has advanced enough to highlight the importance of assessing different factors in temperate forests and grasslands, but there is a lack of agreement about their relative impacts as estimated by different modeling and assessment approaches. Thus, additional research is needed to verify that one or more approaches can be consistently applied in temperate countries and that the results are accurate enough to reflect how policies actually affect greenhouse gases in the atmosphere.

Estimates at smaller scales of activity deployment that can be directly tied to reporting at larger scales (e.g., countries) are important but often costly to deploy. As a result, additional investment in deploying low-cost remote sensing-based monitoring is an essential direction to support mitigation

264 Section | C Case Studies

policies. For example, a high-resolution satellite sensor dedicated to monitoring land-use and land-cover changes across the temperate zone would help provide consistent monitoring across scales, and allow for the detection of "leakage," i.e., actions induced elsewhere as a result of a project activity, that would reduce the anticipated greenhouse gas benefits of the activity. Moving forward, there is increasing acknowledgement that land use and management have additional direct effects on climate that go beyond biogeochemical-induced climate impacts. These biophysical effects include albedo modifications, changes in evapotranspiration, changes in turbulence due to altered surface roughness, and other factors. Their effects on climate may either increase or decrease the cooling effects expected from reducing greenhouse gases, and these effects are highly variable geographically and according to the nature of the activity. While there is currently no agreed-upon accounting framework to merge biogeochemical and biophysical effects into a comprehensive climate response monitoring system, an improved ability to estimate biophysical effects will eventually allow for such an accounting system to be used.

References

Amiotte-Suchet, P., & Probst, J.-L. (1995). A global model for present day atmospheric/soil CO_2 consumption by chemical erosion of continental rocks (GEM-CO_2). *Tellus B*, *47*, 273–2280.

Arneth, A., Sitch, S., Pongratz, J., Stocker, B. D., Ciais, P., Poulter, B., et al. (2017). Historical carbon dioxide emissions caused by land-use changes are possibly larger than assumed. *Nature Geoscience*, *10*, 79–84.

Baldocchi, D. (2003). Assessing the eddy covariance technique for evaluating carbon dioxide exchange rates of ecosystems: Past, present and future. *Global Change Biology*, *9*, 479–492.

Baldocchi, D., Chu, H., & Reichstein, M. (2017). Inter-annual variability of net and gross ecosystem carbon fluxes: A review. *Agricultural and Forest Meteorology*, *249*, 520–533.

Baldocchi, D., Valentini, R., Running, S., Oechel, W., & Dahlman, R. (1996). Strategies for measuring and modelling carbon dioxide and water vapour fluxes over terrestrial ecosystems. *Global Change Biology*, *2*, 159–168.

Ballantyne, A. P., Andres, R., Houghton, R., Stocker, B. D., Wanninkhof, R., Anderegg, W., et al. (2015). Audit of the global carbon budget: Estimate errors and their impact on uptake uncertainty. *Biogeosciences*, *12*(8), 2565–2584.

Barros, N., Cole, J. J., Tranvik, L. J., Prairie, Y. T., Bastviken, D., Huszar, V. L. M., et al. (2011). Carbon emission from hydroelectric reservoirs linked to reservoir age and latitude. *Nature Geoscience*, *4*, 593–596.

Bastos, A., O'Sullivan, M., Ciais, P., Makowski, D., Sitch, S., Friedlingstein, P., et al. (2020). Sources of uncertainty in regional and global terrestrial CO_2 exchange estimates. *Global Biogeochemical Cycles*, *34*, e2019GB006393.

Berner, R. A., Lasaga, A. C., & Garrels, R. M. (1983). The carbonate-silicate geochemical cycle and its effect on atmospheric carbon dioxide over the past 100 million years. *American Journal of Science*, *283*, 641–683.

Birdsey, R., Pregitzer, K., & Lucier, A. (2006). Forest carbon management in the United States. *Journal of Environmental Quality*, *35*, 1461–1469.

Bondeau, A., Smith, P. C., Zaehle, S., Schaphoff, S., Lucht, W., Cramer, W., et al. (2007). Modelling the role of agriculture for the 20th century global terrestrial carbon balance. *Global Change Biology*, *13*, 679–706.

Chen, Y., Gardiner, B., Pasztor, F., Blennow, K., Ryder, J., Valade, A., et al. (2018). Simulating damage for wind storms in the land surface model ORCHIDEE-CAN (revision 4262). *Geoscientific Model Development*, *11*, 771–791.

Chevallier, F., Ciais, P., Conway, T. J., Aalto, T., Anderson, B. E., Bousquet, P., et al. (2010). CO_2 surface fluxes at grid point scale estimated from a global 21 year reanalysis of atmospheric measurements. *Journal of Geophysical Research: Atmospheres*, *115*, D21307.

Cole, J. J., Prairie, Y. T., Caraco, N. F., McDowell, W. H., Tranvik, L. J., Striegl, R. G., et al. (2007). Plumbing the global carbon cycle: Integrating inland waters into the terrestrial carbon budget. *Ecosystems*, *10*, 171–184.

Crutzen, P. J., & Andreae, M. O. (1990). Biomass burning in the tropics: Impact on atmospheric chemistry and biogeochemical cycles. *Science*, *250*, 1669–1678.

Drewniak, B., Song, J., Prell, J., Kotamarthi, V. R., & Jacob, R. (2013). Modeling agriculture in the community land model. *Geoscientific Model Development*, *6*(2), 495–515.

Dürr, H. H., Meybeck, M., & Dürr, S. H. (2005). Lithologic composition of the Earth's continental surfaces derived from a new digital map emphasizing riverine material transfer. *Global Biogeochemical Cycles*, *19*, GB4S10.

Ellis, E. C., & Ramankutty, N. (2008). Putting people in the map: Anthropogenic biomes of the world. *Frontiers in Ecology and the Environment*, *6*, 439–447.

Fan, L., Wigneron, J. P., Ciais, P., Chave, J., Brandt, M., Fensholt, R., et al. (2019). Satellite-observed pantropical carbon dynamics. *Nature Plants*, *5*, 944–951.

Fearnside, P. M. (2000). Global warming and tropical land-use change: Greenhouse gas emissions from biomass burning, decomposition and soils in forest conversion, shifting cultivation and secondary vegetation. *Climatic Change*, *46*, 115–158.

Fisher, R. A., Koven, C. D., Anderegg, W. R. L., Christoffersen, B. O., Dietze, M. C., Farrior, C. E., et al. (2018). Vegetation demographics in earth system models: A review of progress and priorities. *Global Change Biology*, *24*, 35–54.

Gaubert, B., Stephens, B. B., Basu, S., Chevallier, F., Deng, F., Kort, E. A., et al. (2019). Global atmospheric CO_2 inverse models converging on neutral tropical land exchange, but disagreeing on fossil fuel and atmospheric growth rate. *Biogeosciences*, *16*(1), 117–134.

Goodale, C. L., Apps, M. J., Birdsey, R. A., Field, C. B., Heath, L. S., Houghton, R. A., et al. (2002). Forest carbon sinks in the Northern hemisphere. *Ecological Applications*, *12*, 891–899.

Guenther, A. B., Hewitt, C. N., Erickson, D., Fall, R., Geron, C., Graedel, T., et al. (1995). A global model of natural volatile organic compound emissions. *Journal of Geophysical Research-Atmospheres*, *100*, 8873–8892.

Guenther, A. B., Jiang, X., Heald, C. L., Sakulyanontvittaya, T., Duhl, T., Emmons, L. K., et al. (2012). The model of emissions of gases and aerosols from nature version 2.1 (MEGAN2.1): An extended and updated framework for modeling biogenic emissions. *Geoscientific Model Development*, *5*, 1471–1492.

Gurney, K. R., Law, R. M., Denning, A. S., Rayner, P. J., Baker, D., Bousquet, P., et al. (2002). Towards robust regional estimates of CO_2 sources and sinks using atmospheric transport models. *Nature*, *415*, 626–630.

Hansis, E., Davis, S. J., & Pongratz, J. (2015). Relevance of methodological choices for accounting of land use change carbon fluxes. *Global Biogeochemical Cycles*, *29*, 1230–1246.

Hantson, S., Arneth, A., Harrison, S. P., Kelley, D. I., Prentice, I. C., Rabin, S. S., et al. (2012). The status and challenge of global fire modelling. *Biogeosciences*, *13*, 3359–3375.

266 Section | C Case Studies

Harrison, J. A., Seitzinger, S. P., Bouwman, A. F., Caraco, N. F., Beusen, A. H. W., & Vörösmarty, C. J. (2005). Dissolved inorganic phosphorus export to the coastal zone: Results from a spatially explicit, global model. *Global Biogeochemical Cycles*, *19*, GB4S03.

Hartmann, J., Jansen, N., Dürr, H. H., Kempe, S., & Köhler, P. (2009). Global CO_2-consumption by chemical weathering: What is the contribution of highly active weathering regions? *Global Planetary Change*, *69*, 185–194.

Hartmann, J., Lauerwald, R., & Moosdorf, N. (2014). A brief overview of the GLObal RIver chemistry database, GLORICH. *Procedia Earth and Planetary Science*, *10*, 23–27.

Haslwanter, A., Hammerle, A., & Wohlfahrt, G. (2009). Open-path vs. closed-path eddy covariance measurements of the net ecosystem carbon dioxide and water vapour exchange: A long-term perspective. *Agricultural and Forest Meterology*, *149*, 291–302.

Haverd, V., Smith, B., Nieradzik, L. P., & Briggs, P. R. (2014). A stand-alone tree demography and landscape structure module for earth system models: Integration with inventory data from temperate and boreal forests. *Biogeosciences*, *11*, 4039–4055.

Hayes, D. J., Turner, D. P., Stinson, G., McGuire, A. D., Wei, Y., West, T. O., et al. (2012). Reconciling estimates of the contemporary North American carbon balance among terrestrial biosphere models, atmospheric inversions, and a new approach for estimating net ecosystem exchange from inventory-based data. *Global Change Biology*, *18*, 1282–1299.

Houghton, R. A., Hobbie, J. E., Melillo, J. M., Moore, B., Peterson, B. J., Shaver, G. R., et al. (1983). Changes in the carbon content of terrestrial biota and soils between 1860 and 1980: A net release of CO_2 to the atmosphere. *Ecological Monographs*, *53*, 236–262.

Houghton, R. A., & Nassikas, A. A. (2017). Global and regional fluxes of carbon from land use and land cover change 1850–2015. *Global Biogeochemical Cycles*, *31*, 456–472.

Hurtt, G. C., Chini, L., Sahajpal, R., Frolking, S., Bodirsky, B. L., Calvin, K., et al. (2020). Harmonization of global land-use change and management for the period 850–2100 (LUH2) for CMIP6. *Geoscientific Model Development*, *13*, 5425–5464.

Ito, A. (2019). Disequilibrium of terrestrial ecosystem CO_2 budget caused by disturbance-induced emissions and non-CO_2 carbon export flows: A global model assessment. *Earth System Dynamics*, *10*, 685–709.

Jacobson, A. R., Fletcher, S. E. M., Gruber, N., Sarmiento, J. L., & Gloor, M. (2007). A joint atmosphere-ocean inversion for surface fluxes of carbon dioxide: 1. Methods and global-scale fluxes. *Global Biogeochemical Cycles*, *21*, GB1019.

Jung, M., Reichstein, M., Margolis, H. A., Cescatti, A., Richardson, A. D., Arain, M. A., et al. (2011). Global patterns of land-atmosphere fluxes of carbon dioxide, latent heat, and sensible heat derived from eddy covariance, satellite, and meteorological observations. *Journal of Geophysical Research-Biogeosciences*, *116*, 148–227.

Jung, M., Reichstein, M., Schwalm, C. R., Huntingford, C., Sitch, S., Ahlström, A., et al. (2017). Compensatory water effects link yearly global land CO_2 sink changes to temperature. *Nature*, *541*(7638), 516–520.

Jung, M., Schwalm, C., Migliavacca, M., Walther, S., Camps-Valls, G., Koirala, S., et al. (2020). Scaling carbon fluxes from eddy covariance sites to globe: Synthesis and evaluation of the FLUXCOM approach. *Biogeosciences*, *17*, 1343–1365.

Kaiser, J. W., Heil, A., Andreae, M. O., Benedetti, A., Chubarova, N., Jones, L., et al. (2012). Biomass burning emissions estimated with a global fire assimilation system based on observed fire radiative power. *Biogeosciences*, *9*, 527–554.

Kautz, M., Anthoni, P., Meddens, A. J. H., Pugh, T. A. M., & Arneth, A. (2018). Simulating the recent impacts of multiple biotic disturbances on forest carbon cycling across the United States. *Global Change Biology*, *24*, 2079–2092.

Keenan, T. F., Prentice, I. C., Canadell, J. G., Williams, C. A., Wang, H., Raupach, M., et al. (2016). Recent pause in the growth rate of atmospheric CO_2 due to enhanced terrestrial carbon uptake. *Nature Communications, 7*, 13428.

Kim, J. B., Kerns, B. K., Drapek, R. J., Pitts, G. S., & Halofsky, J. E. (2018). Simulating vegetation response to climate change in the Blue Mountains with MC2 dynamic global vegetation model. *Climate Services, 10*, 20–32.

King, A. W., Andres, R. J., Davis, K. J., Hafer, M., Hayes, D. J., Huntzinger, D. N., et al. (2015). North America's net terrestrial CO_2 exchange with the atmosphere 1990–2009. *Biogeosciences, 12*, 399–414.

Klein Goldewijk, K., Beusen, A., Doelman, J., & Stehfest, E. (2017). Anthropogenic land use estimates for the Holocene—HYDE 3.2. *Earth System Science Data, 9*, 927–953.

Kondo, M., Ichii, K., Patra, P. K., Poulter, B., Calle, L., Koven, C., et al. (2018). Plant regrowth as a driver of recent enhancement of terrestrial CO_2 uptake. *Geophysical Research Letters, 45*, 4820–4830.

Kondo, M., Ichii, K., Takagi, H., & Sasakawa, M. (2015). Comparison of the data-driven top-down and bottom-up global terrestrial CO_2 exchanges: GOSAT CO_2 inversion and empirical eddy flux upscaling. *Journal of Geophysical Research-Biogeosciences, 120*, 1226–1245.

Kondo, M., Patra, P. K., Sitch, S., Friedlingstein, P., Poulter, B., Chevallier, F., et al. (2020). State of the science in reconciling top-down and bottom-up approaches for terrestrial CO_2 budget. *Global Change Biology, 26*, 1068–1084.

Koven, C. D., Knox, R. G., Fisher, R. A., Chambers, J. Q., Christoffersen, B. O., Davies, S. J., et al. (2020). Benchmarking and parameter sensitivity of physiological and vegetation dynamics using the functionally assembled terrestrial ecosystem simulator (FATES) at Barro Colorado Island, Panama. *Biogeosciences, 17*, 3017–3044.

Lauerwald, R., Laruelle, G. G., Hartmann, J., Ciais, P., & Regnier, P. A. G. (2015). Spatial patterns in CO_2 evasion from the global river network. *Global Biogeochemical Cycles, 29*(5), 534–554.

Le Quéré, C., Andrew, R. M., Friedlingstein, P., Sitch, S., Hauck, J., Pongratz, J., et al. (2018). Global carbon budget 2018. *Earth System Scientific Data, 10*, 2141–2194.

Levis, S., Hartman, M. D., & Bonan, G. B. (2014). The community land model underestimates land-use CO_2 emissions by neglecting soil disturbance from cultivation. *Geoscientific Model Development, 7*, 613–620.

Liu, Y. Y., van Dijk, A. I. J. M., de Jeu, R. A. M., Canadell, J. G., McCabe, M. F., Evans, J. P., et al. (2015). Recent reversal in loss of global terrestrial biomass. *Nature Climate Change, 5*, 470–474.

Ludwig, W., Amiotte-Suchet, P., Munhoven, G., & Probst, J.-L. (1998). Atmospheric CO_2 consumption by continental erosion: Present-day controls and implications for the last glacial maximum. *Global Planetary Change, 16–17*, 107–120.

Luyssaert, S., Abril, G., Andres, R., Bastviken, D., Bellassen, V., Bergamaschi, P., et al. (2012). The European land and inland water CO_2, CO, CH_4 and N_2O balance between 2001 and 2005. *Biogeosciences, 9*, 3357–3380.

Luyssaert, S., Marie, G., Valade, A., Chen, Y., Djomo, S. N., Ryder, J., et al. (2018). Trade-offs in using European forests to meet climate objectives. *Nature, 562*, 259–262.

Maki, T., Ikegami, M., Fujita, T., Hirahara, T., Yamada, K., Mori, K., et al. (2010). New technique to analyse global distributions of CO_2 concentrations and fluxes from non-processed observational data. *Tellus B: Chemical and Physical Meteorology, 62*, 797–809.

Mayorga, E., Seitzinger, S. P., Harrison, J. A., Dumont, E., Beusen, A. H. W., Bouwman, A. F., et al. (2010). Global nutrient export from WaterSheds 2 (NEWS 2): Model development and implementation. *Environmental Modelling and Software, 25*(7), 837–853.

268 Section | C Case Studies

Meybeck, M. (1993). Riverine transport of atmospheric carbon: Sources, global typology and budget. *Water, Air, and Soil Pollution, 70,* 443–463.

Miles, L., & Kapos, V. (2008). Reducing greenhouse gas emissions from deforestation and forest degradation: Global land-use implications. *Science, 320,* 1454–1455.

Moesinger, L., Dorigo, W., de Jeu, R., van der Schalie, R., Scanlon, T., Teubner, I., et al. (2020). The global long-term microwave vegetation optical depth climate archive (VODCA). *Earth System Science Data, 12,* 177–196.

Moffat, A. M., Reichstein, M., Hollinger, D. Y., Richardson, A. D., Barr, A. G., Beckstein, C., et al. (2007). Comprehensive comparison of gap-filling techniques for eddy covariance net carbon fluxes. *Agricultural and Forest Meterology, 147,* 209–232.

Naudts, K., Chen, Y., McGrath, M. J., Ryder, J., Valade, A., Otto, J., et al. (2016). Europe's forest management did not mitigate climate warming. *Science, 351,* 597–600.

Niwa, Y., & Fujii, Y. (2020). A conjugate BFGS method for accurate estimation of a posterior error covariance matrix in a linear inverse problem. *Quarterly Journal of the Royal Meteorological Society, 146,* 3118–3143.

Niwa, Y., Tomita, H., Satoh, M., Imasu, R., Sawa, Y., Tsuboi, K., et al. (2017). A 4D-Var inversion system based on the icosahedral grid model (NICAM-TM 4D-Var v1.0)—Part 1: Offline forward and adjoint transport models. *Geoscientific Model Development, 10,* 1157–1174.

Olin, S., Schurgers, G., Lindeskog, M., Wårlind, D., Smith, B., Bodin, P., et al. (2015). Modelling the response of yields and tissue C:N to changes in atmospheric CO_2 and N management in the main wheat regions of western Europe. *Biogeosciences, 12,* 2489–2515.

Pan, Y., Birdsey, R. A., Fang, J., Houghton, R., Kauppi, P. E., Kurz, W. A., et al. (2011). A large and persistent carbon sink in the world's forests. *Science, 333,* 988–993.

Papale, D., Reichstein, M., Aubinet, M., Canfora, E., Bernhofer, C., Kutsch, W., et al. (2006). Towards a standardized processing of net ecosystem exchange measured with eddy covariance technique: Algorithms and uncertainty estimation. *Biogeosciences, 3,* 571–583.

Peters, W., Jacobson, A. R., Sweeney, C., Andrews, A. E., Conway, T. J., Masarie, K., et al. (2007). An atmospheric perspective on North American carbon dioxide exchange: CarbonTracker. *Proceedings of the National Academy of Sciences of the United States of America, 104,* 18925–18930.

Peylin, P., Baker, D., Sarmiento, J., Ciais, P., & Bousquet, P. (2002). Influence of transport uncertainty on annual mean and seasonal inversions of atmospheric CO_2 data. *Journal of Geophysical Research, 107*(D19), 4385.

Peylin, P., Law, R. M., Gurney, K. R., Chevallier, F., Jacobson, A. R., Maki, T., et al. (2013). Global atmospheric carbon budget: Results from an ensemble of atmospheric CO_2 inversions. *Biogeosciences, 10,* 6699–6720.

Piao, S. L., Ito, A., Li, S. G., Huang, Y., Ciais, P., Wang, X. H., et al. (2012). The carbon budget of terrestrial ecosystems in East Asia over the last two decades. *Biogeosciences, 9,* 3571–3586.

Prentice, I., Bondeau, A., Cramer, W., Harrison, S., Hickler, T., Lucht, W., et al. (2007). Dynamic global vegetation modeling: Quantifying terrestrial ecosystem responses to large-scale environmental change. In J. Canadell, D. Pataki, & L. Pitelka (Eds.), *Terrestrial ecosystems in a changing world* (pp. 175–192). Berlin, Heidelberg, New York: Springer.

Pugh, T. A. M., Arneth, A., Olin, S., Ahlström, A., Bayer, A. D., Klein Goldewijk, K., et al. (2015). Simulated carbon emissions from land-use change are substantially enhanced by accounting for agricultural management. *Environmental Research Letters, 10*(12), 124008.

Pugh, T. A. M., Lindeskog, M., Smith, B., Poulter, B., Arneth, A., Haverd, V., et al. (2019). Role of forest regrowth in global carbon sink dynamics. *Proceedings of the National Academy of Sciences of the United States of America, 116,* 4382–4387.

Raymond, P. A., Hartmann, J., Lauerwald, R., Sobek, S., McDonald, C., et al. (2013). Global carbon dioxide emissions from inland waters. *Nature, 503*, 355–359.

Renard, K. G., Foster, G. R., Weesies, G. A., & Porter, J. P. (1991). RUSLE, revised universal soil loss equation. *Journal of Soil and Water Conservation, 46*(1), 30–33.

Rödenbeck, C., Houweling, S., Gloor, M., & Heimann, M. (2003). Time-dependent atmospheric CO_2 inversions based on interannually varying tracer transport. *Tellus B: Chemical and Physical Meteorology, 55*, 488–497.

Rödenbeck, C., Zaehle, S., Keeling, R., & Heimann, M. (2018). How does the terrestrial carbon exchange respond to inter-annual climatic variations? A quantification based on atmospheric CO_2 data. *Biogeosciences, 15*, 2481–2498.

Ruhala, S. S., & Zarnetske, J. P. (2017). Using in-situ optical sensors to study dissolved organic carbon dynamics of streams and watersheds: A review. *Science of the Total Environment, 575*, 713–723.

Saatchi, S. S., Harris, N. L., Brown, S., Lefsky, M., Mitchard, E. T., Salas, W., et al. (2011). Benchmark map of forest carbon stocks in tropical regions across three continents. *Proceedings of the National Academy of Sciences of the United States of America, 108*, 9899–9904.

Saeki, T., & Patra, P. K. (2017). Implications of overestimated anthropogenic CO_2 emissions on East Asian and global land CO_2 flux inversion. *Geoscience Letters, 4*, 9.

Seiler, W., & Crutzen, P. J. (1980). Estimates of gross and net fluxes of carbon between the biosphere and the atmosphere from biomass burning. *Climatic Change, 2*, 207–247.

Shafer, S. L., Bartlein, P. J., Gray, E. M., & Pelltier, R. T. (2015). Projected future vegetation changes for the Northwest United States and Southwest Canada at a fine spatial resolution using a dynamic global vegetation model. *PLoS One, 10*, e0138759.

Shi, Y., Matsunaga, T., Saito, M., Yamaguchi, Y., & Chen, X. (2015). Comparison of global inventories of CO_2 emissions from biomass burning during 2002–2011 derived from multiple satellite products. *Environmental Pollution, 206*, 479–487.

Sindelarova, K., Granier, C., Bouarar, I., Guenther, A., Tilmes, S., Stavrakou, T., et al. (2014). Global data set of biogenic VOC emissions calculated by the MEGAN model over the last 30 years. *Atmospheric Chemistry and Physics, 14*, 9317–9341.

Sitch, S., Smith, B., Prentice, I. C., Arneth, A., Bondeau, A., et al. (2003). Evaluation of ecosystem dynamics, plant geography and terrestrial carbon cycling in the LPJ dynamic global vegetation model. *Global Change Biology, 9*, 161–185.

Smith, B., Wårlind, D., Arneth, A., Hickler, T., Leadley, P., Siltberg, J., et al. (2014). Implications of incorporating N cycling and N limitations on primary production in an individual-based dynamic vegetation model. *Biogeosciences, 11*, 2027–2054.

Soloway, A. D., Amiro, B. D., Dunn, A. L., & Wofsy, S. C. (2017). Carbon neutral or a sink? Uncertainty caused by gap-filling long-term flux measurements for an old-growth boreal black spruce forest. *Agricultural and Forest Meterology, 233*, 110–121.

Stephens, B. B., Gurney, K. R., Tans, P. P., Sweeney, C., Peters, W., Bruhwiler, L., et al. (2007). Weak northern and strong tropical land carbon uptake from vertical profiles of atmospheric CO_2. *Science, 316*(5832), 1732–1735.

Syvitski, J. P. M., Morehead, M. D., & Nicholson, M. (1998). HydroTrend: A climate-driven hydrologic-transport model for predicting discharge and sediment load to lakes or oceans. *Computers & Geosciences, 24*, 51–68.

USGCRP. (2018). In N. Cavallaro, G. Shrestha, R. Birdsey, M. A. Mayes, R. G. Najjar, S. C. Reed, P. Romero-Lankao, & Z. Zhu (Eds.), *Second state of the carbon cycle report (SOCCR2): A sustained assessment report*. Washington, DC: U.S. Global Change Research Program. https://doi.org/10.7930/SOCCR2.2018 (878 pp.).

270 Section | C Case Studies

van der Werf, G. R., Randerson, J. T., Giglio, L., Collatz, G. J., Mu, M., Kasibhatla, P. S., et al. (2010). Global fire emissions and the contribution of deforestation, savanna, forest, agricultural, and peat fires (1997–2009). *Atmospheric Chemistry and Physics, 10*, 11707–11735.

Williams, C. A., Gu, H., MacLean, R., Masek, J. G., & Collatz, G. J. (2016). Disturbance and the carbon balance of US forests: A quantitative review of impacts from harvests, fires, insects, and droughts. *Global and Planetary Change, 143*, 66–80.

Wolf, J., West, T. O., Le Page, Y., Kyle, G. P., Zhang, X., Collatz, G. J., et al. (2015). Biogenic carbon fluxes from global agricultural production and consumption. *Global Biogeochemical Cycles, 29*, 1617–1639.

Zaehle, S., & Friend, A. D. (2010). Carbon and nitrogen cycle dynamics in the O-CN land surface model: 1. Model description, site-scale evaluation, and sensitivity to parameter estimates. *Global Biogeochemical Cycles, 24*, GB1005.

Zscheischler, J., Mahecha, M. D., Avitabile, V., Calle, L., Carvalhais, N., Ciais, P., et al. (2017). Reviews and syntheses: An empirical spatiotemporal description of the global surface–atmosphere carbon fluxes: Opportunities and data limitations. *Biogeosciences, 14*, 3685–3703.

Chapter 8

Tropical ecosystem greenhouse gas accounting

Jean Pierre Ometto, Felipe S. Pacheco, Mariana Almeida, Luana Basso, Francisco Gilney Bezerra, Manoel Cardoso, Marcela Miranda, Eráclito Souza Neto, Celso von Randow, Luiz Felipe Rezende, Kelly Ribeiro, and Gisleine Cunha-Zeri

Earth System Science Centre/National Institute for Space Research, Sao Jose dos Campos, São Paulo, Brazil

Key messages

Tropical ecosystems store large amounts of carbon (e.g., tropical forests, peatlands); however, the high rates of ecosystem conversion due to intense land use and land cover changes lead to significant emissions of greenhouse gases (GHGs) to the atmosphere. Additionally, forest degradation, such as unmanaged timber harvest and forest fires, accounts for additional large amount of emissions. The dynamics of economies in transition, in which most of the countries in the tropical belt are included, determine an increasing rate of fossil fuel-related emissions. Nevertheless, deforestation and agriculture still dominate the emissions in this region. Uncertainties in estimates of GHG emissions is normally due to the lack of data and modeling assumptions. Increasing monitoring of GHG emissions, in broader areas, is expected to reduce uncertainty and provide GHG national inventories with more reliable estimates.

1 Introduction and background: Tropical ecosystems

1.1 General description

The tropics are delimited by the Tropic of Cancer (Latitude $23°26'12.1''N$) and Tropic of Capricorn (Latitude $23°26'12.1''S$). With abundant sunlight, the tropical region is under a nonarid climate with a mean annual temperature warmer than $18°C$ ($64°F$). Nevertheless, other climate characteristics are found in the tropical regions associated with deserts and snow-capped mountains. The tropicas comprises 40% of the Earth's surface area, containing approximately 36% of the Earth's land, with 40%–55% of this area covered by forests and savannas (Murphy & Bowman, 2012) and holding the most diverse biomes on the planet (Lewis, Edwards, & Galbraith, 2015). Half of the world's population lives in the tropics, facing challenges in severely inequal societies, and also abundant

Balancing Greenhouse Gas Budgets. https://doi.org/10.1016/B978-0-12-814952-2.00013-7
Copyright © 2022 Elsevier Inc. All rights reserved.

272 Section | C Case Studies

cultural diversity. Approximately 135 nations are distributed among the four continents present in the planet's tropical belt.

1.2 Understanding changes in carbon cycling and storage

The effect of climate change shall pose extra pressure on the infrastructure and society in this region; infrastructure because of depletion of adequate solutions in most countries, and society by the substantial number of poor and vulnerable communities. From natural disasters to sea level rise, extreme events affecting agriculture and food security to the risk of increasing flood events in urban areas will dramatically impact several regions within the tropical region. Therefore, the need for the development of adaptation strategies is critical for most nations, while substantial investments in technologies, infrastructure, and capacity building to mitigate greenhouse gas (GHG) emissions are a priority. On that concern, the identification of key emitter sectors, strategies to mitigate social inclusiveness and involvement, and the development of a strong database and regionally distributed estimates are critical. Nevertheless, the diversity, heterogeneity, broad social context, economic wealth, and basic provisions, among a long list of differences and disparities within tropical nations, affect not only the capability to estimate GHG emissions but also the per capita, country, or sector emissions (Friedlingstein et al., 2019). Therefore, the uncertainty in the country and local emissions estimates may be high.

The forests in the tropical regions are, in general, carbon-dense and highly productive, being both a huge storage of carbon in the vegetation and on soils, as a potential source, considering climate change and intense land-use changes. The expansion of agriculture and conversion of tropical forests to other land uses are still strong drivers of deforestation in the tropics (Aguiar et al., 2016; Drescher et al., 2016). This is one of the major sources of GHG to the atmosphere in the region, especially considering low- and mid-income countries. Most of the global contributions to atmospheric carbon dioxide (CO_2) concentration increasing rates are due to deforestation in the tropical region, overall in South America, Africa, and Southeast Asia (in particular, Indonesia). Climate change, with increase in temperature and extreme droughts, threaten the stability of tropical vegetation. High rates of fire and forest degradation are already being observed (Allen et al., 2017; Aragão et al., 2018) and are expected to increase substantially in the future (Aleman, Jarzyna, & Staver, 2018; Cardoso et al., 2009; Tacconi & Muttaqin, 2019b). Increasing attention is being given to secondary forest growth and dynamics, considering its importance to the carbon cycle, ecosystem restoration, and biodiversity preservation (Silva Junior et al., 2020).

Emissions from agricultural production, including livestock and grain production (in special rice, soy, and corn), are also a major source of GHG emissions in the region. The use of fertilizer is fairly heterogeneous geographically, being highly used in some regions and lacking in others, although the application rates in agricultural production are increasing exponentially, leading to increased emissions of nitrous oxide (N_2O) and ammonia volatilization. Among the large exporters of animal protein for human consumption, tropical regions hold several important exporters of agricultural commodities (e.g., soy, corn, rice, etc.), contributing substantially to the global emissions of methane (CH_4) and (N_2O).

2 GHG budget in the tropics

2.1 Components of the greenhouse gas budget tropical ecosystems

In this chapter, we present the methodologies and results from different landscape footprints, scaling up to regional and global fluxes, related to the tropics. One of the major gaps is the lack of consistent, long term and spatially distributed sampling, and GHG emissions monitoring networks (Fig. 1). Despite the, in general, shorter history and smaller geo graphical representation of GHG emissions data in the tropics, compared to the temperate region (Fig. 2), several methodological approaches have been adapted in the region. Remote sensing and airborne atmospheric sampling have evolved substantially on estimating the GHG fluxes in the tropics; however, some of the samplings are expensive and do not maintain a long-term database. Examples of atmospheric profile sampling are provided for the Amazon region in South America.

Data and methodological approaches to flux measurements considering eddy covariance techniques are also provided, which couple, at a smaller scale, to soil chamber measurements. These two approaches are much more common in the tropics, and are strongly associated with anthropogenic landscapes (e.g., associated to agriculture and pasture), and are less common, but still in

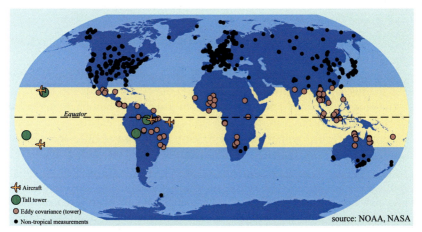

FIG. 1 Global distribution of monitoring sites of GHG fluxes using different methodologies (chamber measurements are not included). *(Source of the sampling site locations: NOAA, NASA.)*

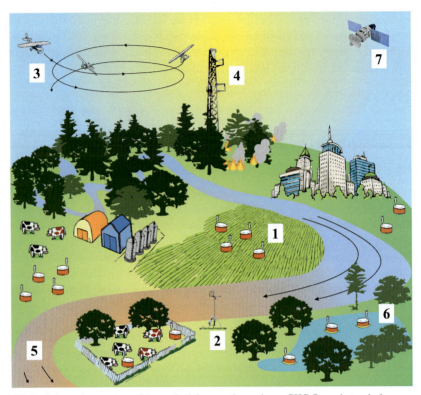

FIG. 2 Schematic overview of the methodology used to estimate GHG fluxes in tropical ecosystems. (1) Chamber measurements in terrestrial ecosystems; (2) micrometeorological eddy covariance; (3) aircraft concentration measurements; (4) tall tower measurements; (5) mass content and streamflow; (6) chamber measurements in aquatic systems; (7) satellite images to calculate land use and land cover change and to support models to estimate gas fluxes.

an important amount, in natural ecosystems in the tropics, including South America, Africa, and SE Asia.

In aquatic systems, carbon balance and associated fluxes are estimated on lotic and lentic systems by mass balance and direct measurements. The complexity of aquatic systems and interactions of a system in motion, with the lateral shores and benches, vertically with the system bottom and the upper surface and atmosphere, and within the system, leads to the use of differences in the chemical species concentration at different points and the flow of the water. The mass balance is used to calculate the transfer of carbon, nitrogen, and other chemical species from one species to another. The use of static chambers is also fairly common in the tropics, but with the majority of studies done in man-made reservoirs (multiple uses, hydropower, water provision, and so on) and in natural lakes. In general, studies done in reservoirs relate GHG emissions to the production of energy or water use for irrigation in agriculture or for human

Tropical ecosystem greenhouse gas accounting **Chapter | 8 275**

supply. On continuous flux measurements, eddy covariance techniques are also used in aquatic systems.

2.2 Methodologies for flux estimation in the tropics

2.2.1 Chamber measurements in tropical terrestrial environments

Microbial processes in soils, sediments, and organic wastes such as manure are a major source of atmospheric GHGs (Butterbach-Bahl, Sander, Pelster, & Díaz-Pinés, 2016). Therefore, a rigorous understanding of the processes and a better quantification of spatiotemporal dynamics of sinks and sources of GHG are pivotal for developing GHG inventories at global, national, and regional scales, identifying hotspots and developing strategies for mitigating GHG emissions from terrestrial systems, whatever in impacted environments or natural systems (Butterbach-Bahl et al., 2016).

Soil-atmosphere exchange fluxes of carbon dioxide (CO_2), methane (CH_4), and nitrous oxide (N_2O) are bidirectional, i.e., what is observed is a net flux of production and consumption processes, though, with regard to CO_2, often only respiratory fluxes are measured (Butterbach-Bahl et al., 2016). One of the most commonly used techniques for measuring fluxes of such gases between terrestrial ecosystems and the atmosphere is enclosure-based (chamber) measurements (Butterbach-Bahl et al., 2016; Livingston & Hutchinson, 1995). Although the technique is widely used in terrestrial environments, chamber measurements in aquatic environments are reported to be the technique that best characterizes large water body flows (Duchemin, Lucotte, & Canuel, 1999; Silva, Lasso, Lubberding, Peña, & Gijzen, 2015).

Chamber measurements of gas fluxes between the land surface and the atmosphere have been conducted for almost a century (Lundegårdh, 1927; Pavelka et al., 2018), have been extensively tested and are generally accepted as providing good results (Christiansen, Korhonen, Juszczak, Giebels, & Pihlatie, 2011; Pumpanen et al., 2004). This technique allows measurements of GHG fluxes at fine scales, with chambers usually covering small soil areas (generally $<1\,m^2$), and they can be operated manually or automatically (Breuer, Papen, & Butterbach-Bahl, 2000; Butterbach-Bahl et al., 2016). The chamber measurement technique is a simple and common method for GHG measurements because it is simple to operate and adaptable to a wide variety of studies (Pavelka et al., 2018), and gas samples can be stored for future analysis and do not require a power supply at the site (with the exception of automated systems) (Butterbach-Bahl et al., 2016). Furthermore, chambers are suitable for measuring treatment effects (e.g., fertilizer and crop trials) or effects of land use, land cover, or topography on GHG exchange (Butterbach-Bahl et al., 2016).

Different chamber techniques, including static and dynamic techniques, have been used with varying degrees of success in estimating GHG fluxes. However, all of these have certain disadvantages that have either prevented

276 Section | C Case Studies

them from providing an adequate estimate of GHG exchange or restricted them to be used under limited conditions (Pavelka et al., 2018). Static chambers are gastight, without forced exchange of the headspace gas volume, and are usually vented to allow pressure equalization between the chamber's headspace and the ambient air pressure (e.g., Xu et al., 2006). The volume of the "vent tube" should be greater than the gas volume taken at each sampling time (Butterbach-Bahl et al., 2016). Manual chamber measurements allow users to investigate the interannual variations of soil GHGs and the influence of environmental factors on them; however, they may not be consistent throughout the year and may miss specific weather events (Pavelka et al., 2018). In addition, attention must be paid to obtaining accurate data, since the installation of the chamber disturbs environmental conditions and measured fluxes might not necessarily reflect fluxes at adjacent sites if some precautions are not considered (Butterbach-Bahl et al., 2016). In aquatic environments, the limitation of the technique is related to chamber-measured fluxes over air-water interfaces, which appear to be subject to considerable uncertainty, since depending on the chamber design, there is a lack of air mixing in the chamber and concentration gradient changes during deployment (Silva et al., 2015).

On the other hand, automated chambers have the advantage of being able to measure continuously for long periods, regardless of the weather and time of day, and allow accurate measurements, with minimal disturbance of the soil surface, and high-resolution datasets for extended periods of time (e.g., Korkiakoski et al., 2017; Pavelka et al., 2018). However, it usually has higher installation and operating costs and more specialized labor (Pavelka et al., 2018). For dynamic chambers, the headspace air is exchanged at a high rate (>1–2 times the chamber's volume per minute), and fluxes are calculated from the difference in gas concentrations at the inlet and outlet of the chambers multiplied by the gas volume flux, thereby considering the area covered by the chamber (Butterbach-Bahl, Papen, & Rennenberg, 1997). Although the existence of equipment that measures fluxes of GHG faster and precisely compared to chamber measurements, the latter is cheaper and more practical and still provides the majority of information being the method most used worldwide (Denmead, 2008; Pavelka et al., 2018).

2.2.2 Micrometeorological eddy covariance measurements in terrestrial ecosystems

With strong institutional developments and international collaboration projects that were stimulated during research programs such as the LBA (Keller et al., 2004) and GoAmazon (Fuentes et al., 2016), the majority of tropical micrometeorological eddy covariance studies of GHG exchange are focused on the Amazon region, but general questions about how tropical forests exchange carbon and other trace gases, variability from seasonal to interannual scales and how a change in land use affects carbon balance are currently being addressed for all tropical regions (Ciais et al., 2011; Gloor et al., 2012; Patra et al., 2013; Valentini et al., 2014).

The establishment of a network of eddy flux towers in the Amazon is one of the legacies of the LBA program, providing important knowledge about the characteristics of water and carbon fluxes throughout the region (Araújo et al., 2002; Borma et al., 2009; Saleska et al., 2003; von Randow et al., 2004). While researchers historically believed that the Amazon forest would respond to dry seasons similar to other water-stressed ecosystems, predicting declines in evapotranspiration and photosynthesis during the dry season, measurements in the towers seem to show very little decline and even a slight increase in evapotranspiration and photosynthesis (related to greater availability of solar radiation) in the dry season (Restrepo-Coupe et al., 2013). However, towers in the southern part of the Amazon, with semideciduous forests or transition areas to the savanna, and towers in deforested areas show clear declines in the fluxes during the dry season and signs of water stress, which are also related to a more intense dry season in these places (Da Rocha et al., 2009; von Randow et al., 2004).

Seasonal patterns of ecosystem metabolism are also reflected in the interannual variations in fluxes, but although the fluxes in the Amazonian ecosystem are expected to be coupled to regional climate conditions, the dynamic mechanisms associated with their interannual variability remain unclear (Nobre, Obregon, Marengo, Fu, & Poveda, 2009). Although observations of a gas exchange over the basin are dynamically linked with anomalies in annual precipitation, which are associated with the ENSO or oscillations in the Atlantic sea surface temperature (SST) (Fu, Dickinson, Chen, & Wang, 2001; Marengo, 1992; Marengo, Tomasella, & Uvo, 1998; Poveda, Waylen, & Pulwarty, 2006), the combined tropical Pacific and Atlantic SST variability explain little more than 50% of interannual precipitation variance over Amazonia, and not much is known about other mechanisms, internal or external to the region, responsible for the remaining unexplained interannual variability (Nobre et al., 2009). The proportions of interannual changes in net carbon or water fluxes directly related to variability in climate drivers remain an open question, and a detailed assessment of the relative roles of changes in climate vs changes in vegetation response on the variability of fluxes is still needed (de Araujo, von Randow, & Restrepo-Coupe, 2016).

2.2.3 Concentration measurements from aircraft and tall towers

Quantifying regional to continental scale carbon sources based on atmospheric data is essential for understanding the global carbon balance and its response to climate and environmental forcing (Gerbig et al., 2003). Two different methodologies of GHG measurements with regional representativeness have been made in tropical regions, tall towers, and aircraft vertical profiles. Around the world, several aircraft networks have been made to monitor GHGs through

278 Section | C Case Studies

TABLE 1 Amazon fluxes estimated by aircraft vertical profiles.

	Annual mean flux	Period
CO_2 emission (PgC yr^{-1})[a]	0.48 ± 0.18	2010 (drought year)
	0.06 ± 0.10	2011 (wet year)
CH_4 emission (TgCH$_4$ yr^{-1})[b]	42.7 ± 5.6	Mean 2010–2013

[a]By Gatti et al. (2014).
[b]By Pangala et al. (2017).

the boundary layer, troposphere, and stratosphere (Sweeney et al., 2015). These observational data have been useful to estimate regional fluxes, helping to validate models and satellite observations (Sweeney et al., 2015). Vertical profiles with higher altitudes are more influenced by more distant regions, being more representative of larger areas than at altitudes closer to the surface (Tans, Bakwin, & Guenther, 1996). A few small aircraft vertical profiles have been made in tropical regions, for example, in the Amazon, with long-term regular measurements of CO_2, CH_4, and N_2O to estimate fluxes at regional scales (approximately 10^5–10^6 km^2, Basso et al., 2016; D'Amelio, Gatti, Miller, & Tans, 2009; Gatti et al., 2010, 2014; Miller et al., 2007). Table 1 shows some examples of the Amazon estimated balance based on CO_2 and CH_4 aircraft vertical profile measurements. These fluxes represent the results of all sources (natural and anthropogenic) and sinks of the study area.

In addition, the National Oceanic and Atmospheric Administration (NOAA) has made GHG measurements with small aircraft around the world, with one site over the South Pacific near Rarotonga (Cook Islands). The seasonal and annual mean profiles are the results of surface fluxes and atmospheric influences (Stephens et al., 2007), requiring careful analysis to extract information on regional sources and sinks (Chou et al., 2002). Comparisons between tower and aircraft estimates showed discrepancies in Amazon CO_2 fluxes. This could be due to the influence of the combination of surface sources/sinks (Chou et al., 2002) and a consequence of the much larger surface coverage from the aircraft budget (Lloyd et al., 2007). The area of influence in the measurements (footprint) is largely related to the sample altitude. Footprints of aircraft vertical profiles are representative of larger areas, with an order of 10^6 km (Gloor et al., 2001), while flux tower measurements that have lower altitude (few tens of meters above the canopy), in general, have a few kilometers of footprint (Araújo et al., 2002), representing better local processes. With tall towers, studies addressing continuous measurements of GHGs in forests, aerosols and suspended particles that promote the formation of clouds at several heights are possible. Its results can reflect local processes at the lowest levels and regional influences at the highest levels (Andrews et al., 2014; Bakwin, Tans, Hurst, & Zhao, 1998) and investigate the transport of air masses over hundreds of kilometers. Measurements in tall towers are commonly used in atmospheric science, and the best known is the ATTO (Amazon Tall Tower Observatory), located in the Brazilian Amazon with a footprint of continental area of $\sim 1.5 \times 10^6$ km^2, which is the tallest structure in South America (325 m, Pöhlker et al., 2019). ATTO was developed to study the interaction between the forest and climate and to better understand the Amazon influences on the global climate. A global network of these towers exists; however, most are located in temperate regions, such as ZOTTO (located in Siberian Tundra), with ATTO being the first tall tower to study tropical forests. Measurements from both small aircraft and tall towers are important ways to represent an integration of local and regional processes and could be helpful to validate satellite measurements and useful for global inversions.

2.2.4 Freshwater/fluvial flux (DOC/DIC/POC) measurements

Freshwater systems, including lakes, reservoirs, rivers, and streams, occupy only a small fraction of the Earth's surface; however, they play a large role in the global C cycle (Cole et al., 2007; Tranvik et al., 2009). Globally, there is a significant latitudinal gradient in C fluxes from these systems (Kosten et al., 2010). Tropical systems are hotspots for C emissions and account for 34% of global inland water emissions (Raymond et al., 2013), and tropical lentic systems emit 3 and 5 times more CH_4 and CO_2 than nontropical systems, respectively (Barros et al., 2011; St Louis, Kelly, Duchemin, Rudd, & Rosenberg, 2000; Tranvik et al., 2009). These systems are dynamic and have a large diversity of fluxes, making the quantification of GHG fluxes in freshwater systems more challenging than in terrestrial environments. The reason is that freshwater GHG exchange includes either vertical or lateral fluxes. Vertically, CO_2 and CH_4 can be emitted into the atmosphere, whereas laterally, carbon is transported as dissolved organic carbon (DOC), dissolved inorganic carbon (DIC), particulate organic carbon (POC) and particulate inorganic carbon (PIC). The latter is generally not considered as less data are available and low attention is given to this form of carbon because it has a lesser contribution to the carbon flux and is gradually trapped in lowlands, floodplains, lakes, estuaries, and on the continental shelf before reaching the coast (Ciais et al., 2008).

The vertical flows in the freshwater system can be estimated by diffusion (CO_2, CH_4, and N_2O) and ebullition (CH_4). The principal methods to measure the emission of GHG from aquatic systems by diffusion are (1) the eddy covariance (EC); (2) the chamber and; (3) the boundary layer method.

The EC method allows direct measurement of the turbulent flux on a continuous basis. The technique is nonintrusive, and the measurement is representative of a region upwind of the measuring tower, which is on the order of a few

280 Section | C Case Studies

hundred meters, called the "footprint" (Schuepp, Leclerc, MacPherson, & Desjardins, 1990). The chamber method and the boundary layer method are relatively simple to apply, being the main measuring techniques used in lakes and reservoir studies about GHG emissions in the tropics (Abril et al., 2014; Ometto et al., 2013; Pacheco et al., 2015; Rosa et al., 2003; Zhu, Chen, Zhu, Wu, & Wu, 2012). The floating chamber method, where the vertical flux at the air-water interface is calculated from the concentration increase within the chamber during the measurement period, has been criticized for causing a modification of the flow at the water-air interface due to the turbulence created by the chamber or to the inhibition of the wind effect on the surface turbulence (Richey, Melack, Aufdenkampe, Ballester, & Hess, 2002; Schubert, Diem, & Eugster, 2012). The mass transfer approach, which uses wind speed (via gas transfer velocity k) and concentration gradient between the air and surface water, requires measuring the partial pressure of CO_2 in water and air, usually by means of an analytical water-air equilibration approach. The limitation of the chamber and mass transfer method is that point measurements in space and time need to be extrapolated for the whole water body and are sometimes used to obtain estimates of long-term average GHG fluxes (Paranaíba et al., 2018; Podgrajsek et al., 2014; Roland et al., 2010; Teodoru, Prairie, & del Giorgio, 2011; Vesala et al., 2006). However, both techniques are inexpensive and extensive data postprocessing is not needed. On the other hand, the EC method estimates GHG flux over a larger area than the other two methods and can continuously collect data; however, the measurements are expensive, require extensive data postprocessing and there are still fill sites worldwide with long data sets.

The emission of CH_4 by ebullition can be estimated by bottom-moored funnels connected to water-filled glass bottles; the gas that accumulates in the bottles is measured in a GHG analyzer, and CH_4 emission is then calculated (Rosa et al., 2003).

From multiple measurements collected in several regions of the globe using different methods, the vertical carbon emissions from tropical freshwater systems have been estimated in the literature (Barros et al., 2011; Bastviken, Tranvik, Downing, Crill, & Enrich-Prast, 2011; Cole et al., 2007; Stanley et al., 2016; Tranvik et al., 2009). In addition, to a lesser extent, the emission of N_2O (Ivens, Tysmans, Kroeze, Löhr, & van Wijnen, 2011; Seitzinger, Kroeze, & Styles, 2000; Soued, del Giorgio, & Maranger, 2016). However, there is a large uncertainty because of the incomplete spatial coverage of GHG emissions, especially in tropical freshwater systems. Global carbon emissions are estimated to be in the range of 1.5–2.1 $PgC yr^{-1}$, and approximately 65%–75% is emitted in the tropics. Streams and rivers emit most of the carbon among lakes, reservoirs, streams and rivers, and the Amazon rivers represent approximately 50% of the carbon evasion in the tropics (0.5 $Pg yr^{-1}$; Richey et al., 2002). In the case of N_2O, emissions in the tropics may represent approximately 80% of the total N_2O emissions from freshwater systems (Soued et al., 2016).

Wetlands are generally defined as ecosystems that are flooded by water, either permanently or seasonally, where the water saturation characteristic of the soil, or peat, determines the composition of the species and the soil biogeochemistry (Keddy, 2010). In these systems, hydrology, temperature, and substrate availability are the main factors that influence methanogenic microorganism activity (anaerobic decomposers), which degrades organic matter slowly in an anoxic environment (Mitsch et al., 2010) and produces a large amount of CH_4. As important methane emitters, these systems emit 22%–30% of the total methane considering all sources globally (Saunois et al., 2020). From the total CH_4 emissions from wetlands, approximately 70% are from tropical regions (Table 2, Mitsch et al., 2013). The largest wetland areas in the tropics are in Amazonia, the Congo Basin and Indonesia

TABLE 2 Global and tropical vertical carbon transport from freshwater ecosystems (including wetlands).

	Global	Tropical
	$PgCyr^{-1}$	
CO_2 emission		
Stream and rivers	1.5–2.1[a]	1.0–1.6[b]
Lakes and reservoirs	0.3–0.6[a]	–
Wetlands	−0.83[c]	−0.56[c]
CH_4 emission		
Stream and rivers	0.026[d]	0.016
Lakes and reservoirs	0.091[e]	0.045[e]
Wetlands	0.080–0.280[f]	0.062–0.218[c]
	Global	**Tropical**
	$TgCyr^{-1}$	
N_2O emission[g]		
Stream and rivers	0.194	0.110
Lakes and reservoirs	0.583	0.510

Positive values are net carbon emissions, and negative values are net carbon uptake.
[a]*Raymond et al. (2013).*
[b]*Butman and Raymond (2011).*
[c]*Mitsch et al. (2013).*
[d]*Stanley et al. (2016).*
[e]*Bastviken et al. (2011).*
[f]*Bridgham, Cadillo-Quiroz, Keller, and Zhuang (2013).*
[g]*Soued et al. (2016).*

282 Section | C Case Studies

(Gumbricht et al., 2017; Xu, Morris, Liu, & Holden, 2018); however, large uncertainties in estimates of CH_4 emissions from tropical wetlands are related mainly to the uncertainties of wetland extent, difficulties in defining wetland CH_4-producing areas, and in parameterizing terrestrial anaerobic conditions that drive sources and the oxidative conditions leading to sinks (Melton et al., 2013; Poulter et al., 2017; Wania et al., 2013).

The lateral transport in freshwater systems can be estimated using two approaches: (1) one uses the carbon content of the river water, and the other (2) considers the mass balance conceptual model. The first approach calculates carbon transport by measuring carbon concentration in the water (usually monthly) and water discharge (Coynel, Seyler, Etcheber, Meybeck, & Orange, 2005; Tong et al., 2015). The second approach uses empirical data to propose simple equations or complex models considering the relationship between the carbon fluxes and climatic, biologic, and geologic factors (Li et al., 2017; Ludwig, Probst, & Kempe, 1996). Measuring the carbon concentration in river water is fieldwork dependent and would be the most accurate way to estimate lateral carbon fluxes on a local and regional scale but may require a large dataset to estimate on a continental to global scale. Additionally, fieldwork may not be the best option for most projects because they require substantial resources, material and financial support to cover relatively large areas. This approach has been used in several tropical regions to estimate carbon fluxes from rivers to the ocean in tropical America (Richey et al., 1990, 2002), Africa (Coynel et al., 2005), and Asia (Huang et al., 2017; Tong et al., 2015).

Despite the lower accuracy, empirical analysis and models have been developed to estimate lateral carbon fluxes at several tropical regions and different scales. These approaches attempt to generalize the mechanisms that drive carbon dynamics in aquatic systems to explain temporal and spatial variations in lateral carbon fluxes. The advantages of these models are the capacity to predict carbon fluxes from regions where data are not available and predict the effects of disturbances in the watershed, such as climate and land-use change, on the carbon balance. The lack of systematic measurements in tropical rivers is generally true, and the use of empirical analysis and models is a useful tool to address lateral transport from main tropical rivers. However, complex processes and several model parameters may temporally bring additional difficulties when applying the model to a large scale (Li et al., 2017).

A mix of both approaches has been used to estimate carbon discharge by rivers worldwide. The total carbon delivered to the ocean by rivers is $0.8–1.3 \, PgC \, yr^{-1}$ (Cole et al., 2007; IPCC, 2007; Li et al., 2017; Meybeck, 1993). From this amount, approximately $0.53 \, PgC$, more than half, is delivered to the ocean by tropical rivers (Huang, Fu, Pan, & Chen, 2012). However, the amount and proportion of DOC, DIC, and POC transported by rivers differ regionally (Fig. 3). For instance, DOC transported by South American and Asian rivers is relatively higher than POC, whereas the opposite is observed

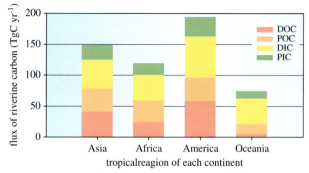

FIG. 3 Total carbon laterally delivered to the ocean by rivers from tropical rivers of different continents. *(Data for the tropical region (30°N–30°S) were extracted from Li, M., Peng, C., Wang, M., Xue, W., Zhang, K., Wang, K., Shi, G., & Zhu, Q. (2017). The carbon flux of global rivers: A reevaluation of amount and spatial patterns. Ecological Indicators, 80, 40–51. https://doi.org/10.1016/j.ecolind.2017.04.049.)*

in rivers from Africa and Oceania. High DOC yields in those regions are related to the massive DOC release from organic soils, such as large areas of peatlands in Southeast Asia and Amazonia (Gumbricht et al., 2017), and the regarded characteristics of Amazonian rainforests as important sources of global riverine DOC export (Medeiros et al., 2015).

2.3 Top-down and bottom-up methods for estimating carbon fluxes in the tropics (modeling)

Analyses on GHGs are also performed in studies assuming different assumptions to integrate measurements and models. In one class of study, knowledge is built from local measurements, which quantify relevant ecosystem processes, including GHG sources and sinks. These studies support detailed inventories and the development of mathematical models able to reproduce the dynamics of these sites with respect to the variables explored (Kondo, Ichii, Takagi, & Sasakawa, 2015). From these inventories and models, "bottom-up" regional estimates of emissions and sinks can be produced, assuming that local values represent the regional dynamics and can be extrapolated for larger areas (Bastos et al., 2018).

In another approach, GHGs are studied from a large-scale atmospheric sampling of the effects of sources and sinks, informing their concentration and distribution over large areas. Typically, these measurements are combined with the knowledge on atmospheric circulation and are used to provide "top-down" estimates of the intensity of sources and sinks for the evaluated GHGs over the whole study region (Cheewaphongphan, Chatani, & Saigusa, 2019). Additionally, corresponding detections from multiple satellites can also be used in large-scale top-down emissions estimations (Pechony, Shindell, & Faluvegi, 2013).

284 Section | C Case Studies

Bottom-up modeling has the advantage of directly representing and quantifying processes for specific locations on the land surface, but the assumption of broad spatial extrapolation may not be appropriate in all cases or when there is a lack of input data. Top-down inverse modeling, on the other hand, evaluates the major effects of sources and sinks and connects their estimates to biomes and other large-scale quantifications but does not report information for specific locations. Below, we discuss examples of studies that have applied these methods in tropical areas.

Using a top-down method in the Amazon Basin, Gatti et al. (2014) sampled vertical profiles of CO_2, SF_6, and CO over four locations in 2010 and 2011, chosen to allow quantifying sources and sinks of these gases according to dominant mid-low tropospheric airflow across the study region. By combining these measurements with modeled air-mass back-trajectories, Gatti et al. (2014) estimated total C fluxes of $0.48 \pm 0.18\,PgC\,yr^{-1}$ and $0.06 \pm 0.1\,PgC\,yr^{-1}$ in 2010 and 2011, respectively, and indicate that the region was a net source of C in a dry year and C neutral in a wet year. The authors also found that the sink caused by the vegetation in 2011 was equal to $0.25 \pm 0.14\,PgC\,yr^{-1}$.

Based on ground observations on biomass change and tree mortality in several different long-term sites across Amazonia, Phillips et al. (2009) estimated that the 2005 drought in the region had a total impact on biomass carbon equal to a 1.2–1.6 Pg loss. For that, the authors also estimated moisture stress from meteorological datasets and scaled up their plot results based on the total drought area. Additionally, using a bottom-up approach, Hashimoto et al. (2015) estimated CO_2 emissions from soil respiration by scaling up local observations from a global database using model relations between climate (temperature, precipitation) and soil respiration. The authors estimated an annual average of $91\,PgC\,yr^{-1}$ globally from total soil respiration from 1965 to 2012, from which 64% comes from the tropics.

As already indicated by other studies, GHG budget estimates from bottom-up and top-down modeling are complementary, and it is important to explore the capabilities of both methods in contributing information to emissions policies (DeCola et al., 2019). For example, top-down methods can suggest maximum regional values and a potential need to constrain bottom-up estimates, which in turn are able to identify specific regions and processes (Saunois et al., 2020). Comparison between the two methods for China helped to identify that rice cultivation and coal mining are categories for which bottom-up CH_4 emission inventories for the country have higher uncertainties (Cheewaphongphan et al., 2019).

Combining the results from bottom-up and top-down studies, Bastos et al. (2018) provide a comprehensive analysis of the impact of El Niño events in 2015–16 on the terrestrial carbon cycle, and indicate that overall, the carbon sink on land was reduced by $0.4–0.7\,PgC\,yr^{-1}$ (based on top-down atmospheric inversions), and by $1\,PgC\,yr^{-1}$ (based on bottom-up land-surface models) in 2015–16. Specifically, for the tropical areas, the authors found that less productive ecosystems rather than higher respiration was a relevant factor for the net sink anomalies associated with drier conditions.

2.4 Terrestrial ecosystem and land surface processes in the tropics

The transformations that the terrestrial system has undergone, mainly with the advent of industrialization and market globalization, have significantly impacted the components that are responsible for environmental stability. For example, the compromise of biogeochemical cycles, loss of biodiversity, regulation of the hydrological and climatic regime corroborate the impoverishment of ecosystems, especially due to the loss of the ability to provide fundamental environmental services to the individuals who depend on them, such as water, clean air, fertile soils, food, etc., and consequently, this affects socioeconomic relations at different scales. Although the way people interact with the earth changes significantly according to the region of the planet, the final consequence is the degradation of environmental conditions (Foley et al., 2005). In this context, it has been observed that land available for agricultural expansion has become an increasingly scarce resource in several regions of the globe (Lambin et al., 2013; Lambin & Meyfroidt, 2011). In practice, the expansion of the agricultural frontier is already concentrated primarily in tropical regions (Defries, 2013; Gibbs et al., 2010).

In these regions, the increase in cultivated areas had an average increase of 34% (Fig. 4) when we analyzed the period from 1960 to 2018. This increase occurred mainly by the increase in the areas of tree nuts that grew 569%, followed by the areas of citrus fruit (387%) and oil crops (237%). When we analyzed the production, an increase in the quantity produced was observed of 249% over that same period, with the cultivations of tree nuts (604%), oil crops

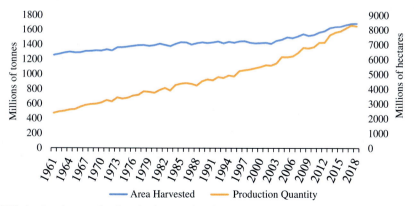

FIG. 4 Area harvested and production quantity of crops (cereals, citrus, coarse grain, fiber crops, fruits, seed-oil crops, pulses, roots and tubers, tree nuts, and vegetables). *(Source: Based on FAO statistics (FAO. (2020). FAOSTAT—Food and Agriculture Organization of the United Nations).*

286 Section | C Case Studies

(579%) and citrus fruit (508%) standing out over the other cultures (FAO, 2020). In part, this significant increase in cultivated areas and the quantity produced is due to the set of technological innovations disseminated by the Green Revolution from the 1950s and 1960s all over the world, which consolidated agriculture as one of the main activities for land use (Heistermann, Müller, & Ronneberger, 2006). Although this increase in areas and agricultural production has contributed to some countries consolidating themselves as the main commodity-exporting countries worldwide and presenting economic growth, as well as maintaining the trade balance, this growth model has been opposed to the conservation and maintenance of natural resources, given the impact that land-use change and land cover have on natural forests, especially the Amazon rainforest.

This transition from natural areas to anthropic systems contributes mainly to the increase in the volume of GHGs. If we analyze the land use, land-use change, and forestry (LULUCF) sector, we can see that in the tropics, between 1990 and 2017, there was an increase in CO_2 emissions from cultivation and pasture areas of 10% and 2%, respectively, to the detriment of a reduction in emissions from forestland (40%) and burning of biomass (17%). In this region, there was a reduction of approximately 24% in the net CO_2 emissions in this sector (FAO, 2020, Fig. 5).

Understanding the factors that influence land use and land cover changes (LULC) is essential (Aguiar, 2006; Geist & Lambin, 2001; Lambin & Geist, 2001, 2003), especially to define reliable indicators to guide policy makers to establish sustainable development strategies. Thus, the models of LULC developed in the last decades are relevant tools to analyze current and possible interactions between human beings and nature. In addition to making it possible to analyze and evaluate the dynamics of the factors that influence changes, as well as the economic and environmental impacts of these changes, they also make it possible to verify the influence of public policies on the current trajectories of transformations on the Earth's surface associated with human activities (Heistermann et al., 2006; Pijanowski, Brown, Shellito, & Manik, 2002; Verburg, Kok, Pontius Jr, & Veldkam, 2006). Due to its complexity, the study of LULC has required from science the ability to integrate a diversity of themes, scales and units of analysis, in addition to different methodological and theoretical approaches (Brondizio, 2014). This integration allows aggregating both quantitative and qualitative information from different sources, such as remote sensing, secondary databases, field research, participatory methodologies, etc. This makes it possible to analyze and explain LULC not only by understanding a single factor but also by considering a complex network of biophysical and socioeconomic factors that interact in time and space in different historical and geographical contexts (Aguiar, 2006).

Despite the diversity of models that explore LULC, dynamic spatial models stand out, as they allow integrating different factors and models, in addition to projecting and visualizing changes, their intensity, and location. In general, the

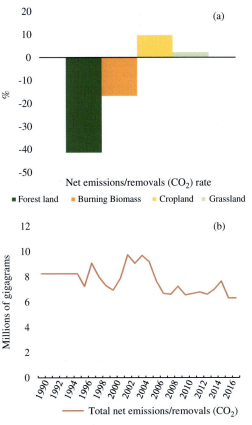

FIG. 5 (A) Net emissions/removals (CO_2) and (B) rate. *(Source: Based on FAO statistics (FAO. (2020). FAOSTAT—Food and Agriculture Organization of the United Nations).)*

design of these models is structured as follows: (1) definition of the object of study, temporal and spatial scales, and identification of possible related factors; (2) preparation and organization of information for the preparation of a database containing all relevant information; (3) analysis of the collected information, quantitative and/or qualitative, using different methodological approaches; (4) parameterization of the models using the information from the previous steps; and (5) execution, calibration, and validation of the developed models. Note that these steps can be reviewed at any time during the development and analysis of the models, as well as the results generated. In addition, the process of modeling LULC should be understood as an approach that aims to understand the interrelationships between man and the environment. The model should not be the end, but the means by which we understand how the transformation process of the terrestrial system takes place and how we can make it more sustainable.

The use of these models integrated with the models of estimation of emissions of GHGs can help in the understanding of the factors intrinsic to the sector of LULUCF, as well as in the estimation of emissions coming from this sector, which would contribute to the definition of national reference levels of forest emissions (FREL) and, thus, subsidize strategies related to efforts to reduce emissions Reducing Emissions from Deforestation and Degradation (REDD+).

2.5 GHG emissions from tropical forest deforestation and degradation

Tropical forests store large amounts of carbon, both in the structure of vegetation and in the soil. In comparison with temperate climate forests, tropical forests are denser and have less seasonal fluctuations in carbon flow, constituting themselves as important carbon stocks that contribute to the stability of the global climate. Tropical forests are still home to approximately 50% of terrestrial biodiversity, playing a fundamental role in regulating the supply of water resources and in soil conservation. Deforestation and forest degradation (e.g., timber harvest and forest fire) emit large amounts of GHG (IPCC, 2014). These emissions are estimated to account for 7%–14% of the total CO_2 emissions from human activities (from 0.57 to 1.22 $PgC\,yr^{-1}$, Harris et al., 2012). Of the total emissions from deforestation and forest degradation, deforestation in tropical regions accounted for 75% (Harris et al., 2012). Only two countries, Brazil and Indonesia, accounted for approximately 55% of the total emissions produced between 2000 and 2005 due to deforestation (0.34 and 0.11 $PgC\,yr^{-1}$, respectively). In Brazil, most deforestation occurs on land in the Amazon, Cerrado, and Pantanal biomes, where livestock expansion has caused extensive deforestation to meet the increasing demand for beef (Barona, Ramankutty, Hyman, & Coomes, 2010; Boucher et al., 2011; Houghton, 2010) and other commodities. In Indonesia, deforestation has been rapidly increasing in peatlands since the 1990s due to clearing land for palm oil and pulp plantations, and by 2010, it was estimated that only 6% of peatlands were covered by primary and secondary peat swamp forests (Miettinen, Shi, & Liew, 2012).

The reducing emissions from deforestation and forest degradation program (REDD+) under the United Nations Framework Convention on Climate Change (UNFCCC) have focused mostly on deforestation, which is easier to detect and thus more readily measured and monitored than forest degradation. However, forest degradation is perceived to be an important contributor both to GHG emissions and to regional economic development. Pearson, Brown, Murray, and Sidman (2017) estimated annual emissions from tropical forest degradation of 0.57 $PgC\,yr^{-1}$, of which 53% were derived from timber harvest, 30% from woodfuel harvest and 17% from forest fire. Again, Brazil and Indonesia, large forested countries, led the ranking of tropical countries with the highest emissions; however, notable emissions from forest fires also occur in

Congo, Bolivia, and Argentina. Additionally, the authors show that the proportion of total forest degradation emissions by degrading activity differs by region. For instance, timber harvest was as high as 69% in South and Central America and just 31% in Africa, while woodfuel harvest was 35% in Asia and just 10% in South and Central America; fire ranged from 33% in Africa to only 5% in Asia (Pearson et al., 2017).

Facing the political framework in Brazil in 2020, where the government has taken serious measures that negatively affect environmental policies (Viola & Gonçalves, 2019), emissions from deforestation and forest degradation may stand in high levels over the upcoming years in the Amazon, Cerrado, and Pantanal biomes. In 2019, approximately $318,000 \, km^2$ of forest area was consumed by fire in Brazil, almost 2 times higher than in 2018. The Pantanal biome suffered the highest increase, where $20,800 \, km^2$ were degraded by fire, approximately 570% higher than in 2018 (http://queimadas.dgi.inpe.br/queimadas/aq1km/). In Indonesia, the emission reductions pledged in its Nationally Determined Contributions (NDC) propose in the Paris Agreement are strictly related to the reduction of deforestation and forest degradation, particularly those affecting peatland, which was approximately 48.5% of total emissions in the country (ID, 2016). However, the proposed activities in Indonesia's regulatory architecture of emissions reduction fall short of the emissions reduction committed in the NDCs (Tacconi & Muttaqin, 2019a). Countries that proposed GHG emission reduction from forests, such as Brazil and Indonesia, will need to implement several policies and activities to achieve the aims of the Paris Agreement (Tacconi & Muttaqin, 2019b). Forest restoration, control measures and incentive schemes will be essential and, at the same time, a complex problem that will require a significant research contribution.

2.6 Review of tropical GHG estimates by sector

Although there are large uncertainties inherent to historical estimates of GHG by country and at the global scale, such estimates are key for observing changes and tendencies over time, as well as for assessing the effectiveness of emissions reductions and overall compliance with the terms of international treaties (Jonas et al., 2019; Marland, Cantrell, Kiser, Marland, & Shirley, 2014). The identification of key emitter sectors is crucial for planning mitigation strategies and adaptation responses to local and regional environmental changes and for analyzing scenarios of future emissions (Ometto, Bun, Jonas, Nahorski, & Gusti, 2014).

In the following discussion, historical GHG emissions from tropical countries are organized by sectoral sources presenting carbon dioxide (CO_2), methane (CH_4), and nitrous oxide (N_2O), both individually and collectively as total GHG emissions. All data are sourced from the Climate Analysis Indicators Toll (CAIT), a data analysis tool on global climate change developed by the World

Resources Institute (WRI) (CAIT, 2019; Damassa, 2014). Sectoral sources are grouped as follows (WRI, 2015):

- **Agriculture**: CH_4 from enteric fermentation and rice cultivation; N_2O from agricultural soils (synthetic fertilizers, manure applied to soils, manure left on pasture, crop residues, cultivation of organic soils); CH_4 and N_2O from livestock manure management and other agricultural sources (burning crop residues, burning savanna).
- **Energy**: data from five subsectors: electricity/heat, manufacturing/construction, transportation (domestic aviation, road, rail, domestic navigation, and other transportation), other fuel combustion, and fugitive emissions. Most energy emissions are related to CO_2 from fossil fuel combustion, but emissions from CH_4 and N_2O may also be significant in the fugitive emissions subsector.
- **Industrial processes**: CO_2 emissions from cement manufacture; N_2O emissions from adipic and nitric acid production; N_2O and CH_4 emissions from other industries (nonagriculture); F-gases (HFCs, PFCs, and SF_6).
- **Land-use change and forestry (LUCF)**: from the net conversion of the forest; forestland (CO_2, CH_4, N_2O); cropland (CO_2); grassland (CO_2); biomass burning (CO_2, CH_4, N_2O).
- **Waste**: CH_4 from landfills (solid waste) and from wastewater treatment; N_2O from human sewage; CH_4 and N_2O from other waste.

The CAIT data are available for 101 tropical countries for the period from 1990 to 2016, organized by region, according to Table 3.

Historical emissions by sector for all GHGs and for each gas individually are presented in Fig. 6. GHG emissions are measured in tons of carbon dioxide equivalents (CO_2e). According to the graphics, GHG and CO_2 emissions come mainly from the energy and LUCF sectors, which also dominate global trends (IPCC, 2014), while agriculture is the greater emitter sector of CH_4 and N_2O.

Emissions from the energy sector are related to electricity production and fossil fuel combustion due to economic growth in the last few years, especially in India and other developing countries (Figueres et al., 2018; Le Quéré et al., 2018). LUCF and agriculture are important drivers of GHG emissions in tropical countries, mostly caused by deforestation and land degradation (Pendrill et al., 2019; van der Werf et al., 2009) driven by increasing exports of agricultural commodities (DeFries, Rudel, Uriarte, & Hansen, 2010; Henders, Persson, & Kastner, 2015; Pendrill et al., 2019).

In 2016, the last year of CAIT available data for countries, most GHG emissions come from the energy sector, followed by LUCF and agriculture; the latter were waste and industrial processes (Fig. 7), according to the trend observed over the past decades.

In comparison with other nontropical regions, GHG emissions from tropical countries are rising over time, approaching those from East Asia (particularly

TABLE 3 List of tropical countries.

Regions	Countries
East Asia and Pacific (23)	Brunei Darussalam, Cambodia, Cook Islands, Fiji, Indonesia, Kiribati, Laos, Malaysia, Marshall Islands, Micronesia, Nauru, Niue, Palau, Papua New Guinea, Philippines, Samoa, Singapore, Solomon Islands, Thailand, Tonga, Tuvalu, Vanuatu, Vietnam.
Latin America and Caribbean (28)	Antigua and Barbuda, Barbados, Belize, Bolivia, Brazil, Colombia, Costa Rica, Cuba, Dominica, Dominican Republic, Ecuador, El Salvador, Grenada, Guatemala, Guyana, Haiti, Honduras, Jamaica, Nicaragua, Panama, Paraguay, Peru, Saint Kitts and Nevis, Saint Lucia, Saint Vincent and the Grenadines, Suriname, Trinidad and Tobago, Venezuela.
Middle East and North Africa (3)	Djibouti, Oman, Yemen
South Asia (3)	India, Maldives, Sri Lanka
Sub-Saharan Africa (44)	Angola, Benin, Botswana, Burkina Faso, Burundi, Cameroon, Cape Verde, Central African Republic, Chad, Comoros, Cote d'Ivoire, Democratic Republic of Congo, Equatorial Guinea, Eritrea, Ethiopia, Gabon, Gambia, Guinea-Bissau, Kenya, Liberia, Madagascar, Malawi, Mali, Mauritania, Mauritius, Mozambique, Namibia, Niger, Nigeria, Republic of Congo, Rwanda, São Tome and Principe, Senegal, Seychelles, Sierra Leone, Somalia, Sudan, Tanzania, Togo, Uganda, Zambia, Zimbabwe.

China) in more recent years (Fig. 8). Within the tropics, the largest emitters are India, Indonesia, and Brazil. These countries are also among the top 10 emitters in the world, according to the CAIT dataset (CAIT, 2019).

Deforestation has long been identified as the main driver for this boost in emissions in tropical countries (Gibbs & Herold, 2007; Houghton, 2005). However, it is important to highlight that each tropical country has special characteristics, ranging from climatic differences (equatorial, monsoon) to dynamics of land-use change and social, political, demographic, cultural, and economic structures (Henry, 2005; Lambin, Geist, & Lepers, 2003). These peculiar characteristics may influence the amount of GHG emissions by sector when countries are analyzed separately instead of regionally, and they should be considered for international environmental agreements and policy planning.

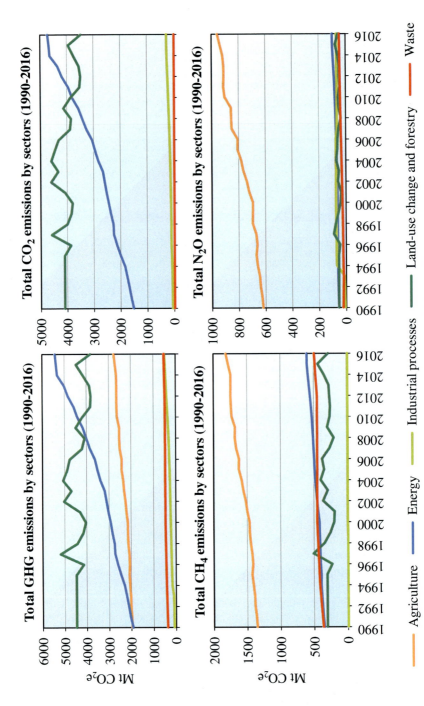

FIG. 6 Emissions by gases (GHG, CH$_4$, CO$_2$, N$_2$O) organized by sectors. (*Source: CAIT/WRI via Climate Watch.*)

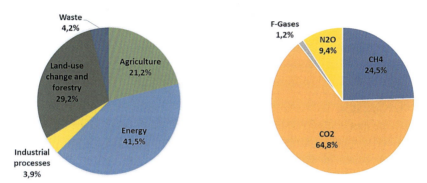

FIG. 7 Total GHG emissions from tropical countries in 2016 (%). *(Source: CAIT/WRI via Climate Watch.)*

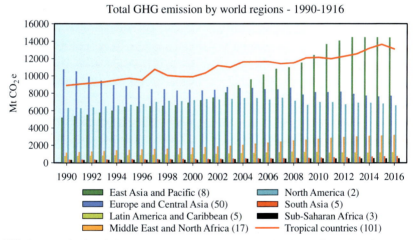

FIG. 8 Total GHG emissions by world regions. *(Source: CAIT/WRI via Climate Watch.)*

3 Uncertainty and reducing uncertainty

Uncertainty can be conceptualized as the value that defines the accuracy level of a reported value. This can be due to measurement error, lack of available data, modeling assumptions, or future estimation (Marland et al., 2014). Uncertainties are present in the estimates of GHG emissions, and in some cases and situations, they are higher, and in others, a reduction has been achieved over the years through the improvement of the methods used (Jia et al., 2019; Restrepo-Coupe et al., 2013). One of the damages due to the uncertainties of the estimates is that they can make it difficult to assess compliance with the country's emission reduction targets and international agreements (Marland et al., 2014; Oda et al., 2019).

For example, in relation to CO_2, it is important to estimate the amount of carbon released and the amount sequestered. This is because if the uncertainty in carbon sequestration during an evaluation period is greater than the uncertainty of emissions, it is difficult to extract reliable and objective information about the carbon balance (Marland et al., 2014). This is often called the ambiguity of uncertainty. Suppose a country that was emitting 100 tons of CO_2 equivalent and committed to reducing by 10%, which means that if its goals are met, emissions should be 90 tons of CO_2 equivalent. If the uncertainty was 20% (which is a plausible situation to happen), it would be difficult to assess whether the goal was met (Marland et al., 2014).

Methods that estimate GHG emissions have had some challenges in addressing some important issues in tropical regions, such as high rates of photosynthesis, thick cloud cover over forests and a lack of data on land-use change (LUC) (Ometto, Aguiar, et al., 2014; Ryu, Berry, & Baldocchi, 2019; Valentini et al., 2014). Cloud cover, more specifically, has been a difficulty for remote sensing applications. One example was an error of incoming photosynthetically active radiation (PAR) in tropical regions that was observed in MODIS-derived global land surface radiation products (Ryu et al., 2019). This error led to a misinterpretation of the 2005 drought in the Amazon. Higher cloud optical thickness, water vapor contents (Pinker & Laszlo, 1992), and less well-known aerosol properties (Martin et al., 2010) result in the imprecise mapping of PAR and diffuse PAR. For this reason, a green-up of Amazon rainforests during the 2005 drought was reported (Ryu et al., 2019; Saleska, Didan, Huete, & Da Rocha, 2007). However, this was contested by a study based on the analysis of a data set that excluded data contaminated by clouds, which showed no green-up in 2005 (Samanta et al., 2010).

Another issue that contributes to errors and uncertainties in tropical regions, as the models that estimate productivity cannot capture it, is the seasonality of irradiance. It is often observed that even in dry seasons, there is no decrease in gross primary productivity (GPP) (which may even occur), which is attributed to a reduction in cloud cover during the dry season (Restrepo-Coupe et al., 2013; Ryu et al., 2019).

For eddy covariance measurements, the problems originate from errors of systematic measurement bias causing uncertainties in characterizing the mean seasonality of photosynthetic patterns (Restrepo-Coupe et al., 2013); yet another complicated canopy structure (Von Randow, Kruijt, Holtslag, & de Oliveira, 2008; Yi et al., 2010); advection errors caused by complex terrain (Aubinet et al., 2005; Feigenwinter et al., 2008); errors in the energy balance (Foken, 2008; Massman & Lee, 2002), errors of the stochastic nature of turbulence (Hollinger & Richardson, 2005; Moncrieff, Malhi, & Leuning, 1996;

Yi et al., 2010) and low coverage of flow towers that affect tropical regions (Ometto, Aguiar, et al., 2014; Schimel et al., 2015).

Surface models coupled with general circulation models of the atmosphere are used to assess stocks and CO_2 emissions, and GHG emissions also have uncertainties as a result of the lack of data available for calibration and validation in tropical regions. Nevertheless, there are problems with the forcing data also from tropical regions that are used as input for the models due to the few flow towers to cover large areas (Sörensson & Ruscica, 2018; Valentini et al., 2014).

Another factor that contributes to uncertainties in tropical regions is LUC in the face of the lack of data to elaborate GHG estimates (Ometto, Bun, et al., 2014; Valentini et al., 2014). The still source of uncertainty is the conversion of primary forests into plantations, and this is shown for West Ghana, in Africa, where land-use change leads to a significant loss of carbon stock (Valentini et al., 2014).

Uncertainties are also high for other gases in tropical regions, as studies are still quite scarce (Silva, Carreiras, Rosa, & Pereira, 2011; Valentini et al., 2014). In addition to the emission of CO_2, Silva et al. (2011) did work for other gases such as CH_4, carbon monoxide (CO), N_2O and nitrogen oxides (NOx) in tropical regions (Africa, America and Asia) and found high standard deviation (SD) for all estimates of these gases on all continents, thus finding high uncertainty (Fig. 9).

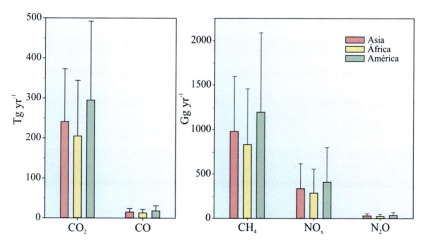

FIG. 9 Total annual greenhouse gas emissions from shifting cultivation in tropical Asia, Africa, and America. Yearly mean and standard deviation values were calculated using Monte Carlo-based uncertainty analysis. *(Source: From Silva, J. M. N., Carreiras, J. M. B., Rosa, I., & Pereira, J. M. C. (2011). Greenhouse gas emissions from shifting cultivation in the tropics, including uncertainty and sensitivity analysis. Journal of Geophysical Research: Atmospheres. https://doi.org/10.1029/2011JD016056.)*

296 Section | C Case Studies

One of the strategies that have been adopted to reduce uncertainties is the combination of several methods. FLUXCOM, for example, uses approaches based on machine learning methods such as artificial neural net (ANN) that integrate FLUXNET site-level observations (eddy covariance measurements), satellite remote sensing [fluxes estimated from mean seasonal cycles of Moderate Resolution Imaging Spectroradiometer (MODIS)—with techniques of Sun-Induced Fluorescence (SIF) and atmospheric inversion results] and meteorological data (daily meteorological information) (Jung et al., 2020).

Statistical and artificial intelligence methods such as the Bayesian approach, maximum likelihood analysis, Monte Carlo technique, and artificial neural network have also been adopted to estimate, derive, and reduce uncertainties (Lo, Ma, & Lo, 2005; Marland et al., 2014; Silva et al., 2011).

Increasing the coverage of flow towers to cover larger areas is expected to reduce uncertainty and feed GHG inventories with more reliable data. (Ometto, Aguiar, et al., 2014; Valentini et al., 2014).

4 Perspective and future opportunities

The increase, stabilization, or even decrease in GHG emissions depends on which scenario of the Intergovernmental Panel on Climate Change (IPCC)—Representative Concentration Pathways (RCPs, Arora et al., 2011)—the civilization will target in the coming years. Carbon dioxide, one of these GHGs, contributes to the increase in temperature in the atmosphere and acts as a fertilizer for vegetation. Thus, carbon dioxide is sequestered and removed from the atmosphere by vegetation (Chen et al., 2019; Phillips et al., 2017).

One of the consequences of this increase in carbon sequestration that has been observed is what is called greening, that is, the increase in biomass in semi-arid lands or tropical dry forests (TDFs) (Donohue, Roderick, McVicar, & Farquhar, 2013; Sheffield, Wood, & Roderick, 2012). It has also been observed that TFDs such as Chaco (Argentina, Paraguay, and Bolivia), Beni savannas (northern Bolivia), Llanos (Colombia and Venezuela), Caatinga (Brazil), and Cerrado (Brazil) have great carbon storage and recovery capacity above ground biomass and nutrient recycling (Portillo-Quintero, Sanchez-Azofeifa, Calvo-Alvarado, Quesada, & do Espirito Santo, 2015), which points to a high degree of resilience and carbon sink for these biomes in the tropical region. Nevertheless, the growth of native vegetation on the land from which it has been removed by human activity has been considered a good way of mitigating emissions (Canadell & Schulze, 2014).

Another point that is added is that with greater availability of CO_2 in the atmosphere, there is a stomatal reduction in plants; therefore, a decrease in transpiration, that is, vegetation reduces water vapor rates (also a GHG) it launches into the atmosphere. On the other hand, still addressing global changes in relation to the climate, it was observed that an intense drought in the Amazon, such as that of 2005, which lasted until 2008, showed evidence of a decrease in the carbon sink (Yang et al., 2018).

Land is both a source and a sink for various GHGs; however, it is difficult to separate natural and anthropic processes (Jia et al., 2019). The processes responsible for land flows were also divided by the IPCC (2010) into three categories: (1) the direct effects of human activity due to changes in land cover and management (LUC); (2) the indirect effects of anthropic environmental changes, such as climate change, fertilization with carbon dioxide (CO_2), and nitrogen deposition; and (3) natural climate variability and natural disturbances (e.g., forest fires and diseases) (Jia et al., 2019).

Land-use change has been a significant source of emissions, being responsible for approximately 22% of global anthropogenic GHG emissions over the period 2003–12 (Ciais, Willem, Friedlingstein, & Munhoven, 2013; Jia et al., 2019; Smith & Dukes, 2013). Promoting agricultural practices that reduce GHG emissions is important, but it is also necessary to improve the farmer's production and income; for some regions, such as South Asia and Latin America, which are large areas with diverse agroclimatic regions, adaptation and mitigation strategies must be location specific and cost effective (Jat et al., 2016).

Improved agricultural practices can offer opportunities for emission reductions with better land use. Technologies that reduce emissions of CH_4 and N_2O bring the added value of no risk of reversion (i.e., avoided emissions are permanent). Significant mitigation activities include better quality food, meeting growing biomass demands, better regulation of flooding regimes in rice paddies with drainage in the middle of the growing season, reduced burning of agricultural waste, and better distribution of nitrogen fertilizers (Canadell & Schulze, 2014).

Nevertheless, some research and development components, such as algae-based biofuels and bioinspired catalytic systems, can contribute significantly to mitigating the effects of GHG emissions and reduce the pressure that currently occurs on land (Canadell & Schulze, 2014). Algae, for example, are able to accumulate large quantities of lipids for biodiesel production, with little or no requirements for use of productive lands. To date, the bioenergy produced by algae has not reached a level of sustainability for its production compared to biodiesel, such as the large consumption of water and nutrients that are needed. To address these issues, future research looks for ways to reduce nutrient demand by recycling wastewater containing significant amounts of nitrate and phosphates; production of new strains of algae with higher growth rates and accumulation of lipids and greater tolerance to contaminants (Canadell & Schulze, 2014).

Dihydrogen (H_2) is a bioinspired catalytic system, a renewable energy carrier, the oxidation of which produces heat and water. Dihydrogen is also important to many industrial processes, such as the production of N fertilizer, and in combination with the use of H_2 to reduce CO_2 to CO, gas mixtures would be

298 Section | C Case Studies

suitable for the production of fuels such as methanol (Canadell & Schulze, 2014).

Some advances in satellite technologies and remote sensing will allow us to increase the accuracy of GHG estimates in tropical regions that have cloud cover issues. Geostationary satellites that the global remote sensing of photosynthesis has relied on at most one or two snapshots per day, currently provide multichannel images at very high revisit frequencies ($\sim$10 min) with a moderate spatial resolution ($\sim$1 km) (Bessho et al., 2016; Ryu et al., 2019).

CubeSat, another technology, is a low-cost, miniaturized satellite that is composed of multiples of cubic units ($10 \times 10 \times 10$ cm). They offer maps of red, green, blue, and visible to near-infrared (NIR) spectral reflectance channels at 3 m resolution, with daily revisit frequency over the entire globe (Hand, 2015; Ryu et al., 2019).

Finally, hyperspectral satellites, planned for the next decade, will provide information on vegetation, photosynthetic processes and hyperspectral reflectance maps (DuBois et al., 2018; Ryu et al., 2019).

These new technologies will offer a large amount of information (big data) at high spatial, temporal, and spectral resolutions. Therefore, large computational resources and several platforms are already available, including the Google Earth Engine (Gorelick et al., 2017), NASA Earth Exchange (Nemani, Votava, Michaelis, Melton, & Milesi, 2011), and other cloud computing services (Agarwal et al., 2011; Ryu et al., 2019).

Acknowledgments

The authors thank the Sao Paulo State Foundation (FAPESP) for their support. Grant no. 2017/22269-2; L.S.B. was funded by FAPESP Grant no. 2018/14006-4.

References

Abril, G., et al. (2014). Amazon River carbon dioxide outgassing fuelled by wetlands. *Nature, 505* (7483), 395–398. https://doi.org/10.1038/nature12797.

Agarwal, D., et al. (2011). Data-intensive science: The Terapixel and MODISAzure projects. *International Journal of High Performance Computing Applications, 25*(3), 304–316. https://doi.org/10.1177/1094342011414746.

Aguiar, A. P. D. (2006). *Modelagem de mudança do uso da terra na amazônia: Explorando a heterogeneidade intra-regional* (p. 182). São José dos Campos.

Aguiar, A. P. D., et al. (2016). Land use change emission scenarios: Anticipating a forest transition process in the Brazilian Amazon. *Global Change Biology, 22*(5), 1821–1840. https://doi.org/10.1111/gcb.13134.

Aleman, J. C., Jarzyna, M. A., & Staver, A. C. (2018). Forest extent and deforestation in tropical Africa since 1900. *Nature Ecology & Evolution, 2*(1), 26–33. https://doi.org/10.1038/s41559-017-0406-1.

Allen, K., et al. (2017). Will seasonally dry tropical forests be sensitive or resistant to future changes in rainfall regimes? *Environmental Research Letters, 12*(2). https://doi.org/10.1088/1748-9326/aa5968, 023001.

Andrews, A. E., et al. (2014). CO_2, CO, and CH_4 measurements from tall towers in the NOAA Earth System Research Laboratory's Global Greenhouse Gas Reference Network: Instrumentation, uncertainty analysis, and recommendations for future high-accuracy greenhouse gas monitoring efforts. *Atmospheric Measurement Techniques, 7*(2), 647–687. https://doi.org/10.5194/amt-7-647-2014.

Aragão, L. E. O. C., et al. (2018). 21st century drought-related fires counteract the decline of Amazon deforestation carbon emissions. *Nature Communications, 9*(1), 536. https://doi.org/10.1038/s41467-017-02771-y.

Araújo, A. C., et al. (2002). Comparative measurements of carbon dioxide fluxes from two nearby towers in a central Amazonian rainforest: The Manaus LBA site. *Journal of Geophysical Research: Atmospheres, 107*(D20). https://doi.org/10.1029/2001jd000676. LBA 58-51-LBA 58-20.

Arora, V. K., Scinocca, J. F., Boer, G. J., Christian, J. R., Denman, K. L., Flato, G. M., et al. (2011). Carbon emission limits required to satisfy future representative concentration pathways of greenhouse gases. *Geophysical Research Letters*. https://doi.org/10.1029/2010GL046270.

Aubinet, M., et al. (2005). Comparing CO_2 storage and advection conditions at night at different carboeuroflux sites. *Boundary-Layer Meteorology, 116*(1), 63–93. https://doi.org/10.1007/s10546-004-7091-8.

Bakwin, P. S., Tans, P. P., Hurst, D. F., & Zhao, C. (1998). Measurements of carbon dioxide on very tall towers: Results of the NOAA/CMDL program. *Tellus B: Chemical and Physical Meteorology, 50*(5), 401–415. https://doi.org/10.3402/tellusb.v50i5.16216.

Barona, E., Ramankutty, N., Hyman, G., & Coomes, O. T. (2010). The role of pasture and soybean in deforestation of the Brazilian Amazon. *Environmental Research Letters, 5*(2). https://doi.org/10.1088/1748-9326/5/2/024002, 024002.

Barros, N., Cole, J. J., Tranvik, L. J., Prairie, Y. T., Bastviken, D., Huszar, V. L. M., et al. (2011). Carbon emission from hydroelectric reservoirs linked to reservoir age and latitude. *Nature Geoscience, 4*(9), 593–596. https://doi.org/10.1038/Ngeo1211.

Basso, L. S., Gatti, L. V., Gloor, M., Miller, J. B., Domingues, L. G., Correia, C. S., et al. (2016). Seasonality and interannual variability of CH(4) fluxes from the eastern Amazon Basin inferred from atmospheric mole fraction profiles. *Journal of Geophysical Research: Atmospheres, 121*(1), 168–184. https://doi.org/10.1002/2015jd023874.

Bastos, A., et al. (2018). Impact of the 2015/2016 El Niño on the terrestrial carbon cycle constrained by bottom-up and top-down approaches. *Philosophical Transactions of the Royal Society of London. Series B, Biological Sciences, 373*(1760). https://doi.org/10.1098/rstb.2017.0304.

Bastviken, D., Tranvik, L. J., Downing, J. A., Crill, P. M., & Enrich-Prast, A. (2011). Freshwater methane emissions offset the continental carbon sink. *Science, 331*(6013), 50. https://doi.org/10.1126/science.1196808.

Bessho, K., et al. (2016). An introduction to Himawari-8/9—Japan's new-generation geostationary meteorological satellites. *Journal of the Meteorological Society of Japan*. https://doi.org/10.2151/jmsj.2016-009.

Borma, L. S., et al. (2009). Atmosphere and hydrological controls of the evapotranspiration over a floodplain forest in the Bananal Island region, Amazonia. *Journal of Geophysical Research Biogeosciences, 114*(G01003), 1–12. https://doi.org/10.1029/2007JG000641.

Boucher, D., Elias, P., Lininger, K., May-Tobin, C., Roquemore, S., & Saxon, E. (2011). *The root of the problem: What's driving tropical deforestation today?*. Cambridge: Union of Concerned Scientists.

Breuer, L., Papen, H., & Butterbach-Bahl, K. (2000). N_2O emission from tropical forest soils of Australia. *Journal of Geophysical Research: Atmospheres, 105*(D21), 26353–26367. https://doi.org/10.1029/2000JD900424.

300 Section | C Case Studies

Bridgham, S. D., Cadillo-Quiroz, H., Keller, J. K., & Zhuang, Q. (2013). Methane emissions from wetlands: Biogeochemical, microbial, and modeling perspectives from local to global scales. *Global Change Biology, 19*(5), 1325–1346. https://doi.org/10.1111/gcb.12131.

Brondizio, E. S. (2014). In I. C. G. Vieira, P. M. Toledo, & R. A. O. Santos Jr., (Eds.), *Abordagens teóricas e metodológicas para o estudo de mudança de usos da terra* (p. 504). Rio de Janeiro: Garamond.

Butman, D., & Raymond, P. A. (2011). Significant efflux of carbon dioxide from streams and rivers in the United States. *Nature Geoscience, 4*(12), 839–842. https://doi.org/10.1038/ngeo1294.

Butterbach-Bahl, K., Papen, H., & Rennenberg, H. (1997). Impact of gas transport through rice cultivars on methane emission from rice paddy fields. *Plant Cell & Environment, 20*(9), 1175–1183.

Butterbach-Bahl, K., Sander, B. O., Pelster, D., & Díaz-Pinés, E. (2016). Quantifying greenhouse gas emissions from managed and natural soils. In T. S. Rosenstock, M. C. Rufino, K. Butterbach-Bahl, L. Wollenberg, & M. Richards (Eds.), *Methods for measuring greenhouse gas balances and evaluating mitigation options in smallholder agriculture* (pp. 71–96). Cham: Springer International Publishing. https://doi.org/10.1007/978-3-319-29794-1_4.

CAIT. (2019). *CAIT climate data explorer*. Washington, DC, USA: World Resources Institute (WRI).

Canadell, J. G., & Schulze, E. D. (2014). Global potential of biospheric carbon management for climate mitigation. *Nature Communications*. https://doi.org/10.1038/ncomms6282.

Cardoso, M., Nobre, C., Sampaio, G., Hirota, M., Valeriano, D., & Câmara, G. (2009). Long-term potential for tropical-forest degradation due to deforestation and fires in the Brazilian Amazon. *Biologia, 64*(3), 433–437. https://doi.org/10.2478/s11756-009-0076-9.

Cheewaphongphan, P., Chatani, S., & Saigusa, N. (2019). Exploring gaps between bottom-up and top-down emission estimates based on uncertainties in multiple emission inventories: A case study on CH_4 emissions in China. *Sustainablility, 11*(7), 2054.

Chen, J. M., Ju, W., Ciais, P., Viovy, N., Liu, R., Liu, Y., et al. (2019). Vegetation structural change since 1981 significantly enhanced the terrestrial carbon sink. *Nature Communications*. https://doi.org/10.1038/s41467-019-12257-8.

Chou, W. W., Wofsy, S. C., Harriss, R. C., Lin, J. C., Gerbig, C., & Sachse, G. W. (2002). Net fluxes of CO_2 in Amazonia derived from aircraft observations. *Journal of Geophysical Research: Atmospheres, 107*(D22). https://doi.org/10.1029/2001jd001295. ACH 4-1-ACH 4-15.

Christiansen, J., Korhonen, J., Juszczak, R., Giebels, M., & Pihlatie, M. (2011). Assessing the effects of chamber placement, manual sampling and headspace mixing on CH_4 fluxes in a laboratory experiment. *Plant and Soil, 343*, 171–185. https://doi.org/10.1007/s11104-010-0701-y.

Ciais, P., Bombelli, A., Williams, M., Piao, S. L., Chave, J., Ryan, C. M., et al. (2011). The carbon balance of Africa: Synthesis of recent research studies. *Philosophical Transactions of the Royal Society A: Mathematical, Physical and Engineering Sciences*. https://doi.org/10.1098/rsta.2010.0328.

Ciais, P., Borges, A. V., Abril, G., Meybeck, M., Folberth, G., Hauglustaine, D., et al. (2008). The impact of lateral carbon fluxes on the European carbon balance. *Biogeosciences, 5*(5), 1259–1271. https://doi.org/10.5194/bg-5-1259-2008.

Ciais, P., Willem, J., Friedlingstein, P., & Munhoven, G. (2013). Carbon and other biogeochemical cycles. In *Climate change 2013*, ISBN:978-1-107-66182-0.

Cole, J. J., et al. (2007). Plumbing the global carbon cycle: Integrating inland waters into the terrestrial carbon budget. *Ecosystems, 10*(1), 171–184. https://doi.org/10.1007/s10021-006-9013-8.

Coynel, A., Seyler, P., Etcheber, H., Meybeck, M., & Orange, D. (2005). Spatial and seasonal dynamics of total suspended sediment and organic carbon species in the Congo River. *Global Biogeochemical Cycles, 19*(4). https://doi.org/10.1029/2004GB002335.

Da Rocha, H. R., et al. (2009). Patterns of water and heat flux across a biome gradient from tropical forest to savanna in Brazil. *Journal of Geophysical Research: Biogeosciences, 114*(1). https://doi.org/10.1029/2007JG000640.

Damassa, T. (2014). *Climate analysis indicators tool (CAIT)* (pp. 949–951). Dordrecht, Netherlands: Springer. https://doi.org/10.1007/978-94-007-0753-5_403.

D'Amelio, M. T. S., Gatti, L. V., Miller, J. B., & Tans, P. (2009). Regional N_2O fluxes in Amazonia derived from aircraft vertical profiles. *Atmospheric Chemistry and Physics, 9*(22), 8785–8797. https://doi.org/10.5194/acp-9-8785-2009.

de Araujo, A. C., von Randow, C., & Restrepo-Coupe, N. (2016). In L. Nagy, B. R. Forsberg, & P. Artaxo (Eds.), *Ecosystem–atmosphere exchanges of CO_2 in dense and open 'Terra Firme' rainforests in Brazilian Amazonia* (pp. 149–169). Springer-Verlag.

DeCola, P., et al. (2019). *An integrated global greenhouse gas information system (IG3IS) science implementation plan* (p. 245). World Meteorological Organization (WMO).

Defries, R. (2013). In M. C. H. Frederic Achard (Ed.), *Why forest monitoring matters for people and the panet* (p. 354). CRC Press.

DeFries, R. S., Rudel, T., Uriarte, M., & Hansen, M. (2010). Deforestation driven by urban population growth and agricultural trade in the twenty-first century. *Nature Geoscience, 3*(3), 178–181. https://doi.org/10.1038/ngeo756.

Denmead, O. T. (2008). Approaches to measuring fluxes of methane and nitrous oxide between landscapes and the atmosphere. *Plant and Soil, 309*(1), 5–24. https://doi.org/10.1007/s11104-008-9599-z.

Donohue, R. J., Roderick, M. L., McVicar, T. R., & Farquhar, G. D. (2013). Impact of CO_2 fertilization on maximum foliage cover across the globe's warm, arid environments. *Geophysical Research Letters*. https://doi.org/10.1002/grl.50563.

Drescher, J., et al. (2016). Ecological and socio-economic functions across tropical land use systems after rainforest conversion. *Philosophical Transactions of the Royal Society B: Biological Sciences, 371*(1694), 20150275. https://doi.org/10.1098/rstb.2015.0275.

DuBois, S., et al. (2018). Using imaging spectroscopy to detect variation in terrestrial ecosystem productivity across a water-stressed landscape. *Ecological Applications*. https://doi.org/10.1002/eap.1733.

Duchemin, E., Lucotte, M., & Canuel, R. (1999). Comparison of static chamber and thin boundary layer equation methods for measuring greenhouse gas emissions from large water bodies. *Environmental Science & Technology, 33*(2), 350–357. https://doi.org/10.1021/es9800840.

FAO. (2020). *FAOSTAT—Food and Agriculture Organization of the United Nations.*

Feigenwinter, C., et al. (2008). Comparison of horizontal and vertical advective CO_2 fluxes at three forest sites. *Agricultural and Forest Meteorology*. https://doi.org/10.1016/j.agrformet.2007.08.013.

Figueres, C., Le Quéré, C., Mahindra, A., Bäte, O., Whiteman, G., Peters, G., et al. (2018). Emissions are still rising: Ramp up the cuts. *Nature, 564*(7734), 27–30. https://doi.org/10.1038/d41586-018-07585-6.

Foken, T. (2008). The energy balance closure problem: An overview. *Ecological Applications*. https://doi.org/10.1890/06-0922.1.

Foley, J. A., et al. (2005). Global consequences of land use. *Science, 309*(5734), 570–574. https://doi.org/10.1126/science.1111772.

302 Section | C Case Studies

Friedlingstein, P., et al. (2019). Global carbon budget 2019. *Earth System Science Data, 11*(4), 1783–1838. https://doi.org/10.5194/essd-11-1783-2019.

Fu, R., Dickinson, R., Chen, M., & Wang, H. (2001). How do tropical sea surface temperatures influence the seasonal distribution of precipitation in the equatorial Amazon? *Journal of Climate.* https://doi.org/10.1175/1520-0442(2001)014<4003:HDTSST>2.0.CO;2.

Fuentes, J. D., et al. (2016). Linking meteorology, turbulence, and air chemistry in the amazon rain forest. *Bulletin of the American Meteorological Society, 97*(12). https://doi.org/10.1175/BAMS-D-15-00152.1.

Gatti, L. V., Miller, J. B., D'amelio, M. T. S., Martinewski, A., Basso, L. S., Gloor, M. E., et al. (2010). Vertical profiles of CO_2 above eastern Amazonia suggest a net carbon flux to the atmosphere and balanced biosphere between 2000 and 2009. *Tellus B, 62*(5), 581–594. https://doi.org/10.1111/j.1600-0889.2010.00484.x.

Gatti, L. V., et al. (2014). Drought sensitivity of Amazonian carbon balance revealed by atmospheric measurements. *Nature, 506*(7486), 76–80. https://doi.org/10.1038/nature12957.

Geist, H. J., & Lambin, E. F. (2001). *What drives tropical deforestation? A meta-analysis ofproximate and underlying causes of deforestation based on sub-national case studyevidence, LUCC Report* (p. 116). Belgium: LUCC International Project Office.

Gerbig, C., Lin, J. C., Wofsy, S. C., Daube, B. C., Andrews, A. E., Stephens, B. B., et al. (2003). Toward constraining regional-scale fluxes of CO_2 with atmospheric observations over a continent: 1. Observed spatial variability from airborne platforms. *Journal of Geophysical Research: Atmospheres, 108*(D24). https://doi.org/10.1029/2002JD003018.

Gibbs, H. K., & Herold, M. (2007). Tropical deforestation and greenhouse gas emissions. *Environmental Research Letters, 2*(4), 045021. https://doi.org/10.1088/1748-9326/2/4/045021.

Gibbs, H. K., Ruesch, A. S., Achard, F., Clayton, M. K., Holmgren, P., Ramankutty, N., et al. (2010). Tropical forests were the primary sources of new agricultural land in the 1980s and 1990s. *Proceedings of the National Academy of Sciences of the United States of America, 107*(38), 16732–16737. https://doi.org/10.1073/pnas.0910275107.

Gloor, M., et al. (2012). The carbon balance of South America: A review of the status, decadal trends and main determinants. *Biogeosciences, 9*, 5407–5430. https://doi.org/10.5194/bg-9-5407-2012.

Gloor, M., Bakwin, P., Hurst, D., Lock, L., Draxler, R., & Tans, P. (2001). What is the concentration footprint of a tall tower? *Journal of Geophysical Research: Atmospheres, 106*(D16), 17831–17840. https://doi.org/10.1029/2001JD900021.

Gorelick, N., Hancher, M., Dixon, M., Ilyushchenko, S., Thau, D., & Moore, R. (2017). Google Earth engine: Planetary-scale geospatial analysis for everyone. *Remote Sensing of Environment.* https://doi.org/10.1016/j.rse.2017.06.031.

Gumbricht, T., Roman-Cuesta, R. M., Verchot, L., Herold, M., Wittmann, F., Householder, E., et al. (2017). An expert system model for mapping tropical wetlands and peatlands reveals South America as the largest contributor. *Global Change Biology, 23*(9), 3581–3599. https://doi.org/10.1111/gcb.13689.

Hand, E. (2015). Startup liftoff. *Science.* https://doi.org/10.1126/science.348.6231.172.

Harris, N. L., Brown, S., Hagen, S. C., Saatchi, S. S., Petrova, S., Salas, W., et al. (2012). Baseline map of carbon emissions from deforestation in tropical regions. *Science, 336*(6088), 1573–1576.

Hashimoto, S., Carvalhais, N., Ito, A., Migliavacca, M., Nishina, K., & Reichstein, M. (2015). Global spatiotemporal distribution of soil respiration modeled using a global database. *Biogeosciences, 12*(13), 4121–4132. https://doi.org/10.5194/bg-12-4121-2015.

Tropical ecosystem greenhouse gas accounting **Chapter | 8** **303**

Heistermann, M., Müller, C., & Ronneberger, K. (2006). Land in sight? Achievements, deficits and potentials of continental to global scale land-use modeling. *Agriculture, Ecosystems & Environment, 114*(2–4), 141–158. https://doi.org/10.1016/j.agee.2005.11.015.

Henders, S., Persson, U. M., & Kastner, T. (2015). Trading forests: Land-use change and carbon emissions embodied in production and exports of forest-risk commodities. *Environmental Research Letters, 10*(12), 125012. https://doi.org/10.1088/1748-9326/10/12/125012.

Henry, J. (2005). In J. E. Oliver (Ed.), *Tropical and equatorial climates* (pp. 742–750). Dordrecht, Netherlands: Springer. https://doi.org/10.1007/1-4020-3266-8_212.

Hollinger, D. Y., & Richardson, A. D. (2005). Uncertainty in eddy covariance measurements and its application to physiological models. *Tree Physiology, 25*(7), 873–885. https://doi.org/10.1093/treephys/25.7.873.

Houghton, R. A. (2005). In P. Moutinho, & S. Schwartzman (Eds.), *Tropical deforestation as a source of greenhouse gas emissions* (pp. 13–21). Washington, DC: IPAM (Instituto de Pesquisa Ambiental da Amazonia).

Houghton, R. A. (2010). How well do we know the flux of CO_2 from land-use change? *Tellus B, 62*(5), 337–351. https://doi.org/10.1111/j.1600-0889.2010.00473.x.

Huang, T.-H., Fu, Y.-H., Pan, P.-Y., & Chen, C.-T. A. (2012). Fluvial carbon fluxes in tropical rivers. *Current Opinion in Environmental Sustainability, 4*(2), 162–169. https://doi.org/10.1016/j.cosust.2012.02.004.

Huang, T. H., et al. (2017). Riverine carbon fluxes to the South China Sea. *Journal of Geophysical Research: Biogeosciences, 122*(5), 1239–1259. https://doi.org/10.1002/2016JG003701.

ID. (2016). *Republic of Indonesia: First nationally determined contribution.*

IPCC, Parry, M. L., Canziani, O. F., Palutikof, J. P., van der Linden, P. J., & Hanson, C. E. (Eds.). (2007). Climate change 2007: Impacts, adaptation and vulnerability. In *Contribution of working group II to the fourth assessment report of the IPCC* (p. 976). Cambridge, UK: Cambridge University Press.

IPCC. (2010). *Revisiting the use of managed land as a proxy for estimating national anthropogenic emissions and removals.*, ISBN:9784887880610.

IPCC. (2014). Climate change 2014: Mitigation of climate change. In *Contribution of working group III to the fifth assessment report of the Intergovernmental Panel on Climate Change* (p. 1454). Cambridge, United Kingdom and New York, NY, USA: Cambridge University Press.

Ivens, W. P. M. F., Tysmans, D. J. J., Kroeze, C., Löhr, A. J., & van Wijnen, J. (2011). Modeling global N_2O emissions from aquatic systems. *Current Opinion in Environmental Sustainability, 3*(5), 350–358. https://doi.org/10.1016/j.cosust.2011.07.007.

Jat, M. L., et al. (2016). Climate change and agriculture: Adaptation strategies and mitigation opportunities for food security in South Asia and Latin America. In D. L. Sparks (Ed.), *Advances in agronomy* (pp. 127–235). Academic Press. https://doi.org/10.1016/bs.agron.2015.12.005 (Chapter 3).

Jia, G., et al. (2019). Land-climate interactions. In P. R. Shukla, et al. (Eds.), *Climate change and land: An IPCC special report on climate change, desertification, land degradation, sustainable land management, food security, and greenhouse gas fluxes in terrestrial ecosystems* IPCC.

Jonas, M., Bun, R., Nahorski, Z., Marland, G., Gusti, M., & Danylo, O. (2019). Quantifying greenhouse gas emissions. *Mitigation and Adaptation Strategies for Global Change, 24*(6), 839–852. https://doi.org/10.1007/s11027-019-09867-4.

Jung, M., et al. (2020). Scaling carbon fluxes from eddy covariance sites to globe: Synthesis and evaluation of the FLUXCOM approach. *Biogeosciences.* https://doi.org/10.5194/bg-17-1343-2020.

304 Section | C Case Studies

Keddy, P. A. (2010). *Wetland ecology: Principles and conservation.* Cambridge University Press.

Keller, M., et al. (2004). Ecological research in the large-scale biosphere-atmosphere experiment in Amazonia: Early results. *Ecological Applications*, *14*(sp4), 3–16. https://doi.org/10.1890/03-6003. ESA—Ecological Society of America.

Kondo, M., Ichii, K., Takagi, H., & Sasakawa, M. (2015). Comparison of the data-driven top-down and bottom-up global terrestrial CO_2 exchanges: GOSAT CO_2 inversion and empirical eddy flux upscaling. *Journal of Geophysical Research: Biogeosciences*, *120*(7), 1226–1245. https://doi.org/10.1002/2014jg002866.

Korkiakoski, M., Tuovinen, J. P., Aurela, M., Koskinen, M., Minkkinen, K., Ojanen, P., et al. (2017). Methane exchange at the peatland forest floor—Automatic chamber system exposes the dynamics of small fluxes. *Biogeosciences*, *14*(7), 1947–1967. https://doi.org/10.5194/bg-14-1947-2017.

Kosten, S., Roland, F., Marques, D. M. L. D., Van Nes, E. H., Mazzeo, N., Sternberg, L. D. L., et al. (2010). Climate-dependent CO_2 emissions from lakes. *Global Biogeochemical Cycles*, *24*. https://doi.org/10.1029/2009gb003618. Artn Gb2007.

Lambin, E. F., & Geist, H. J. (2001). Global land-use and land-cover change: What have we learned so far? *Global Change Newsletter*, *46*(6), 27–30.

Lambin, E. F., & Geist, H. J. (2003). Regional differences in tropical deforestation. *Environment: Science and Policy for Sustainable Development*, *45*(6), 22–36. https://doi.org/10.1080/00139157.2003.10544695.

Lambin, E. F., Geist, H. J., & Lepers, E. (2003). Dynamics of land-use and land-cover change in tropical regions. *Annual Review of Environment and Resources*, *28*(1), 205–241. https://doi.org/10.1146/annurev.energy.28.050302.105459.

Lambin, E. F., Gibbs, H. K., Ferreira, L., Grau, R., Mayaux, P., Meyfroidt, P., et al. (2013). Estimating the world's potentially available cropland using a bottom-up approach. *Global Environmental Change*, *23*(5), 892–901. https://doi.org/10.1016/j.gloenvcha.2013.05.005.

Lambin, E. F., & Meyfroidt, P. (2011). Global land use change, economic globalization, and the looming land scarcity. *Proceedings of the National Academy of Sciences of the United States of America*, *108*(9), 3465–3472. https://doi.org/10.1073/pnas.1100480108.

Le Quéré, C., et al. (2018). Global carbon budget 2018. *Earth System Science Data*, *10*(4), 2141–2194. https://doi.org/10.5194/essd-10-2141-2018.

Lewis, S. L., Edwards, D. P., & Galbraith, D. (2015). Increasing human dominance of tropical forests. *Science*, *349*(6250), 827. https://doi.org/10.1126/science.aaa9932.

Li, M., Peng, C., Wang, M., Xue, W., Zhang, K., Wang, K., et al. (2017). The carbon flux of global rivers: A re-evaluation of amount and spatial patterns. *Ecological Indicators*, *80*, 40–51. https://doi.org/10.1016/j.ecolind.2017.04.049.

Livingston, G. P., & Hutchinson, G. L. (1995). Enclosure-based measurement of trace gas exchange: Applications and sources of error. In P. A. Matson, & R. C. Harris (Eds.), *Biogenic trace gases: Measuring emissions from soil and water* (pp. 14–51). Oxford, UK: Blackwell Science.

Lloyd, J., et al. (2007). An airborne regional carbon balance for Central Amazonia. *Biogeosciences*, *4*(5), 759–768. https://doi.org/10.5194/bg-4-759-2007.

Lo, S. C., Ma, H. W., & Lo, S. L. (2005). Quantifying and reducing uncertainty in life cycle assessment using the Bayesian Monte Carlo method. *Science of the Total Environment.* https://doi.org/10.1016/j.scitotenv.2004.08.020.

Ludwig, W., Probst, J.-L., & Kempe, S. (1996). Predicting the oceanic input of organic carbon by continental erosion. *Global Biogeochemical Cycles*, *10*(1), 23–41. https://doi.org/10.1029/95GB02925.

Lundegårdh, H. (1927). Carbon dioxide evolution of soil and crop growth. *Soil Science*, *23*(6).

Marengo, J. A. (1992). Interannual variability of surface climate in the Amazon basin. *International Journal of Climatology*, *12*(8), 853–863. https://doi.org/10.1002/joc.3370120808.

Marengo, J. A., Tomasella, J., & Uvo, C. R. (1998). Trends in streamflow and rainfall in tropical South America: Amazonia, eastern Brazil, and northwestern Peru. *Journal of Geophysical Research: Atmospheres*. https://doi.org/10.1029/97JD02551.

Marland, E., Cantrell, J., Kiser, K., Marland, G., & Shirley, K. (2014). Valuing uncertainty part I: The impact of uncertainty in GHG accounting. *Carbon Management*, *5*(1), 35–42. https://doi.org/10.4155/cmt.13.75.

Martin, S. T., et al. (2010). Sources and properties of Amazonian aerosol particles. *Reviews of Geophysics*. https://doi.org/10.1029/2008RG000280.

Massman, W. J., & Lee, X. (2002). Eddy covariance flux corrections and uncertainties in long-term studies of carbon and energy exchanges. *Agricultural and Forest Meteorology*. https://doi.org/10.1016/S0168-1923(02)00105-3.

Medeiros, P. M., Seidel, M., Ward, N. D., Carpenter, E. J., Gomes, H. R., Niggemann, J., et al. (2015). Fate of the Amazon River dissolved organic matter in the tropical Atlantic Ocean. *Global Biogeochemical Cycles*, *29*(5), 677–690. https://doi.org/10.1002/2015GB005115.

Melton, J. R., et al. (2013). Present state of global wetland extent and wetland methane modelling: Conclusions from a model inter-comparison project (WETCHIMP). *Biogeosciences*, *10*(2), 753–788. https://doi.org/10.5194/bg-10-753-2013.

Meybeck, M. (1993). Riverine transport of atmospheric carbon: Sources, global typology and budget. *Water, Air, & Soil Pollution*, *70*(1), 443–463. https://doi.org/10.1007/BF01105015.

Miettinen, J., Shi, C., & Liew, S. C. (2012). Two decades of destruction in Southeast Asia's peat swamp forests. *Frontiers in Ecology and the Environment*, *10*(3), 124–128. https://doi.org/10.1890/100236.

Miller, J. B., Gatti, L. V., d'Amelio, M. T. S., Crotwell, A. M., Dlugokencky, E. J., Bakwin, P., et al. (2007). Airborne measurements indicate large methane emissions from the eastern Amazon basin. *Geophysical Research Letters*, *34*(10). https://doi.org/10.1029/2006gl029213.

Mitsch, W. J., Bernal, B., Nahlik, A. M., Mander, Ü., Zhang, L., Anderson, C. J., et al. (2013). Wetlands, carbon, and climate change. *Landscape Ecology*, *28*(4), 583–597. https://doi.org/10.1007/s10980-012-9758-8.

Mitsch, W. J., Nahlik, A., Wolski, P., Bernal, B., Zhang, L., & Ramberg, L. (2010). Tropical wetlands: Seasonal hydrologic pulsing, carbon sequestration, and methane emissions. *Wetlands Ecology and Management*, *18*(5), 573–586.

Moncrieff, J. B., Malhi, Y., & Leuning, R. (1996). The propagation of errors in long-term measurements of land-atmosphere fluxes of carbon and water. *Global Change Biology*. https://doi.org/10.1111/j.1365-2486.1996.tb00075.x.

Murphy, B. P., & Bowman, D. M. J. S. (2012). What controls the distribution of tropical forest and savanna? *Ecology Letters*, *15*(7), 748–758. https://doi.org/10.1111/j.1461-0248.2012.01771.x.

Nemani, R., Votava, P., Michaelis, A., Melton, F., & Milesi, C. (2011). Collaborative supercomputing for global change science. *Eos*. https://doi.org/10.1029/2011EO130001.

Nobre, C. A., Obregon, G., Marengo, J. A., Fu, R., & Poveda, G. (2009). In M. Keller, M. Bustamante, J. H. C. Gash, & P. Silva Dias (Eds.), *Characteristics of Amazonian climate: Main features* (pp. 149–162). Washington, DC: American Geophysical Union.

Oda, T., et al. (2019). Errors and uncertainties in a gridded carbon dioxide emissions inventory. *Mitigation and Adaptation Strategies for Global Change*, *24*(6), 1007–1050. https://doi.org/10.1007/s11027-019-09877-2.

Ometto, J. P., Aguiar, A. P., Assis, T., Soler, L., Valle, P., Tejada, G., et al. (2014). Amazon forest biomass density maps: Tackling the uncertainty in carbon emission estimates. *Climatic Change*. https://doi.org/10.1007/s10584-014-1058-7.

306 Section | C Case Studies

Ometto, J. P., Bun, R., Jonas, M., Nahorski, Z., & Gusti, M. I. (2014). Uncertainties in greenhouse gases inventories—Expanding our perspective. *Climatic Change, 124*(3), 451–458. https://doi.org/10.1007/s10584-014-1149-5.

Ometto, J. P., Cimbleris, A. C. P., dos Santos, M. A., Rosa, L. P., Abe, D., Tundisi, J. G., et al. (2013). Carbon emission as a function of energy generation in hydroelectric reservoirs in Brazilian dry tropical biome. *Energy Policy, 58*, 109–116. https://doi.org/10.1016/j.enpol.2013.02.041.

Pacheco, F. S., Soares, M. C. S., Assireu, A. T., Curtarelli, M. P., Roland, F., Abril, G., et al. (2015). The effects of river inflow and retention time on the spatial heterogeneity of chlorophyll and water–air CO_2 fluxes in a tropical hydropower reservoir. *Biogeosciences, 12*(1), 147–162. https://doi.org/10.5194/bg-12-147-2015.

Pangala, S. R., et al. (2017). Large emissions from floodplain trees close the Amazon methane budget. *Nature, 552*(7684), 230–234. https://doi.org/10.1038/nature24639.

Paranaíba, J. R., Barros, N., Mendonça, R., Linkhorst, A., Isidorova, A., Roland, F., et al. (2018). Spatially resolved measurements of CO_2 and CH_4 concentration and gas-exchange velocity highly influence carbon-emission estimates of reservoirs. *Environmental Science & Technology, 52*(2), 607–615. https://doi.org/10.1021/acs.est.7b05138.

Patra, P. K., et al. (2013). The carbon budget of South Asia. *Biogeosciences*. https://doi.org/10.5194/bg-10-513-2013.

Pavelka, M., et al. (2018). Standardisation of chamber technique for CO_2, N_2O and CH_4 fluxes measurements from terrestrial ecosystems. *International Agrophysics, 32*(4), 569–587. https://doi.org/10.1515/intag-2017-0045.

Pearson, T. R. H., Brown, S., Murray, L., & Sidman, G. (2017). Greenhouse gas emissions from tropical forest degradation: An underestimated source. *Carbon Balance and Management, 12*(1), 3. https://doi.org/10.1186/s13021-017-0072-2.

Pechony, O., Shindell, D. T., & Faluvegi, G. (2013). Direct top-down estimates of biomass burning CO emissions using TES and MOPITT versus bottom-up GFED inventory. *Journal of Geophysical Research: Atmospheres, 118*(14), 8054–8066. https://doi.org/10.1002/jgrd.50624.

Pendrill, F., Persson, U. M., Godar, J., Kastner, T., Moran, D., Schmidt, S., et al. (2019). Agricultural and forestry trade drives large share of tropical deforestation emissions. *Global Environmental Change, 56*, 1–10. https://doi.org/10.1016/j.gloenvcha.2019.03.002.

Phillips, O. L., et al. (2009). Drought sensitivity of the Amazon rainforest. *Science, 323*(5919), 1344–1347. https://doi.org/10.1126/science.1164033.

Phillips, O. L., et al. (2017). Carbon uptake by mature Amazon forests has mitigated Amazon nations' carbon emissions. *Carbon Balance and Management*. https://doi.org/10.1186/s13021-016-0069-2.

Pijanowski, B. C., Brown, D. G., Shellito, B. A., & Manik, G. A. (2002). Using neural networks and GIS to forecast land use changes: A land transformation model. *Computers, Environment and Urban Systems, 26*(6), 553–575. https://doi.org/10.1016/S0198-9715(01)00015-1.

Pinker, R. T., & Laszlo, I. (1992). Modeling surface solar irradiance for satellite applications on a global scale. *Journal of Applied Meteorology*. https://doi.org/10.1175/1520-0450(1992)031<0194:MSSIFS>2.0.CO;2.

Podgrajsek, E., Sahlée, E., Bastviken, D., Holst, J., Lindroth, A., Tranvik, L., et al. (2014). Comparison of floating chamber and eddy covariance measurements of lake greenhouse gas fluxes. *Biogeosciences, 11*(15), 4225–4233. https://doi.org/10.5194/bg-11-4225-2014.

Pöhlker, C., et al. (2019). Land cover and its transformation in the backward trajectory footprint region of the Amazon Tall Tower Observatory. *Atmospheric Chemistry and Physics, 19*(13), 8425–8470. https://doi.org/10.5194/acp-19-8425-2019.

Portillo-Quintero, C., Sanchez-Azofeifa, A., Calvo-Alvarado, J., Quesada, M., & do Espirito Santo, M. M. (2015). The role of tropical dry forests for biodiversity, carbon and water conservation in the neotropics: Lessons learned and opportunities for its sustainable management. *Regional Environmental Change*. https://doi.org/10.1007/s10113-014-0689-6.

Poulter, B., et al. (2017). Global wetland contribution to 2000–2012 atmospheric methane growth rate dynamics. *Environmental Research Letters*, *12*(9). https://doi.org/10.1088/1748-9326/aa8391, 094013.

Poveda, G., Waylen, P. R., & Pulwarty, R. S. (2006). Annual and inter-annual variability of the present climate in northern South America and southern Mesoamerica. *Palaeogeography, Palaeoclimatology, Palaeoecology*. https://doi.org/10.1016/j.palaeo.2005.10.031.

Pumpanen, J., et al. (2004). Comparison of different chamber techniques for measuring soil CO_2 efflux. *Agricultural and Forest Meteorology*, *123*(3), 159–176. https://doi.org/10.1016/j.agrformet.2003.12.001.

Raymond, P. A., et al. (2013). Global carbon dioxide emissions from inland waters. *Nature*, *503* (7476), 355–359. https://doi.org/10.1038/nature12760.

Restrepo-Coupe, N., et al. (2013). What drives the seasonality of photosynthesis across the Amazon basin? A cross-site analysis of eddy flux tower measurements from the Brasil flux network. *Agricultural and Forest Meteorology*. https://doi.org/10.1016/j.agrformet.2013.04.031.

Richey, J. E., Hedges, J. I., Devol, A. H., Quay, P. D., Victoria, R., Martinelli, L., et al. (1990). Biogeochemistry of carbon in the Amazon River. *Limnology and Oceanography*, *35*(2), 352–371. https://doi.org/10.4319/lo.1990.35.2.0352.

Richey, J. E., Melack, J. M., Aufdenkampe, A. K., Ballester, V. M., & Hess, L. L. (2002). Outgassing from Amazonian rivers and wetlands as a large tropical source of atmospheric CO_2. *Nature*, *416* (6881), 617–620.

Roland, F., Vidal, L. O., Pacheco, F. S., Barros, N. O., Assireu, A., Ometto, J. P., et al. (2010). Variability of carbon dioxide flux from tropical (Cerrado) hydroelectric reservoirs. *Aquatic Sciences*, *72*(3), 283–293.

Rosa, L. P., Dos Santos, M. A., Matvienko, B., Sikar, E., Lourenço, R. S. M., & Menezes, C. F. (2003). Biogenic gas production from major Amazon reservoirs, Brazil. *Hydrological Processes*, *17*(7), 1443–1450. https://doi.org/10.1002/hyp.1295.

Ryu, Y., Berry, J. A., & Baldocchi, D. D. (2019). What is global photosynthesis? History, uncertainties and opportunities. *Remote Sensing of Environment*, *223*, 95–114. https://doi.org/10.1016/j.rse.2019.01.016.

Saleska, S. R., Didan, K., Huete, A. R., & Da Rocha, H. R. (2007). Amazon forests green-up during 2005 drought. *Science*. https://doi.org/10.1126/science.1146663.

Saleska, S. R., et al. (2003). Carbon in Amazon forests: Unexpected seasonal fluxes and disturbance-induced losses. *Science Magazine*, *302*(5650), 1554–1557. https://doi.org/10.1126/science.1091165.

Samanta, A., Ganguly, S., Hashimoto, H., Devadiga, S., Vermote, E., Knyazikhin, Y., et al. (2010). Amazon forests did not green-up during the 2005 drought. *Geophysical Research Letters*. https://doi.org/10.1029/2009GL042154.

Saunois, M., et al. (2020). The global methane budget 2000–2017. *Earth System Science Data*, *12*(3), 1561–1623. https://doi.org/10.5194/essd-12-1561-2020.

Schimel, D., Pavlick, R., Fisher, J. B., Asner, G. P., Saatchi, S., Townsend, P., et al. (2015). Observing terrestrial ecosystems and the carbon cycle from space. *Global Change Biology*. https://doi.org/10.1111/gcb.12822.

Schubert, C. J., Diem, T., & Eugster, W. (2012). Methane emissions from a small wind shielded lake determined by eddy covariance, flux chambers, anchored funnels, and boundary model calculations: A comparison. *Environmental Science and Technology*, *46*(8), 4515–4522.

308 Section | C Case Studies

Schuepp, P., Leclerc, M., MacPherson, J., & Desjardins, R. (1990). Footprint prediction of scalar fluxes from analytical solutions of the diffusion equation. *Boundary-Layer Meteorology, 50* (1–4), 355–373.

Seitzinger, S. P., Kroeze, C., & Styles, R. V. (2000). Global distribution of N_2O emissions from aquatic systems: Natural emissions and anthropogenic effects. *Chemosphere—Global Change Science, 2*(3), 267–279. https://doi.org/10.1016/S1465-9972(00)00015-5.

Sheffield, J., Wood, E. F., & Roderick, M. L. (2012). Little change in global drought over the past 60 years. *Nature.* https://doi.org/10.1038/nature11575.

Silva, J. M. N., Carreiras, J. M. B., Rosa, I., & Pereira, J. M. C. (2011). Greenhouse gas emissions from shifting cultivation in the tropics, including uncertainty and sensitivity analysis. *Journal of Geophysical Research: Atmospheres.* https://doi.org/10.1029/2011JD016056.

Silva Junior, C. H. L., Heinrich, V. H. A., Freire, A. T. G., et al. (2020). Benchmark maps of 33 years of secondary forest age for Brazil. *Scientific Data, 7,* 269. https://doi.org/10.1038/s41597-020-00600-4.

Silva, J. P., Lasso, A., Lubberding, H. J., Peña, M. R., & Gijzen, H. J. (2015). Biases in greenhouse gases static chambers measurements in stabilization ponds: Comparison of flux estimation using linear and non-linear models. *Atmospheric Environment, 109,* 130–138. https://doi.org/10.1016/j.atmosenv.2015.02.068.

Smith, N. G., & Dukes, J. S. (2013). Plant respiration and photosynthesis in global-scale models: Incorporating acclimation to temperature and CO_2. *Global Change Biology.* https://doi.org/10.1111/j.1365-2486.2012.02797.x.

Sörensson, A. A., & Ruscica, R. C. (2018). Intercomparison and uncertainty assessment of nine evapotranspiration estimates over South America. *Water Resources Research.* https://doi.org/10.1002/2017WR021682.

Soued, C., del Giorgio, P. A., & Maranger, R. (2016). Nitrous oxide sinks and emissions in boreal aquatic networks in Québec. *Nature Geoscience, 9*(2), 116–120. https://doi.org/10.1038/ngeo2611.

St Louis, V. L., Kelly, C. A., Duchemin, E., Rudd, J. W. M., & Rosenberg, D. M. (2000). Reservoir surfaces as sources of greenhouse gases to the atmosphere: A global estimate. *Bioscience, 50*(9), 766–775.

Stanley, E. H., Casson, N. J., Christel, S. T., Crawford, J. T., Loken, L. C., & Oliver, S. K. (2016). The ecology of methane in streams and rivers: Patterns, controls, and global significance. *Ecological Monographs, 86*(2), 146–171. https://doi.org/10.1890/15-1027.

Stephens, B. B., et al. (2007). Weak northern and strong tropical land carbon uptake from vertical profiles of atmospheric CO_2. *Science, 316*(5832), 1732–1735. https://doi.org/10.1126/science.1137004.

Sweeney, C., et al. (2015). Seasonal climatology of CO_2 across North America from aircraft measurements in the NOAA/ESRL global greenhouse gas reference network. *Journal of Geophysical Research: Atmospheres, 120*(10), 5155–5190. https://doi.org/10.1002/2014jd022591.

Tacconi, L., & Muttaqin, M. Z. (2019a). Policy forum: Institutional architecture and activities to reduce emissions from forests in Indonesia. *Forest Policy and Economics, 108,* 101980.

Tacconi, L., & Muttaqin, M. Z. (2019b). Reducing emissions from land use change in Indonesia: An overview. *Forest Policy and Economics, 108,* 101979. https://doi.org/10.1016/j.forpol.2019.101979.

Tans, P. P., Bakwin, P. S., & Guenther, D. W. (1996). A feasible global carbon cycle observing system: A plan to decipher today's carbon cycle based on observations. *Global Change Biology, 2*(3), 309–318. https://doi.org/10.1111/j.1365-2486.1996.tb00082.x.

Teodoru, C. R., Prairie, Y. T., & del Giorgio, P. A. (2011). Spatial heterogeneity of surface CO_2 fluxes in a newly created Eastmain-1 reservoir in Northern Quebec, Canada. *Ecosystems*, *14*(1), 28–46. https://doi.org/10.1007/s10021-010-9393-7.

Tong, Y., Zhao, Y., Zhen, G., Chi, J., Liu, X., Lu, Y., et al. (2015). Nutrient loads flowing into coastal waters from the main rivers of China (2006–2012). *Scientific Reports*, *5*(1), 16678. https://doi.org/10.1038/srep16678.

Tranvik, L. J., et al. (2009). Lakes and reservoirs as regulators of carbon cycling and climate. *Limnology and Oceanography*, *54*(6), 2298–2314.

Valentini, R., et al. (2014). A full greenhouse gases budget of Africa: Synthesis, uncertainties, and vulnerabilities. *Biogeosciences*. https://doi.org/10.5194/bg-11-381-2014.

van der Werf, G. R., Morton, D. C., DeFries, R. S., Olivier, J. G. J., Kasibhatla, P. S., Jackson, R. B., et al. (2009). CO_2 emissions from forest loss. *Nature Geoscience*, *2*(11), 737–738. https://doi.org/10.1038/ngeo671.

Verburg, P. H., Kok, K., Pontius, R. G., Jr., & Veldkam, A. (2006). In E. F. Lambin, & H. Geist (Eds.), *Modeling land-use and land-cover change* (pp. 117–135). Berlin: Springer.

Vesala, T., Huotari, J., Rannik, Ü., Suni, T., Smolander, S., Sogachev, A., et al. (2006). Eddy covariance measurements of carbon exchange and latent and sensible heat fluxes over a boreal lake for a full open-water period. *Journal of Geophysical Research: Atmospheres*, *111*(D11101).

Viola, E., & Gonçalves, V. K. (2019). Brazil ups and downs in global environmental governance in the 21st century. *Revista Brasileira de Política Internacional*, *62*.

Von Randow, C., Kruijt, B., Holtslag, A. A. M., & de Oliveira, M. B. L. (2008). Exploring eddy-covariance and large-aperture scintillometer measurements in an Amazonian rain forest. *Agricultural and Forest Meteorology*. https://doi.org/10.1016/j.agrformet.2007.11.011.

von Randow, C., et al. (2004). Comparative measurements and seasonal variations in energy and carbon exchange over forest and pasture in South West Amazonia. *Theoretical and Applied Climatology*, *78*(1–3). https://doi.org/10.1007/s00704-004-0041-z.

Wania, R., et al. (2013). Present state of global wetland extent and wetland methane modelling: Methodology of a model inter-comparison project (WETCHIMP). *Geoscientific Model Development*, *6*(3), 617–641. https://doi.org/10.5194/gmd-6-617-2013.

WRI. (2015). *CAIT country greenhouse gas emissions: Sources & methods.*

Xu, L., Furtaw, M. D., Madsen, R. A., Garcia, R. L., Anderson, D. J., & McDermitt, D. K. (2006). On maintaining pressure equilibrium between a soil CO_2 flux chamber and the ambient air. *Journal of Geophysical Research: Atmospheres*, *111*(D8). https://doi.org/10.1029/2005JD006435.

Xu, J., Morris, P. J., Liu, J., & Holden, J. (2018). PEATMAP: Refining estimates of global peatland distribution based on a meta-analysis. *Catena*, *160*, 134–140. https://doi.org/10.1016/j.catena.2017.09.010.

Yang, Y., Saatchi, S. S., Xu, L., Yu, Y., Choi, S., Phillips, N., et al. (2018). Post-drought decline of the Amazon carbon sink. *Nature Communications*. https://doi.org/10.1038/s41467-018-05668-6.

Yi, C., Li, R., Wolbeck, J., Xu, X., Nilsson, M., et al. (2010). Climate control of terrestrial carbon exchange across biomes and continents. *Environmental Research Letters*. https://doi.org/10.1088/1748-9326/5/3/034007.

Zhu, D., Chen, H., Zhu, Q.a., Wu, Y., & Wu, N. (2012). High carbon dioxide evasion from an Alpine Peatland Lake: The central role of terrestrial dissolved organic carbon input. *Water, Air, & Soil Pollution*, *223*(5), 2563–2569. https://doi.org/10.1007/s11270-011-1048-6.

Chapter 9

Semiarid ecosystems

Ana Bastos[a], Victoria Naipal[b,c], Anders Ahlström[d], Natasha MacBean[e], William Kolby Smith[f], and Benjamin Poulter[g]

[a]Department of Biogeochemical Integration, Max Planck Institute for Biogeochemistry, Jena, Germany, [b]Departement of Geosciences, École Normale Supérieure, Paris, France, [c]Laboratory for Climate and Environmental Sciences (LSCE), Gif-sur-Yvette, France, [d]Department of Physical Geography and Ecosystem Science, Lund University, Lund, Sweden, [e]Department of Geography, Indiana University, Bloomington, IN, United States, [f]School of Natural Resources and the Environment, University of Arizona, Tucson, AZ, United States, [g]Biospheric Sciences Laboratory, NASA Goddard Space Flight Center, Greenbelt, MD, United States

1 Introduction and background: Global drylands and semiarid ecosystems

Drylands, comprising subhumid, semiarid, and arid areas, cover 41% of the land surface and are home to 38% of the global population (Reynolds et al., 2007). Semiarid ecosystems are located in transitional zones between arid and humid biomes and cover about 15%–25% of the land surface (Ahlström et al., 2015; Huang et al., 2016; Smith et al., 2019) (Fig. 1). Semiarid regions are characterized by low annual precipitation but differ from arid regions in that they can receive substantial rainfall during parts of the year, which allows relatively higher levels of vegetation growth and biomass. Semiarid regions can be further divided into hot and cold regions based on mean annual temperature normals. Since they are defined in comparison with arid and humid biomes, their distribution varies between studies, depending on their definition and the underlying climate datasets (Beck et al., 2018; Greve et al., 2014; Kriticos et al., 2012; Peel, Finlayson, & McMahon, 2007). As examples, hot semiarid regions can be found in Australia, Southwestern United States, and Mexico, parts of the Mediterranean region, Southern Africa and the Sahel, Eastern Brazil, and parts of India and Pakistan (Fig. 1). Cold semiarid ecosystems are mostly located at high latitudes or high elevation regions in Western North America and Eurasia, but can also be found in Southern South America and Southern Australia (Fig. 1).

Seasonal or interannual variability in rainfall in several semiarid regions is strongly influenced by coupled ocean-atmosphere variability patterns. For example, rainfall is strongly coupled with the monsoons in parts of southern Asia and eastern Africa; the El-Niño/Southern Oscillation (ENSO) in North

---

Balancing Greenhouse Gas Budgets. https://doi.org/10.1016/B978-0-12-814952-2.00012-5
Copyright © 2022 Elsevier Inc. All rights reserved.

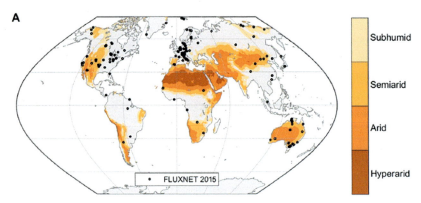

FIG. 1 Distribution of global semiarid and dry lands, from Smith et al. (2019). The distribution of eddy-covariance monitoring sites is shown in *circles*.

America, South America, Southern Africa, and Australia (Bastos, Running, Gouveia, & Trigo, 2013; Poulter et al., 2014); and the Indian Ocean Dipole in Southern Asia and Australia (Ashok, Guan, & Yamagata, 2003). Across these regions long-lasting and intense droughts recur every few years or decades and are interspersed with recurring wet periods.

1.1 Ecology

Semiarid ecosystems are characterized by relatively low mean annual, yet highly seasonally variable precipitation (Biederman et al., 2017; Noy-Meir, 1973; Reynolds et al., 2007). Dry periods are punctuated by precipitation pulses that lead to a cascade of pulsed ecosystem responses, ultimately expressed in rapid changes in the land-atmosphere exchanges of energy, water, and carbon (Huxman et al., 2004; Scott, Jenerette, Potts, & Huxman, 2009). Within a given area, semiarid ecosystems consist of multiple vegetation types with different strategies for enduring the seasonal or long-term droughts, i.e., grasslands, schlerophyllous shrubs, and/or trees and savannas. Additionally, biological soil crusts, or biocrusts—communities of photosynthetic mosses, lichens, and/or cyanobacteria found at the soil surface—are common across semiarid ecosystems worldwide and fill a critical semiarid niche in their ability to endure prolonged drought and rapidly respond to surficial wetting (Eldridge et al., 2020; Tucker et al., 2017). Semiarid regions are thus characterized by diverse community assemblage, complex ecosystem structure, and variable plant functional characteristics each sensitive to different climate drivers (Biederman et al., 2017; Cable, Ogle, Williams, Weltzin, & Huxman, 2008; Smith et al., 2019; Swetnam, Brooks, Barnard, Harpold, & Gallo, 2017). Many semiarid plants have developed key additional strategies to survive under reoccurring drought, such as deep roots to access ground-water during dry seasons (Barron-Gafford et al., 2017; Fan, Miguez-Macho, Jobbágy, Jackson, & Otero-Casal, 2017) or small leaves sometimes covered with wax to minimize water-loss

(Smith, 1978). Due to the above-described unique challenges, semiarid regions are dominated by water stress-tolerant species usually associated with lower biomass and soil carbon (C) densities than humid or temperate regions (Ruesch & Gibbs, 2008).

Even though semiarid regions hold less C than more humid biomes, they are large in areal extent and located in regions with strong pantropical climate variability (see above), and are thought to contribute to 35%–50% of the interannual variability in the atmospheric CO_2 growth rate (Ahlström et al., 2015; Piao et al., 2020; Poulter et al., 2014), 55% of the variability in tropical biomass C variability (Fan et al., 2019), and 60% of the long-term trend (Ahlström et al., 2015). The variability and drivers of the global atmospheric CO_2 growth rate and tropical C sinks are major uncertainties in future climate projections (Cox et al., 2013; Friedlingstein et al., 2014). Therefore, quantifying the C balance of semiarid regions and its variability is key to evaluate the future dynamics of the C cycle and carbon-climate feedbacks.

1.2 Threats

Because of their low tree cover, semiarid ecosystems are particularly vulnerable to multiple threats including erosion and desertification (Eldridge et al., 2020; Reynolds et al., 2007; Rodríguez-Caballero et al., 2018). Biocrusts cover most vegetation interspace across semiarid regions and provide critical stabilizing ecosystems services including soil water, erosion, climate, and air quality regulation (Eldridge et al., 2020; Rodríguez-Caballero et al., 2018). However, biocrusts are very sensitive to grazing, disturbance, and drought (Tucker, Antoninka, Day, Poff, & Reed, 2020), and loss of biocrust can lead to significant increases in erosion, which can feedback to drive persistent ecosystem degradation (Cannell & Weeks, 1979). The progressive loss of carbon and nutrients due to erosion puts semiarid ecosystems at risk of difficult-to-reverse desertification (Tucker et al., 2020). Some regions might receive dust from nearby desert regions, which can partly compensate for this loss (Stewart, Unger, & Jones, 1985).

In addition to erosion, semiarid hot biomes are characterized by recurring fires during hot and dry periods, which are also favored by the more flammable dry vegetation. Fires result in immediate losses of biocrusts and accumulated carbon stocks, followed by recovery of grasses, trees, and shrubs, which in these regions are generally adapted to fires (Pellegrini et al., 2017). The fire regime characteristics (recurrence and intensity) shape ecosystem composition and distribution (Staver, Archibald, & Levin, 2011). Furthermore, semiarid ecosystem are particularly sensitive to human pressure from increasing population, population migration and conversion of pastoralism to agricultural settlements (Stewart, Unger, & Jones, 1985) and urban expansion (Huang et al., 2016).

Finally, semiarid regions are particularly sensitive to climate change, given that warming increases evaporative demand, while precipitation trends might increase or reduce the water supply. Several semiarid regions have progressively becoming drier (Dannenberg, Wise, & Smith, 2019; Greve et al.,

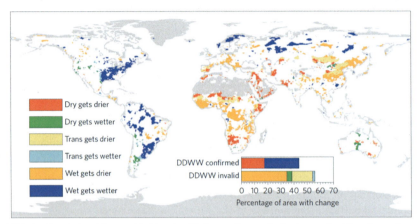

FIG. 2 Trends in aridity worldwide, classified according to the background aridity classes from Greve et al. (2014). *DDWW*, dry gets drier and wet gets wetter.

2014) (Fig. 2), a trend that is expected to persist in the coming decades. According to Huang et al. (2016), semiarid regions have expanded by 7% between the mid and the late 20th century. On the other hand, increasing atmospheric CO_2 is expected to increase plant water-use efficiency and therefore partly alleviate the negative impacts of progressive dryness, though it is not clear to what extent, as other factors might progressively become more important (Donohue, Roderick, McVicar, & Farquhar, 2013; Obermeier et al., 2017; Vicente-Serrano, McVicar, Miralles, Yang, & Tomas-Burguera, 2020). For example, forest expansion and woody encroachment in water-limited savannas in Australia, North America, and central and western Africa has been reported (Archer et al., 2017; Mitchard & Flintrop, 2013; Song et al., 2018) though with variable outcomes regarding carbon sequestration potential (Archer et al., 2017). In spite of the general drying trend, some semiarid regions, have received higher rainfall, for example, the Sahel where increased tree cover has been found (Song et al., 2018). Increasing spring temperatures are expected to prolong the growing season in the cold semiarid regions and lead to increased vegetation growth (Barichivich et al., 2013), which, in turn, might increase water depletion and contribute to extreme summer heat-drought events and, therefore, result in negative impacts on ecosystem C balance (Buermann et al., 2018; Piao et al., 2014).

2 Methodologies

2.1 Components of the greenhouse gas budget of semiarid ecosystems

The ecosystem carbon budget for a given region can be estimated as the change of C stock over time, and decomposed into changes in the different pools.

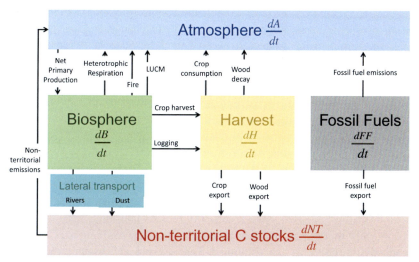

FIG. 3 Regional carbon pools and fluxes. *(Modified from Haverd, V., Raupach, M.R., Briggs, P.R., Canadell, J.G., Davis, S. J., Law, R.M., Meyer, C.P., Peters, G.P., Pickett-Heaps, C., & Sherman, B. (2013). The Australian terrestrial carbon budget. Biogeosciences, 10(2), 851–869. http://www.biogeosciences.net/10/851/2013/.)*

Here, we follow the framework proposed by Haverd et al. (2013) for the Australian carbon budget (Fig. 3):

$$\frac{dC}{dt} = \frac{dB}{dt} + \frac{dFF}{dt} + \frac{dH}{dt} + \frac{dNT}{dt} = \frac{-dA}{dt} \qquad (1)$$

where C is the total C stock, B is the amount of carbon stored in the biosphere (vegetation and soils), FF is the carbon in fossil fuels, and H is the carbon in harvest products (wood and cropland). For regional fluxes, the balance between inputs and outputs from other regions (changes in nonterritorial carbon stock, NT) also needs to be considered. The total change in C over time corresponds to the inverse of the change in the C stored in the atmosphere due to a given region (A).

The changes in live and dead biospheric carbon stocks correspond to the net atmosphere-to-land flux, or net biospheric productivity (F_{NBP}) and can be further decomposed into:

$$\frac{dB}{dt} = F_{NBP} = F_{NPP} + F_{RH} + F_{LUCM} + F_{Fire} + F_{LT} \qquad (2)$$

where NPP is the net primary productivity, RH corresponds to heterotrophic respiration (decomposition of carbon in the soil and litter), LUCM refers to changes in land-cover, land-use, and management practices (including deforestation fires), F_{Fire} corresponds to emissions from natural fires and LT refers to lateral transport of carbon as dissolved carbon in rivers and groundwater or in dust.

316 Section | C Case Studies

The changes in FF stocks can be decomposed into emissions from fossil fuel burning (FF_E) and transport by trade (FF_T) and the changes in harvested products correspond to the balance between harvested wood and crop products (F_{WH}, F_{CH}) and their consumption (F_{CW}, F_{CC}) (Fig. 3).

2.2 In situ based methodologies for flux estimation in semiarid ecosystems

Quantification of ecosystem carbon fluxes at local scales can be effectively quantified using the closed chamber technique, in which bottomless containers are used to cover the surface and measure changes in CO_2 concentration over a given time frame, allowing to quantify surface fluxes. Chamber measurements are relatively low cost and portable, which allows measuring surface fluxes in several locations, including remote sites (Matson & Goldstein, 2000). The combined use of transparent and dark chambers allows determining gross primary productivity (GPP) as the difference between net ecosystem productivity (NEP, corresponding to the net land-atmosphere C flux excluding human and natural disturbances) and total ecosystem respiration (TER = RH + autotropic respiration).

Automated chambers allow measuring CO_2 fluxes continuously over long periods at high frequency; therefore, they are well suited to analyze CO_2 flux dynamics from subdaily to annual or interannual time-scales. However, because of their small footprint, chamber measurements are not necessarily representative of surface fluxes at ecosystem scale. Moreover, closed chambers alter the micrometeorology inside the chamber, eliminating atmospheric turbulence, artificially changing pressure gradients and altering the long-wave radiation balance, which result in higher fluxes compared to eddy-covariance (EC) measurements (Riederer, Serafimovich, & Foken, 2014).

The EC flux measurement technique circumvents some of the limitations of chamber measurements. EC measurements allow directly measuring net CO_2 fluxes between the canopy and the atmosphere without influencing the environment. EC measurements allow estimating CO_2 fluxes at the ecosystem scale over large footprints (hundreds of meters to few kilometers) and across different temporal scales from hourly to interannual (Baldocchi, 2008; Baldocchi & Meyers, 1998). Therefore, EC measurements are well suited to evaluate changes in ecosystem CO_2 fluxes in response to environmental perturbations and disturbances. The EC method requires a flat terrain, steady state environmental conditions, and a horizontal homogeneous surface (Foken, Aubinet, & Leuning, 2012; Riederer et al., 2014). Technological improvements in the 1990s allowed for continuous EC measurements, when the first towers of the FLUXNET network were installed (Baldocchi et al., 2001; Foken et al., 2012). The most recent FLUXNET2015 dataset compiles 1500 site-years of data from 212 EC measurement sites extending over a range of multiple climate conditions and biomes (Pastorello et al., 2017). The wide geographical and temporal coverage of the FLUXNET datasets allows bridging results from field observations and large-scale datasets such as remote-sensing data, atmospheric inversions, or climate and ecosystem models. FLUXNET sites cover several semiarid ecosystem types (Fig. 1), although with limited coverage in Africa and South America (Smith et al., 2019).

EC data have been upscaled globally using machine learning methods by Tramontana et al. (2016) and recently updated by Jung et al. (2019), so that they can be directly compared to global gridded datasets. This method shows good consistency with observations in capturing the spatial variability of ecosystem carbon fluxes and their seasonality, but interannual variability is highly underestimated (Tramontana et al., 2016). In semiarid regions, the errors of interannual variability in net surface CO_2 fluxes are lower compared to other regions (Jung et al., 2019). Uncertainties in the upscaled fluxes are due to representativeness issues, errors in observations, limitations in the driving predictors used by the upscaling methods, and in the methods themselves (Jung et al., 2019).

Several studies have performed analysis of carbon fluxes and their responses to environmental changes estimated by these different methods, some of them comparing both approaches. Hutley, Leuning, Beringer, and Cleugh (2005) estimated NBP and GPP at a semiarid savanna site in Australia over two growing seasons in 2001/02 and 2002/03. Over this period, average NEP was $35\,gC\,m^{-2}\,year^{-1}$ and GPP $379\,gC\,m^{-2}\,year^{-1}$, though with high variability between years. For a Mediterranean beech forest in Europe, Valentini et al. (2000) reported annual NEP of $660\,gC\,m^{-2}\,year^{-1}$, with GPP values of $1300\,gC\,m^{-2}\,year^{-1}$ and TER of $645\pm51\,gC\,m^{-2}\,year^{-1}$ (Janssens et al., 2001; Matteucci, 1998), possibly because of summer drought stress. Van Gorsel, Leuning, Cleugh, Keith, and Suni (2007) found good agreement between daytime and nighttime measurements of NEE and TER from chamber measurements and an EC site over an eucalyptus forest in southeastern Australia. In a semiarid grassland in Southwest Idaho, United States, Myklebust, Hipps, and Ryel (2008) have shown that there is good consistency between nighttime RH estimates from EC and chamber measurements, except for the period of high canopy growth. They estimated average annual RH of $406\pm73\,gC\,m^{-2}\,year^{-1}$, consistent with the estimates of Tang and Baldocchi (2005), $394\,gC\,m^{-2}\,year^{-1}$, for the open section of a savanna ecosystem in California. For the closed area (oak trees), Tang and Baldocchi (2005) estimate RH values of $616\,gC\,m^{-2}\,year^{-1}$, and the average for the savanna ecosystem was $488\,gC\,m^{-2}\,year^{-1}$. More recently, Wang et al. (2019) performed a metaanalysis of ecosystem carbon fluxes and their responses to warming from in situ based studies, including different types of semiarid ecosystems. The average values of the considered studies for NEP, GPP, and TER were 165, 780, and $740\,gC\,m^{-2}\,year^{-1}$, respectively. Based on the FLUXNET upscaling,

318   Section | C Case Studies

Tramontana et al. (2016) estimated global NEP over dry and arid ecosystems of $150-168\,gC\,m^{-2}\,year^{-1}$, and for savannas of $252-259\,gC\,m^{-2}\,year^{-1}$. GPP and TER estimates were, respectively, $219-285$ and $179-245\,gC\,m^{-2}\,year^{-1}$ for dry and arid ecosystems, and $296-318$ and $248-314\,gC\,m^{-2}\,year^{-1}$ for savannas.

2.3   Atmospheric inversion monitoring of semiarid ecosystems

Atmospheric inversions combine in situ CO_2 and aircraft concentration measurements and atmospheric transport model simulations to derive land-atmosphere surface fluxes (Peylin et al., 2013). Because they are based on CO_2 concentration measurements, estimates from atmospheric inversions are expected to be consistent with the global CO_2 growth rate and provide robust constraints of the global biosphere and ocean fluxes (Kondo et al., 2020). However, uncertainty in in situ measurement-based inversion estimates increases as the spatial scale of the estimated fluxes decreases (Kaminski & Heimann, 2001), and therefore uncertainty is still large in the regional partitioning of the surface fluxes (Kondo et al., 2020). In addition, some regions are poorly constrained by observations, which explains large disagreements between inversion estimates, especially in the tropics and the Southern Hemisphere (Peylin et al., 2013). Williams et al. (2007) have compared the carbon balance for African ecosystems from multiple inversions, showing a large spread in estimates at both the continental scale, and especially in Northern Africa (arid and semiarid biomes). Haverd et al. (2013) pointed out that global atmospheric inversions do not constrain the Australian carbon budget meaningfully, because the observational network is not representative of the continent (Law, Rayner, & Wang, 2004).

Satellite-based total column CO_2 measurements are more recently providing global coverage including in the interior of continents, allowing more detail on the regional patterns of surface CO_2 fluxes in those regions poorly constrained by in situ observations (Chevallier et al., 2019). For example, Palmer et al. (2019) have shown that NBP estimated by satellite-based inversions presents much higher seasonal variability in Northern Tropical Africa than that of in situ based inversions. Ma et al. (2016) analyzed the response of Australian semiarid ecosystems to the drought following the 2010/11 La Niña using another satellite-based inversion. Satellite-based inversions are, therefore, a promising method to estimate CO_2 fluxes from regions poorly constrained by observational networks, as is the case of many semiarid biomes.

It should be noted that net land-atmosphere surface fluxes estimated by inversions include a "sink" term of carbon that is taken up by ecosystems, then transported as dissolved inorganic and dissolved organic carbon (DIC/DOC) in rivers and groundwater and finally released in the oceans or lakes. Therefore, in order to compare NBP from inversions with other estimates, a correction for this term should be applied (Bastos et al., 2020; Zscheischler et al., 2017).

2.4 Remote sensing

Satellite-based measurements of land surface reflectance allow estimating the properties of the vegetated canopy, which can be used to derive changes in ecosystem function and structure (Smith et al., 2019). Due in part to favorable atmospheric conditions, many remote sensing techniques have been pioneered in semiarid ecosystems, including the first Landsat retrievals of surface reflectance (Kowalik, Marsh, & Lyon, 1982; Marsh & Lyon, 1980) and derived estimates of vegetation condition based on the Landsat red and near-infrared surface reflectance bands (Rouse, Haas, Schell, Deering, & Harlan, 1974), which would later become the Normalized Difference Vegetation Index (NDVI) (Myneni et al., 1995; Tucker, 1979; Zhou et al., 2001). Tucker, Fung, Keeling, and Gammon (1986) have shown that variations in NDVI were related with variations in CO_2 concentration and proposed that NDVI could be a good proxy for photosynthesis. NDVI and related vegetation indices remain widely used as proxies of fraction of Absorbed Photosynthetically Active Radiation (fAPAR) to calculate GPP from remote-sensing data (Nemani et al., 2003), and combined with meteorological data to model NPP (Nemani et al., 2003; Smith et al., 2016; Zhao, Running, & Nemani, 2006).

Studies based on vegetation indices from remote sensing (Nemani et al., 2003; Seddon, Macias-Fauria, Long, Benz, & Willis, 2016) have shown that a large fraction of land ecosystems are limited by water. Because NDVI and related vegetation indices are known to saturate at high leaf-area index values, they have limitations in monitoring dense forests, especially in the tropics (Zhou et al., 2014). While semiarid ecosystems have relatively lower Leaf Area Index (LAI) values and are less prone to NDVI saturation, they present their own challenges to remote sensing, including low vegetation signal-to-noise ratios, high soil background reflectance, high spatial heterogeneity, and irregular growing seasons due to unpredictable seasonal rainfall and frequent periods of drought (Smith et al., 2019). Despite these challenges, there have been many significant achievements in semiarid remote sensing, many of which leverage new and emerging remote sensing resources and techniques (Nemani et al., 2003; Seddon et al., 2016) (Fig. 4).

Long-term NDVI and related vegetation index records indicate a generalized greening trend over most of the globe, in particular in high-latitude ecosystems, but also in semiarid regions, where the increase in atmospheric CO_2 concentrations are expected to alleviate water stress (Donohue et al., 2013; Piao et al., 2020), but may increase vulnerability to high temperatures (Obermeier et al., 2017). In semiarid ecosystems greening has been explained in part by woody encroachment, i.e., increase in tree and shrub cover replacing herbaceous vegetation (mainly in the Sahel and Southern Africa), but it should be noted that several semiarid regions also displayed "browning" trends with short vegetation being replaced by bare ground (Southwestern United States, Southern Argentina, Inland Asia, and Inland Australia), which indicates land

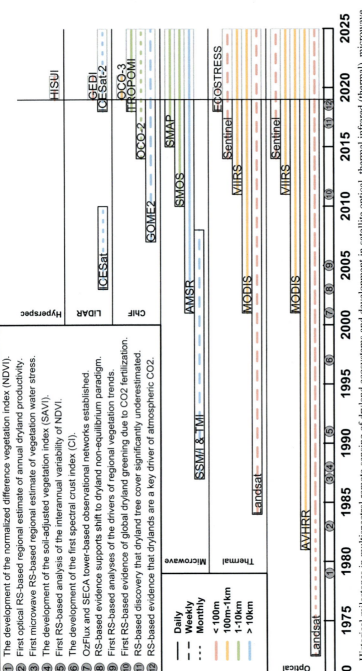

FIG. 4 Historical milestones in multispectral remote sensing of dryland ecosystems and development in satellite optical, thermal infrared (thermal), microwave, chlorophyll fluorescence (ChlF), light detection and ranging (LiDAR), and hyperspectral (Hyperspec) capabilities from Smith et al. (2019). Sensor timelines are provided for context and are not meant to provide a comprehensive overview of available sensors. The timeline showing the development of satellite capabilities from optical to hyperspectral demonstrates both the increasing spatiotemporal resolution of satellite data as well as the expanding diversity of remote sensing techniques. References associated with each historical milestone: (1) Tucker (1979), (2) Tucker, Vanpraet, Sharman, and Van Ittersum (1985), (3) Tucker and Choudhury (1987), (4) Huete (1988), (5) Tucker, Dregne, and Newcomb (1991), (6) Karnieli (1997), (7) Leuning, Cleugh, Zegelin, and Hughes (2005), Scott et al. (2008), and Scott et al. (2009), (8) Briske, Fuhlendorf, and Smeins (2003), (9) Anyamba and Tucker (2005), Donohue, Mcvicar, and Roderick (2009), and Fensholt and Rasmussen (2011), (10) Helldén and Tottrup (2008), Fensholt et al. (2012), and Donohue et al. (2013), (11) Bastin et al. (2017), (12) Poulter et al. (2014), Ahlström et al. (2015), Biederman et al. (2017), and Humphrey et al. (2018). (Seddon et al., 2016).

degradation or desertification (Song et al., 2018). The recently developed NIRv index (NDVI × NIR reflectance) has been found to be relatively less sensitive to soil background brightness and could thus significantly improve our ability to track long-term semiarid vegetation function. NIRv correlates closely with GPP from EC measurements across broad ecosystem types (Badgley, Field, & Berry, 2017) and has been recently used to scale up GPP from EC sites to the globe (Badgley, Anderegg, Berry, & Field, 2018). More detailed evaluations across semiarid regions continue to highlight the promise of this new surface reflectance-based index (Turner et al., 2020).

In addition to NDVI and related surface reflectance-based indices, satellite-based passive microwave measurements of vegetation optical depth (VOD) allow estimating vegetation water content (Tian et al., 2016) and in turn woody canopy cover density (Brandt et al., 2018). While the X-band VOD datasets saturate at high biomass values, the more recent L-band VOD measurements show a strong linear relationship with aboveground biomass carbon stocks, especially in the drylands (Brandt et al., 2018). Tian, Brandt, Liu, Rasmussen, and Fensholt (2017) have analyzed trends of the leaf and woody components of tropical drylands using VOD data, showing indeed an increase in woody cover in the Sahel and South Africa, mainly associated with shrub encroachment, and in eastern Australia, driven by changes in rainfall. Decreases are found in eastern Africa, the Gran Chaco region, driven by human activity, and in western Australia, linked to an increase in dry conditions. In some regions, such as the Gran Chaco and Southern India, where extensive cropland expansion over forest has occurred, VOD and NDVI show diverging trends, since NDVI does not capture canopy structure changes (Tian et al., 2017). Based on aboveground biomass estimates derived from L-band VOD measurements. Fan et al. (2019) estimated that semiarid biomes control about 55% of the interannual variability of carbon stocks in the tropical regions, supporting results from models and observation-based datasets indicating that semiarid ecosystems contribute significantly to the variability in the global land CO_2 sink (Ahlström et al., 2015; Bastos et al., 2013; Humphrey et al., 2018; Poulter et al., 2014).

Recent advances in the remote sensing of Sun-induced chlorophyll fluorescence (SIF) from satellites have proposed SIF as a promising approach to estimate global patterns of photosynthesis (Frankenberg et al., 2011). SIF is one of the pathways that light absorbed by the chloroplasts can follow, and has therefore been proposed to be directly related to photosynthesis (Guanter et al., 2014). However, the mechanistic link between satellite-based SIF and actual photosynthesis is complex and not yet well understood (Porcar-Castell et al., 2014; Wohlfahrt et al., 2018). As a proxy for GPP, SIF has shown improved performance to greenness-based indices (Wieneke et al., 2016), and comparable results to the NIRv. Specifically, SIF has shown good skill in capturing seasonal to interannual GPP of dryland ecosystems (Badgley et al., 2018; Smith et al., 2018). Moreover, SIF has shown promising results in constraining land—surface model estimates of global GPP (MacBean et al., 2018; Parazoo et al.,

2014). Colocation of SIF and thermal infrared (TIR) observations may further enable new insights into variability and trends in vegetation water use efficiency (GPP/ET)—a factor predicted to change most rapidly across dryland ecosystems (Stavros et al., 2017). Importantly, SIF is not always linearly related to GPP, for instance, Wohlfahrt et al. found that SIF only captured 35% of GPP variability during a heatwave in a Mediterranean forest (Wohlfahrt et al., 2018). More studies are currently needed to evaluate the ability of SIF to quantify GPP variability at interannual timescales (Smith et al., 2018), particularly in semiarid ecosystems that are frequently affected by drought and high-temperature conditions (Smith et al., 2019).

Additional emerging remote sensing techniques including hyperspectral imaging spectroscopy—which can provide new information on a variety of functional traits including foliar nitrogen, chlorophyll, and carotenoid concentrations—and LiDAR—which can be used to measure foliar vertical profiles, vegetation height, and total aboveground biomass—is rapidly providing new insights into semiarid ecosystem structure and physiological function (Jetz et al., 2016; Sankey et al., 2017; Schneider et al., 2017; Stavros et al., 2017). We expect that as these data become more widely available, integration of SIF, TIR, hyperspectral, and LiDAR techniques will lead to revolutionized understanding of semiarid ecosystem structural and functional dynamics under a changing climate (Fig. 4).

2.5   Land surface modeling of semiarid ecosystems

Carbon budgets in ecosystems can be additionally estimated by process-based land-surface models (LSMs) either forced by observed climate (offline) or integrated within Earth system models (ESMs). Land-surface models estimate the main water, energy, and biogeochemical exchanges between the surface and the atmosphere and the ecosystem processes driving these exchanges: photosynthesis, carbon allocation, growth, maintenance respiration and decomposition, and mortality. The LSMs contributing regularly to the global carbon budgets also simulate the effects of human activities on ecosystems' C balance, including fluxes from land-use/land-cover change, harvest, and some management practices. Some LSMs, but not all, simulate explicitly vegetation dynamics, fire disturbances, and nutrient cycling (Friedlingstein et al., 2019). None of the models integrating the global carbon budgets simulate lateral transport of C by water or wind, or the effects of other disturbances explicitly. However, developments have been made in including DOC and DIC fluxes in LSMs (Bowring et al., 2020; Camino-Serrano et al., 2018) as well as carbon fluxes due to water erosion (Naipal, Lauerwald, Ciais, Guenet, & Wang, 2020), though not specifically for semiarid regions.

LSMs consistently attribute most of the variability in the global carbon sink to semiarid ecosystems (Ahlström et al., 2015; Poulter et al., 2014; Tagesson et al., 2020), but different studies differ on their contribution to the global trend

(Ahlström et al., 2015; Tagesson et al., 2020). Models estimate that strong contribution to the global sink variability is mainly driven by variations in GPP to warm/dry and cool/wet conditions, largely controlled by ENSO (Bastos et al., 2013; Poulter et al., 2014), although they estimate stronger sensitivity of net and gross productivity to water availability than that reported by observation based-datasets (Bastos et al., 2018; Piao et al., 2013). The disagreements between modeled and observed GPP sensitivity to water may be due to biases in climate forcing (van Erik et al., 2018) or the simulated climate anomalies, when analyzing carbon fluxes from ESMs (Ahlström, Schurgers, Arneth, & Smith, 2012). van Erik et al. (2018) further showed that uncertainties in the sensitivity of GPP to soil-moisture anomalies during one El-Niño event can be related with the large uncertainty of plant rooting depth, especially in the seasonally dry rainforest. An important characteristic of semiarid ecosystems is the presence of plants with deep roots, which can tap into groundwater during the dry periods (Fan et al., 2017; Fan, Li, & Miguez-Macho, 2013), which is not well represented in LSMs. According to an estimate by Fan et al. (2017), in some semiarid regions rooting depth can reach 10–20 m, with the water-table depth reaching 20–40 m (Fan et al., 2013), while the LSMs typically simulate water-table depths up to 2–7 m and typically lack representation of groundwater. Additionally, many tall shrubs in semiarid regions have extensive shallow, lateral root structures that are not implemented in LSMs. Semiarid phenology schemes in LSMs have been adapted from their temperate and boreal counterparts; however, new models may be needed to simulate the complex plant strategies for maintaining growth during periods of water stress (Renwick et al., 2019).

2.6 Soil erosion

Soil erosion in semiarid regions occurs mainly in three different forms: by rainfall and runoff (water erosion), by wind (wind erosion), or by grazing, which promotes both water and wind erosion (Lawton & Wilke, 1979; Ravi, Breshears, Huxman, & D'Odorico, 2010). Even though precipitation rates are relatively low in semiarid regions, the low vegetation density, in combination with steep slopes in mountainous regions and occasional high-intensity storms make these regions vulnerable to water erosion. Wind erosion is one of the principal processes in semiarid regions and can be even the dominant form of soil erosion (Duniway et al., 2019; Ravi et al., 2010), leading to large dust emissions and carbon and nutrient losses and resulting in desertification. Both water and wind erosion can be measured; however, observations at larger spatial scales are missing. To estimate the overall extent of erosion for semiarid regions, models are needed. During the past few decades substantial work has been done on developing global water and wind erosion models, which have been extensively validated with observations (Borrelli et al., 2017; Chappell et al., 2019; Chappell, Baldock, & Sanderman, 2016; Naipal et al., 2020; Naipal, Reick, Pongratz, & Van Oost, 2015; Van Oost et al., 2007).

324 Section | C Case Studies

TABLE 1 Estimates of soil and soil C erosion from water for semiarid regions and the globe averaged over the period 2000–05 as estimated by CE-DYNAM-v1 (Naipal et al., 2018, 2020).

	Semiarid (Bsk and Bsh)			World (all climates)		
	Min	Default	Max	Min	Default	Max
Soil total (Pg)	2.00	2.91	4.23	28.73	37.67	54.76
Soil average ($t\,ha^{-1}$)	1.20	1.74	2.53	2.39	3.18	4.62
C total (TgC)	13.93	18.36	26.45	413.25	536.73	767.78
C average ($kg\,C\,ha^{-1}$)	8.64	11.65	16.79	35.60	46.41	66.38

Min, default and max are the minimum, maximum and default soil erosion scenario's as described in Naipal et al. (2018).

However, quantifying the effects of soil erosion on the C and nutrient cycles has been limited. In the rest of this section, we present some recent novel results on the effects of both water and wind erosion on the removal of soil C for semiarid regions.

Naipal et al. (2018) combined a global soil-erosion model with the process-based land surface model ORCHIDEE to estimate gross soil carbon losses from soil due to water erosion globally. The results for semiarid ecosystems are shown in Table 1.

Semiarid regions contributed between 6.96% and 7.72% to the total global water erosion rates, and between 3.37% and 3.44% to the total global C erosion rates during the period 2000–05. The C concentration in eroded soil is on average 0.67% for semiarid regions, while the global average is 1.46% for this time period. This is a result of the low soil organic carbon (SOC) stocks in especially low-latitude semiarid regions, while soil erosion rates are usually high here due to a higher rainfall erosivity.

Fig. 5 shows the spatial distribution of C erosion rates for the semiarid regions. Although the overall C erosion rates are much smaller compared to tropical and temperate regions, there are sites with high C erosion by water ($>20\,kg\,C\,ha^{-1}$), for example in mountainous areas, but also in low vegetation areas such as in Mongolia. The same observation can be made for wind erosion, where SOC removal by wind erosion is found to be generally small for drylands due to the low C content in the soils in these regions. However, C erosion rates by wind erosion can be large in certain areas, such as in Mongolia and in the Sahel (Chappell et al., 2019).

Fig. 6 shows the temporal variability in C erosion rates due to water erosion over the period 1850–2005 for a set of different sites in global semiarid regions.

Semiarid ecosystems **Chapter | 9** **325**

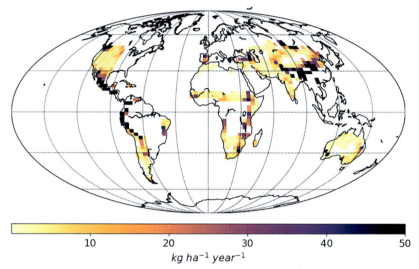

FIG. 5 C rates by water erosion for semiarid regions (units=kg ha^{-1} year^{-1}), as estimated by CE-DYNAM (Naipal et al., 2018).

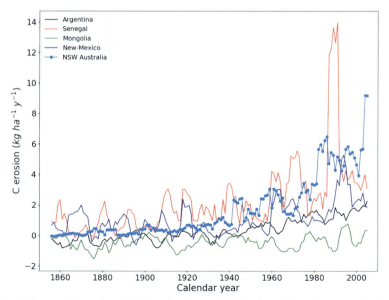

FIG. 6 Temporal dynamics of SOC erosion due to rainfall and runoff for the period 1850–2005 in selected countries. The values shown are 5-year averages.

The C erosion by water increases over the abovementioned period for most of the sites except for Mongolia, where a slight decrease is observed. The increase in C erosion is mostly due to the increased atmospheric CO_2 concentrations, which promote increased biomass C and soil C stocks leading to larger SOC erosion rates over time. This is especially the case for the tropical Senegal. For low vegetation areas like Mongolia, this is not the case; here, the soil erosion rates dominate both the short and long-term trends. Although the overall long-term trends of the different sites gradually increase, the short-term variability can be large due to large precipitation events. It should be noted that the erosion-driven feedbacks on the net primary productivity are not yet included in CE-DYNAM.

Chappell et al. (2019), who developed and applied a wind erosion model globally, also found increasing C erosion rates by wind for drylands. However, in contrast to water erosion, similar sites as in Fig. 6 under wind erosion show abrupt changes in SOC removal due to changes in soil surface shear stress or wind speed.

The above results indicate that lateral fluxes due to both water and wind erosion will play an important role in the near future in semiarid regions and are likely to increase under future climate change and land use change.

3   Future perspectives

As transition regions, semiarid regions are particularly susceptible to global change. As temperatures and atmospheric evaporative demand increase, semiarid and arid regions are prone to experience more severe droughts, with impacts on agricultural drought and ecosystems during periods without rainfall (Vicente-Serrano et al., 2020), although these impacts might be compensated to some extent by increased stomatal regulation of ecosystems adapted to periodic drought (Konings & Gentine, 2017). At the same time, several semiarid regions (Mediterranean, Central America, and southern Africa) are projected to receive lower rainfall under future climate scenarios, further contributing to the occurrence of droughts (Martin, 2018). The predicted increase in aridity in drylands and increase in extreme droughts and extreme precipitation events will not only increase soil erosion rates directly but also indirectly through shifts in vegetation (Feng & Fu, 2013; Ravi et al., 2010).

Quantifying the carbon budgets of semiarid regions is important to understand the dynamics of the carbon cycle, evaluate the resistance and resilience of these ecosystems to disturbances, and estimate their vulnerability to global change. Such understanding could help in guiding sustainable management practices and erosion control measures are needed to prevent the further increase in C losses in these regions, and to mitigate climate change by preserving soil C stocks and vegetation density.

Acknowledgment

This chapter is dedicated to the memory of Vanessa Haverd and her outstanding contributions to our understanding of semiarid ecosystems, the carbon cycle, and the Earth system.

References

Ahlström, A., Raupach, M. R., Schurgers, G., Smith, B., Arneth, A., Jung, M., et al. (2015). The dominant role of semi-arid ecosystems in the trend and variability of the land CO2 sink. *Science*, *348*(6237), 895–899.

Ahlström, A., Schurgers, G., Arneth, A., & Smith, B. (2012). Robustness and uncertainty in terrestrial ecosystem carbon response to CMIP5 climate change projections. *Environmental Research Letters*, *7*(4), 044008. http://stacks.iop.org/1748-9326/7/i=4/a=044008.

Anyamba, A., & Tucker, C. J. (2005). Analysis of Sahelian vegetation dynamics using NOAA-AVHRR NDVI data from 1981–2003. *Journal of Arid Environments*, *63*, 596–614. https://doi.org/10.1016/J.JARIDENV.2005.03.007.

Archer, S. R., Andersen, E. M., Predick, K. I., Schwinning, S., Steidl, R. J., & Woods, S. R. (2017). Woody plant encroachment: Causes and consequences. In *Rangeland systems* (pp. 25–84). Cham: Springer.

Ashok, K., Guan, Z., & Yamagata, T. (2003). Influence of the Indian Ocean dipole on the Australian winter rainfall. *Geophysical Research Letters*, *30*(15). https://doi.org/10.1029/2003GL017926, 1821.

Badgley, G., Anderegg, L. D., Berry, J. A., & Field, C. B. (2018). Terrestrial gross primary production: Using NIRV to scale from site to globe. *Global Change Biology*, *25*, 3731–3740.

Badgley, G., Field, C. B., & Berry, J. A. (2017). Canopy near-infrared reflectance and terrestrial photosynthesis. *Science Advances*, *3*(3). https://doi.org/10.1126/sciadv.1602244, e1602244.

Baldocchi, D. (2008). TURNER REVIEW No. 15 "Breathing" of the terrestrial biosphere: Lessons learned from a global network of carbon dioxide flux measurement systems. *Australian Journal of Botany*, *56*(1), 1–26. https://doi.org/10.1071/BT07151.

Baldocchi, D., Falge, E., Gu, L., Olson, R., Hollinger, D., Running, S., et al. (2001). FLUXNET: A new tool to study the temporal and spatial variability of ecosystem–scale carbon dioxide, water vapor, and energy flux densities. *Bulletin of the American Meteorological Society*, *82*(11), 2415–2434. https://doi.org/10.1175/1520-0477(2001)082<2415:FANTTS>2.3.CO;2.

Baldocchi, D., & Meyers, T. (1998). On using eco-physiological, micrometeorological and biogeochemical theory to evaluate carbon dioxide, water vapor and trace gas fluxes over vegetation: A perspective. *Agricultural and Forest Meteorology*, *90*(1-2), 1–25. https://doi.org/10.1016/S0168-1923(97)00072-5.

Barichivich, J., Briffa, K. R., Myneni, R. B., Osborn, T. J., Melvin, T. M., Ciais, P., et al. (2013). Large-scale variations in the vegetation growing season and annual cycle of atmospheric CO2 at high northern latitudes from 1950 to 2011. *Global Change Biology*, *19*(10), 3167–3183.

Barron-Gafford, G. A., Sanchez-Cañete, E. P., Minor, R. L., Hendryx, S. M., Lee, E., Sutter, L. F., et al. (2017). Impacts of hydraulic redistribution on grass-tree competition vs facilitation in a semi-arid savanna. *The New Phytologist*, *215*(4), 1451–1461. https://doi.org/10.1111/nph.14693.

Bastin, J., Berrahmouni, N., Grainger, A., Maniatis, D., Mollicone, D., Moore, R., et al. (2017). The extent of forest in dryland biomes. *Science*, *638*, 1–5. https://doi.org/10.1126/science.aam6527.

Bastos, A., Friedlingstein, P., Sitch, S., Chen, C., Mialon, A., Wigneron, J.-P., et al. (2018). Impact of the 2015/2016 El Niño on the terrestrial carbon cycle constrained by bottom-up and top-down

328 Section | C Case Studies

approaches. *Philosophical Transactions of the Royal Society, B: Biological Sciences, 373* (1760), 20170304.

Bastos, A., O'Sullivan, M., Ciais, P., Makowski, D., Sitch, S., Friedlingstein, P., et al. (2020). Sources of uncertainty in regional and global terrestrial CO_2-exchange estimates. *Global Biogeochemical Cycles, 34*. https://doi.org/10.1029/2019GB006393, e2019GB006393.

Bastos, A., Running, S. W., Gouveia, C., & Trigo, R. M. (2013). The global NPP dependence on ENSO: La Niña and the extraordinary year of 2011. *Journal of Geophysical Research – Biogeosciences, 118*, 1247–1255. https://doi.org/10.1002/jgrg.20100.

Beck, H. E., Zimmermann, N. E., McVicar, T. R., Vergopolan, N., Berg, A., & Wood, E. F. (2018). Present and future Köppen-Geiger climate classification maps at 1-km resolution. *Scientific Data, 5*(1), 180214. https://doi.org/10.1038/sdata.2018.214.

Biederman, J. A., Scott, R. L., Bell, T. W., Bowling, D. R., Dore, S., Garatuza-Payan, J., et al. (2017). CO_2 exchange and evapotranspiration across dryland ecosystems of southwestern North America. *Global Change Biology, 23*(10), 4204–4221. https://doi.org/10.1111/gcb.13686.

Borrelli, P., Robinson, D. A., Fleischer, L. R., Lugato, E., Ballabio, C., Alewell, C., et al. (2017). An assessment of the global impact of 21st century land use change on soil erosion. *Nature Communications, 8*(1), 2013. https://doi.org/10.1038/s41467-017-02142-7.

Bowring, S. P., Lauerwald, R., Guenet, B., Zhu, D., Guimberteau, M., Regnier, P., et al. (2020). ORCHIDEE MICT-LEAK (r5459), a global model for the production, transport, and transformation of dissolved organic carbon from Arctic permafrost regions–part 2: Model evaluation over the Lena River basin. *Geoscientific Model Development, 13*(2), 507–520. https://doi.org/10.5194/gmd-13-507-2020.

Brandt, M., Wigneron, J.-P., Chave, J., Tagesson, T., Penuelas, J., Ciais, P., et al. (2018). Satellite passive microwaves reveal recent climate-induced carbon losses in African drylands. *Nature Ecology & Evolution, 2*(5), 827.

Briske, D. D., Fuhlendorf, S. D., & Smeins, F. E. (2003). Vegetation dynamics on rangelands: a critique of the current paradigms. *Journal of Applied Ecology, 40*, 601–614. https://doi.org/10.1046/j.1365-2664.2003.00837.x.

Buermann, W., Forkel, M., O'Sullivan, M., Sitch, S., Friedlingstein, P., Haverd, V., et al. (2018). Widespread seasonal compensation effects of spring warming on northern plant productivity. *Nature, 562*(7725), 110.

Cable, J. M., Ogle, K., Williams, D. G., Weltzin, J. F., & Huxman, T. E. (2008). Soil texture drives responses of soil respiration to precipitation pulses in the Sonoran Desert: Implications for climate change. *Ecosystems, 11*(6), 961–979. https://doi.org/10.1007/s10021-008-9172-x.

Camino-Serrano, M., Guenet, B., Luyssaert, S., Ciais, P., Bastrikov, V., De Vos, B., et al. (2018). ORCHIDEE-SOM: Modeling soil organic carbon (SOC) and dissolved organic carbon (DOC) dynamics along vertical soil profiles in Europe. *Geoscientific Model Development, 11*(3), 937–957.

Cannell, G. H., & Weeks, L. V. (1979). Erosion and its control in semi-arid regions. In A. E. Hall, G. H. Cannell, & H. W. Lawton (Eds.), *Agriculture in semi-arid environments* (pp. 238–256). Berlin, Heidelberg: Springer. https://doi.org/10.1007/978-3-642-67328-3_10.

Chappell, A., Baldock, J., & Sanderman, J. (2016). The global significance of omitting soil erosion from soil organic carbon cycling schemes. *Nature Climate Change, 6*(2), 187–191. https://doi.org/10.1038/nclimate2829.

Chappell, A., Webb, N. P., Leys, J. F., Waters, C. M., Orgill, S., & Eyres, M. J. (2019). Minimising soil organic carbon erosion by wind is critical for land degradation neutrality. *Environmental Science & Policy, 93*, 43–52. https://doi.org/10.1016/j.envsci.2018.12.020.

Chevallier, F., Remaud, M., O'Dell, C. W., Baker, D., Peylin, P., & Cozic, A. (2019). Objective evaluation of surface-and satellite-driven carbon dioxide atmospheric inversions. *Atmospheric Chemistry and Physics*, *19*(22), 14233–14251. https://doi.org/10.5194/acp-19-14233-2019.

Cox, P. M., Pearson, D., Booth, B. B., Friedlingstein, P., Huntingford, C., Jones, C. D., et al. (2013). Sensitivity of tropical carbon to climate change constrained by carbon dioxide variability. *Nature*. https://doi.org/10.1038/nature11882. advance online publication.

Dannenberg, M. P., Wise, E. K., & Smith, W. K. (2019). Reduced tree growth in the semiarid United States due to asymmetric responses to intensifying precipitation extremes. *Science Advances*, *5*(10). https://doi.org/10.1126/sciadv.aaw0667, eaaw0667.

Donohue, R. J., Mcvicar, T. R., & Roderick, M. L. (2009). Climate-related trends in Australian vegetation cover as inferred from satellite observations, 1981–2006. *Global Change Biology*, *15*, 1025–1039. https://doi.org/10.1111/j.1365-2486.2008.01746.x.

Donohue, R. J., Roderick, M. L., McVicar, T. R., & Farquhar, G. D. (2013). Impact of CO_2 fertilization on maximum foliage cover across the globe's warm, arid environments. *Geophysical Research Letters*, *40*(12), 3031–3035. https://doi.org/10.1002/grl.50563.

Duniway, M. C., Pfennigwerth, A. A., Fick, S. E., Nauman, T. W., Belnap, J., & Barger, N. N. (2019). Wind erosion and dust from \textlessspan style="font-variant:Small-caps;"\textgreaterUS\textless/span\textgreater drylands: A review of causes, consequences, and solutions in a changing world. *Ecosphere*, *10*(3). https://doi.org/10.1002/ecs2.2650, e02650.

Eldridge, D. J., Reed, S., Travers, S. K., Bowker, M. A., Maestre, F. T., Ding, J., et al. (2020). The pervasive and multifaceted influence of biocrusts on water in the world's drylands. *Global Change Biology*, *26*(10), 6003–6014. https://doi.org/10.1111/gcb.15232.

Fan, Y., Li, H., & Miguez-Macho, G. (2013). Global patterns of groundwater table depth. *Science*, *339*(6122), 940–943. https://doi.org/10.1126/science.1229881.

Fan, Y., Miguez-Macho, G., Jobbágy, E. G., Jackson, R. B., & Otero-Casal, C. (2017). Hydrologic regulation of plant rooting depth. *Proceedings of the National Academy of Sciences*, *114*(40), 10572–10577.

Fan, L., Wigneron, J.-P., Ciais, P., Chave, J., Brandt, M., Fensholt, R., et al. (2019). Satellite-observed pantropical carbon dynamics. *Nature Plants*, *5*(9), 944–951.

Feng, S., & Fu, Q. (2013). Expansion of global drylands under a warming climate. *Atmospheric Chemistry and Physics*, *13*(19), 10081–10094. https://doi.org/10.5194/acp-13-10081-2013.

Fensholt, R., Langanke, T., Rasmussen, K., Reenberg, A., Prince, S. D., Tucker, C., et al. (2012). Greenness in semi-arid areas across the globe 1981–2007—An Earth Observing Satellite based analysis of trends and drivers. *Remote Sensing of Environment*, *121*, 144–158. https://doi.org/10.1016/j.rse.2012.01.017.

Fensholt, R., & Rasmussen, K. (2011). Analysis of trends in the Sahelian 'rain-use efficiency' using GIMMS NDVI, RFE and GPCP rainfall data. *Remote Sensing of Environment*, *115*, 438–451. https://doi.org/10.1016/J.RSE.2010.09.014.

Foken, T., Aubinet, M., & Leuning, R. (2012). The eddy covariance method. In *Eddy covariance* (pp. 1–19). Springer.

Frankenberg, C., Fisher, J. B., Worden, J., Badgley, G., Saatchi, S. S., Lee, J., et al. (2011). New global observations of the terrestrial carbon cycle from GOSAT: Patterns of plant fluorescence with gross primary productivity. *Geophysical Research Letters*, *38*(17). https://doi.org/10.1029/2011GL048738, L17706.

Friedlingstein, P., Jones, M., O'Sullivan, M., Andrew, R., Hauck, J., Peters, G., et al. (2019). Global carbon budget 2019. *Earth System Science Data*, *11*(4), 1783–1838.

Friedlingstein, P., Meinshausen, M., Arora, V. K., Jones, C. D., Anav, A., Liddicoat, S. K., et al. (2014). Uncertainties in CMIP5 climate projections due to carbon cycle feedbacks. *Journal of Climate*, *27*(2), 511–526. https://doi.org/10.1175/JCLI-D-12-00579.1.

330 Section | C Case Studies

Greve, P., Orlowsky, B., Mueller, B., Sheffield, J., Reichstein, M., & Seneviratne, S. I. (2014). Global assessment of trends in wetting and drying over land. *Nature Geoscience*, *7*(10), 716–721.

Guanter, L., Zhang, Y., Jung, M., Joiner, J., Voigt, M., Berry, J. A., et al. (2014). Global and time-resolved monitoring of crop photosynthesis with chlorophyll fluorescence. *Proceedings of the National Academy of Sciences*, *111*(14), E1327–E1333.

Haverd, V., Raupach, M. R., Briggs, P. R., Canadell, J. G., Davis, S. J., Law, R. M., et al. (2013). The Australian terrestrial carbon budget. *Biogeosciences*, *10*(2), 851–869. http://www.biogeosciences.net/10/851/2013/.

Helldén, U., & Tottrup, C. (2008). Regional desertification: a global synthesis. *Global and Planetary Change*, *64*, 169–176. https://doi.org/10.1016/J.GLOPLACHA.2008.10.006.

Huang, J., Ji, M., Xie, Y., Wang, S., He, Y., & Ran, J. (2016). Global semi-arid climate change over last 60 years. *Climate Dynamics*, *46*(3–4), 1131–1150.

Huete, A. R. (1988). A soil-adjusted vegetation index (SAVI). *Remote Sensing of Environment*, *25*, 295–309. https://doi.org/10.1016/0034-4257(88)90106-X.

Humphrey, V., Zscheischler, J., Ciais, P., Gudmundsson, L., Sitch, S., & Seneviratne, S. I. (2018). Sensitivity of atmospheric CO_2 growth rate to observed changes in terrestrial water storage. *Nature*, *560*(7720), 628.

Hutley, L. B., Leuning, R., Beringer, J., & Cleugh, H. A. (2005). The utility of the eddy covariance techniques as a tool in carbon accounting: Tropical savanna as a case study. *Australian Journal of Botany*, *53*(7), 663–675.

Huxman, T. E., Snyder, K. A., Tissue, D., Leffler, A. J., Ogle, K., Pockman, W. T., et al. (2004). Precipitation pulses and carbon fluxes in semiarid and arid ecosystems. *Oecologia*, *141*(2), 254–268. https://doi.org/10.1007/s00442-004-1682-4.

Janssens, I., Lankreijer, H., Matteucci, G., Kowalski, A., Buchmann, N., Epron, D., et al. (2001). Productivity overshadows temperature in determining soil and ecosystem respiration across European forests. *Global Change Biology*, *7*(3), 269–278.

Jetz, W., Cavender-Bares, J., Pavlick, R., Schimel, D., Davis, F. W., Asner, G. P., et al. (2016). Monitoring plant functional diversity from space. *Nature Plants*, *2*(3), 16024. https://doi.org/10.1038/nplants.2016.24.

Jung, M., Schwalm, C., Migliavacca, M., Walther, S., Camps-Valls, G., Koirala, S., et al. (2019). Scaling carbon fluxes from eddy covariance sites to globe: Synthesis and evaluation of the FLUXCOM approach. *Biogeosciences Discussions*, *2019*, 1–40. https://doi.org/10.5194/bg-2019-368.

Kaminski, T., & Heimann, M. (2001). Inverse modeling of atmospheric carbon dioxide fluxes. *Science*, *294*(5541), 259. http://www.sciencemag.org/content/294/5541/259.short.

Karnieli, A. (1997). Development and implementation of spectral crust index over dune sands. *International Journal of Remote Sensing*, *18*, 1207–1220. https://doi.org/10.1080/0143 11697218368.

Kondo, M., Patra, P. K., Sitch, S., Friedlingstein, P., Poulter, B., Chevallier, F., et al. (2020). State of the science in reconciling top-down and bottom-up approaches for terrestrial CO2 budget. *Global Change Biology*, *26*(3), 1068–1084.

Konings, A. G., & Gentine, P. (2017). Global variations in ecosystem-scale isohydricity. *Global Change Biology*, *23*(2), 891–905.

Kowalik, W. S., Marsh, S. E., & Lyon, R. J. P. (1982). A relation between landsat digital numbers, surface reflectance, and the cosine of the solar zenith angle. *Remote Sensing of Environment*, *12*(1), 39–55. https://doi.org/10.1016/0034-4257(82)90006-2.

Kriticos, D. J., Webber, B. L., Leriche, A., Ota, N., Macadam, I., Bathols, J., et al. (2012). CliMond: Global high-resolution historical and future scenario climate surfaces for bioclimatic modelling. *Methods in Ecology and Evolution*, *3*(1), 53–64.

Law, R., Rayner, P., & Wang, Y. (2004). Inversion of diurnally varying synthetic CO_2: Network optimization for an Australian test case. *Global Biogeochemical Cycles*, *18*(1), GB104410.1029/2003GB002136.

Lawton, P. J., & Wilke, H. W. (1979). Ancient agricultural systems in dry regions. In *Vol. 34. Agriculture in semi-arid environments* Springer. https://doi.org/10.1007/978-3-642-67328-3_1.

Leuning, R., Cleugh, H. A., Zegelin, S. J., & Hughes, D. (2005). Carbon and water fluxes over a temperate Eucalyptus forest and a tropical wet/dry savanna in Australia: measurements and comparison with MODIS remote sensing estimates. *Agricultural and Forest Meteorology*, *129*, 151–173. https://doi.org/10.1016/j.agrformet.2004.12.004.

Ma, X., Huete, A., Cleverly, J., Eamus, D., Chevallier, F., Joiner, J., et al. (2016). Drought rapidly diminishes the large net CO_2 uptake in 2011 over semi-arid Australia. *Scientific Reports*, *6*, 37747.

MacBean, N., Maignan, F., Bacour, C., Lewis, P., Peylin, P., Guanter, L., et al. (2018). Strong constraint on modelled global carbon uptake using solar-induced chlorophyll fluorescence data. *Scientific Reports*, *8*(1), 1973.

Marsh, S. E., & Lyon, R. J. P. (1980). Quantitative relationships of near-surface spectra to Landsat radiometric data. *Remote Sensing of Environment*, *10*(4), 241–261. https://doi.org/10.1016/0034-4257(80)90085-1.

Martin, E. R. (2018). Future projections of global pluvial and drought event characteristics. *Geophysical Research Letters*, *45*(21), 11,913–11,920. https://doi.org/10.1029/2018GL079807.

Matson, P., & Goldstein, A. (2000). Biogenic trace gas exchanges. In *Methods in ecosystem science* (pp. 235–248). Springer.

Matteucci, G. (1998). *Bilancio del carbonio in una faggeta dell'Italia Centro-Meridionale: Determinanti ecofisiolo-gici, integrazione a livello di copertura e simulazione dell'impatto dei cambiamenti ambientali*. Padova: Universita Degli Studi Di Padova.

Mitchard, E. T., & Flintrop, C. M. (2013). Woody encroachment and forest degradation in sub-Saharan Africa's woodlands and savannas 1982–2006. *Philosophical Transactions of the Royal Society, B: Biological Sciences*, *368*(1625), 20120406.

Myklebust, M., Hipps, L., & Ryel, R. J. (2008). Comparison of eddy covariance, chamber, and gradient methods of measuring soil CO2 efflux in an annual semi-arid grass, Bromus tectorum. *Agricultural and Forest Meteorology*, *148*(11), 1894–1907.

Myneni, R., Maggion, S., Iaquinta, J., Privette, J., Gobron, N., Pinty, B., et al. (1995). Optical remote sensing of vegetation: Modeling, caveats, and algorithms. *Remote Sensing of Environment*, *51*(1), 169–188.

Naipal, V., Ciais, P., Wang, Y., Lauerwald, R., Guenet, B., & Van Oost, K. (2018). Global soil organic carbon removal by water erosion under climate change and land use change during AD 1850–2005. *Biogeosciences*, *15*(14), 4459–4480. https://doi.org/10.5194/bg-15-4459-2018.

Naipal, V., Lauerwald, R., Ciais, P., Guenet, B., & Wang, Y. (2020). CE-DYNAM (v1): A spatially explicit process-based carbon erosion scheme for use in earth system models. *Geoscientific Model Development*, *13*(3), 1201–1222.

Naipal, V., Reick, C., Pongratz, J., & Van Oost, K. (2015). Improving the global applicability of the RUSLE model—Adjustment of the topographical and rainfall erosivity factors. *Geoscientific Model Development Discussion*, *8*(3), 2991–3035. https://doi.org/10.5194/gmdd-8-2991-2015.

332 Section | C Case Studies

Nemani, R. R., Keeling, C. D., Hashimoto, H., Jolly, W. M., Piper, S. C., Tucker, C. J., et al. (2003). Climate-driven increases in global terrestrial net primary production from 1982 to 1999. *Science*, *300*(5625), 1560–1563. https://doi.org/10.1126/science.1082750.

Noy-Meir, I. (1973). Desert ecosystems: Environment and producers. *Annual Review of Ecology and Systematics*, *4*(1), 25–51. https://doi.org/10.1146/annurev.es.04.110173.000325.

Obermeier, W., Lehnert, L., Kammann, C., Müller, C., Grünhage, L., Luterbacher, J., et al. (2017). Reduced CO_2 fertilization effect in temperate C3 grasslands under more extreme weather conditions. *Nature Climate Change*, *7*(2), 137.

Palmer, P. I., Feng, L., Baker, D., Chevallier, F., Bösch, H., & Somkuti, P. (2019). Net carbon emissions from African biosphere dominate pan-tropical atmospheric CO 2 signal. *Nature Communications*, *10*(1), 1–9.

Parazoo, N. C., Bowman, K., Fisher, J. B., Frankenberg, C., Jones, D. B. A., Cescatti, A., et al. (2014). Terrestrial gross primary production inferred from satellite fluorescence and vegetation models. *Global Change Biology*, *20*(10), 3103–3121. https://doi.org/10.1111/gcb.12652.

Pastorello, G., Papale, D., Chu, H., Trotta, C., Agarwal, D., Canfora, E., et al. (2017). A new data set to keep a sharper eye on land-air exchanges. *Eos, Transactions American Geophysical Union*, *98*(8). https://doi.org/10.1029/2017EO071597 (Online).

Peel, M. C., Finlayson, B. L., & McMahon, T. A. (2007). Updated world map of the Köppen-Geiger climate classification. *Hydrology and Earth System Sciences*, *11*, 1633–1644. https://doi.org/10.5194/hess-11-1633-2007.

Pellegrini, A. F. A., Anderegg, W. R. L., Paine, C. E. T., Hoffmann, W. A., Kartzinel, T., Rabin, S. S., et al. (2017). Convergence of bark investment according to fire and climate structures ecosystem vulnerability to future change. *Ecology Letters*, *20*(3), 307–316. https://doi.org/10.1111/ele.12725.

Peylin, P., Law, R. M., Gurney, K. R., Chevallier, F., Jacobson, A. R., Maki, T., et al. (2013). Global atmospheric carbon budget: Results from an ensemble of atmospheric CO_2 inversions. *Biogeosciences*, *10*(10), 6699–6720. https://doi.org/10.5194/bg-10-6699-2013.

Piao, S., Nan, H., Huntingford, C., Ciais, P., Friedlingstein, P., Sitch, S., et al. (2014). Evidence for a weakening relationship between interannual temperature variability and northern vegetation activity. *Nature Communications*, *5*. https://doi.org/10.1038/ncomms6018.

Piao, S., Sitch, S., Ciais, P., Friedlingstein, P., Peylin, P., Wang, X., et al. (2013). Evaluation of terrestrial carbon cycle models for their response to climate variability and to CO2 trends. *Global Change Biology*, *19*(7), 2117–2132. https://doi.org/10.1111/gcb.12187.

Piao, S., Wang, X., Wang, K., Li, X., Bastos, A., Canadell, J. G., et al. (2020). Interannual variation of terrestrial carbon cycle: Issues and perspectives. *Global Change Biology*, *26*(1), 300–318.

Porcar-Castell, A., Tyystjärvi, E., Atherton, J., van der Tol, C., Flexas, J., Pfündel, E. E., et al. (2014). Linking chlorophyll a fluorescence to photosynthesis for remote sensing applications: Mechanisms and challenges. *Journal of Experimental Botany*, *65*(15), 4065–4095. https://doi.org/10.1093/jxb/eru191.

Poulter, B., Frank, D., Ciais, P., Myneni, R. B., Andela, N., Bi, J., et al. (2014). Contribution of semi-arid ecosystems to interannual variability of the global carbon cycle. *Nature*. https://doi.org/10.1038/nature13376. advance online publication.

Ravi, S., Breshears, D. D., Huxman, T. E., & D'Odorico, P. (2010). Land degradation in drylands: Interactions among hydrologic–aeolian erosion and vegetation dynamics. *Geomorphology*, *116* (3–4), 236–245. https://doi.org/10.1016/j.geomorph.2009.11.023.

Renwick, K. M., Fellows, A., Flerchinger, G. N., Lohse, K. A., Clark, P. E., Smith, W. K., et al. (2019). Modeling phenological controls on carbon dynamics in dryland sagebrush ecosystems. *Agricultural and Forest Meteorology*, *274*, 85–94. https://doi.org/10.1016/j.agrformet.2019.04.003.

Reynolds, J. F., Smith, D. M. S., Lambin, E. F., Turner, B. L., Mortimore, M., Batterbury, S. P. J., et al. (2007). Global desertification: Building a science for dryland development. *Science*, *316* (5826), 847–851. https://doi.org/10.1126/science.1131634.

Riederer, M., Serafimovich, A., & Foken, T. (2014). Net ecosystem CO_2 exchange measurements by the closed chamber method and the eddy covariance technique and their dependence on atmospheric conditions. *Atmospheric Measurement Techniques*, *7*(4), 1057–1064.

Rodríguez-Caballero, E., Castro, A. J., Chamizo, S., Quintas-Soriano, C., Garcia-Llorente, M., Cantón, Y., et al. (2018). Ecosystem services provided by biocrusts: From ecosystem functions to social values. *Ecosystem Services in Dryland Systems of the World*, *159*, 45–53. https://doi.org/10.1016/j.jaridenv.2017.09.005.

Rouse, J. W., Haas, R. H., Schell, J. A., Deering, D. W., & Harlan, J. C. (1974). *Monitoring the vernal advancement and retrogradation (greenwave effect) of natural vegetation. NASA/GSFC type III final report.*

Ruesch, A., & Gibbs, H. K. (2008). *New IPCC Tier-1 global biomass carbon map for the year 2000. Available online from the carbon dioxide information analysis Center [Http://Cdiac.Ornl.Gov].* Oak Ridge, TN: Oak Ridge National Laboratory.

Sankey, T. T., McVay, J., Swetnam, T. L., McClaran, M. P., Heilman, P., & Nichols, M. (2017). UAV hyperspectral and lidar data and their fusion for arid and semi-arid land vegetation monitoring. *Remote Sensing in Ecology and Conservation*. https://doi.org/10.1002/rse2.44.

Schneider, F. D., Morsdorf, F., Schmid, B., Petchey, O. L., Hueni, A., Schimel, D. S., et al. (2017). Mapping functional diversity from remotely sensed morphological and physiological forest traits. *Nature Communications*, *8*(1). https://doi.org/10.1038/s41467-017-01530-3.

Scott, R. L., Cable, W. L., Huxman, T. E., Nagler, P. L., Hernandez, M., & Goodrich, D. C. (2008). Multiyear riparian evapotranspiration and groundwater use for a semiarid watershed. *Journal of Arid Environments*, *72*, 1232–1246. https://doi.org/10.1016/j.jaridenv.2008.01.001.

Scott, R. L., Jenerette, G. D., Potts, D. L., & Huxman, T. E. (2009). Effects of seasonal drought on net carbon dioxide exchange from a woody-plant-encroached semiarid grassland. *Journal of Geophysical Research*, *114*. https://doi.org/10.1029/2008JG000900, G04004.

Seddon, A. W., Macias-Fauria, M., Long, P. R., Benz, D., & Willis, K. J. (2016). Sensitivity of global terrestrial ecosystems to climate variability. *Nature*, *531*(7593), 229–232.

Smith, W. K. (1978). Temperatures of desert plants: Another perspective on the adaptability of leaf size. *Science*, *201*(4356), 614. https://doi.org/10.1126/science.201.4356.614.

Smith, W. K., Biederman, J. A., Scott, R. L., Moore, D. J. P., He, M., Kimball, J. S., et al. (2018). Chlorophyll fluorescence better captures seasonal and interannual gross primary productivity dynamics across dryland ecosystems of Southwestern North America. *Geophysical Research Letters*, *45*(2), 748–757. https://doi.org/10.1002/2017GL075922.

Smith, W. K., Dannenberg, M. P., Yan, D., Herrmann, S., Barnes, M. L., Barron-Gafford, G. A., et al. (2019). Remote sensing of dryland ecosystem structure and function: Progress, challenges, and opportunities. *Remote Sensing of Environment*, *233*, 111401. https://doi.org/10.1016/j.rse.2019.111401.

Smith, W. K., Reed, S. C., Cleveland, C. C., Ballantyne, A. P., Anderegg, W. R., Wieder, W. R., et al. (2016). Large divergence of satellite and earth system model estimates of global terrestrial CO2 fertilization. *Nature Climate Change*, *6*(3), 306–310.

Song, X.-P., Hansen, M. C., Stehman, S. V., Potapov, P. V., Tyukavina, A., Vermote, E. F., et al. (2018). Global land change from 1982 to 2016. *Nature*, *560*, 639–643. https://doi.org/10.1038/s41586-018-0411-9.

Staver, A. C., Archibald, S., & Levin, S. A. (2011). The global extent and determinants of Savanna and forest as alternative biome states. *Science*, *334*(6053), 230–232. https://doi.org/10.1126/science.1210465.

334 Section | C Case Studies

Stavros, E. N., Schimel, D., Pavlick, R., Serbin, S., Swann, A., Duncanson, L., et al. (2017). ISS observations offer insights into plant function. *Nature Ecology and Evolution, 1*(7), 1–4. https://doi.org/10.1038/s41559-017-0194.

Stewart, B., Unger, P., & Jones, O. (1985). Soil and water conservation in semiarid regions. *Soil erosion and conservation*. Ankeny, IA: Soil Science Society of America.

Swetnam, T. L., Brooks, P. D., Barnard, H. R., Harpold, A. A., & Gallo, E. L. (2017). Topographically driven differences in energy and water constrain climatic control on forest carbon sequestration. *Ecosphere, 8*(4). https://doi.org/10.1002/ecs2.1797, e01797.

Tagesson, T., Schurgers, G., Horion, S., Ciais, P., Tian, F., Brandt, M., et al. (2020). Recent divergence in the contributions of tropical and boreal forests to the terrestrial carbon sink. *Nature Ecology & Evolution, 4*, 202–209. https://doi.org/10.1038/s41559-019-1090-0.

Tang, J., & Baldocchi, D. D. (2005). Spatial–temporal variation in soil respiration in an oak–grass savanna ecosystem in California and its partitioning into autotrophic and heterotrophic components. *Biogeochemistry, 73*(1), 183–207.

Tian, F., Brandt, M., Liu, Y. Y., Rasmussen, K., & Fensholt, R. (2017). Mapping gains and losses in woody vegetation across global tropical drylands. *Global Change Biology, 23*(4), 1748–1760.

Tian, F., Brandt, M., Liu, Y. Y., Verger, A., Tagesson, T., Diouf, A. A., et al. (2016). Remote sensing of vegetation dynamics in drylands: Evaluating vegetation optical depth (VOD) using AVHRR NDVI and in situ green biomass data over west African Sahel. *Remote Sensing of Environment, 177*, 265–276. https://doi.org/10.1016/j.rse.2016.02.056.

Tramontana, G., Jung, M., Schwalm, C. R., Ichii, K., Camps-Valls, G., Ráduly, B., et al. (2016). Predicting carbon dioxide and energy fluxes across global FLUXNET sites with regression algorithms. *Biogeosciences, 13*, 4291–4313. https://doi.org/10.5194/bg-13-4291-2016.

Tucker, C. J. (1979). Red and photographic infrared linear combinations for monitoring vegetation. *Remote Sensing of Environment, 8*(2), 127–150.

Tucker, C., Antoninka, A., Day, N., Poff, B., & Reed, S. (2020). Biological soil crust salvage for dryland restoration: An opportunity for natural resource restoration. *Restoration Ecology, 28* (S2), S9–S16. https://doi.org/10.1111/rec.13115.

Tucker, C., Fung, I. Y., Keeling, C., & Gammon, R. (1986). Relationship between atmospheric CO_2 variations and a satellite-derived vegetation index. *Nature, 319*(6050), 195–199.

Tucker, C. J., & Choudhury, B. J. (1987). Satellite remote sensing of drought conditions. *Remote Sensing of Environment, 23*(2), 243–251. https://doi.org/10.1016/0034-4257(87)90040-X.

Tucker, C. J., Dregne, H. E., & Newcomb, W. W. (1991). Expansion and contraction of the sahara desert from 1980 to 1990. *Science, 253*, 299–300. https://doi.org/10.1126/science. 253.5017.299.

Tucker, C. J., Vanpraet, C. L., Sharman, M. J., & Van Ittersum, G. (1985). Satellite remote sensing of total herbaceous biomass production in the Senegalese Sahel: 1980–1984. *Remote Sensing of Environment, 17*, 233–249. https://doi.org/10.1016/0034-4257(85) 90097-5.

Tucker, C. L., McHugh, T. A., Howell, A., Gill, R., Weber, B., Belnap, J., et al. (2017). The concurrent use of novel soil surface microclimate measurements to evaluate CO2 pulses in biocrusted interspaces in a cool desert ecosystem. *Biogeochemistry, 135*(3), 239–249. https://doi.org/10.1007/s10533-017-0372-3.

Turner, A. J., Köhler, P., Magney, T. S., Frankenberg, C., Fung, I., & Cohen, R. C. (2020). A double peak in the seasonality of California's photosynthesis as observed from space. *Biogeosciences, 17*(2), 405–422. https://doi.org/10.5194/bg-17-405-2020.

Valentini, R., Matteucci, G., Dolman, A., Schulze, E.-D., Rebmann, C., Moors, E., et al. (2000). Respiration as the main determinant of carbon balance in European forests. *Nature, 404* (6780), 861–865.

Semiarid ecosystems **Chapter | 9** **335**

van Erik, S., Lars, K., Smith Naomi, E., Gerbrand, K., van Beek, L. P. H., Wouter, P., et al. (2018). Changes in surface hydrology, soil moisture and gross primary production in the Amazon during the 2015/2016 El Niño. *Philosophical Transactions of the Royal Society, B: Biological Sciences, 373*(1760), 20180084. https://doi.org/10.1098/rstb.2018.0084.

Van Gorsel, E., Leuning, R., Cleugh, H. A., Keith, H., & Suni, T. (2007). Nocturnal carbon efflux: Reconciliation of eddy covariance and chamber measurements using an alternative to the u.-threshold filtering technique. *Tellus Series B: Chemical and Physical Meteorology, 59*(3), 397–403.

Van Oost, K., Quine, T. A., Govers, G., De Gryze, S., Six, J., Harden, J. W., et al. (2007). The impact of agricultural soil Erosion on the global carbon cycle. *Science, 318*(5850), 626–629. https://doi.org/10.1126/science.1145724.

Vicente-Serrano, S. M., McVicar, T. R., Miralles, D. G., Yang, Y., & Tomas-Burguera, M. (2020). Unraveling the influence of atmospheric evaporative demand on drought and its response to climate change. *Wiley Interdisciplinary Reviews: Climate Change, 11*(2), e632.

Wang, N., Quesada, B., Xia, L., Butterbach-Bahl, K., Goodale, C. L., & Kiese, R. (2019). Effects of climate warming on carbon fluxes in grasslands—A global meta-analysis. *Global Change Biology, 25*(5), 1839–1851. https://doi.org/10.1111/gcb.14603.

Wieneke, S., Ahrends, H., Damm, A., Pinto, F., Stadler, A., Rossini, M., et al. (2016). Airborne based spectroscopy of red and far-red sun-induced chlorophyll fluorescence: Implications for improved estimates of gross primary productivity. *Remote Sensing of Environment, 184*, 654–667.

Williams, C. A., Hanan, N. P., Neff, J. C., Scholes, R. J., Berry, J. A., Denning, A. S., et al. (2007). Africa and the global carbon cycle. *Carbon Balance and Management, 2*(1), 3.

Wohlfahrt, G., Gerdel, K., Migliavacca, M., Rotenberg, E., Tatarinov, F., Müller, J., et al. (2018). Sun-induced fluorescence and gross primary productivity during a heat wave. *Scientific Reports, 8*(1), 14169. https://doi.org/10.1038/s41598-018-32602-z.

Zhao, M., Running, S. W., & Nemani, R. R. (2006). Sensitivity of moderate resolution imaging Spectroradiometer (MODIS) terrestrial primary production to the accuracy of meteorological reanalyses. *Journal of Geophysical Research – Biogeosciences, 111*(G1). https://doi.org/10.1029/2004JG000004, G01002.

Zhou, L., Tian, Y., Myneni, R. B., Ciais, P., Saatchi, S., Liu, Y. Y., et al. (2014). Widespread decline of Congo rainforest greenness in the past decade. *Nature, 509*, 86–90.

Zhou, L., Tucker, C. J., Kaufmann, R. K., Slayback, D., Shabanov, N. V., & Myneni, R. B. (2001). Variations in northern vegetation activity inferred from satellite data of vegetation index during 1981 to 1999. *Journal of Geophysical Research, 106*(D17), 20069–20083. https://doi.org/10.1029/2000JD000115.

Zscheischler, J., Mahecha, M. D., Avitabile, V., Calle, L., Carvalhais, N., Ciais, P., et al. (2017). Reviews and syntheses: An empirical spatiotemporal description of the global surface-atmosphere carbon fluxes: Opportunities and data limitations. *Biogeosciences, 14*(15), 3685–3703.

Chapter 10

Urban environments and trans-boundary linkages

Kangkang Tong[a] and Anu Ramaswami[a,b,c]

[a]China-UK Low Carbon College, Shanghai Jiao Tong University, Pudong New District, Shanghai, China, [b]High Meadows Environmental Institute, Princeton University, Princeton, NJ, United States, [c]M.S. Chadha Center for Global India, Princeton University, Princeton, NJ, United States

1 From science to policy for urban carbon accounting

Globally, cities house about 55% of the world's population and produce about 80% of GDP (UN-Habitat, 2016). Meanwhile, consumption, production, and transportation activities in cities contribute to about 70% of global anthropogenic greenhouse gas (GHG) emissions (UN-Habitat, 2016). The global urban population is predicted to increase by 50% in the next 30 years, from about 4.2 billion currently to 6.6 billion in 2050 (United Nations, Department of Economic and Social Affairs, Population division, 2015). The contribution of cities to global GHG emissions is expected to increase due to this ongoing urbanization if the emission intensity associated with urban activities remains the same. Meanwhile, about 60% of urban areas that will exist in 2050 have yet to be built (UNEP, 2013), which signals that low-carbon urban development is a critical opportunity for reducing global GHG emissions.

Cities have begun to integrate low-carbon development into their agendas to address climate change. For example, in 2005, the U.S. Conference of Mayors issued the Climate Protection Agreement to prioritize GHG reduction (U.S. Conference of Mayors, 2005). The World Mayors Council on Climate Change issued "The Mexico City Pact" in 2010 (World Mayors Summit on Climate, 2010), which demonstrates the consensus among cities worldwide on voluntarily reducing carbon emissions. In addition, global nongovernmental organizations, e.g., the World Resources Institute (WRI), ICLEI-Local Governments for Sustainability (ICLEI), and the C40 Cities Climate Leadership Group (C40), actively work with cities, national governments, and researchers to develop GHG inventory protocols specifically supporting urban climate policies.

Urban climate policies range from behavior nudges, improving energy (e.g., electricity and fossil fuel) efficiency in buildings regardless of where energy is

Balancing Greenhouse Gas Budgets. https://doi.org/10.1016/B978-0-12-814952-2.00005-8
Copyright © 2022 Elsevier Inc. All rights reserved.

generated, travel demand reduction, waste-to-energy (e.g., food waste to energy), and local renewable energy adoption. These urban policies are linked with national deep-decarbonization pathways, such as zero-carbon electricity grid, electrification, and efficiency improvement (Deep-Decarbonization Pathways Project, 2015). These urban policies collectively have the potential to reduce GHG emissions globally, in addition to reducing emissions within the city. This indicates urban climate actions and policies require a systems approach to evaluate their local and trans-boundary mitigation effects.

During the past decade, the science of urban carbon accounting has been advanced in response to policy considerations. Four broad carbon accounting approaches have emerged at the city scale. Atmospheric and climate scientists follow the IPCC national carbon accounting guidelines (IPCC, 2006) to account for carbon emissions from each source within a city's boundary. This is a source-based approach and is called the *purely territorial carbon accounting (PTA) approach*. However, this approach does not recognize that cities rely heavily on goods and service supply chains that extend well beyond the city boundary (Kennedy et al., 2009; Ramaswami, Hillman, Janson, Reiner, & Thomas, 2008). Horizontal material and energy flows that are critical for urban activities result in GHG emissions elsewhere. Therefore, a second broad approach, *community-wide trans-boundary infrastructure supply-chain carbon footprinting (CIF)*, quantifies GHG emissions associated with seven key infrastructure and food provisioning sectors (i.e., energy, transportation and communication, water, waste, shelter materials, public/green spaces, and food). These key provisioning sectors have been shown to globally contribute to about 90% of anthropogenic GHGs, as well as premature mortality (Ramaswami, Russell, Culligan, Sharma, & Kumar, 2016). The CIF approach accounts for both direct carbon emissions and emissions embodied in the supply chain of these key infrastructural provisioning sectors in a city. The third broad approach, *consumption-based footprinting (CBF)*, considers the consumption of other services and goods primarily by households in addition to the key infrastructure sectors (Larsen & Hertwich, 2009). CBF is not a community-wide approach because it excludes GHGs from local industries and businesses that export goods and services elsewhere. The last broad approach, *total community-wide carbon footprinting (TCF)*, quantifies direct carbon emissions from both infrastructural and noninfrastructural use activities in cities as well as embodied emissions in their supply chains (Chavez & Ramaswami, 2013).

The differences and overlaps of these accounting approaches have been discussed in many scientific articles (Chavez & Ramaswami, 2013; Ramaswami & Chavez, 2013). Increasingly, GHG accounting protocols developed for cities, such as ICLEI-USA's community protocol and the British Standards Institution (BSI)'s Publicly Available Specification (PAS) 2070, recognize the value of multiple accounting perspectives, because they inform different policies and actions for different social actors. In this chapter, we highlight the *value* that each approach offers to *urban climate policies*, emphasizing that each approach

Urban environments and trans-boundary linkages Chapter | 10 **339**

is complete for its designed purposes and is valuable for different stakeholders (e.g., homes, businesses, industry, local government, etc.).

The four city carbon accounting approaches mentioned above have primarily focused on GHGs associated with fossil fuel combustion, waste management, and industrial processes, as well as agricultural activities when applicable. These approaches have less emphasis on vertical biogenic carbon from urban green spaces and are related to land-use change. Recently, scholars and city networks have advocated for nature-based solutions to take advantage of address the carbon sequestration potential of urban green space. Yet science on the contribution of urban green space and land-use change to cities' carbon budgets is still nascent. In Section 3, emerging efforts to quantify biogenic carbon fluxes from urban areas are discussed, including current approaches for estimating biogenic carbon emissions and their implications for urban policies.

Another emerging research area is how to align urban emissions with the national total to guide effective multilevel collaboration for local and national GHG mitigation. For example, Gurney et al. applied the PTA approach to create a fossil fuel carbon emission inventory for all urban areas in the United States (Gurney et al., 2009). Ramaswami et al. developed and implemented an innovative methodology using the CIF approach to construct a carbon footprint database focusing on the energy use in all urban areas in China (Ramaswami et al., 2017a). This database was used to quantify the collective carbon mitigation potential of new cross-sectoral urban mitigation strategies at the national level (Ramaswami et al., 2017a). Moran et al. downscaled household carbon footprints using the CBF approach from a national/subnational dataset of 13,000 cities globally based on population density and a purchasing power index (Moran et al., 2018). In the future, new data sources and data analytic techniques (e.g., machine learning, deep learning, artificial intelligence, etc.) will contribute to further developing policy-relevant databases covering all urban areas nationally or globally (Ramaswami et al., 2021).

This chapter provides a review of these broad urban carbon accounting approaches for individual cities, including their policy relevance and benchmarking metrics in Section 2. We discuss emerging efforts for quantifying biogenic carbon from urban land use and land-use change in Section 3. Lastly, we summarize new initiatives for carbon accounting for all urban areas/cities in a nation or globally, as well as how new data and analysis techniques can contribute to these initiatives.

2 Four carbon accounting approaches for individual cities

This section details four broad carbon accounting approaches for individual cities: (1) the purely territorial GHG accounting approach (PTA), (2) the community-wide trans-boundary infrastructure supply-chain carbon footprinting approach (CIF), (3) the consumption-based carbon footprinting approach (CBF), and (4) the total community-wide carbon footprinting approach

340 Section | C Case Studies

(TCF). These four approaches are not mutually exclusive, as they account for GHG emissions associated with different activities in cities (see Fig. 1). Each approach provides different benchmarking metrics to track the effectiveness of *urban climate policies* (see Box 1–4). The analytical steps and data required for each approach are discussed next, as well as uncertainties in data.

Not all urban mitigation actions reduce direct carbon emissions, and not all direct emissions are related to urban mitigation policies. This is a science-policy conundrum for urban carbon accounting approaches in informing urban climate policies. For example, cities can reduce electricity use in buildings through efficiency rebates, but the carbon reduction of this policy happens both inside and outside of a city. Thus, it is necessary to clarify the value that each approach provides to urban climate policies. The *value* of each approach to *urban climate policies* is evaluated based on the following three criteria:

- what it offers to systemic urban climate policies;
- whether it captures the direction (increase or decrease) of emission change associated with urban climate policies; and
- what uncertainties exist in the results of each approach and the sensitivity to urban policies.

The following sections focus on emissions related to fossil fuel use, waste management, industrial production, and agrifood. GHG species in these accounting approaches primarily include anthropogenic carbon dioxide (CO_2), methane (CH_4), and nitrous oxide (N_2O). In Section 3, we discuss accounting approaches for biogenic carbon from urban green spaces and related to land-use change.

2.1 Purely territorial carbon accounting approaches

Direct GHG emissions from urban areas can be quantified either by the urban metabolic inventory method or through atmospheric science-based measurement. The urban metabolic method, following the IPCC national GHG accounting guidelines, focuses on emission sources within a city boundary. Cities have implemented this method to estimate GHG emissions (see Box 1). Researchers have also implemented this method to construct a gridded urban fossil fuel CO_2 emission dataset with high spatial and temporal resolution (Gurney et al., 2019). Meanwhile, atmospheric scientists have developed a method to infer fossil fuel carbon emission flux from urban areas based on CO_2 concentration. This atmospheric science-based approach has been tested by scientists in several cities to explore the application of this method at the city scale (Box 1).

Method overview of territorial urban metabolic inventory method: The territorial urban metabolic inventory method focuses on sources that directly release GHGs in a city. The overall analytical procedure is to implement a material and energy flow analysis for individual cities and apply emission factors to these sources resulting in direct GHG emissions from cities. In-boundary

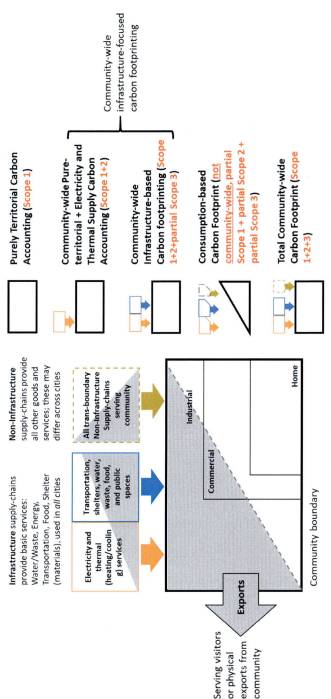

FIG. 1 Relationships among four carbon accounting approaches: purely territorial GHG accounting (PTA), community-wide infrastructure-based carbon footprinting (CIF), consumption-based carbon footprinting (CBF), and total community-wide carbon footprinting (TCF). *(Figure revised from Chavez, A., & Ramaswami, A. (2013). Articulating a trans-boundary infrastructure supply chain greenhouse gas emission footprint for cities: Mathematical relationships and policy relevance. Energy Policy, 54, 376–384. https://doi.org/10.1016/j.enpol.2012.10.037.)*

342 Section | C Case Studies

> **BOX 1 Purely territorial carbon accounting: City cases and benchmarking metrics.**
>
> **City Practice Examples:**
> - *Territorial urban metabolic inventory:* About 80 cities in voluntary reporting platforms report Scope 1 emissions, using the urban metabolic based inventory method.
> - *Atmospheric science-based method:* Not adopted by practitioners in cities.
>
> **Scientific Studies:**
> - *Territorial urban metabolic inventory:* e.g., Indianapolis (Gurney et al., 2012), Los Angeles (Gurney et al., 2019)
> - *Atmospheric science-based method:*
>   - In situ *tower measurement:* e.g., Boston (Sargent et al., 2018), Salt Lake City (Mitchell et al., 2018), Indianapolis (Turnbull et al., 2019)
>   - *Aircraft mass balance measurement:* e.g., Indianapolis (Turnbull et al., 2019), London (Pitt et al., 2019)
>   - CO_2-*satellite data:* e.g., Los Angeles, Cairo (Ye et al., 2017)
>
> The total direct emissions estimated from in situ tower and aircraft mass balance measurement is consistent with the Scope 1 urban metabolic inventory approach in Indianapolis, Indiana, United States (Turnbull et al., 2019).
>
> **Benchmarking metrics:**
> - CO_2 emission flux, g-CO_2/m^2/s
> - Fossil fuel-based CO_2 concentration, ppm

emission sources include stationary and mobile fuel combustion, waste management, industrial process and product use (IPPU), and nonenergy-related agricultural activities. Other than the source of fuel combustion, the economic structure of a city determines whether it has the other three sources. These direct GHGs from a city's boundary are also referred to as Scope 1 emissions when aligning with the scope concepts in GHG inventories (Fong et al., 2014).

Greenhouse gas emissions from energy combustion, including stationary and mobile sources, can be calculated based on the following equation.

$$GHG_{fuel} = \sum_{i,j}^{k} fuel_{i,j} \times EF_{i,j}^{k} \times GWP^{k}, \tag{1}$$

in which $fuel_{i,j,k}$ is the ith type of fuel combusted in the jth end-use sector. Fossil fuels, including coal, natural gas, liquefied petroleum gas, gasoline, etc. are used in buildings for heating and cooking, industrial manufacturing, agricultural production, and transportation. Biomass, such as crop residues, firewood, and animal manure, can also be used by households for cooking and/or heating in some cities. $EF_{i,j}^{k}$ is the emission factor of the ith type of fuel in the jth sector for the kth type of GHGs (mainly CO_2, CH_4, and N_2O). GWP^{k} is the global warming potential for the kth GHG.

BOX 2 Community-wide trans-boundary infrastructure supply chain carbon footprints.

A. Community-wide Purely Territorial plus Imported Electricity and Thermal Supply (Scope 1 + 2) GHG Accounting:

City Practice Examples:

- Reported by 260 cities through CDP and Bonn Center Local Climate Action and Reporting platforms globally (of which 83 only report Scope 1, and 36 cities do not separate Scope 1 and 2) (Nangini et al., 2019)

Scientific Review:

- Kennedy et al. reviewed 50 global cities (Kennedy, Ramaswami, Carney, & Dhakal, 2011)
- Nangini et al. reviewed and compiled an emission database including 343 global cities (Nangini et al., 2019)

B. Community-wide Infrastructure-based Carbon Footprinting (Scope 1 + 2 + 3 GHG Accounting):

City Practice Examples:

- More than 27 cities using ICLEI's Community Protocol have integrated trans-boundary supply of additional sectors according to the Carbon Disclosure Project database.

Scientific Review:

- Fuel: Kennedy et al. added the life-cycle emissions embodied in fuel in 10 global cities (Kennedy et al., 2010)
- Fuel, cement use, food and trans-boundary water supply/treatment GHG:
  - Ramaswami et al. applied this approach in 20 US cities (Ramaswami & Chavez, 2013);
  - Chavez and Ramaswami implanted in Delhi, India (Chavez, Ramaswami, Nath, Guru, & Kumar, 2012);
  - Lin et al. implemented in Xiamen, China (Lin, Liu, Meng, Cui, & Xu, 2013); Tong et al. implemented in four Chinese cities (Tong et al., 2016).

Benchmarking metrics are is now about <u>use activities</u> (as emission factors primarily drawn from national data) by infrastructure sectors. Several typical benchmarking metrics are:

- Energy use per household (with fossil fuel and electricity separated)
- Energy use per industrial economic output/employee
- Vehicle miles traveled (VMT) per capita
- Water use per household and water use per industrial economic output
- Municipal waste per household
- Food consumption per household
- Cement use per newly added floor area
- Carbon footprints per economic output

The benchmarking is available for eight US cities (Hillman & Ramaswami, 2010), four Chinese cities (Tong et al., 2016), and one Indian city (Chavez et al., 2012). The use activities data have been shown to be consistent with fuel sales data and state data in the transportation sector (Hillman et al., 2011).

344 Section | C Case Studies

BOX 3 Consumption-based carbon footprinting: City cases and benchmarking metrics.

City practice examples:
- C40 reported consumption-based carbon footprinting of 79 cities (C40, 2018).
- London released its consumption-based carbon footprints.

Scientific studies: CBF has been implemented in cities where input-output data are available.
- Chen et al. calculated CBF for Sydney Australia (Chen, Wiedmann, Wang, & Hadjikakou, 2016).
- Feng et al. quantified CBF in four megacities in China (Feng, Hubacek, Sun, & Liu, 2014).
- Larsen and Hertwich calculated CBF in Trondheim, Norway (Larsen & Hertwich, 2009).
- Ramaswami et al. quantified CBF in 20 US cities (Ramaswami & Chavez, 2013) and compared this with CIF results.

Benchmarking metrics are on household consumption activities. In addition to the consumption metrics listed in CIF approach, this approach provides the following metrics:
- Expenditure on nondurable goods per household
- Carbon footprints per household

BOX 4 Total community-wide carbon footprinting (infrastructure + noninfrastructure).

City Practice Examples:
- No cities have implemented this approach.

Scientific Studies:
- Chen et al. implemented in four Chinese municipalities (Chen, Long, Chen, Feng, & Hubacek, 2019); Hu et al. implemented in eight Chinese cities (Hu, Lin, Cui, & Khanna, 2016)
- Murakami et al. implemented in six Japanese cities (Murakami, Kaneko, Dhakal, & Sharifi, 2020)
- Chavez implemented in three US cities

These studies explore the numeric relationships between various city-level carbon accounting approaches to reveal the impact of setting accounting boundaries.

Benchmarking metrics:
- This approach can provide the same metrics, using infrastructure provisioning metrics that would have been provided by CIF.
- Per capita CO_2 or CO_2-e emissions should be calculated only using emissions related to household consumption.

When a city has solid waste and/or wastewater treatment plants within its boundary, GHG emissions released from these plants are counted in Scope 1 emissions. If a city does not have such facilities within its boundary, the purely territorial accounting approach does not account for any emissions related to waste management. A simplified equation below shows the calculation of GHG emissions from waste management facilities within a city's boundary.

$$GHG_{waste} = \sum_{m}^{k} waste\ amount_m \times EF_m^k \times GWP^k, \tag{2}$$

in which *waste amount$_m$* is the amount of waste (e.g., municipal waste and industrial waste) being treated by the mth treatment technology (e.g., landfill, incineration, open disposal, etc.). EF_m^k is the emission factor for the kth type of GHGs for the corresponding technology.

When a city has industrial manufacturing releasing GHGs from production processes, Eq. (3) is applied to calculate the emissions. GHG emissions from industrial processes and product use (excluding fossil fuel combustion) are calculated based on the amount of product produced, the technology of production, and emission factors from the production process, as shown in the following equation:

$$GHG_{IPPU} = \sum_{i}^{k} product_i \times EF_i^k \times GWP^k, \tag{3}$$

in which *product$_i$* is the amount of the ith industrial product. There may be multiple processes releasing carbon emissions in a manufacturing plant. Here, the product does not necessarily need to be the final product from a factory. EF_i^k is the emission factor for the kth type of GHGs released from producing the ith industrial product.

Livestock and manure management are two primary nonenergy-related GHG sources from agricultural activities, which can be calculated using Eq. (4). Livestock's enteric fermentation generates CH_4. And both CH_4 and N_2O are released from manure management.

$$GHG_{agri} = M_{livestock} * EF_{livestock} * GWP^{CH_4} + M_{manure} * EF_{manure}^{CH_4\ or\ N_2O} \\ * GWP^{CH_4\ or\ N_2O}, \tag{4}$$

in which $M_{livestock}$ is the amount of livestock and $EF_{livestock}$ is the emission factor for the corresponding livestock. GHGs from livestock should be calculated based on the type of livestock. M_{manure} is the total amount of manure processed and $EF_{manure}^{CH_4\ or\ N_2O}$ is the emission factor for manure management. When manure is burned as fuel, emissions are counted in GHG_{fuel}, not in GHG_{agri}. GHG emissions from agricultural land use and land-use change in a city are included in Section 3.

Data sources for urban metabolic inventory approach: Collecting data at the city level requires coordination with local energy utilities, landfill operators, and transportation agencies, as well as gathering local industrial fuel use data that may not be provided by utilities. Energy use and production-related data in the industrial sector may not be available in many cities. In some countries, cities collect data on fossil fuel use in buildings, industrial manufacturing, and transportation activities and release these data in their statistical yearbooks. These yearbooks become a primary data source. Where data cannot be collected through utilities or from city statistical handbooks, they are estimated based on scaling-up (i.e., data collected from a small sample in a city is used to estimate the city total) or scaling-down (i.e., national or subnational data are scaled to the city-level based on socioeconomic proxy) methods.

There are more and more reliable data sources reporting fossil fuel use (e.g., utilities, tax records, statistical yearbooks, and planning documents). In comparison, data on biomass use are less frequently reported and have higher uncertainties (Bailis, Drigo, Ghilardi, & Masera, 2015). Data on the amount of fuel combusted in industrial manufacturing are also less available in some countries (e.g., the United States). The amount of motor fuel use can be calculated from local fuel sales or downscaled from state-level motor fuel sales to cities based on local traffic volume or road segments. If such data are not available, vehicle-miles traveled (VMT) is a commonly used parameter to back-calculate the amount of motor fuel combusted. Fuel use data from VMT estimation have been found to be aligned with motor fuel sale data at the state level (Hillman, Janson, & Ramaswami, 2011). Data on waste treatment, if such facilities are within the city boundary, are less available in public datasets. Researchers can contact local facilities to obtain data on the amount of treated waste and the waste treatment technology. Alternatively, a scaling-up or scaling-down method can be adapted to estimate the amount of waste treated, when such data are not available at the city scale. This data collection method is also applicable for industrial and agricultural production. Regarding emission factors, the regional or national default values (e.g., IPCC's national GHG guidelines provide default emission factors for fossil fuel, biomass burning, etc.) can be used to calculate Scope 1 emissions. Computational simulation methods, such as Monte Carlo simulation, can be used to demonstrate the uncertainty in estimated direct emissions (Shan, Liu, Liu, Shao, & Guan, 2019).

Method overview of atmospheric science-based measurement for fossil fuel CO_2: Atmospheric scientists apply inverse modeling to estimate the anthropogenic fossil fuel carbon flux from measured CO_2 elevated from an urban area. Three methods measuring CO_2 concentration from urban areas include in situ tower measurement, aircraft observation, and satellite-based CO_2 data. The in situ tower measurement method collects CO_2 concentration data through monitors at several locations continuously over a period of time. This method has a high temporal resolution, but low spatial density. The aircraft observation method collects CO_2 concentration data by flying an aircraft throughout a city to measure CO_2 mole fraction at different locations. This method needs carefully designed flying routes with consideration of local meteorological conditions, given that the temporal resolution of this measurement is low. There are studies using data from in situ tower measurements and aircraft observations to estimate fossil fuel emission flux in a city (Turnbull et al., 2019). Satellite data on CO_2 concentration are currently available only for larger cities.

Data sources for territorial atmospheric science-based measurement: Data for inverse modeling include prior fossil fuel CO_2 emission flux, monitored CO_2 concentration, transport model parameters, and urban biogenic carbon flux. The territorial urban metabolic inventory provides the prior fossil fuel CO_2 emission flux. CO_2 concentration is collected by one of the measurement methods listed above. Biogenic carbon fluxes from urban areas are modeled through Vegetation Photosynthesis and Respiration Models. Weather Research and Forecasting (WRF)-Chem is a commonly used transport model for inverse modeling at the city level.

Inverse modeling has several challenges when applied to urban areas. First, biogenic carbon should be separated from total emissions. Studies have found that separating biogenic carbon can reduce the uncertainty in results (Wu et al., 2018). It is a challenge to separate background biogenic CO_2 emissions when using satellite data and aircraft measurements, due to temporal constraints when sampling. Second, the measurement site density within the designated area is also important for reducing uncertainty in the final estimated carbon flux (Wu et al., 2018), but studies in urban areas have less measurement coverage (using the in situ tower method). Third, inverse modeling needs prior flux data as the input, and high quality in these data can reduce uncertainty in the modeled results (Wu et al., 2018). The Scope 1 emission inventory is the prior emission data. Thus, the quality of results from this approach relies on improving the quality of data from the traditional inventory-based method. Fourth, there are errors in atmospheric transport models in urban areas (Wu et al., 2018; Ye et al., 2017). Thus, the uncertainties in the results also should be carefully evaluated.

Value to Urban Climate Policies: This purely territorial carbon accounting approach quantifies vertical carbon fluxes. It does not evaluate the carbon impact of urban activities from the systems perspective, as it does not account for carbon emissions embodied in trans-boundary supply chains. With its sole focus on direct emissions, this approach can only inform limited urban climate policies (e.g., reducing emissions from direct fossil fuel combustions and banning petrol vehicles with low efficiency). However, low carbon development in cities encompasses more actions than what this approach can inform, such as energy efficiency rebate programs for electrical appliances, electrifying heating and mobility, and transit-oriented planning.

Because this territorial carbon accounting approach does not evaluate the systemic carbon impact of urban climate actions, it cannot assess the direction of carbon emission change associated with these actions. Energy efficiency rebate programs can reduce direct fossil fuel use and electricity use. This approach does not account for the full carbon mitigation potential of these rebate programs. Food waste-to-energy initiatives, where cities can collect food

348 Section | C Case Studies

waste and convert it into methane for local electricity generation, are another urban climate action that will be difficult to evaluate using this approach. Furthermore, this action may even increase direct local emissions, while it can reduce direct emissions from waste decomposition elsewhere.

The uncertainties in total direct carbon emissions from urban areas can be evaluated by comparing results from two approaches. The atmospheric science-based approach can help to reduce the uncertainties in the metabolic inventory method by providing locally relevant emission factors. And a high-quality direct emission inventory can reduce uncertainties in results from the atmospheric science-based approach. In addition to fully utilizing two approaches, multiple research groups have independently applied the metabolic inventory method to develop direct fossil fuel CO_2 emission datasets for cross-comparison. This comparison reveals uncertainties in data and methods when accounting for direct carbon emissions from urban areas.

2.2 Community-wide trans-boundary infrastructure supply-chain carbon footprinting approaches

Urban activities heavily rely on energy and materials produced outside of their geographic boundaries. For example, many cities do not have power plants located within their boundaries to self-supply all electricity consumed. In China, about 40% of cities have power plants within their administrative boundaries (Tong et al., 2018). In the United States, about 93% of counties import some share of their electricity (Ramaswami & Chavez, 2013). Additionally, travel never stops at a city's geographical boundary. The purely territorial carbon accounting approach does not quantify carbon emissions embodied in trans-boundary supply chains. Therefore, a series of new methods have been developed to quantify carbon emissions embodied in trans-boundary supply chains. These approaches account for carbon emissions directly from in-boundary use activities and carbon associated with their supply chains when assessing urban carbon footprints.

It is critical to explain the *differences between use activities and GHG sources*. Table 1 details the difference between GHG sources and use activities in all key infrastructure sectors. GHG sources and use activities are the same in some circumstances. For example, natural gas use in households is both a use activity and an emission source. However, use activities differ from emission sources in many other infrastructure provisioning sectors. Electricity use, food consumption, and building materials used in a city are not necessarily direct emission sources. Instead, GHGs associated with these use activities are usually outside of a city's boundary. Take transportation as an example. In Scope 1 emissions, the source of emissions is exhaust gases from vehicles running within a city. In comparison, the use of transportation for personal mobility can be calculated based on the origin and destination of household trips. Focusing on the use of the transportation system can support evaluating the impact of

Urban environments and trans-boundary linkages **Chapter | 10 349**

TABLE 1 Examples of matching emission sources and use activities in seven key infrastructure sectors.

Emissions type		Source or activity?
Built environment		
Use of fuel in residential and commercial stationary combustion equipment		Source AND Activity
Industrial stationary combustion sources		Source
Electricity	Fossil fuel combustion for electric power generation in the community	Source
	Use of electricity by the community	Activity
District heating/cooling	Fossil fuel combustion in district heating/ cooling facilities in the community	Source
	Use of district heating/cooling by the community	Activity
GHG emissions from industrial process in the community		Source
Refrigerant leakage in the community		Source
Transportation and other mobile sources		
On-road passenger vehicles	On-road passenger vehicles operating within the community boundary	Source
	On-road passenger vehicle travel associated with community land uses	Activity
On-road freight vehicles	On-road freight and service vehicles operating within the community boundary	Source
	On-road freight and service vehicle travel associated with community land uses	Activity
On-road transit vehicles operating within the community boundary		Source
Transit rail	Transit rail vehicles operating within the community boundary	Source
	Use of transit rail travel by the community	Activity
Intercity passenger rail vehicles operating within the community boundary		Source
Freight rail vehicles operating within the community boundary		Source

Continued

350 Section | C Case Studies

TABLE 1 Examples of matching emission sources and use activities in seven key infrastructure sectors—cont'd

Emissions type		Source or activity?
Marine	Marine vessels operating within the community boundary	Source
	Use of ferries by the community	Activity
Off-road surface vehicles and other mobile equipment operating within the community boundary		Source
Use of air travel by the community		Activity
Solid waste		
Solid waste	Operation of solid waste disposal facilities in the community	Source
	Generation and disposal of solid waste by the community	Activity
Water and wastewater		
Potable water—energy use	Operation of water delivery facilities in the community	Source
	Use of energy associated with use of potable water by the community	Activity
Use of energy associated with generation of wastewater by the community		Activity
Centralized wastewater systems—process emissions	Process emissions from operation of wastewater treatment facilities located in the community	Source
	Process emissions associated with generation of wastewater by the community	Activity
Use of septic systems in the community		Source AND activity
Agriculture		
Domesticated animal production in the community		Source
Manure decomposition and treatment in the community		Source
Food consumed in household in the community		Activity

Urban environments and trans-boundary linkages Chapter | 10 351

TABLE 1 Examples of matching emission sources and use activities in seven key infrastructure sectors—cont'd

Emissions type	Source or activity?
Upstream impacts of community-wide activities	
Upstream impacts of fuels used in stationary applications by the community	Activity
Upstream and transmission and distribution (T&D) impacts of purchased electricity used by the community	Activity
Upstream impacts of fuels used for transportation in trips associated with the community	Activity
Upstream impacts of fuels used by water and wastewater facilities for water used and wastewater generated within the community boundary	Activity
Upstream impacts of select materials (concrete, food, paper, carpets, etc.) used by the whole community. Note: Additional community-wide flows of goods and services will create significant double counting issues.	Activity

Revised from ICLEI U.S. Community-Wide GHG Accounting Protocol (ICLEI-USA. (2012). *U.S. community protocol for accounting and reporting of greenhouse gas emissions.* http://www.icleiusa.org/tools/ghg-protocol/community-protocol).

urban transportation policies, such as transit-oriented planning, on reducing automobile use and related emissions.

Two community-wide infrastructure accounting methods have been developed with the increasing inclusion of trans-boundary supply from electricity and heating/cooling services (when applicable) to seven key infrastructure plus food provisioning sectors. Data requirements vary in these approaches, due to the differing inclusion of trans-boundary supply chains. Additionally, each method has different implications for urban climate policies (Table 2).

a. *Community-wide purely territorial plus trans-boundary electricity and heating/cooling supply carbon accounting*

Method overview: This method accounts for GHGs embodied in imported electricity and heating/cooling services, in addition to purely territorial carbon emissions. GHGs embodied in imported electricity and heating/cooling services (for buildings) are also defined as Scope 2 emissions. The method and data for quantifying Scope 1 emissions are the same as what is detailed in the territorial

TABLE 2 Implication for urban climate policies of different carbon accounting approaches.

Desired policy attributes	Utility of greenhouse gas accounting methods to policy attribute *** represents greatest relevance; [Explanations] are provided for reduced relevance				
	Purely territorial accounting (PAT)	Purely geographic + imported electricity	Community-wide infrastructure footprint (CIF)	Consumption-based footprint (CBF)	Total community-wide carbon footprint (TCF)
Informs future city and regional infrastructure (multilevel) planning and policy for sustainable transitions	* [Most infrastructures transcend city boundaries]	** [Only including electricity, while other major infrastructures transcend city boundaries]	*** [Most relevant]	* [Excludes infrastructures serving local businesses and industries that export goods]	*** [Most relevant]
Linkage of energy use to local urban heat islands, local air quality, and public health	*** [Most relevant]	*** [Most relevant and considering the trans-boundary electricity supply impact]	** [Energy use in key infrastructures is allocated based on use, not location]	* [Energy use in all industries and businesses are allocated based on consumption, not location]	*** [Most relevant]
Linkage of local consumption and social inequality	* [Fossil fuel use inequality in households and businesses]	* [Fossil fuel and electricity use inequality in households and businesses]	** [Infrastructure use and inequality in households and businesses]	** [Infrastructure and noninfrastructure use inequality in households]	*** [Infrastructure and noninfrastructure use inequality in households and businesses]

Informs supply-chain vulnerability for future planning	* [Most infrastructure transcend city boundaries]	** [Only including electricity, while other major infrastructures transcend city boundaries]	*** [Most relevant]	* [Allocates GHG after consumption occurs, but does not address future planning for local supply vulnerability]	*** [Most relevant]
Enables intercity comparisons using per capita metrics to inform residents	* [Per capita metric can only apply to carbon emissions from households]	* [Per capita metric can only apply to carbon footprints of household energy use]	*** [Per capita metric can only apply to household carbon footprints of infrastructure use]	*** [Most relevant]	*** [Most relevant when only using household carbon footprints]
Enables intercity comparisons using economic productivity metrics	* [Most infrastructures transcend city boundaries]	** [Included trans-boundary electricity Most infrastructures transcend city boundaries]	*** [Most relevant]	N/A [Per economic output metric is incorrectly applied]	*** [Most relevant]
Enables aligning with national total	*** [Can be added up directly to national total]	** [The portion of in-boundary emissions can be added up directly to national total. Electricity use and local production should be processed carefully to avoid double counting]	** [The portion of in-boundary emissions can be added up directly to national total. Infrastructure use and local provision should be processed carefully to avoid double counting]	*** [Household carbon footprints can be added up directly to total national household carbon footprints]	* [Infrastructure and noninfrastructure use and local provision should be processed carefully to avoid double counting]

Continued

TABLE 2 Implication for urban climate policies of different carbon accounting approaches—cont'd

Desired policy attributes	Utility of greenhouse gas accounting methods to policy attribute *** represents greatest relevance; [Explanations] are provided for reduced relevance				
	Purely territorial accounting (PAT)	*Purely geographic + imported electricity*	*Community-wide infrastructure footprint (CIF)*	*Consumption-based footprint (CBF)*	*Total community-wide carbon footprint (TCF)*
Data availability, quality, and ability to benchmark or verify energy use and GHG emissions data	[Remote sensing, in situ tower, aircraft sampling may enable independent verification]	[Electricity use and emission factors of imported electricity]	[Emission factors of trans-boundary supply; Use activities in food, materials]	[IO models are calibrated to personal consumption and other data, not separately verifiable. Uncertainty is difficult to verify]	[IO tables are less available at the city level. Uncertainty is difficult to verify]

Modified based on Table 2 in Chavez, A., & Ramaswami, A. (2013). Articulating a trans-boundary infrastructure supply chain greenhouse gas emission footprint for cities: Mathematical relationships and policy relevance. *Energy Policy, 54*, 376–384. https://doi.org/10.1016/j.enpol.2012.10.037.

Urban environments and trans-boundary linkages **Chapter | 10** **355**

urban metabolic carbon accounting approach. Total Scope 2 GHG emissions can be calculated based on the following equation.

$$GHG_{Scope\ 2} = Electricity^{net\ import} \times EF_{electricity}^{CO_2-eq} + Heating/cooling^{net\ import}$$
$$\times EF_{heating/cooling}^{CO_2-eq}, \tag{5}$$

in which $Electricity^{net\ import}$ is the amount of electricity imported to serve households, indusial production, and businesses in a city. $EF_{elecriticy}^{CO_2-eq}$ is the emission factor of imported electricity. $Heating/cooling^{net\ import}$ and $EF_{heating/cooling}^{CO_2-eq}$ refer to buildings' heating/cooling service use and the emission factors of these thermal service provisions, respectively.

Cities' Scope 1+2 emissions are the sum of emissions from Eqs. (1) to (5), as shown in Eq. (6). The part in parenthesis is optional, depending on whether a city has these sources.

$$GHG_{Scope\ 1+2} = GHG_{Scope\ 1,\ fuel} + GHG_{Scope\ 2}$$
$$+ \left(GHG_{Scope\ 1,waste} + GHG_{Scope\ 1,\ IPPU} + GHG_{Scope\ 1,\ agri}\right). \tag{6}$$

When accounting for GHGs embodied in the trans-boundary energy supply, a possible double counting can occur if cities also have in-boundary electricity generation. This double counting can be controlled by applying the ratio of the amount of electricity used locally versus generated locally.

Data sources for direct emissions within a city's boundary are the same as those detailed in the territorial urban metabolic inventory method. Additional data on electricity and heating/cooling energy use can be collected by contacting local utilities. Official statistical handbooks or official planning documents are another data source for electricity or heating/cooling energy use. When these city-scale data are not available, the scaling-up or scaling-down technique is applied to estimate electricity and heating/cooling service use. Some utility companies routinely report their grid-level GHG emission factors. When the emission factor of trans-boundary electricity supply is not reported, researchers can calculate it based on grid-wide energy use and total generation. This grid-level electricity generation and fuel use data are either publicly available or can be collected through contacting transmission companies.

Value to Urban Climate Policies: This method only applies the systems perspective to imported electricity, as well as district heating/cooling services, but not to other trans-boundary infrastructure supplies. This accounting approach can inform urban policies related to in-boundary fuel and electricity use. For example, cities can use rebate programs to improve energy efficiency in the building sector, regardless of using electricity or fossil fuel.

Results provided by this method can only evaluate the directionality (carbon mitigation or increase) of limited urban climate actions. It cannot assess the carbon reduction of electrifying personal mobility and heating services from a systems perspective, because it does not account for the life-cycle well-to-pump GHG of the displaced petroleum. Furthermore, this approach does not consider

356 Section | C Case Studies

the life-cycle carbon emissions of nonelectricity service provisions (e.g., building material use, food provision, and waste treatment) outside a city's boundary. Thus, it cannot be used to inform urban cross-sectoral actions, such as the food-energy-water nexus (e.g., converting mixed residential-commercial food waste to electricity), or systems approaches (e.g., compact urban mixed-use planning) that influence land-use change. However, cities are actively exploring these choices to reduce their carbon impacts. Therefore, the inclusion of other trans-boundary infrastructure supply chains is critical to inform more urban climate policies.

Uncertainty exists in data about local electricity use and emission factors of imported electricity and heating/cooling service supply. More utilities are releasing data on local electricity use and grid-level emission factors. These reliable at-scale data can reduce the uncertainties in Scope 2 GHG emissions. Additionally, the power system is a critical infrastructure for a nation. Electricity generation and fuel use at the grid level are regularly tracked, and data are released by many countries. Thus, uncertainties in emission factors are comparably low for electricity use. Data on imported heating and cooling services need to be collected from utilities, who provide reliable city-level data.

b. *Community-wide infrastructure-based carbon footprinting*

Method Overview: Community-wide infrastructure-based carbon footprinting accounts for direct and embodied emissions in seven key infrastructure plus food provisioning sectors for a city. These key provisioning sectors are energy, transportation and communication, water, waste, shelter materials, public/green spaces, and food supply. These sectors are critical for supporting urban activities. CIF quantifies both direct and life-cycle carbon emissions associated with the use of these infrastructural sectors. When aligning CIF with GHG inventory "scopes," direct emissions from the use phase and locally supplied services are Scope 1, embodied emissions in imported electricity and heating/cooling energy are Scope 2, and emissions embodied in other trans-boundary infrastructure supplies are Scope 3 emissions. CIF, with its focus on use activities and trans-boundary infrastructure supply chains, provides data to inform local mitigation actions from a systems perspective.

CIF begins with collecting data on the use activities in key infrastructure and food sectors. The equation for this approach (Hillman et al., 2011) is presented below.

$$\mathrm{CIF}^{GHG} = \underbrace{\sum_{NG, petrol, coal, biomass} Fuel * EF_{fuel} + M_{waste} * EF_{waste} + M_{IPPU} * EF_{IPPU} + M_{agri} * EF_{agri}}_{Scope\ 1\ (if\ WWT\ and\ MSW\ faciliteis\ in\ boundary)}$$

$$+ \underbrace{\sum_{ele\ or\ heating/coolilng} Energy * EF_{ele\ or\ heating/cooling}}_{Scope\ 2}$$

$$+ \underbrace{MFA_{key\ materials} * EF_{LCA} + M_{fuel} * EF_{WTW}}_{Scope\ 3}.$$

(7)

in which *Fuel* is the total amount of fuel combusted and EF_{fuel} is the emission factor of fuel combustion. M_{waste} is the amount of waste treated in facilities within a city's boundary, and EF_{waste} is the emission factor of in-boundary waste treatment facilities. Here Eq. (7) details the condition when a community has in-boundary solid waste and wastewater treatment facilities. When a community's waste is processed outside of its boundary, $M_{waste} * EF_{waste}$ should be included in Scope 3 emissions. $M_{IPPU} * EF_{IPPU}$ calculates GHG emissions from industrial products and the production process. $M_{agri} * EF_{agri}$ is the simplified version of calculating emissions from livestock and manure management if these activities are within a city's boundary. The part of Scope 2 is the same as Eq. (5). $MFA_{key\ materials}$ is the amount of key materials (e.g., building materials, food) supplied to a community. EF_{LCA} is the life-cycle emission factor of supplying these materials to a city. M_{fuel} is the amount of fuel supplied to a community, and EF_{WTW} is the well-to-wheel emission factor.

There is double counting when locally supplied versus trans-boundary supplied material and energy use is not separated. Detailed trade data can track material and energy flows among different use sectors. These data can be used to further separate local production from local consumption. However, this information is rarely available at the city level for building materials and food supply. A commonly used approach is to take the net ratio between local use activities and local production to control double-counting.

Data sources: Additional data for this method include the use of trans-boundary travel, water, waste management, food, and building materials, as well as life-cycle emission factors of these infrastructure provisioning sectors. Data for these use activities are less available compared to data for fossil fuel and electricity use at the city level. Scaling-up or scaling-down is commonly used to estimate the amount of local use in these additional sectors. Furthermore, new methodologies or new local surveys may be developed to estimate the trans-boundary food or material supply. The accurate emission factors of trans-boundary service provisions can be calculated based on detailed information tracking the supply chains (Ramaswami et al., 2017b). But this type of data is rare at the city level. Regarding emission factors of trans-boundary infrastructure supply, national- or subnational-level life-cycle emission factors are used to quantify the emissions embodied in trans-boundary service provisions.

The inclusion of more trans-boundary infrastructure supply introduces further sources of uncertainties in data. And the estimation of use activities, especially for regional transportation, food, and building materials, can be very uncertain, due to the lack of city-specific data. For example, the amount of building materials used in cities each year is challenging to track. Data on food consumption and supply at the city level also are estimated based on several assumptions, which introduce uncertainties in the results. Tracking trans-boundary food and material provisions is also very difficult for cities, and data are rarely available to comprehensively reveal the supply network. Although

358 Section | C Case Studies

conducting CIF needs more data on use activity and supply chain, this approach has been adopted by local policymakers to support cities' climate action plans (Box 2).

Value to Urban Climate Policy: CIF applies systems thinking when accounting for GHGs associated with seven key infrastructure and food provisioning sectors in a city. It can be used to evaluate the direction of GHG emissions changes of urban climate strategies related to these infrastructure sectors. Additional urban strategies include food system transitions with the linkage to the food-energy-water nexus (Boyer & Ramaswami, 2017), urban policies on sustainable buildings and materials, and land-use change (e.g., compact development, zoning, and green space planning). Furthermore, CIF can inform emerging city-level circular economy strategies involving material and energy exchanges among multiple end-use sectors, i.e., residential, commercial, and industrial sectors (Ramaswami et al., 2017a). Therefore, it is valuable for multisector urban infrastructure systems planning. Moreover, the focus on infrastructure use and provisioning allows for addressing both GHG mitigation and cobenefits of infrastructure transitions, e.g., inequality in access to clean water, energy, and healthy food; air pollution; and climate mitigation and adaptation related to these seven key sectors.

CIF does not account for carbon footprints embodied in noninfrastructure goods and service provisions. These noninfrastructure sectors include clothing, furniture, shoes, and other nonfood-related service provisions. Thus, CIF cannot inform international or regional trade policies to reduce emissions embodied in these noninfrastructure services and goods. For example, the impact of consumers' behavior change related to the fashion industry on GHGs cannot be addressed by this approach. Therefore, the total community-wide footprint (TCF), adding the noninfrastructure supply chain into CIF, can provide data to support policies addressing the consumption of noninfrastructure goods.

This community-wide trans-boundary supply chain carbon footprinting approach evaluates systemic carbon impacts and supports the directionality of urban climate policies. However, there can be significant uncertainties in trans-boundary nonelectricity infrastructure use and supply chains, due to lack of city-level data. New data and analytical techniques are expected to provide more data tracking trans-boundary commodity flow to reduce uncertainties in both use activity estimates and emission factors (Ramaswami et al., 2021).

2.3 Consumption-based carbon footprinting approaches

The consumption-based GHG footprinting (CBF) approach accounts for emissions embodied in the consumption activities of households and governments within a city. CBF is not a community-wide carbon accounting approach because it only focuses on consumption activities of households and government within the city, excluding consumption of goods and services by nonresidents (e.g., visitors and tourists), as well as goods and services exported from

Urban environments and trans-boundary linkages **Chapter | 10** **359**

the city. When mapping CBF to the three different emission inventory scopes, CBF is a mix of partial Scope 1 (emissions from energy combustions in *households and governmental buildings, and locally provided services to households and government*) plus partial Scope 2 (emissions embodied in electricity and/or thermal services imported to *homes and governmental buildings*) plus partial Scope 3 (emissions embodied in other trans-boundary goods and services consumed by *households and governments*). Two different methods have been developed to calculate consumption-based carbon footprinting at the city level. Examples of applying this approach at the city level can be found in Box 3.

Method overview of economic input-output-based CBF: This method relies on cities' input-output tables to quantify GHGs embodied in the final demand (households and government) columns in a city's economic input-output table (Chen et al., 2019). Consumption-based GHG emissions can be calculated based on the following equation when a city-level economic input-output table is available:

$$GHG^{CBF} = ([F] + [M_F]) \times [EF^{use}] + [B][L] \times ([F] + [M_F]), \qquad (8)$$

in which $[F]$ is the amount of final consumption met by local production in cities; $[M_F]$ is the amount of final consumption met by imports; $[EF^{use}]$ is the emission factor of fuel combustion; and $[B]$ is the direct emission intensity of city-level production (mt-CO_2e/$-output). $[L]$ is the Leontif matrix derived from the input-output table. This method assumes that the production structure of imports to final demand is the same as the local production structure. Eq. (8) is used when a single-region input-output table is available.

With the development of the input-output method, more data are available to support multiregional input-output modeling. Multiregional input-output tables detail global flows of goods and services among different regions (national or subnational areas). CBF can be calculated using Eq. (9) when a multiregion input-output table is available.

$$GHG^{CBF} = ([F] + [M_F]) \times [EF^{use}] + [B][L][F] + \sum_i [B_i][L_i] \times ([M_{F_i}]), \qquad (9)$$

i indicates the ith region other than local production. This method requires the further breakdown of imports from different regions or countries.

Data sources for input-output data-based CBF: Citywide input-output tables and local emission intensity are critical data for this method, but not many cities have such data readily available. Even in the United States, the county-level input-output table constructed by IMPLAN is downscaled from the national level (Chavez & Ramaswami, 2013). Four Chinese megacities (provincial-level municipalities) regularly release their economic input-output tables. When the economic input-output data are available, a challenge is to separate total imports to household consumption and intermediate industrial production. The U.S. Bureau of Economic Analysis (BEA) has developed a method to allocate national imports, but it may not be applicable to individual

360 Section | C Case Studies

cities due to the high level of trans-boundary activities. Tracking the flows of all imported goods is complicated at the city level, and new analysis methods are needed. In general, the economic input-output method is less applicable to many cities globally due to the lack of input-output tables.

The uncertainty in input-output tables is complex and difficult to evaluate (Heijungs & Lenzen, 2014; Lenzen, 2000). While several studies have examined the uncertainty in national economic input-output data by tracing the raw data and methods used to generate the table, few studies detail uncertainties in city-level input-output data. Chavez and Ramaswami compared data from city input-output tables with energy flow data reported at the city scale, and they found more than 100% difference (Chavez & Ramaswami, 2013). This indicates a considerable uncertainty when using the downscaled city-level input-output table. Although many studies used economic input-output tables to estimate cities' CBF (Box 3), few have discussed uncertainties when applying this approach to cities or how to address these uncertainties. The level of uncertainty in city-level input-output data is not known and is difficult to verify (Chen et al., 2020).

Method overview of hybrid life-cycle assessment-based CBF: The other method for estimating urban consumption-based carbon footprints uses household consumption data and the life-cycle emissions factors to quantify the amount of emissions directly from household fuel combustion and the embodied emissions in fuel, food, and other goods/service consumption at the city level (Jones & Kammen, 2011). This hybrid method can also be applied to quantify the GHGs associated with governmental consumption in cities. This method requires two sets of data: household consumption surveys and the life-cycle emission factors of supplying these goods/services. Household consumption-based carbon footprints can be calculated based on the following equation.

$$GHG_{HH}^{CBF} = \sum_i F_i \times EF_i^{use} + F_i \times EF_i^{LCA}, \tag{10}$$

in which F_i is the expenditure on the ith goods/service in a household. EF_i^{use} is the emission factor of directly using these goods/services, e.g., natural gas and fuel oil for heating and cooling in homes. EF_i^{LCA} is the life-cycle supply chain emission factor of ith good and services. The life-cycle emission factor can be drawn from a national level economic input-output database.

Data sources for hybrid life-cycle assessment-based CBF: Household consumption survey data are available in many countries. The local or national government may release the household expenditure/consumption survey data. This method also needs detailed life-cycle emission factors for different final consumption items. These life-cycle emission factors can be extracted from the national-level environmental extended economic input-output analysis.

When using the hybrid method to calculate the CBF of households, uncertainties in the survey data and the life-cycle emission factors should be evaluated. The uncertainties in consumption activities are mainly evaluated based on the statistical sampling method. In comparison, uncertainties in life-cycle emission factors of services and goods production can be more complicated than in the expenditure data. Jones and Kammen estimated errors according to data from the literature. Their analysis found the uncertainty in life-cycle emission factors is about 15%–20% (Jones & Kammen, 2011). In comparison, the estimated error of direct fuel combustion emission factors is about 1% (Jones & Kammen, 2011). Harmonizing between final consumption items and production industries may also introduce uncertainties to the results. Comparing the life-cycle emission factors reported in different studies can uncover potential uncertainties.

Value to Urban Climate Policies: CBF cannot inform community-wide urban planning, which must consider homes, industrial manufacturing, and businesses together. GHG emissions associated with services provided to visitors and exports are not counted in this method. Many cities in developing countries are experiencing rapid industrialization, which makes manufacturing a critical activity in a city. Even for cities in developed countries, many commercial activities, such as tourism and commercial services, are pillar industries. CBF does not account for the carbon impacts associated with these activities in a city. Thus, CBF cannot inform many urban policy levers, such as infrastructure planning, because these actions influence both residents and nonresidents.

CBF informs households and governments about their consumption choices beyond the seven key infrastructure sectors and durable goods already covered by CIF. These additional sectors include products like shoes, clothing, furniture, services, etc. CBF links urban consumption activities with national and global supply chains. It informs consumption-production in a multilevel transnational perspective, beyond the scope of urban planning policy. For example, the fashion industry and electronic appliances (phones, computers, laptops, etc.) have significant environmental impacts. CBF can inform consumers about the carbon impacts of their consumption behaviors. Cities can establish educational programs to inform consumers, but these programs for reducing consumption do not necessarily directly reduce cities' Scope 1 and Scope 2 emissions.

The uncertainties in input-output tables and the modeling approach are very high and hard to evaluate, as discussed above. When the raw data used to construct the table are not provided, it is difficult to evaluate the uncertainty of the input-output table. One way to understand the level of uncertainty is to compare CBF results with results from urban material and energy flow analysis. When applying the hybrid life-cycle assessment approach, the uncertainties are comparably easier to evaluate than using the input-output table.

362 Section | C Case Studies

2.4 Total community-wide carbon footprinting

The total community-wide carbon footprinting (TCF) approach quantifies direct emissions (Scope 1) from local activities and the embodied emissions (Scope 2 and Scope 3) in trans-boundary infrastructure and noninfrastructure provisions for a city. This approach provides the full scope of direct and indirect carbon associated with urban activities. It can inform multiple policies, influencing both production and consumption practices in cities. In addition, TCF contributes to identifying city typologies to advance our understanding of urban activities and carbon impacts. TCF can build upon CIF to include trans-boundary noninfrastructure service provisions. Chavez and Ramaswami derived the equation below for quantifying TCF (Chavez & Ramaswami, 2013):

$$GHG^{Total\ GHG\ Footprint} = [B][Z + F + E] + [EF^{use}] \times ([F] + [M_F])$$
$$+ [B][L]\left[M_F^{infra} + M_Z^{infra}\right] + [B][L]\left[M_F^{noninfra} + M_Z^{noninfra}\right],$$

$$(11)$$

in which Z is the local interindustry transaction across the city economy represented in the input-output table. F is local household and governmental consumption supplied by local production. E refers to exports. M_F is the amount of imports for households and governmental consumption, which can be separated into infrastructure-related (M_F^{infra}) and noninfrastructure-related ($M_F^{noninfra}$). M_Z is the amount of imports for local industrial and commercial activities. Similarly to M_F, M_Z also can be separated into infrastructure-related and noninfrastructure-related. In Eq. (11), $[B][Z+F+E]+[EF^{use}] \times ([F]+[M_F])$ is total Scope 1 emissions, and CIF footprints are equivalent to $[B][Z+F+E]+[EF^{use}] \times ([F]+[M_F])+[B][L][M_F^{infra}+M_Z^{infra}]$.

Chavez and Ramaswami rewrote Eq. (11) as below to specify the consumption-based carbon footprint and the nonconsumption-based carbon footprints in TCF (Chavez & Ramaswami, 2013).

$$GHG^{Total\ GHG\ Footprint} = [B][Z_F + F] + [EF^{use}] \times ([F] + [M_F])$$
$$+ [B][L]\left[M_F^{infra} + M_F^{noninfra}\right] + [B][Z_E + E]$$
$$+ [B][L]\left[M_Z^{infra} + M_Z^{noninfra}\right].$$

$$(12)$$

Z_F is the total local production serving local consumption, Z_E is the amount of local production exported. $[B][Z_F+F]+[EF^{use}] \times ([F]+[M_F])+[B][L][M_F^{infra}+M_F^{noninfra}]$ is the consumption-based carbon footprint, which is equivalent to Eq. (9).

Value to Urban Climate Policies: TCF informs urban policies and actions by revealing systemic impacts. Furthermore, the directionality of urban actions on reducing or increasing carbon emissions can be fully revealed through total consumption-based footprinting approach. This approach relies on local input-

Urban environments and trans-boundary linkages **Chapter | 10 363**

output tables, which delineate monetary flows among different economic sectors, as well as monetary flows to final consumption in households and governmental consumption. However, only a limited number of cities construct their own input-output tables. The alternative source is to downscale national or regional input-output to the city level. But this alternative approach introduces high uncertainties in the data. Due to constraints in data availability and uncertainty, TCF has been developed but not yet been integrated into practical protocols. Scientific studies implemented TCF in several cities where data were available (Box 4).

These four approaches vary in the calculation process, data requirement, and policy implications. The complexity of urban systems drives the development of these methods. Each approach is complete to provide the data that this approach is developed for. The purely territorial approach (PTA) is designed to quantify all emissions directly released from a city's boundary; CIF is developed with a focus on quantifying direct and embodied emissions in seven key infrastructure and provisioning systems, and CBF is developed with a focus on households and governmental consumption. Practitioners and scientists can choose the appropriate approaches according to their purposes when developing a city-level carbon inventory, instead of using completeness as a gold standard. The policy inferences of these approaches are compared and summarized in Table 2, along with different benchmarking metrics.

3 Accounting biogenic carbon from land use and land-use change in individual cities

Land use, land-use change, and forestry (LULUCF) at the global scale have long been recognized as key components in carbon mitigation. The role of LULUCF has been specified in global carbon budgets, while its role has been less quantified at the city scale. Part of the reason is that the spatial scale of models and available data for global carbon budgets are not suitable for urban-scale analysis. Emerging efforts have developed two accounting methods to quantify urban biogenic carbon associated with local LULUCF and implemented them in several cities. The first is the inventory-based method, which accounts for biogenic emissions based on the carbon stock change due to land-use change over time, as well as emission fluxes from existing urban green spaces. The second is the atmospheric observation-based method, which quantifies biogenic carbon flux from urban areas based on CO_2 concentration. Both approaches estimate direct emissions related to land-use change and current land use.

Method overview of inventory-based method: The inventory-based method follows the processes detailed in the IPCC national GHG accounting guidelines. City-level carbon protocols, such as GPC Basic Plus (Fong et al., 2014), follow the same procedures detailed in the IPCC guidelines. The generic equation below calculates the carbon stock change due to land-use change.

$$\Delta C_{LU} = \Delta C_{FL} + \Delta C_{CL} + \Delta C_{GL} + \Delta C_{WL} + \Delta C_{SL} + \Delta C_{OL}, \qquad (13)$$

ΔC means the change in carbon stock. FL refers to forestland. CL is cropland, GL is grassland, WL is wetlands, SL is settlements, and OL refers to other land-use types. Within each land use category, carbon is sequestrated in aboveground biomass, belowground biomass, soil organic matter, deadwood, and litter, as well as harvested wood products. The estimate of carbon stock change at the city level depends on data availability and may not count change in every carbon pool within a land-use category.

In addition to carbon stock change caused by land-use change, carbon can be sequestrated and released from existing urban green spaces. The amount of biogenic carbon sequestrated is calculated based on biogenic activity intensity of the corresponding land-use type. Eq. (14) is presented in the GPC protocol (Fong et al., 2014).

$$CO_2 = \sum_{LU} [Flux_{LU} \times Area_{LU}] \times 44/12, \qquad (14)$$

LU refers to different land-use category; $Flux_{LU}$ is the net annual rate of change in carbon stocks per hectare (in tonnes of C per hectare per year) of the corresponding land-use type; $Area_{LU}$ is the land area of this land-use type in a city.

Data sources for inventory-based method: The first set of data for this method include a current land-cover map and land-cover change over a period of time. City-level analysis needs land-use maps with a high spatial resolution. For example, in the United States, vegetation land maps are released regularly at the spatial resolution of 30 meters (U.S. Geological Survey USGS Gap Analysis Project GAP, 2016), which are suitable for studying land-use change and land cover at the city level, especially for large cities. The spatial resolution of the U.S. National Agriculture Imagery Program (NAIP) products is 5 meters, or even 1 meter in some states. NAIP products can be processed through deep learning algorithms to extract land-use types. Although these new datasets provide high-quality data for city-level analysis, land-use maps over time may not have the same spatial resolution. In this case, spatial aggregation is needed to calculate land-use change over time. This spatial aggregation can introduce uncertainties into the estimate of land-use change.

Carbon intensity is the second set of data needed for estimating carbon emissions related to land-use change. These data can be extracted from various scientific studies. Researchers have conducted surveys of urban green spaces, including trees, soil systems, grass, etc., to estimate citywide biogenic carbon stock, considering urban microclimate features and land use management. For example, Nowak et al. surveyed trees in 10 US cities and estimated carbon stock in urban trees across the nation (Nowak & Crane, 2002). A similar method has been adopted to estimate organic carbon in urban soil systems (Pouyat, Yesilonis, & Nowak, 2006). In addition to evaluating individual ecological systems in urban areas, Hutyra et al. mapped carbon stocks in multiple aboveground carbon pools, including trees, grassland, and biomass litter, from urban to suburban areas in the Seattle metropolitan area (Hutyra, Yoon, & Alberti, 2011). These studies all conducted surveys in their analyses to collect relevant local information.

The net carbon uptake in urban green spaces differs from that of the natural environment. For example, one study found that urban trees grow faster and die younger than trees in suburban areas (Smith, Dearborn, & Hutyra, 2019). Thus, the annual carbon sequestration rate of urban trees can be higher than the average value of all trees. Researchers have begun to integrate urban microclimate parameters into the Vegetation Photosynthesis and Respiration Model (VPRM) to estimate urban biogenic carbon fluxes. For example, the VPRM model was applied in the Boston region to estimate biogenic carbon fluxes with consideration of urban microclimate characteristics (Hardiman et al., 2017). These urban-specific estimates can contribute to reducing uncertainties in biogenic carbon from urban land use.

Method overview of atmospheric observation-based method: This method is similar to the method inversing fossil fuel carbon flux using the atmospheric science-based method. It estimates biogenic carbon emissions from measured carbon concentration in urban areas. This atmospheric observation-based method applies to cities where carbon concentration measurement and direct fossil fuel carbon inventory data are available at the city level. This method provides local biogenic carbon flux, which can also be used to reduce uncertainties in the inventory-based estimate. This method is still under development and has not yet been integrated into practical urban carbon accounting protocols.

Data sources for atmospheric observation-based method: Data needed for this method include high spatial resolution vegetation maps, measured carbon concentration, direct fossil fuel carbon emission inventories, and transport models. The process is similar to inferring fossil fuel CO_2 flux from concentration measurements in urban areas. Researchers either set up in situ towers or use aircraft to measure CO_2 concentration in the atmosphere. The observed carbon dioxide concentration is processed to correct the background CO_2 concentration. Researchers then separate carbon emissions fluxes from fossil fuel combustion and biogenic activities through inverse modeling (Lopez-Coto, Prasad, & Whetstone, 2017). The in situ tower measurements have a high temporal resolution because the equipment can take measurements continuously once it is set up. However, observation sites are very limited in cities. For example, the Indianapolis region has 13 observation sites (Turnbull et al., 2019), Boston's core area only has 2 sites (Sargent et al., 2018), and the Salt Lake City area has 6 (Mitchell et al., 2018). To obtain data from a larger area, researchers use aircraft to collect data, but the temporal resolution of this measurement is low. Researchers can gain high temporal and broad spatial coverage through combining in situ tower measurements and aircraft sampling. This combination has been used to study biogenic carbon flux from Indianapolis (Lopez-Coto et al., 2017).

The uncertainties of this atmospheric CO_2-based method come from the coverage of CO_2 concentration measurement, uncertainties in the fossil fuel emission inventory, and transport models. Sampling sites in cities can be limited. Studies of inverse modeling the fossil fuel emissions from CO_2 concentration have indicated that a higher density of monitoring sites would improve the accuracy of results (Wu et al., 2018). Considering that the inverse modeling theory is the same for fossil fuel and biogenic carbon, it is expected that a higher monitoring density reduces uncertainties in biogenic carbon flux. Besides, a high-quality fossil fuel carbon emission inventory is necessary for this approach to separate the contribution of fossil fuel carbon from total carbon concentration. Regarding transport models, there are errors in atmospheric transport models, which have been discussed in extracting fossil fuel carbon flux from the concentration.

Policy relevance: The requirement for reporting biogenic carbon from land use and land-use change is lower compared to reporting fossil fuel carbon. For example, GPC Basic Plus requires cities to report biogenic carbon from land use, unlike GPC Basic. Fewer cities quantify and report biogenic carbon, comparing to fossil fuel carbon emissions. However, cities are exploring how much urban green spaces, as a key part of nature-based solutions, can offset fossil fuel carbon emissions. Advancements in data and methodology can provide more empirical data on this issue. When comparing biogenic carbon flux versus fossil fuel carbon flux, net biogenic carbon sequestration flux is about 0.6% of fossil fuel carbon flux in the city of Boston (Hardiman et al., 2017). Due to the small proportion of biogenic carbon stock in urban settlements, the role of urban green space is limited in directly offsetting carbon emissions from fossil fuel combustion in cities. Furthermore, the effectiveness of urban green spaces in offsetting fossil fuel emissions should be examined and compared with other policy interventions, such as energy use efficiency improvement, food-to-energy strategies, and mixed land use.

4 From individual cities to initiatives for all urban areas' carbon accounting

When advancements in individual cities' carbon accounting approaches inform different urban climate policies, fewer efforts either at the national or at the global level are made to control carbon emissions associated with activities in all urban areas. For example, countries committed to the Paris Agreement have released their intended nationally determined contributions, but few of them specify actions to control or reduce GHGs from all urban areas. And while individual cities have voluntarily taken efforts to evaluate and control their GHG emissions, fewer studies have quantified the collective impacts of these actions on reducing national carbon emissions.

One reason is that we lack policy-relevant carbon emission databases that cover all urban areas in a nation. Most of the time, national or global anthropogenic GHG inventories are constructed based on activity sectors (i.e., residential sector, commercial sector, industrial sector, and transportation) rather than urban versus rural areas. Several studies have begun to estimate the contribution from all urban areas, but the value is highly uncertain due to different accounting scopes and urban definitions (Seto et al., 2014). New approaches are needed to develop databases that cover all cities/urban areas with a focus on energy use and strong alignment with national totals. Recent studies have compiled city-level databases covering 200–300 cities globally (Creutzig, Baiocchi, Bierkandt, Pichler, & Seto, 2015; Nangini et al., 2019). Although these studies have advanced the science of city typology (Creutzig et al., 2015), their datasets could neither add up to the national total for revealing uncertainty in the database nor demonstrate collective contribution of urban mitigation actions from all cities in a nation.

How to define urban areas or cities is the first question when developing such databases. There are differences between urban agglomeration areas and cities. Urban areas can be defined based on population size and a visible built-up settlement boundary. In comparison, cities are defined as an administrative jurisdiction with its own local governing body. These two definitions are not the same but can overlap. For example, the US census defines urbanized areas as a continuously built-up area housing 50,000 or more people in its boundary. A US city, as an administrative unit having an executive governmental system, can occupy part of this urbanized area. Countries across the world define cities differently (United Nations, Department of Economic and Social Affairs & Population Division, 2019). Four general criteria used to define cities include administrative, economic, population size/density, and urban characteristics, according to World Urbanization Prospects (United Nations, Department of Economic and Social Affairs & Population Division, 2019). Cities are defined based either on one criterion or a combination of multiple criteria. When quantifying GHGs from all urban areas at the global level, it is not easy to use a universal definition of cities across counties. Scholars adopt definitions of cities/urban areas, depending on data types and research questions.

Several studies have quantified direct GHG emissions from fossil fuel combustion in all urban areas nationwide or globally. Marcotullio et al. downscaled EDGAR emission inventory (constructed based on IPCC's national GHG guidelines) to urban areas based on population and economic output data covering urban areas at the global level (Marcotullio, Sarzynski, Albrecht, Schulz, & Garcia, 2013). In 2000, direct emissions related to energy use from urban areas accounted for about 41.5% of total energy emissions. The Open-source Data Inventory for Anthropogenic CO_2 (ODIAC) is developed at the spatial scale of 1 km, allocating national emission inventories using satellite night data and significant emission point sources (such as power plants) (Oda et al., 2019). Gridded emissions can be aggregated to urban agglomeration

368 Section | C Case Studies

areas at the global level. In the United States, the Vulcan database includes detailed local information on power plants, large emission facilities, and county-level travel data (Parshall et al., 2010). This study found that direct energy use in all urban counties in the United States contributed to 37% of nationwide carbon emissions in 2002, while this proportion is sensitive to how we define urban areas (Parshall et al., 2010). Due to the rapid development of Scope 1 emission inventories, these data products have been compared with each other to reveal uncertainty in different databases. These inventories are valuable for linking urban areas' carbon emission fluxes with global climate change modeling. As discussed above, a purely Scope 1 emission inventory has limited policy relevance to urban climate policies and databases, and a focus on use activities in cities is needed.

Researchers have explored how to develop carbon emission inventories of all urban areas with a focus on energy-use activities instead of direct fossil fuel CO_2 fluxes. A hybrid bottom-up and top-down approach has been developed to construct the Chinese-City-Industrial-Infrastructure database (Ramaswami et al., 2017a; Tong et al., 2018). This database includes CO_2 emissions embodied in imported electricity use, in addition to Scope 1 fossil fuel emissions in Chinese cities (Ramaswami et al., 2017a; Tong et al., 2018). Chinese cities collectively contributed 62% (Scope 1) to 67% (Scope 1+2) of national energy-related CO_2 emissions in 2010 (Tong et al., 2018). This difference between Scope 1 and Scope 1+2 indicates that the accounting scopes influence our understanding of collective carbon contributions from all urban areas. Moreover, urban-industrial symbiosis strategies to maximize material and energy exchanges in urban areas can contribute to an additional 40% reduction in carbon by improving nationwide efficiency in China (Ramaswami et al., 2017a). This database was also used to quantify local-level carbon mitigation and health cobenefits at the national scale (Ramaswami et al., 2017a). The U.S. Department of Energy and National Renewable Energy Laboratory developed a State and Local Energy Database that includes fossil fuel and electricity used in residential, commercial, industrial sectors, and transportation at the city level (USDOE & NREL, 2015). This database helps cities to evaluate the carbon mitigation of local policies (Aznar, Day, Doris, Mathur, & Donohoo-Vallett, 2015). These examples demonstrate how a policy-relevant database focusing on use activities can be used to evaluate systemic carbon mitigation actions. Currently, nations and cities are making efforts to significantly reduce GHG emissions. Deep-decarbonization actions are expected to shape energy systems and transportation infrastructure significantly. Urban infrastructure use and provision databases can support both urban and national infrastructure policies for deep decarbonization.

The consumption-based carbon footprinting approach also has been implemented at the national level, covering all urban areas focusing on household consumption. Minx et al. quantified household carbon footprints in 434 municipalities in the United Kingdom (Minx et al., 2013). Jones and Kammen quantified household carbon footprints covering all urban areas in the continental United States (Jones & Kammen, 2014). Moran et al. have downscaled national or regional household carbon footprints to 13,000 urban systems (Moran et al., 2018). This purely top-down method used the purchasing power index and urban population information to allocate carbon footprints from the national or subnational level (depending on data availability) (Moran et al., 2018). These consumption-based carbon footprint databases have been used to study the impacts of urban form, local climate factors, and lifestyles on urban residents' carbon footprints (Jones & Kammen, 2014; Minx et al., 2013; Moran et al., 2018).

Regarding biogenic carbon emissions, researchers have adopted surveys plus scaling-up methods to quantify carbon sequestrated in urban green spaces, including trees, urban soil systems, grassland, etc. These studies combined field survey data and land-cover maps extracted from remote sensing or imagery data. Nowak et al. surveyed trees in 28 cities and derived national urban tree coverage from Google Earth's pictures using aerial photograph interpretation (Nowak, Greenfield, Hoehn, & Lapoint, 2013). A similar method is also adopted to estimate the organic carbon in urban soil systems in sampled cities, which is then scaled to the national level (Pouyat et al., 2006). It is estimated that urban soil systems capture about 1940 million metric tonnes of carbon (Pouyat et al., 2006), and the urban forest can capture 643 million metric tonnes of carbon in all US urban communities (Nowak et al., 2013). These studies contribute to identifying the capacity of carbon sequestration in urban green spaces. A potential future research direction could be to reveal the total potential of biogenic carbon in urban carbon budgets for all urban areas.

Various remote sensing data can contribute to developing carbon emission and sequestration databases covering all urban areas. Remote sensing data have been used for improving the quality of city-level carbon emission inventories. Nighttime light data are used to construct high spatial resolution gridded emissions with other fossil fuel emission data (Oda, Maksyutov, & Andres, 2018). Satellite imagery can be used to track building footprints in cities over time (Mahtta, Mahendra, & Seto, 2019), providing new data for estimating building material use. Urban morphology can be detected using satellite data (Stokes & Seto, 2019). Researchers can further examine how urban activities, such as mobility and energy use, are influenced by urban form (Stokes & Seto, 2019). Land use and vegetation imagery products are widely used to reveal land-use patterns. These data are foundational for calculating biogenic carbon emissions from land-use change. Remote sensing data will be used more often when developing future all-encompassing urban-area carbon emission databases.

New data sources and analytical techniques, in addition to remote sensing data, can either enlarge data coverage or reduce uncertainty in the estimates of use activities. Smartphones, smart meters, and social media expand the types of data that can be extracted, including location, text, images, video, and even

370   Section | C Case Studies

personal preferences. These data can provide metrics for evaluating city-level energy and carbon emissions using machine learning, deep learning, or artificial intelligence techniques. For example, all rooftop solar panels were identified in the continental United States using deep learning (Yu, Wang, Majumdar, & Rajagopal, 2018). A Microsoft team used deep learning to process images and generated a building footprint database covering all states in the United States (Microsoft, 2018). StreetLight Data, a private company, collects GPS and phone signal tower data to estimate daily travel demand and other travel-related metrics using machine learning techniques. New data also can improve the accuracy of use activities. Mobile phone usage data were used to correct the occupancy rate of buildings to better estimate energy consumption in buildings in Boston (Barbour et al., 2019). "Big data" was used to map direct and indirect emissions of a city district, covering transportation and building energy use (Yamagata, Yoshida, Murakami, Matsui, & Akiyama, 2018). These new data sources and analytical techniques can be used to develop national urban-area carbon emission databases in the future.

References

Aznar, A., Day, M., Doris, E., Mathur, S., & Donohoo-Vallett, P. (2015). *City-level energy decision making: data use in energy planning, implementation, and evaluation in US cities*. Golden, CO: National Renewable Energy Laboratory (NREL).

Bailis, R., Drigo, R., Ghilardi, A., & Masera, O. (2015). The carbon footprint of traditional woodfuels. *Nature Climate Change, 5*, 266. https://doi.org/10.1038/nclimate2491.

Barbour, E., et al. (2019). Planning for sustainable cities by estimating building occupancy with mobile phones. *Nature Communications, 10*, 3736. https://doi.org/10.1038/s41467-019-11685-w.

Boyer, D., & Ramaswami, A. (2017). What is the contribution of city-scale actions to the overall food system's environmental impacts?: Assessing water, greenhouse gas, and land impacts of future urban food scenarios. *Environmental Science & Technology, 51*, 12035–12045. https://doi.org/10.1021/acs.est.7b03176.

C40. (2018). *Consumption-based GHG emissions of C40 cities*.

Chavez, A., & Ramaswami, A. (2013). Articulating a trans-boundary infrastructure supply chain greenhouse gas emission footprint for cities: Mathematical relationships and policy relevance. *Energy Policy, 54*, 376–384. https://doi.org/10.1016/j.enpol.2012.10.037.

Chavez, A., Ramaswami, A., Nath, D., Guru, R., & Kumar, E. (2012). Implementing trans-boundary infrastructure-based greenhouse gas accounting for Delhi, India. *Journal of Industrial Ecology, 16*, 814–828. https://doi.org/10.1111/j.1530-9290.2012.00546.x.

Chen, S., Chen, B., Feng, K., Liu, Z., Fromer, N., Tan, X., et al. (2020). Physical and virtual carbon metabolism of global cities. *Nature Communications, 11*(182), 3–11. https://doi.org/10.1038/s41467-019-13757-3.

Chen, S., Long, H., Chen, B., Feng, K., & Hubacek, K. (2019). Urban carbon footprints across scale: Important considerations for choosing system boundaries. *Applied Energy*, 114201. https://doi.org/10.1016/j.apenergy.2019.114201.

Chen, G., Wiedmann, T., Wang, Y., & Hadjikakou, M. (2016). Transnational city carbon footprint networks—Exploring carbon links between Australian and Chinese cities. *Applied Energy, 184*, 1082–1092. https://doi.org/10.1016/j.apenergy.2016.08.053.

Urban environments and trans-boundary linkages **Chapter | 10** **371**

Chen, G., et al. (2019). Review on city-level carbon accounting. *Environmental Science & Technology, 53*, 5545–5558. https://doi.org/10.1021/acs.est.8b07071.

Creutzig, F., Baiocchi, G., Bierkandt, R., Pichler, P.-P., & Seto, K. C. (2015). Global typology of urban energy use and potentials for an urbanization mitigation wedge. *Proceedings of the National Academy of Sciences, 112*, 6283–6288. https://doi.org/10.1073/pnas.1315545112.

Deep-Decarbonization Pathways Project. (2015). *Pathways to deep decarbonization 2015 report*. SDSN and IDDRI.

Feng, K., Hubacek, K., Sun, L., & Liu, Z. (2014). Consumption-based CO2 accounting of China's megacities: The case of Beijing, Tianjin, Shanghai and Chongqing. *Ecological Indicators, 47*, 26–31. https://doi.org/10.1016/j.ecolind.2014.04.045.

Fong, W. K., et al. (2014). *Global protocol for community-scale greenhouse gas emission inventories: An accounting and reporting standard for cities*. World Resources Institute; C40 Cities Climate Leadership Group; ICLEI—Local Governments for Sustainability.

Gurney, K. R., et al. (2009). High resolution fossil fuel combustion CO2 emission fluxes for the United States. *Environmental Science & Technology, 43*, 5535–5541. https://doi.org/10.1021/es900806c.

Gurney, K. R., et al. (2012). Quantification of fossil fuel CO2 emissions on the building/street scale for a large U.S. city. *Environmental Science & Technology, 46*, 12194–12202. https://doi.org/10.1021/es3011282.

Gurney, K. R., et al. (2019). The Hestia fossil fuel CO2 emissions data product for the Los Angeles megacity (Hestia-LA). *Earth System Science Data, 11*, 1309–1335. https://doi.org/10.5194/essd-11-1309-2019.

Hardiman, B. S., et al. (2017). Accounting for urban biogenic fluxes in regional carbon budgets. *Science of the Total Environment, 592*, 366–372. https://doi.org/10.1016/j.scitotenv.2017.03.028.

Heijungs, R., & Lenzen, M. (2014). Error propagation methods for LCA—A comparison. *The International Journal of Life Cycle Assessment, 19*, 1445–1461. https://doi.org/10.1007/s11367-014-0751-0.

Hillman, T., Janson, B., & Ramaswami, A. (2011). Spatial allocation of transportation greenhouse gas emissions at the city scale. *Journal of Transportation Engineering, 137*, 416–425. https://doi.org/10.1061/(ASCE)TE.1943-5436.0000136.

Hillman, T., & Ramaswami, A. (2010). Greenhouse gas emission footprints and energy use benchmarks for eight US cities. *Environmental Science & Technology, 44*, 1902–1910.

Hu, Y., Lin, J., Cui, S., & Khanna, N. Z. (2016). Measuring urban carbon footprint from carbon flows in the global supply chain. *Environmental Science & Technology, 50*, 6154–6163. https://doi.org/10.1021/acs.est.6b00985.

Hutyra, L. R., Yoon, B., & Alberti, M. (2011). Terrestrial carbon stocks across a gradient of urbanization: A study of the Seattle, WA region. *Global Change Biology, 17*, 783–797. https://doi.org/10.1111/j.1365-2486.2010.02238.x.

IPCC. (2006). In H. S. Eggleston, L. Buendia, K. Miwa, T. Ngara, & K. Tanabe (Eds.), *2006 IPCC Guidelines for National Greenhouse Gas Inventories, Prepared by the National Greenhouse Gas Inventories Programme*. Japan: IGES.

Jones, C. M., & Kammen, D. M. (2011). Quantifying carbon footprint reduction opportunities for U.S. households and communities. *Environmental Science & Technology, 45*, 4088–4095. https://doi.org/10.1021/es102221h.

Jones, C., & Kammen, D. M. (2014). Spatial distribution of U.S. household carbon footprints reveals suburbanization undermines greenhouse gas benefits of urban population density. *Environmental Science & Technology, 48*, 895–902. https://doi.org/10.1021/es4034364.

372 Section | C Case Studies

Kennedy, C., et al. (2009). Greenhouse gas emissions from global cities. *Environmental Science & Technology, 43*, 7297–7302.

Kennedy, C., et al. (2010). Methodology for inventorying greenhouse gas emissions from global cities. *Energy Policy, 38*, 4828–4837.

Kennedy, C. A., Ramaswami, A., Carney, S., & Dhakal, S. (2011). Greenhouse gas emission baselines for global cities and metropolitan regions. *Cities and climate change: Responding to an urgent agenda* (pp. 15–54). World Bank.

Larsen, H. N., & Hertwich, E. G. (2009). The case for consumption-based accounting of greenhouse gas emissions to promote local climate action. *Environmental Science & Policy, 12*, 791–798. https://doi.org/10.1016/j.envsci.2009.07.010.

Lenzen, M. (2000). Errors in conventional and input-output—based life—cycle inventories. *Journal of Industrial Ecology, 4*, 127–148. https://doi.org/10.1162/10881980052541981.

Lin, J., Liu, Y., Meng, F., Cui, S., & Xu, L. (2013). Using hybrid method to evaluate carbon footprint of Xiamen City, China. *Energy Policy, 58*, 220–227. https://doi.org/10.1016/j.enpol. 2013.03.007.

Lopez-Coto, I., Prasad, K., & Whetstone, J. (2017). *Carbon dioxide biogenic vs anthropogenic sectoral contribution for the Indianapolis flux experiment (INFLUX).* https://doi.org/10.6028/ NIST.SP.1237.

Mahtta, R., Mahendra, A., & Seto, K. C. (2019). Building up or spreading out? Typologies of urban growth across 478 cities of 1 million+. *Environmental Research Letters, 14*, 124077. https://doi. org/10.1088/1748-9326/ab59bf.

Marcotullio, P. J., Sarzynski, A., Albrecht, J., Schulz, N., & Garcia, J. (2013). The geography of global urban greenhouse gas emissions: An exploratory analysis. *Climatic Change, 121*, 621–634. https://doi.org/10.1007/s10584-013-0977-z.

Microsoft. (2018). *US building footprints.*

Minx, J., et al. (2013). Carbon footprints of cities and other human settlements in the UK. *Environmental Research Letters, 8*, 035039. https://doi.org/10.1088/1748-9326/8/3/035039.

Mitchell, L. E., et al. (2018). Long-term urban carbon dioxide observations reveal spatial and temporal dynamics related to urban characteristics and growth. *Proceedings of the National Academy of Sciences, 115*, 2912–2917. https://doi.org/10.1073/pnas.1702393115.

Moran, D., et al. (2018). Carbon footprints of 13000 cities. *Environmental Research Letters, 13*, 064041. https://doi.org/10.1088/1748-9326/aac72a.

Murakami, K., Kaneko, S., Dhakal, S., & Sharifi, A. (2020). Changes in per capita CO2 emissions of six large Japanese cities between 1980 and 2000: An analysis using "The Four System Boundaries" approach. *Sustainable Cities and Society, 52*, 101784. https://doi.org/10.1016/j.scs. 2019.101784.

Nangini, C., et al. (2019). A global dataset of CO2 emissions and ancillary data related to emissions for 343 cities. *Scientific Data, 6*, 180280. https://doi.org/10.1038/sdata.2018.280.

Nowak, D. J., & Crane, D. E. (2002). Carbon storage and sequestration by urban trees in the USA. *Environmental Pollution, 116*, 381–389. https://doi.org/10.1016/S0269-7491(01)00214-7.

Nowak, D. J., Greenfield, E. J., Hoehn, R. E., & Lapoint, E. (2013). Carbon storage and sequestration by trees in urban and community areas of the United States. *Environmental Pollution, 178*, 229–236. https://doi.org/10.1016/j.envpol.2013.03.019.

Oda, T., Maksyutov, S., & Andres, R. J. (2018). The Open-source Data Inventory for Anthropogenic CO2, version 2016 (ODIAC2016): A global monthly fossil fuel CO2 gridded emissions data product for tracer transport simulations and surface flux inversions. *Earth System Science Data, 10*, 87–107. https://doi.org/10.5194/essd-10-87-2018.

Oda, T., et al. (2019). Errors and uncertainties in a gridded carbon dioxide emissions inventory. *Mitigation and Adaptation Strategies for Global Change*, *24*, 1007–1050. https://doi.org/10.1007/s11027-019-09877-2.

Parshall, L., et al. (2010). Modeling energy consumption and CO2 emissions at the urban scale: Methodological challenges and insights from the United States. *Energy Policy*, *38*, 4765–4782. https://doi.org/10.1016/j.enpol.2009.07.006.

Pitt, J. R., et al. (2019). Assessing London CO2, CH4 and CO emissions using aircraft measurements and dispersion modelling. *Atmospheric Chemistry and Physics*, *19*, 8931–8945. https://doi.org/10.5194/acp-19-8931-2019.

Pouyat, R. V., Yesilonis, I. D., & Nowak, D. J. (2006). Carbon storage by urban soils in the United States. *Journal of Environmental Quality*, *35*, 1566–1575. https://doi.org/10.2134/jeq2005.0215.

Ramaswami, A., & Chavez, A. (2013). What metrics best reflect the energy and carbon intensity of cities? Insights from theory and modeling of 20 US cities. *Environmental Research Letters*, *8*, 035011. https://doi.org/10.1088/1748-9326/8/3/035011.

Ramaswami, A., Hillman, T., Janson, B., Reiner, M., & Thomas, G. (2008). A demand-centered, hybrid life-cycle methodology for city-scale greenhouse gas inventories. *Environmental Science & Technology*, *42*, 6455–6461.

Ramaswami, A., Russell, A. G., Culligan, P. J., Sharma, K. R., & Kumar, E. (2016). Meta-principles for developing smart, sustainable, and healthy cities. *Science*, *352*, 940–943. https://doi.org/10.1126/science.aaf7160.

Ramaswami, A., et al. (2017a). Urban cross-sector actions for carbon mitigation with local health co-benefits in China. *Nature Climate Change*, *7*, 736–742. https://doi.org/10.1038/nclimate3373.

Ramaswami, A., et al. (2017b). An urban systems framework to assess the trans-boundary food-energy-water nexus: Implementation in Delhi, India. *Environmental Research Letters*, *12*, 025008. https://doi.org/10.1088/1748-9326/aa5556.

Ramaswami, A., Tong, K., Canadell, J., Jackson, R., Stokes, E., Dhakal, S., et al. (2021). Carbon analytics for net-zero emissions sustainable cities. *Nature Sustainability*. https://doi.org/10.1038/s41893-021-00715-5.

Sargent, M., et al. (2018). Anthropogenic and biogenic CO2 fluxes in the Boston urban region. *Proceedings of the National Academy of Sciences of the United States of America*, *115*, 7491–7496. https://doi.org/10.1073/pnas.1803715115.

Seto, K. C., Dhakal, S., Bigio, A., Blanco, H., Delgado, G. C., Dewar, D., et al. (2014). Human settlements, infrastructure and spatial planning. In *Climate change 2014: Mitigation of climate change. Contribution of working group III to the fifth assessment report of the intergovernmental panel on climate change.* Cambridge, United Kingdom and New York, NY, USA: Cambridge University Press.

Shan, Y., Liu, J., Liu, Z., Shao, S., & Guan, D. (2019). An emissions-socioeconomic inventory of Chinese cities. *Scientific Data*, *6*, 190027. https://doi.org/10.1038/sdata.2019.27.

Smith, I. A., Dearborn, V. K., & Hutyra, L. R. (2019). Live fast, die young: Accelerated growth, mortality, and turnover in street trees. *PLoS One*, *14*, e0215846. https://doi.org/10.1371/journal.pone.0215846.

Stokes, E. C., & Seto, K. C. (2019). Characterizing and measuring urban landscapes for sustainability. *Environmental Research Letters*, *14*, 045002. https://doi.org/10.1088/1748-9326/aafab8.

Tong, K., et al. (2016). Greenhouse gas emissions from key infrastructure sectors in larger and smaller Chinese cities: Method development and benchmarking. *Carbon Management*, *7*, 27–39. https://doi.org/10.1080/17583004.2016.1165354.

374 Section | C Case Studies

Tong, K., et al. (2018). The collective contribution of Chinese cities to territorial and electricity-related CO2 emissions. *Journal of Cleaner Production, 189*, 910–921. https://doi.org/10.1016/j.jclepro.2018.04.037.

Turnbull, J. C., et al. (2019). Synthesis of urban CO2 emission estimates from multiple methods from the Indianapolis flux project (INFLUX). *Environmental Science & Technology, 53*, 287–295. https://doi.org/10.1021/acs.est.8b05552.

United Nations, Department of Economic and Social Affairs, Population division. (2015). *World urbanization prospects: The 2014 revision, highlights* (p. 1). New York City, USA: United Nations.

U.S. Conference of Mayors. (2005). *Endorsing the U.S. Mayors climate protection agreement.* https://www.usmayors.org/the-conference/resolutions/?category=c1212&meeting=73rd%20Annual%20Meeting.

UNEP. (2013). *City-level decoupling: Urban resource flows and the governance of infrastructure transitions.* Norway: Birkeland Trykkeri AS.

UN-Habitat. (2016). *Urbanization and development: emerging futures—Key findings and messages.* New York: UN-Habitat, Pub. United Nations.

United Nations, Department of Economic and Social Affairs & Population Division. (2019). *World urbanization prospects: The 2018 revision, methodology.* New York City United Nations.

U.S. Geological Survey (USGS) Gap Analysis Project (GAP). (2016). *GAP/LANDFIRE National Terrestrial Ecosystems 2011.* U.S. Geological Survey.

USDOE, & NREL. (2015). State and local energy use data. *Website*.

World Mayors Summit on Climate. (2010). *The global cities covenant on climate ("Mexico City Pact").* http://mexicocitypact.org.

Wu, K., Lauvaux, T., Davis, K. J., Deng, A., Lopez Coto, I., Gurney, K. R., et al. (2018). Joint inverse estimation of fossil fuel and biogenic CO 2 fluxes in an urban environment: An observing system simulation experiment to assess the impact of multiple uncertainties. *Elementa: Science of the Anthropocene, 6*(17), 1–19.

Yamagata, Y., Yoshida, T., Murakami, D., Matsui, T., & Akiyama, Y. (2018). Seasonal urban carbon emission estimation using spatial micro big data. *Sustainability, 10*, 4472.

Ye, X., et al. (2017). Constraining fossil fuel CO2 emissions from urban area using OCO-2 observations of total column CO2. *Atmospheric Chemistry and Physics Discussions, 2017*, 1–30. https://doi.org/10.5194/acp-2017-1022.

Yu, J., Wang, Z., Majumdar, A., & Rajagopal, R. (2018). DeepSolar: A machine learning framework to efficiently construct a solar deployment database in the United States. *Joule, 2*, 2605–2617. https://doi.org/10.1016/j.joule.2018.11.021.

Chapter 11

Agricultural systems

Stephen M. Ogle[a], Pete Smith[b], Francesco N. Tubiello[c], Shawn Archibeque[d], Miguel Taboada[e], Donovan Campbell[f], and Cynthia Nevison[g]

[a]*Natural Resource Ecology Laboratory and Department of Ecosystem Science and Sustainability, Colorado State University, Fort Collins, CO, United States,* [b]*Institute of Biological & Environmental Sciences, University of Aberdeen, Aberdeen, United Kingdom,* [c]*Statistics Division, Food and Agriculture Organization of the United Nations, Rome, Italy,* [d]*Department of Animal Sciences, Colorado State University, Fort Collins, CO, United States,* [e]*National Agricultural Technology Institute (INTA), Natural Resources Research Center (CIRN), Institute of Soils, Ciudad Autónoma de Buenos Aires, Argentina,* [f]*The University of the West Indies, Jamaica, West Indies,* [g]*Institute for Arctic and Alpine Research, University of Colorado, Boulder, CO, United States*

1 Introduction

Agriculture provides the world's food, feed, and fiber and may become a key component of bioenergy production in the future (FAO, 2018). Agricultural lands cover approximately 37% of the world's land surface (FAO, 2019; Ramankutty, Evan, Monfreda, & Foley, 2008). In general, land use systems are a significant source of anthropogenic GHG emissions including agricultural land management and forestry (Mbow et al., 2019; Smith et al., 2014). For 2006–15, the total net carbon flux from agriculture and forestry has averaged 1.11 $(\pm 0.35)\,GtC\,yr^{-1}$ with a net source associated with the tropics $(1.41 \pm 0.17\,GtC\,yr^{-1})$, a net sink associated with northern mid-latitudes $(-0.28 \pm 0.21\,GtC\,yr^{-1})$, and carbon neutral emissions in the southern mid-latitudes (Houghton & Nassikas, 2017). Land use change is a key driver of the total emissions, particularly deforestation for purposes of crop production and livestock grazing. When all relevant greenhouse gases (GHG) are considered, emissions from associated land use and change have averaged 4.8 $\pm 2.4\,GtCO_2\,eq\,yr^{-1}$ over the 2007–16 period (Tubiello, 2018).

Agriculture alone, defined as the set of within-farm-gate activities, emits approximately 5.4 Gt CO_2 eq yr^{-1} of GHG, about 10%–12% of the total annual anthropogenic GHG emissions (Tubiello, 2018; Tubiello et al., 2015, 2013). Additional GHG emissions linked to food and agriculture are associated with the concept of food systems, i.e., the entire web of actions and actors connecting

Balancing Greenhouse Gas Budgets. https://doi.org/10.1016/B978-0-12-814952-2.00009-5
Copyright © 2022 Elsevier Inc. All rights reserved.

food supply and demand including food supply chains (Fig. 1). In addition to emissions within the farm gate and those from land use and land use change, these additional emissions include those from manufacturing of pesticides, fertilizer, and animal feed; emissions due to the burning of fossil fuels from usage of farm machinery; supply chain emission associated with processing, transport, storage, and packaging of products. When incorporating all of these emission sources along with emissions from land use, land use change, and livestock systems, the total contribution of food systems has been estimated in the range of 30%–40% of total global anthropogenic emissions (Mbow et al., 2019;

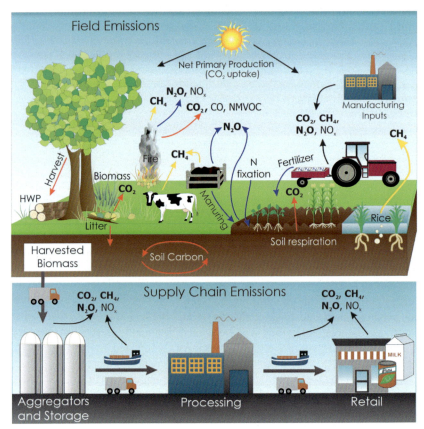

FIG. 1 GHG emissions from the web of activities associated with agricultural food production, including (A) manufacturing product inputs that are used in farming operations, such as fertilizers and pesticides; (B) GHG emissions from within-farm-gate crop and livestock management; and (C) GHG emissions from transport, storage, processing, and selling agricultural products. *(This figure was adapted with permission from the Intergovernmental Panel on Climate Change from Fig. 1.1 in Chapter 1, Volume IV of the 2006 IPCC National Greenhouse Gas Inventory Guidelines. This report was prepared by the IPCC Task Force on National Greenhouse Gas Inventories and published by the Institute for Global Environmental Strategies on behalf of the IPCC.)*

Vermeulen, Campbell, & Ingram, 2012). While all of these estimates provide valuable information about the contribution of agriculture to global emissions, it is important to recognize that estimates do vary depending on the boundary of the study and consequently the sources that are included in the analysis.

To understand the role of agriculture in regional greenhouse gas budgets, it is important to understand the context and role of agriculture in society. The most critical issue surrounding agriculture is food supply, food nutrition, and security, i.e., maintaining safe food supplies at sufficient levels to feed society. Food security is a high priority for policymakers due to a growing global population that is expected to reach 9–10 billion people by 2050, leading to considerably more pressure on agricultural systems to meet the caloric and nutritional needs of society (FAO, 2018; Godfray et al., 2010). With this level of growth, enhancing food security will become even more challenging for governments. There is potential to promote sustainable practices to help meet food security goals while avoiding the negative effects of agriculture on the environment, including air pollution from GHG emissions (Bajželj et al., 2014; Paustian et al., 2016; Smith et al., 2008, 2013). Agronomic strategies that can be devised in support of such challenges will likely include improving crop varieties and intensification of production on existing cropland (Chen et al., 2014; Mueller et al., 2012; Tilman et al., 2011; Tilman, Cassman, Matson, Naylor, & Polasky, 2002).

Other food security strategies aim at reducing food waste at different stages of the food chain (Ericksson, Strid, & Hansson, 2015; FAO, 2019b). In addition, societal demand may also influence total emissions via changes in GHG emission intensities of food products (amount of emission per unit production). For example, meat products have a higher GHG emission intensity than grain products. Consequently, society could effectively reduce GHG emissions from agriculture by increasing demand for grain products to meet nutritional and caloric needs while reducing demand for meat products (Bajželj et al., 2014; Mbow et al., 2019).

Food and agriculture, and food systems in general, will likely be a significant contributor to GHG emissions in the future, but emissions could be reduced depending on policy, management, and societal choice. Regional GHG emission inventories associated with agriculture will be an important source of information for evaluating the outcomes of any actions, such as the recent 4 per mille initiative to decrease GHG emissions through soil C sequestration (Minasny et al., 2017).

Given this context, our objective is to review the methods that are used to conduct regional GHG inventories for crops and livestock systems. This review provides a summary of our understanding of carbon stocks and flows through these systems, as well as some discussion of methane (CH_4) and nitrous oxide (N_2O) emissions, which are major sources of GHG emissions in agricultural systems (Mbow et al., 2019; Smith et al., 2014; Tubiello, 2019; Tubiello et al., 2013). In the latter part of this chapter, we discuss methods for quantifying GHG emissions. The methods fall into three Tiers as defined by the Intergovernmental Panel on Climate Change ,(Ogle et al., 2019; Paustian et al., 2006). Tier 1 includes simple methods with equations and emissions factors provided in the IPCC guidelines. Tier 2 involves development of country-specific emission factors, while typically relying on the same Tier 1 equations provided in the IPCC guidelines to calculate emissions. Tier 3 involves development of country-specific methods, either measurement- or model-based approaches. While the IPCC guidelines (Ogle et al., 2019; Paustian et al., 2006) mostly focus on bottom-up approaches for estimating GHG emissions from source categories, other methods can also be used that utilize top-down approaches with inverse modeling (Ciais et al., 2011; West, Brown, Duren, Ogle, & Moss, 2013). A few examples of global datasets of GHG emissions from agriculture are provided in Table 1.

Several developments could improve quantification of regional GHG emissions from agricultural systems. These improvements include (a) more complete activity data on agricultural statistics, including information on management practices, (b) advancing to higher tier methods if there are adequate data to support the method, (c) developing or expanding monitoring networks that can inform the methods, and (d) incorporation of top-down methodologies into the inventory framework to verify and better constrain

TABLE 1 Existing datasets providing global estimates of agricultural GHG emissions.

Dataset	Emission sources	General approach	Reference/ download site
US EPA global non-CO_2 GHG emissions data	All non-CO_2 GHG sources in agriculture	IPCC Tier 1	https://www.epa.gov/ ghgemissions/global- greenhouse-gas- emissions-data
FAOSTAT GHG emissions database	All non-CO_2 GHG sources in agriculture and land use change, wetland and fire emissions	IPCC Tier 1	Tubiello et al. (2013)
EDGAR database	All non-CO_2 GHG sources in agriculture and land use change, wetland and fire emissions	IPCC Tier 1	https://www.eea. europa.eu/themes/air/ links/data-sources/ emission-database- for-global- atmospheric

Agricultural systems **Chapter | 11 379**

emissions data. In the remainder of this chapter, we provide information on the emission sources in agricultural systems, estimation methods, and key future advancements for estimation of regional GHG emissions from agricultural systems.

2 Carbon stocks, flows, and emissions in agricultural systems

2.1 Cropping systems

Carbon (C) flows through agricultural systems from the atmosphere as CO_2 into crops; crop biomass is harvested or left as residues to decompose in the crop field; and a portion of C in residues becomes soil organic matter. Plant and microbial respiration returns C to the atmosphere, as well as the consumption and breakdown of the food, feed, and fiber that is harvested. There are also lateral flows of C due to erosion and dissolved organic C that flow across the landscape (Kindler et al., 2011; Van Oost et al., 2007). In addition, lateral flows occur with harvested biomass that can be moved short or longer distances in trade and influence the net fluxes in regions where the biomass is produced and used (Ciais, Bousquet, Freibauer, & Naegler, 2007). There are impacts on inorganic C cycles that contribute to net fluxes of CO_2 in croplands and river basins by enhancing the weathering of soils (Raymond, Oh, Turner, & Broussard, 2008) and through application of carbonate limes to soils (Hamilton, Kurzman, Arango, Jin, & Robertson, 2007; Oh & Raymond, 2006). For example, losses of inorganic C from cropland soils in Europe have been estimated at $14.6\,\mathrm{g\,m^2\,yr^{-1}}$ on a per unit area basis (Kindler et al., 2011).

Net primary production (NPP) in agricultural lands can reach high to intermediate levels of CO_2 uptake during the growing season; for example, croplands in the conterminous United States have high to intermediate levels of CO_2 uptake compared to other vegetation types (Xiao et al., 2008). Over the entire globe, Wolf et al. (2015) estimated total NPP in agricultural lands at 4.6 to 5.0 Gt C from 2005 to 2010, which is approximately 10% of global terrestrial NPP (52–53 Gt C; Zhao & Running, 2010). However, the net balance of CO_2 flux associated with NPP in croplands is almost 0 over a few years because agricultural commodities are typically consumed or otherwise used and CO_2 is returned to the atmosphere over short timescales (Smith et al., 2014; West et al., 2013). The exceptions are agroforestry and perennial tree crop systems, where there could be significant impacts on the regional GHG inventory if this type of cropping is significantly increasing or decreasing over time. Agroforestry and perennial crop systems are not discussed further in this chapter, but examples and discussion can be found in Kim, Kirschbaum, and Beedy (2016), Pardon et al. (2017), and Wolz et al. (2018).

Plant residues are another pool of C in agricultural systems. However, residues that are left in the field typically decompose over a few years and do not contribute to long-term storage of C in the biosphere. Burning of residues and residues that are used for feed also tend to return C to the atmosphere over short timescales and contribute little to net storage or release of C over periods of years to decades. However, non-CO_2 GHG emissions can be released in fires due to burning of agricultural residues as well as prescribed burning and wildfires in savannas, and these emissions are included in GHG inventories (Gupta et al., 2004; McCarty, 2011; Smil, 1999).

Due to limited potential for storage of C in biomass and litter, regional C inventories for agricultural croplands tend to focus on soil organic carbon (SOC) in order to quantify the net uptake or release of CO_2 to the atmosphere. This pool can accumulate or lose SOC over years to decades or centuries. For example, Sanderman, Hengl, and Fiske (2017) estimated that land use change to croplands globally has led to losses of 0.13 Gt C yr^{-1} over the last century. Other losses, such as dissolved organic or inorganic C, are also important, but less attention has been given to quantifying the atmospheric CO_2 fluxes associated with these pathways in GHG inventories. Incorporating other flows should be further investigated and possibly incorporated into future C inventory methods.

Considerable research has focused on estimating regional changes in SOC. For example, Vanden Bygaart, Gregorich, Angers, and Stoklas (2004) estimated total net gains of 3.2–8.3 Mt C yr^{-1} in agricultural soils of Canada for the 1990s. Ogle et al. (2010) estimated a total net gain in SOC of 14.6–17.5 Mt C yr^{-1} during the 1990s for croplands in the United States. In contrast, West et al. (2008) estimated considerably higher rates of SOC accumulation in croplands of the Midwestern United States at 70.3 Mt C yr^{-1} during the same time frame. In China, SOC stocks in croplands have increased by 15–20 Mt C yr^{-1} (Piao et al., 2009), and a significant portion of the C uptake is associated with erosion-deposition processes (Yue et al., 2016). Soil C has also increased in Russia by an average of 6.4 Mt C yr^{-1} during the 1990s due to abandonment of crop production (Vuichard, Ciais, Belelli, Smith, & Valentini, 2008). In some studies, fluxes are estimated on a per unit of area basis. For example, Ciais et al. (2010) estimated a net uptake of 8.3–13.3 g C m^{-2} yr^{-1} for European croplands, which was partly associated with an increase in SOC.

Other regions have lost SOC during recent times, such as the lower Amazon Basin in Brazil with estimated loses between 3.7 and 5.4 Mt C yr^{-1} from 1970 to 2002, combining both land-use change and management impacts on SOC pools (Maia, Ogle, Cerri, & Cerri, 2010). Even though there are historical losses of SOC in this region, there is potential to reduce emissions with the adoption of agricultural systems in South America based on Low-Carbon Agriculture strategies (e.g., no till farming, integrated crop-livestock). These strategies could offset the global annual emissions by 22% from 2021 to 2035 and 25% from 2036 to 2050, as well as mitigate 8.24 Pg C for the period

2016–50 (Sá et al., 2017). In other regions, Sleutel, De Neve, and Hofman (2003) estimated losses of SOC from croplands in Belgium during the 1990s, but Meersmans et al. (2011) found insignificant changes in SOC in Belgium from 1960 to 2006.

Additional GHG emissions can occur from croplands, particularly CH_4 and N_2O. Methane is emitted from soils in anaerobic conditions through a process known as methanogenesis. Methane emissions are high in paddy rice production systems, particularly in Asia where a large proportion of the world's rice is grown (Yan, Akiyama, Yagi, & Akimoto, 2009). Smith et al. (2014) synthesized previous studies and concluded that global CH_4 emissions from paddy rice were between 0.5 and 0.7 Gt CO_2 eq. in 2005.

Nitrous oxide is emitted from soils through processes of nitrification and denitrification (Firestone & Davidson, 1989). Emission rates can increase due to management practices that enhance the availability of mineral N in soils, such as fertilization with synthetic fertilizers and/or organic amendments (e.g., manure), and seeding legume species (Galloway et al., 2008; Mosier, Duxbury, Freney, Heinemeyer, & Minami, 1998; Reay et al., 2012). Smith et al. (2014) concluded that between about 1.5 and 2.2 Gt CO_2 eq. of N_2O was emitted in 2005 from agricultural soils based on a synthesis of available data.

2.2 Livestock systems

Livestock systems are a major source of GHG emissions. As with cropping systems, grazing lands (i.e., grasslands) can be a C source or sink depending on management impacts on SOC pools (Conant, Paustian, & Elliott, 2001). For example, in a recent global review of SOC stock changes in grazing lands, Viglizzo, Ricard, Taboada, and Vázquez-Amábile (2019) estimated that C gains outweigh C losses in grasslands under extensive grazing conditions. Grazing systems are also an important source of N_2O emissions (Oenema, Velthof, Yamulki, & Jarvis, 1997). These same processes drive GHG emissions from grazing lands as discussed in the "Cropping Systems" section.

Livestock add another dimension to emissions from grazing systems compared to cropping systems as discussed in the previous section. Methane emissions from livestock enteric fermentation are typically the largest source of GHG emissions in grazing systems. The bulk of the enteric CH_4 is from ruminant animals that use microbial fermentation to digest cellulose and hemicellulose in grasses, which would have essentially no nutritive value for animals or humans without enteric fermentation. This is not necessarily true for grain-fed livestock in feedlots/confined systems because grains such as corn and soybeans can be digested to meet the daily caloric needs of humans (Schader et al., 2015). Even in feedlot/confined systems, grains can be supplemented with by-products that are not consumed by humans, such as bakery waste, brewer's grains, distiller's grains, soybean meal, and yellow grease.

The production of CH_4 by ruminants begins approximately 4 weeks after birth when the neonatal ruminant is exposed to solid feed that is retained within the reticulorumen (Anderson et al., 1987). As the reticulorumen develops, the extent of fermentation and subsequent CH_4 production rates rise rapidly until reaching full adult rates. Due to the caloric density associated with the formation of enteric CH_4, these emissions represent uncaptured energy from the diets fed to these animals. Therefore, it is beneficial to both productivity of livestock and overall livestock efficiency to reduce the formation of CH_4 within the digestive tract of the animal, with approximately 3%–12% losses of the ingested energy as CH_4 (Johnson et al., 1993). For example, Jayanegara, Leiber, and Kreuzer (2012) concluded that increasing the level of dietary tannins leads to a clear decrease in ruminal CH_4 emissions, although food supplements such as tannins must take into account palatability and adverse effects on animal performance.

In addition to enteric fermentation, livestock systems produce CH_4 and N_2O emissions from manure associated with digestive and metabolic waste. Livestock manure contributes as much as 30%–50% to the global N_2O emissions from agriculture, as a function of type and number of livestock, N excretion per animal, and manure management. In particular, dung and urine that is deposited directly in pastures account for 68% of the total N_2O emissions by livestock (Oenema et al., 2005). This waste represents nutrients and organic C that were not used within the livestock digestive system. Volatile solids in manure become an energy source for microbial organisms to drive methanogenesis and CH_4 emissions. Nitrous oxide is emitted through nitrification and denitrification from urea and volatile solids directly deposited on soils in paddock, range, and pasture systems, as well as manure in storage and after application to the land as soil amendments.

Nutrients and undigested organic C in the manure systems are the main determinants of emissions, and can be minimized by enhancing the efficiency of livestock production to utilize more of the energy and nutritional value of the feedstock. One of the key strategies is to reduce the time needed for animals to reach maturity through better quality feeds and forage. Nevertheless, livestock will never be 100% efficient and some loss and associated GHG emissions are inevitable from these production systems.

Overall, around 6%–8% of global anthropogenic greenhouse gas emissions are from livestock production systems, dominated by enteric fermentation and manure management and applications (FAOSTAT, 2019). This figure can be as high as 14.5% if extended to a life cycle analysis, which includes emissions from feedstock production, transport, and other GHG emissions beyond just CH_4 and N_2O from enteric fermentation and manure management (Gerber et al., 2013). For example, Gerber et al. (2013) estimated that there is approximately 3.3 Gt CO_2 eq. associated with the production, processing, and transport of feed, which in turn accounts for approximately 45% of the emissions from livestock systems.

2.3 Lateral transport and supply chains

Another key consideration in regional GHG budgets associated with agriculture is the movement of commodities with trade. In terms of net effect, there is a paucity of evidence of the impact of food supply on GHG emissions (Zimmermann, Benda, Webber, & Jafari, 2018). Poore and Nemecek (2018) estimated that contemporary food supply chains contribute approximately 26% of GHG emissions (approximately 14 Gt CO_2 eq.) with significant regional variation. In the EU for example, Sandström et al. (2018) estimated the average GHG footprint of food supply at 1.1 Mg CO_2-eq. $cap^{-1} yr^{-1}$.

While food trade makes a relatively minor contribution to supply chain emissions (Edwards-Jones et al., 2008; Garnett, 2011), refrigeration is estimated to account for 15% of global energy consumption and is the most energy-intensive component of the food system (Vermeulen et al., 2012). One estimate suggests that emission from transport and refrigeration associated with livestock production can account for up to 24%–32% of total livestock emissions (Weiss & Leip, 2012). In a recent study, Accorsi, Gallo, and Manzini (2017) highlighted the energy-intensive processes involved with food storage and distribution and the need to improve energy efficiency along the 'cold chain' (Ingram et al., 2016). Another estimate suggests that the cold chain accounts for 1% of CO_2 production globally (James & James, 2010). Improving efficiency could reduce food waste and health risks associated with poor storage management practices, as well as contribute to a reduction in GHG emissions (James & James, 2010; Vermeulen et al., 2012).

3 Methodologies

Methodologies for estimating GHG emissions and C stock changes fall into two general categories, bottom-up inventories and top-down methodologies. Bottom-up inventory approaches may range from simple accounting approaches, such as Tier 1 and 2 methods as defined by the IPCC to measurement-based approaches and complex process-based models, which are classified at Tier 3 methods by the (Ogle et al., 2019; Paustian et al., 2006). These approaches estimate the impact of management activities and environmental drivers on individual sources of emissions, such as cropland or livestock, and individual units associated with the source, such as a parcel of land or an individual animal. Emissions from individual units are aggregated to produce a total for the source, and then sources are aggregated to obtain total regional emissions. These approaches tend to provide very detailed information about the spatial patterns of emissions associated with individual sources, but may not produce as accurate an estimation of total regional emissions as top-down approaches. Some sources may not be included or a portion of the individual units may be missing from the analysis either due to an oversight or due to the factoring out of some emissions (Canadell et al., 2007). In addition, regional GHG inventories exclude unmanaged land from the estimation (Bickel, Richards, Köhl, Rodrigues, & Stahl, 2006; Reddy et al., 2019) in order to focus on land areas that are directly impacted by management and, in turn, by changes in policy (Ogle et al., 2018).

Top-down methodologies use a time series of greenhouse gas concentration data from flask and tower networks, aircraft, and other sources to infer a set of fluxes that represent an optimal balance between the prescribed prior estimates and observed patterns in the available atmospheric data (Ciais et al., 2007; Enting, 2002). Top-down inversions arguably are preferred for estimating total regional emissions, although this approach typically does not provide much information on individual sources of emissions, particularly if there are many sources and sinks for the greenhouse gas in a region (e.g., Ogle et al., 2015). Furthermore, top-down inversions necessarily require an atmospheric transport model to translate surface fluxes into atmospheric distributions, which introduces uncertainty (Gurney et al., 2004; Peylin, Baker, Sarmiento, Ciais, & Bousquet, 2002). Inversion results are also sensitive to discretionary assumptions about the relative weight of prior information about fluxes versus new information inferred from available atmospheric data (Michalak et al., 2005).

3.1 Inventories of agricultural soil C stock changes, N_2O and CH_4 emissions

Inventory methods (i.e., bottom-up approaches) to quantify regional greenhouse gas emissions from agricultural soils fall into one of three types: (a) measurement-based, (b) empirical/statistical methods, and (c) process-based models (Smith et al., 2019). In the context of IPCC method tiers, Tiers 1 and 2 are empirical/statistical methods, while Tier 3 can be measurement- or model-based (Ogle et al., 2019; Paustian et al., 2006). All of these inventory methods may produce accurate results, but each has its own strengths and weaknesses (Table 2).

Measurement-based methods are likely to provide the strongest inference on regional GHG emissions from croplands, but taking sufficient measurements can be costly and generally requires considerable amounts of time and labor (Conant, Ogle, Paul, & Paustian, 2011). Bellamy et al. (2005) compiled an example of a regional inventory for SOC based on measurements for England and Wales. However, this study did not quantify the change in the size of the C stock pool, but rather just changes in concentration. While understanding changes in concentration is interesting and potentially useful for policy, it does not provide the necessary information to quantify regional fluxes of CO_2 associated with changes in the SOC pool. Quantifying SOC stocks is challenging given the amount of data needed even for well-studied regions such as Europe (Saby et al., 2008). Measurement-based methods are even more difficult to implement for soil CH_4 and N_2O emissions due to the costs of deploying instruments, but technologies are improving and may be less expensive in

TABLE 2 Soil carbon inventory methods.

Method	Strengths	Weaknesses	Examples
Measurement-based	Direct inference on SOC stock changes	Expensive to collect sufficient measurements/re-measurement	Bellamy, Loveland, Bradley, Lark, and Kirk (2005) and Sleutel et al. (2003)
Model-based			
Empirical/statistical methods	Estimation derived directly from measurements; fewer data requirements; Generally simpler than other model approaches	Often relies on experimental data that may not represent the full set of conditions	Ogle, Breidt, and Paustian (2006), Ogle, Eve, Breidt, and Paustian (2003), Vanden Bygaart et al. (2004), West et al. (2008), Piao et al. (2009), Maia et al. (2010), Meersmans et al. (2009), Mishra, Torn, Masanet, and Ogle (2012)
Process-based	Ability to apply across a broader range of conditions than available measurements; Provides mechanistic understanding of underlying patterns	Complexity and lack of transparency in how estimates are derived; typically requires more data inputs	Smith, Desjardins, and Pattey (2000), Ogle et al. (2010), van Wesemael et al. (2010), Izaurralde et al. (2007), Lugato, Paustian, Panagos, Jones, and Borrelli (2016)

the future. Nevertheless, soil monitoring networks are needed to provide a strong inference on GHG emissions from agricultural lands at regional scales, and this information can be used to support model-based inventories. In some cases, it may be possible to quantify regional changes directly from the measurements (van Wesemael et al., 2011).

Empirical methods are the most straightforward to apply, requiring measurement data that can be analyzed to produce a statistical relationship between SOC stock changes and management practices. The advantage of these methods

386 Section | C Case Studies

is that regional inventories can be compiled with data from experiments, which are available in many countries, to produce empirically derived factors (e.g., Ogle, Breidt, Eve, & Paustian, 2003; Vanden Bygaart et al., 2004) or carbon response curves (West et al., 2008). Geospatial methods can also be incorporated into empirical approaches to estimate regional C inventories (Mishra et al., 2012). These methods typically require a static representation of climate and select edaphic characteristics, such as soil texture. Management data are a key component of these analyses, such as information on crop types and yields, tillage practices, mineral fertilization, organic amendments, cover crops, use of bare fallow, and irrigation.

A key uncertainty in empirical approaches is that they are entirely reliant on the robustness of the underlying measurement data that are used to derive the empirical relationships. In theory, the data supporting these analyses should be a random sample of locations across the region of interest. This is almost never the case because the methods rely on experiments that are conducted, often for convenience, in locations near research and academic institutions. This does not mean that the results from experiments are not useful, but they may not be representative of the entire population of farms in a region, which contributes to uncertainty in estimates of regional GHG emissions derived from these data. Uncertainty may also be associated with large variances in the empirically derived relationships. An example is the direct soil N_2O emission factor provided by the IPCC in which 1% of N additions are emitted as N_2O with an uncertainty of $-70\%/+200\%$ (De Klein, Novoa, Ogle, Smith, et al., 2006). Uncertainty may also be associated with management activity data and environmental conditions, but these sources are likely less important than the uncertainty in empirically derived relationships from the measurement data (Ogle et al., 2003).

Process-based models are arguably the most advanced approaches and can lead to greater accuracy if a model is well tested and the required management activity and environmental data are available to drive the model simulations (Smith et al., 2012). Models can vary in scope and representation of processes from models such as Roth C (Jenkinson & Rayner, 1977) and DayCent (Parton, Hartman, Ojima, & Schimel, 1998) that represent an intermediate level of complexity in processes to a more complex representation of processes such as in DNDC (Li, Frokling, Harriss, & Terry, 1994). In addition, our understanding of soil organic matter dynamics is changing (Lehmann & Kleber, 2015), and recent models are adopting new paradigms such as those found in the newly developed Millennial (Abramoff et al., 2018) and MEMS (Robertson et al., 2019) models.

In theory, process-based models are intended to be generalizable so that they can be applied beyond the domain of the measurement data in which they are developed. Process-based models therefore may overcome the key weakness of empirical models, which are most representative of the conditions (climates, soils, and management) in the empirical dataset. However, a key weakness is

Agricultural systems **Chapter | 11** **387**

that these models may not fully represent the processes driving soil organic matter dynamics, leading to significant uncertainty associated with model structure and parameters (Ogle et al., 2010). The level of detail in management activity and environmental conditions is often greater for process-based models and not always available (Smith et al., 2012), which can also lead to uncertainty if the data are not adequate. Nevertheless, these approaches show promise for improving the accuracy of estimate results at regional scales (e.g., Del Grosso, Ogle, & Parton, 2011; Smith et al., 2019).

3.2 Inventories of emissions from livestock systems

Livestock emissions can be estimated using empirical and model-based methods (Table 3). In theory, measurement-based approaches could also be used, but to our knowledge, this approach has not been applied at regional scales. Often, emissions are inferred from dry matter or gross energy intake of the livestock. This information along with measurements of the methane emissions per unit of dry matter or energy intake, i.e., methane conversion rate, is the basis for several methods, including Tier 1 and 2 methods provided by the IPCC (Gavrilova et al., 2019; Hongmin et al., 2006). There are several process-based methodologies available as well. It should be noted that a substantial majority of these methodologies were developed primarily with the use of data from North American, New Zealand, and European dairy systems. There is a paucity of data and models that focus on grazing and feedlot systems. The CNCPS model is an example of a model that does differentiate between beef and dairy cattle for the estimations of enteric fermentation (Van Amburgh et al., 2015).

In most of these systems, there tends to be a trade-off between the simple and complex method, with more complexity leading to higher accuracy but at the cost of a greater number of inputs. In turn, there tends to be a greater uncertainty in the inputs for some regions or nations, particularly in developing countries, that leads to less accurate estimates with the more complex methods.

3.3 Lateral transport and supply chains

Accounting for GHG emissions in food supply chains involves methodologies designed to capture the complex and interactive networks of small- and medium-sized enterprise operations that characterize food systems (Scholten, Verdouw, Beulens, & van der Vorst, 2016). Three of the most common GHG inventory methodologies utilized for supply chain analysis include (1) the GHG Protocol Corporate Accounting and Reporting Standard (GPCARS); (2) the Scope 3 GHG Protocol Corporate Value Chain (GPCVC); and (3) the GHG Protocol Product Life Cycle Accounting and Reporting Standard (GPLCAR). Both the GPCARS and GPLCAR have been developed by the World Resources Institute (WRI, 2018) and the World Business Council for

TABLE 3 Enteric emission inventory methods.

Method	Model	Variable modeled	Inputs/comments
Empirical/statistical methods			
	HOLOS	Enteric CH_4, manure CH_4	Based on IPCC
	CNCPS (Van Amburgh et al., 2015)		Uses equation of Mills et al. (2003) for dairy and Ellis et al. (2007) for beef. Animal characteristics, diet nutrient composition, feed protein fractions, animal performance, animal management, in situ degradability of feeds
	Integrated farm system model, Rotz, Corson, Chianese, and Coiner (2009)	Enteric CH_4, DMI, nutrient excretion	Uses the Mits 3 equation of Mills et al. (2003) for enteric CH4, (Hongmin et al., 2006) for manure CH_4 and DAYCENT Chianese, Rotz, and Richard (2009) or IPCC for manure N_2O
	Phetteplace, Johnson, and Seidl (2001)	Enteric CH_4, DMI, nutrient excretion, manure CH_4, manure N_2O	Animal characteristics, diet nutrient composition, animal management, manure management
	(Hongmin et al., 2006)	Manure CH_4	Number of animals, type of manure storage, VS excreted per animal, MCF
	(Hongmin et al., 2006)	Manure N_2O	Type of manure storage, N content of manure, number of animals
Process-based methods			
	Kebreab, Clark, Wagner-Riddle, France, and Wagner (2006) and Kebreab, Dijkstra, Bannink, and France (2009)	Enteric CH_4, nutrient excretion	DMI, NDF, degradable NDF, total starch, degradable starch, soluble sugars, diet N, NH_3-N in diet, indigestible protein, rate of degradation of starch and protein
	Sommer and Petersen (2004)	Liquid manure CH_4	Proportion of degradable and undegradable VS in manure, daily VS load, emptying frequency, potential and actual CH_4 yields, residence time, air temperature, average monthly temperature
	RUMINANT (Herrero, Thornton, Kruska, & Reid, 2008)	Enteric CH_4	Estimates intake and supply of nutrients in feed along with nutrient uptake, fermentation kinetics and excretion. The model incorporates protein-energy interactions, pH effects, feeing level effects, supplementation regimes and other drivers of emissions

Sustainable Development (WBCSD). The GPCARS provides a framework for cross-sector assessments with company guidelines for measuring and reporting GHG emissions. Scope 3 is a widely accepted technique used to monitor and track GHG emissions of companies across their value chains for 15 established categories. The GPLCAR is a product-specific method for inventories of GHG emissions and is aligned with the Scope 3 and corporate standard guidelines.

Several GHG calculators exist to support the computation of emissions across the supply chain. Smith (2018) identified the Cool Farm Tool (CFT) as one of the most promising applications. The CFT was originally designed as a farm-level GHG calculator to support farm management functions. Among other features, the tool allows companies to account for GHG emissions in transport and other aspects of the supply chain. The tool has been adopted by some of the largest agribusiness companies such as Pepsi, Unilever, and Heineken (Smith, 2018). In terms of protocol, the emission factors associated with the CFT are based on the IPCC Tier 1 methodology for national greenhouse gas inventory (Paustian et al., 2006). The CFT is freely available to growers but at a cost to supply chain businesses (Richards, 2018).

Other supply chain GHG tools include BigChain, Agricultural Life Cycle Inventory Generator (ALCIG), and GLEAM-I (Global Livestock Environmental Assessment Model—Interactive). The BigChain tool was developed by the South Pole Group and allows suppliers and buyers to monitor the impact of their supply chain on forests. The database sources include the Global Forest Watch, Earthstat, and the FAO. The ALCIG tool utilizes datasets from ecoinvent and the World Food LCA Database (WFLDB), along with farm management data to generate company-specific emissions to support a Life Cycle Analysis. GLEAM-I is a publicly available tool that is designed for life cycle analyses but tailored to livestock production. GLEAM-I uses spatial and temporal datasets from the Gridded Livestock of the World (Robinson et al., 2014) and FAO in a Geographic Information System (GIS) framework to generate spatially explicit GHG emissions for production systems.

3.4 Top-down inversion methods

Top-down inversion methods use observed atmospheric concentrations, in combination with an atmospheric tracer transport model (ATM), to infer the surface fluxes that led to the observed atmospheric concentration pattern. Atmospheric inversions focused on agricultural greenhouse gas fluxes, including CO_2, CH_4, and N_2O, have been run on both global and regional scales (e.g., Ganesan et al., 2015; Nevison et al., 2018; Schuh et al., 2013; Thompson et al., 2014). In most cases, many possible flux combinations could create the observed atmospheric pattern, but the problem can be posed in a tractable manner if certain assumptions are made about the allowable space/time patterns of the fluxes (Enting, 2002).

In a classical Bayesian Inversion, the methodology begins with an a priori guess of the true fluxes based on the available bottom-up information. These "prior" fluxes are used as an input to the ATM in order to create an atmospheric distribution, which is sampled at model grids corresponding to the available atmospheric measurements. An optimal set of "posterior" fluxes are generated from the analysis by minimizing a cost function that assigns discretionary relative weights to the priors versus the model-observed mismatch in the atmospheric distribution (Hu, Andrews, Thoning, et al., 2019; Michalak et al., 2005).

Geostatistical inversions evolved from Bayesian methods with a philosophy of letting the atmospheric data determine the inferred fluxes as much as possible without the bias that can be introduced with the prescribed prior fluxes (Michalak, Bruhwiler, & Tans, 2004). However, geostatistical inversions typically replace the prescribed prior flux with geostatistical fields, e.g., temperature and precipitation, which are posited as explanatory variables that are meaningfully related to the fluxes of interest (Shiga et al., 2018). Geostatistical inversions must still make discretionary assumptions about the allowed spatial correlation structure of the fluxes and the relative importance of uncertainty in fluxes versus model-data mismatch.

As atmospheric data streams achieve higher temporal and spatial resolution, the $n \times m$ matrices (where n is the number of observations and m is the number of fluxes that are solved in the inversion analysis) required in conventional Bayesian or geostatistical approaches become too large to be inverted (Baker, Doney, & Schimel, 2006). To address this problem, inverse modelers are turning to increasingly sophisticated approaches such as 4Dvar and Kalman filter methods to handle these large new datasets, e.g., column CO_2 data from the OCO_2 satellite (Bruhwiler, Michalak, Peters, Baker, & Tans, 2005; Crowell et al., 2019; Meirink, Bergamaschi, & Krol, 2008; Peters et al., 2005). To date, these new approaches have generally not focused on regional quantification of agricultural emissions.

4 Improving regional GHG inventories for agriculture

Agriculture is a significant emitter of GHGs, and emissions are likely to increase with a growing population in the coming decades. At the same time, GHG emissions intensity has decreased in recent decades and there is evidence to suggest that this trend will continue into the future (Bennetzen, Smith, & Porter, 2016; FAOSTAT, 2019). Regional inventories will be important tools for monitoring and assessing emissions that may be reduced through management and policy, or societal choice in the future. Frameworks are becoming more accurate for estimating emissions and there is considerable potential for further advancement in the near future (e.g., Smith et al., 2012, 2019).

Activity data collection associated with management of crops and livestock on agricultural lands is critical for accurate regional GHG inventories. A

Agricultural systems **Chapter | 11   391**

sophisticated model will not provide useful results if the management activity data are not available to drive the model. Furthermore, measurement-based methods could provide useful results for regional GHG budgets even without activity data, but the emissions data will not be useful for policy and management decisions without the context of the underlying management activity.

Activity data collection may involve a census, survey, or expert knowledge (Goodwin et al., 2006). Survey and census data collection associated with agricultural management are taking new forms with recent advancements, particularly remote sensing technologies and crowdsourcing data using cellular technologies. Remote sensing is increasingly used for monitoring land use and land cover change (e.g., Friedl et al., 2002; Homer et al., 2007), but there are also methods under development to classify agricultural management practices from remote sensing imagery, such as tillage management (Zheng, Campbell, Serbin, & Galbraith, 2014). Use of remote sensing technologies can be used as a census, with wall-to-wall coverage of the management practice, or as a survey with data collection based on a sample of images covering a region, such as may occur with aerial photos or possibly high-resolution satellite data.

Crowdsourcing with cellular technology is also an efficient way to enhance data collection (See et al., 2013), and in practice could be implemented as a survey or census type of data collection. Given the widespread deployment of cellular networks, managers are accessible across many countries with applications designed for cellular phones. However, there are still major challenges to overcome, such as implementation in areas with limited cellular networks, before these technologies will provide the data streams needed to improve the accuracy of regional GHG inventories in agriculture.

Methods are also becoming more accurate at estimating GHG emissions. In particular, Tier 3 methods have been developed for estimating regional GHG emissions with process-based models to estimate GHG emissions from agriculture (e.g., Kebreab, Dijkstra, Bannink, and France (2009); Lugato et al., 2016; Ogle et al., 2010). As basic understanding of processes evolve, these models are likely to become even more accurate for estimating GHG emissions (e.g., Abramoff et al., 2018; Robertson et al., 2019). The advantage of process-based models, assuming that they are generalizable, is that they are applicable across a broader range of environmental conditions that typically occur in regions beyond the data that are used in their development. Furthermore, recent developments with sophisticated parameterization and testing methods make these models more robust for regional inventories, including optimization (Necpálová et al., 2015) and Bayesian calibration methods (e.g., Gurung, Ogle, Breidt, Williams, & Parton, 2020; Hararuk, Shaw, & Kurz, 2017; Hararuk, Xia, & Luo, 2014).

Data collection is essential for further advancement of regional GHG inventories associated with agriculture. Empirical methods can be advanced with additional measurements, particularly in regions where it is not feasible to apply

392 Section | C Case Studies

Tier 3 methods due to lack of driver data for model applications. Tier 3 methods will provide a stronger inference on regional emissions with monitoring networks of measurement sites underlying the parameterization and verification. New technologies could improve efficiency in data collection, such as spectroscopy for SOC measurements (Nocita, Stevens, Noon, & van Wesemael, 2013) and SF6 or micrometeorological methods for livestock enteric CH_4 (Storm, Hellwing, Nielsen, & Madsen, 2012), allowing for rapid expansion of monitoring sites within existing networks and deployment of new networks in undersampled regions. In some regions, it may be possible to develop networks of monitoring sites that are large enough to infer emissions directly from the measurements (van Wesemael et al., 2010).

Incorporation of top-down methodologies into inventory frameworks is another key advancement that has been used to estimate CH_4 (Henne et al., 2016), N_2O (Nevison et al., 2018), and CO_2 emissions (Ogle et al., 2015) from regions dominated by agriculture. These approaches can verify and possibly constrain total emissions associated with bottom-up inventories. Methane and N_2O emissions are likely more robust for evaluation of agricultural GHG inventories with top-down methods because agricultural activities are key sources of these gases. In contrast, CO_2 fluxes from agricultural activities are more challenging to evaluate because of the large number of sources. Changes in SOC, in particular, are difficult to assess with these technologies given the small net fluxes associated with this pool compared to other sources (Ogle et al., 2015). Nevertheless, top-down methodologies provide another line of evidence that will likely be useful for verifying and constraining regional GHG inventories associated with agriculture in the future.

5 Conclusions

Future GHG emissions from agricultural systems will largely depend on land use and management, livestock management, and management of supply chains for agricultural commodities. Accurate estimation of GHG emissions in regional inventories is needed to monitor and track progress in reducing the anthropogenic impact of agriculture on the climate system. Bottom-up inventories provide the basis for monitoring GHG emissions at regional scales, and may be based on measurements, modeling, or some combination of the two methods (Ogle et al., 2019; Paustian et al., 2006). Inventories may be improved with more complete datasets on management practices, in addition to improved models with a better understanding of the processes driving emissions that are developed with more rigorous parameterization methods. In addition, top-down inversion modeling may be used to verify GHG emissions at regional scales where agriculture is the dominant land use influencing emissions (Henne et al., 2016; Nevison et al., 2018; Ogle et al., 2015), and it could potentially provide a constraint on GHG inventories in the future.

Agricultural systems **Chapter | 11** **393**

Acknowledgments

We thank Amy Swan who adapted Fig. 1 from the 2006 IPCC National Greenhouse Gas Inventory Guidelines with an illustration of emissions from the manufacturing of farm inputs, such as fertilizer and pesticides, along with supply chain emissions associated with transport, storage, processing, and selling of agricultural products.

References

Abramoff, R., Xu, X., Hartman, M., O'Brien, S., Feng, W., Davidson, E., et al. (2018). The millennial model: In search of measurable pools and transformations for modeling soil carbon in the new century. *Biogeochemistry, 137*(1), 51–71.

Accorsi, R., Gallo, A., & Manzini, R. (2017). A climate driven decision-support model for the distribution of perishable products. *Journal of Cleaner Production, 165*, 917–929.

Anderson, K. L., Nagaraja, T. G., Morrill, J. L., Avery, T. B., Galitzer, S. J., & Boyer, J. E. (1987). Ruminal microbial development in conventionally or early-weaned calves. *Journal of Animal Science, 64*, 1215–1226.

Bajželj, B., et al. (2014). Importance of food-demand management for climate mitigation. *Nature Climate Change, 4*(10), 924–929.

Baker, D. F., Doney, S. C., & Schimel, D. S. (2006). Variational data assimilation for atmospheric CO_2. *Tellus Series B: Chemical and Physical Meteorology, 58*(5), 359–365.

Bellamy, P. H., Loveland, P. J., Bradley, R. I., Lark, R. M., & Kirk, G. J. D. (2005). Carbon losses from all soils across England and Wales 1978-2003. *Nature, 437*(7056), 245–248.

Bennetzen, E. H., Smith, P., & Porter, J. R. (2016). Decoupling of greenhouse gas emissions from global agricultural production: 1970–2050. *Global Change Biology, 22*(2), 763–781.

Bickel, K., Richards, G., Köhl, M., Rodrigues, R. L. V., & Stahl, G. (2006). Chapter 3: Consistent representation of lands. In *2006 IPCC Guidelines for National Greenhouse Gas Inventories, Volume IV*. Japan: Intergovernmental Panel on Climate Change, IGES.

Bruhwiler, L. M. P., Michalak, A. M., Peters, W., Baker, D. F., & Tans, P. (2005). An improved Kalman Smoother for atmospheric inversions. *Atmospheric Chemistry and Physics Discussions, 5*, 1891–1923.

Canadell, J. G., Kirschbaum, M. U. F., Kurz, W. A., Sanz, M.-J., Schlamadinger, B., & Yamagata, Y. (2007). Factoring out natural and indirect human effects on terrestrial carbon sources and sinks. *Environmental Science & Policy, 10*(4), 370–384.

Chen, X., Cui, S., Fan, M., Vitousek, P., Zhao, M., et al. (2014). Producing more grain with lower environmental costs. *Nature, 514*, 486–491.

Chianese, D. S., Rotz, C. A., & Richard, T. (2009). Simulation of methane emissions from dairy farms to assess greenhouse gas reduction strategies. *Transactions of the ASABE, 52*, 1313–1323. https://doi.org/10.13031/2013.27781.

Ciais, P., Bousquet, P., Freibauer, A., & Naegler, T. (2007). Horizontal displacement of carbon associated with agriculture and its impacts on atmospheric CO_2. *Global Biogeochemical Cycles, 21*(2). GB2014 2011-2012.

Ciais, P., Rayner, P., Chevallier, F., Bousquet, P., Logan, M., Peylin, P., et al. (2011). Atmospheric inversions for estimating CO_2 fluxes: Methods and perspectives. In M. Jonas, Z. Nahorski, S. Nilsson, & T. Whiter (Eds.), *Greenhouse gas inventories: Dealing with uncertainty* (pp. 69–92). Dordrecht, Netherlands: Springer.

Ciais, P., Wattenbach, M., Vuichard, N., Smith, P., Piao, S. L., Don, A., et al. (2010). The European carbon balance. Part 2: Croplands. *Global Change Biology, 16*(5), 1409–1428.

Conant, R. T., Ogle, S. M., Paul, E. A., & Paustian, K. (2011). Measuring and monitoring soil organic carbon stocks in agricultural lands for climate mitigation. *Frontiers in Ecology and the Environment, 9*(3), 169–173.

Conant, R. T., Paustian, K., & Elliott, E. T. (2001). Grassland management and conversion into grassland: Effects on soil carbon. *Ecological Applications, 11*, 343–355.

Crowell, S., Baker, D., Schuh, A., Basu, S., Jacobson, A. R., Chatterjee, A., et al. (2019). The 2015-2016 carbon cycle as seen from OCO-2 and the global in situ network. *Atmospheric Chemistry and Physics Discussions*. https://doi.org/10.5194/acp-2019-87.

De Klein, C., Novoa, R. S. A., Ogle, S., Smith, K. A., et al., Eggleston, S. (Eds.). (2006). Chapter 11: N_2O emissions from managed soil, and CO_2 emissions from lime and urea application. In *Agriculture, forestry and other land use: Vol. 4. IPCC guidelines for national greenhouse gas inventories*. Kanagawa, Japan: IGES.

Del Grosso, S. J., Ogle, S. M., & Parton, W. J. (2011). Soil organic matter cycling and greenhouse gas accounting methodologies. In L. Guo, A. Gunasekara, & L. McConnell (Eds.), *Understanding greenhouse gas emissions from agricultural management*. Washington, D.C.: American Chemical Society.

Edwards-Jones, G., et al. (2008). Testing the assertion that "local food is best": The challenges of an evidence-based approach. *Trends in Food Science and Technology, 19*, 265–274.

Ellis, J. L., Kebreab, E., Odongo, N. E., McBride, B. W., Okine, E. K., & France, J. (2007). Prediction of methane production from dairy and beef cattle. *Journal of Dairy Science, 90*, 3456–3466. https://doi.org/10.3168/jds.2006-675.

Enting, I. G. (2002). *Inverse problems in atmospheric constituent transport*. New York: Cambridge University Press.

Ericksson, M., Strid, I., & Hansson, P. A. (2015). Carbon footprint of food waste management options in the waste hierarchy—A Swedish case study. *Journal of Cleaner Production, 93*, 15–125.

FAO. (2018). *The future of food and agriculture—Alternative pathways to 2050*. Rome: Food and Agriculture Organization of the United Nations. 224 pp. License: CC BY-NC-SA 3.0 IGO.

FAO. (2019). *FAOSTAT land use database*. Rome: Food and Agriculture Organization of the United Nations. http://www.fao.org/faostat/.

FAO. (2019b). *The state of food and agriculture 2019. Moving forward on food loss and waste reduction*. Rome: Food and Agriculture Organization of the United Nations. Licence: CC BY-NC-SA 3.0 IGO.

Firestone, M. K., & Davidson, E. A. (1989). *Microbiological basis of NO and N_2O production and consumption in soil*. New York: John Wiley & Sons.

Friedl, M. A., McIver, D. K., Hodges, J. C. F., Zhang, X. Y., Muchoney, D., Strahler, A. H., et al. (2002). Global land cover mapping from MODIS: Algorithms and early results. *Remote Sensing of Environment, 83*(1), 287–302.

Galloway, J. N., Townsend, A. R., Erisman, J. W., Bekunda, M., Cai, Z., Freney, J. R., et al. (2008). Transformation of the nitrogen cycle: Recent trends, questions, and potential solutions. *Science, 320*(5878), 889–892.

Ganesan, A. L., Manning, A. J., Grant, A., Young, D., Oram, D. E., Sturges, W. T., et al. (2015). Quantifying methane and nitrous oxide emissions from the UK and Ireland using a national-scale monitoring network. *Atmospheric Chemistry and Physics, 15*, 6393–6406. https://doi.org/10.5194/acp-15-6393-2015.

Garnett, T. (2011). Where are the best opportunities for reducing greenhouse gas emissions in the food system? *Food Policy, 36*(Suppl. 1), S23–S32.

Agricultural systems **Chapter | 11** **395**

Gavrilova, O., Leip, A., Dong, H., MacDonald, J. D., Bravo, C. A. G., Amon, B., et al. (2019). Emissions from livestock and manure management. In E. Calvo Buendia, K. Tanabe, A. Kranjc, J. Baasansuren, M. Fukuda, S. Ngarize, et al. (Eds.), *Refinement to the 2006 IPCC Guidelines for National Greenhouse Gas Inventories, Volume IV*. Switzerland: Intergovernmental Panel on Climate Change. IPCC.

Gerber, P. J., Steinfeld, H., Henderson, B., Mottet, A., Opio, C., Dijkman, J., et al. (2013). *Tackling climate change through livestock: A global assessment of emissions and mitigation opportunities*. Rome: Food and Agriculture Organization of the United Nations.

Godfray, H. C. J., et al. (2010). Food security: The challenge of feeding 9 billion people. *Science*, *327*(5967), 812–818.

Goodwin, J., Woodfield, M., Ibnoaf, M., Koch, M., Yan, H., Frey, C., et al. (2006). Chapter 2: Approaches to data collection. In *2006 IPCC Guidelines for National Greenhouse Gas Inventories, Volume I*. Japan: Intergovernmental Panel on Climate Change, IGES.

Gupta, P. K., Sahai, S., Singh, N., Dixit, C. K., Singh, D. P., Sharma, C., et al. (2004). Residue burning in rice–wheat cropping system: Causes and implications. *Current Science*, *87*(12), 1713–1717.

Gurney, K. R., Law, R. M., Denning, A. S., Rayner, P. J., Pak, B. C., Baker, D., et al. (2004). Transcom3 inversion intercomparison: Model mean results for the estimation of seasonal carbon sources and sinks. *Global Biogeochem. Cycle*, *18*. https://doi.org/10.1029/2003GB002111, GB1010.

Gurung, R., Ogle, S. M., Breidt, F. J., Williams, S., & Parton, W. J. (2020). Bayesian calibration methods of a process-based ecosystem model to simulate soil organic carbon dynamics. *Geoderma*, *376*, 114529.

Hamilton, S. K., Kurzman, A. L., Arango, C., Jin, L., & Robertson, G. P. (2007). Evidence for carbon sequestration by agricultural liming. *Global Biogeochemical Cycles*, *21*(2). GB2021 2021-2012.

Hararuk, O., Shaw, C., & Kurz, W. A. (2017). Constraining the organic matter decay parameters in the CBM-CFS3 using Canadian National Forest Inventory data and a Bayesian inversion technique. *Ecological Modelling*, *364*, 1–12.

Hararuk, O., Xia, J., & Luo, Y. (2014). Evaluation and improvement of a global land model against soil carbon data using a Bayesian Markov chain Monte Carlo method. *Journal of Geophysical Research – Biogeosciences*, *119*(3), 403–417.

Henne, S., Brunner, D., Oney, B., Leuenberger, M., Eugster, W., Bamberger, I., et al. (2016). Validation of the Swiss methane emission inventory by atmospheric observations and inverse modelling. *Atmospheric Chemistry and Physics*, *16*(6), 3683–3710.

Herrero, M., Thornton, P. K., Kruska, R., & Reid, R. S. (2008). Systems dynamics and the spatial distribution of methane emissions from African domestic ruminants to 2030. *Agriculture, Ecosystems & Environment*, *126*(1), 122–137.

Homer, C., Dewitz, J., Fry, J., Coan, M., Hossain, C., Larson, C., et al. (2007). Completion of the 2001 National Land Cover Database for the conterminous United States. *Photogrammetric Engineering and Remote Sensing*, *73*(4), 337–341.

Hongmin, D., Mangino, J., McAllister, T. A., Hatfield, J. L., Johnson, D. E., Lassey, K. R., et al. (2006). Chapter 10: Emissions from livestock and manure management. In *2006 IPCC Guidelines for National Greenhouse Gas Inventories, Volume IV*. Japan: Intergovernmental Panel on Climate Change, IGES.

Houghton, R. A., & Nassikas, A. A. (2017). Global and regional fluxes of carbon from land use and land cover change 1850–2015. *Global Biogeochemical Cycles*, *31*, 456–472.

Hu, L., Andrews, A. E., Thoning, K. W., et al. (2019). Enhanced North American carbon uptake associated with El Niño. *Science Advances*, *5*, eaaw0076.

396 Section | C Case Studies

Ingram, J., Dyball, R., Howden, M., Vermeulen, S., Ganett, T., Redlingshöfer, B., et al. (2016). Food security, food systems, and environmental change. *Solutions, 7*(3), 63–73.

Izaurralde, R., Williams, J., Post, W., Thomson, A., McGill, W., Owens, L., et al. (2007). Long-term modeling of soil C erosion and sequestration at the small watershed scale. *Climatic Change, 80* (1), 73–90.

James, S. J., & James, C. (2010). The food cold-chain and climate change. *Food Research International, 43,* 1944–1956.

Jayanegara, A., Leiber, F., & Kreuzer, M. (2012). Meta-analysis of the relationship between dietary tannin level and methane formation in ruminants from in vivo and in vitro experiments. *Journal of Animal Physiology and Animal Nutrition, 96*(3), 365–375.

Jenkinson, D. S., & Rayner, J. H. (1977). The turnover of soil organic matter in some of the Rothamsted classical experiments. *Soil Science, 123,* 298–305.

Johnson, D. E., Hill, T. M., Ward, G. M., Johnson, K. A., Branine, M. E., Carmean, B. R., et al. (1993). Principle factors varying methane emissions from ruminants and other animals. In M. A. K. Khalil (Ed.), *NATO ADI series: Vol. 113. Atmospheric methane: Sources, sinks, and role in global change.* Berlin, Germany: Springer-Verlag.

Kebreab, E., Clark, K., Wagner-Riddle, C., France, J., & Wagner, K. (2006). Methane and nitrous oxide emissions from Canadian animal agriculture: A review. *The Canadian Veterinary Journal, 86,* 135–158. https://doi.org/10.4141/A05-010.

Kebreab, E., Dijkstra, J., Bannink, A., & France, J. (2009). Recent advances in modeling nutrient utilization in ruminants. *Journal of Animal Science, 87*(Suppl_14), E111–E122. https://doi.org/10.2527/jas.2008-1313.

Kim, D.-G., Kirschbaum, M. U. F., & Beedy, T. L. (2016). Carbon sequestration and net emissions of CH4 and N2O under agroforestry: Synthesizing available data and suggestions for future studies. *Agriculture, Ecosystems & Environment, 226,* 65–78.

Kindler, R., Siemens, J. A. N., Kaiser, K., Walmsley, D. C., Bernhofer, C., Buchmann, N., et al. (2011). Dissolved carbon leaching from soil is a crucial component of the net ecosystem carbon balance. *Global Change Biology, 17*(2), 1167–1185.

Lehmann, J., & Kleber, M. (2015). The contentious nature of soil organic matter. *Nature, 528,* 60–68.

Li, C., Frokling, S. E., Harriss, R. C., & Terry, R. E. (1994). Modeling nitrous oxide emissions from agriculture: A Florida case study. *Chemosphere, 28,* 1401–1415.

Lugato, E., Paustian, K., Panagos, P., Jones, A., & Borrelli, P. (2016). Quantifying the erosion effect on current carbon budget of European agricultural soils at high spatial resolution. *Global Change Biology, 22*(5), 1976–1984.

Maia, S. M. F., Ogle, S. M., Cerri, C. E. P., & Cerri, C. C. (2010). Soil organic carbon stock change due to land use activity along the agricultural frontier of the southwestern Amazon, Brazil, between 1970 and 2002. *Global Change Biology, 16*(10), 2775–2788.

Mbow, C., Rosenzweig, C., Barioni, L. G., Benton, T. G., Herrero, M., Krishnapillai, M., et al. (2019). Chapter 5: Food security. In *Climate change and land: An IPCC special report on climate change, desertification, land degradation, sustainable land management, food security, and greenhouse gas fluxes in terrestrial ecosystems* (p. 1542). Intergovernmental Panel on Climate Change, World Meterological Organization and United Nations Environmental Programme.

McCarty, J. L. (2011). Remote sensing-based estimates of annual and seasonal emissions from crop residue burning in the contiguous United States. *Journal of the Air & Waste Management Association, 61*(1), 22–34.

Meersmans, J., Van, B. W., De, F. R., Fallas, M. D., De, S. B., & Van, M. M. (2009). Changes in organic carbon distribution with depth in agricultural soils in northern Belgium, 1960–2006. *Global Change Biology*, *15*(11), 2739–2750.

Meersmans, J., Wesemael, B. V., Goidts, E., Molle, M. V., Baets, S. D., & Ridder, F. D. (2011). Spatial analysis of soil organic carbon evolution in Belgian croplands and grasslands, 1960–2006. *Global Change Biology*, *17*(17), 466–479.

Meirink, J. F., Bergamaschi, P., & Krol, M. C. (2008). Four-dimensional variational data assimilation for inverse modelling of atmospheric methane emissions: Method and comparison with synthesis inversion. *Atmospheric Chemistry and Physics*, *8*, 6341–6353.

Michalak, A. M., Bruhwiler, L., & Tans, P. P. (2004). A geostatistical approach to surface flux estimation of atmospheric trace gases. *Journal of Geophysical Research*, *109*. https://doi.org/10.1029/2003JD004422, D14109.

Michalak, A. M., et al. (2005). Maximum likelihood estimation of covariance parameters for Bayesian atmospheric trace gas surface flux inversions. *Journal of Geophysical Research*, *110*, D24107.

Mills, J. A. N., Kebreab, E., Yates, C. M., Crompton, L. A., Cammell, S. B., Dhanoa, M. S., et al. (2003). Alternative approaches to predicting methane emissions from dairy cows. *Journal of Animal Science*, *81*, 3141–3150. https://doi.org/10.2527/2003.81123141x.

Minasny, B., Malone, B. P., McBratney, A. B., Angers, D. A., Arrouays, D., Chambers, A., et al. (2017). Soil carbon 4 per mille. *Geoderma*, *292*, 59–86.

Mishra, U., Torn, M. S., Masanet, E., & Ogle, S. M. (2012). Improving regional soil carbon inventories: Combining the IPCC carbon inventory method with regression kriging. *Geoderma*, *189*, 288–295.

Mosier, A. R., Duxbury, J. M., Freney, J. R., Heinemeyer, O., & Minami, K. (1998). Assessing and mitigating N_2O emissions from agricultural soils. *Climatic Change*, *40*, 7–38.

Mueller, N. D., et al. (2012). Closing yield gaps through nutrient and water management. *Nature*, *490*(7419), 254–257.

Necpálová, M., Anex, R. P., Fienen, M. N., Del Grosso, S. J., Castellano, M. J., Sawyer, J. E., et al. (2015). Understanding the Day Cent model: Calibration, sensitivity, and identifiability through inverse modeling. *Environmental Modelling & Software*, *66*, 110–130.

Nevison, C., Andrews, A., Thoning, K., Dlugokencky, E., Sweeney, C., Miller, S., et al. (2018). Nitrous oxide emissions estimated with the Carbon Tracker-Lagrange North American regional inversion framework. *Global Biogeochemical Cycles*, *32*, 463–485.

Nocita, M., Stevens, A., Noon, C., & van Wesemael, B. (2013). Prediction of soil organic carbon for different levels of soil moisture using Vis-NIR spectroscopy. *Geoderma*, *199*, 37–42.

Oenema, O., Velthof, G. L., Yamulki, S., & Jarvis, S. C. (1997). Nitrous oxide emissions from grazed grassland. *Soil Use and Management*, *13*, 288–295.

Oenema, O., Wrage, N., Velthof, G. L., van Groenigen, J. W., Dolfing, J., & Kuikman, P.-J. (2005). Trends in global nitrous oxide emissions from animal production systems. *Nutrient Cycling in Agroecosystems*, *72*, 51–65.

Ogle, S. M., Breidt, F. J., Easter, M., Williams, S., Killian, K., & Paustian, K. (2010). Scale and uncertainty in modeled soil organic carbon stock changes for US croplands using a process-based model. *Global Change Biology*, *16*, 810–820.

Ogle, S. M., Breidt, F. J., Eve, M. D., & Paustian, K. (2003). Uncertainty in estimating land use and management impacts on soil organic carbon storage for US agricultural lands between 1982 and 1997. *Global Change Biology*, *9*(11), 1521–1542.

398 Section | C Case Studies

Ogle, S. M., Breidt, F. J., & Paustian, K. (2006). Bias and variance in model results associated with spatial scaling of measurements for parameterization in regional assessments. *Global Change Biology, 12*, 516–523.

Ogle, S. M., Davis, K., Lauvaux, T., Schuh, A., Cooley, D., West, T. O., et al. (2015). An approach for verifying biogenic greenhouse gas emissions inventories with atmospheric CO_2 concentration data. *Environmental Research Letters, 10*, 034012.

Ogle, S. M., Domke, G., Kurz, W. A., Rocha, M. T., Huffman, T., Swan, A., et al. (2018). Delineating managed land for reporting national greenhouse gas emissions and removals to the United Nations framework convention on climate change. *Carbon Balance and Management, 13*(1), 9.

Ogle, S. M., Eve, M. D., Breidt, F. J., & Paustian, K. (2003). Uncertainty in estimating land use and management impacts on soil organic carbon storage for U.S. agroecosystems between 1982 and 1997. *Global Change Biology, 9*, 1521–1542.

Ogle, S. M., Kurz, W. A., Green, C., Brandon, A., Baldock, J., Domke, G., et al. (2019). Chapter 2: General methodologies applicable to multiple land-use categories. In E. Calvo Buendia, K. Tanabe, A. Kranjc, J. Baasansuren, M. Fukuda, S. Ngarize, et al. (Eds.), *Refinement to the 2006 IPCC Guidelines for National Greenhouse Gas Inventories, Volume IV*. Switzerland: Intergovernmental Panel on Climate Change.

Oh, N.-H., & Raymond, P. A. (2006). Contribution of agricultural liming to riverine bicarbonate export and CO_2 sequestration in the Ohio River basin. *Global Biogeochemical Cycles, 20* (3). https://doi.org/10.1029/2005GB002565.

Pardon, P., Reubens, B., Reheul, D., Mertens, J., De Frenne, P., Coussement, T., et al. (2017). Trees increase soil organic carbon and nutrient availability in temperate agroforestry systems. *Agriculture, Ecosystems & Environment, 247*, 98–111.

Parton, W. J., Hartman, M., Ojima, D., & Schimel, D. (1998). DAYCENT and its land surface submodel: Description and testing. *Global and Planetary Change, 19*, 35–48.

Paustian, K., Lehmann, J., Ogle, S., Reay, D., Robertson, G. P., & Smith, P. (2016). Climate-smart soils. *Nature, 532*(7597), 49–57.

Paustian, K., Ravindranath, N. H., van Amstel, A., Aalde, H., Gonzalez, P., Gytarsky, M., et al. (2006). Chapter 2: Generic methodologies applicable to multiple land-use categories. In *2006 IPCC Guidelines for National Greenhouse Gas Inventories, Volume IV*. Japan: Intergovernmental Panel on Climate Change, IGES.

Peters, W., Miller, J. B., Whitaker, J., Denning, A. S., Hirsch, A., Krol, M. C., et al. (2005). An ensemble data assimilation system to estimate CO_2 surface fluxes from atmospheric trace gas observations. *Journal of Geophysical Research, 110*. https://doi.org/10.1029/2005JD006157, D24304.

Peylin, P., Baker, D., Sarmiento, J., Ciais, P., & Bousquet, P. (2002). Influence of transport uncertainty on annual mean and seasonal inversions of atmospheric CO_2 data. *Journal of Geophysical Research, 107*(D19), 4385. https://doi.org/10.1029/2001JD000857.

Phetteplace, H. W., Johnson, D. E., & Seidl, A. F. (2001). Greenhouse gas emissions from simulated beef and dairy livestock systems in the United States. *Nutrient Cycling in Agroecosystems, 60*, 99–102. https://doi.org/10.1023/A:1012657230589.

Piao, S., Fang, J., Ciais, P., Peylin, P., Huang, Y., Sitch, S., et al. (2009). The carbon balance of terrestrial ecosystems in China. *Nature, 458*(7241), 1009–1013.

Poore, J., & Nemecek, T. (2018). Reducing food's environmental impacts through producers and consumers. *Science, 360*(6392), 987–992.

Agricultural systems Chapter | 11 **399**

Ramankutty, N., Evan, A. T., Monfreda, C., & Foley, J. A. (2008). Farming the planet: 1. Geographic distribution of global agricultural lands in the year 2000. *Global Biogeochemical Cycles, 22*. GB1003 1001-1019.

Raymond, P. A., Oh, N.-H., Turner, R. E., & Broussard, W. (2008). Anthropogenically enhanced fluxes of water and carbon from the Mississippi River. *Nature, 451*(7177), 449–452.

Reay, D. S., Davidson, E. A., Smith, K. A., Smith, P., Melillo, J. M., Dentener, F., et al. (2012). Global agriculture and nitrous oxide emissions. *Nature Climate Change, 2*, 410–416.

Reddy, S., Panichelli, L., Waterworth, R. M., Federici, S., Green, C., Jonckheere, I., et al. (2019). Chapter 3: Consistent representation of lands. In E. Calvo Buendia, K. Tanabe, A. Kranjc, J. Baasansuren, M. Fukuda, S. Ngarize, et al. (Eds.), *Refinement to the 2006 IPCC Guidelines for National Greenhouse Gas Inventories, Volume IV*. Switzerland: Intergovernmental Panel on Climate Change.

Richards, M. (2018). *Measure the chain: Tools for assessing GHG emissions in agricultural supply chains*. CGIAR Research Program on Climate Change, Agriculture and Food Security; Rubenstein School for Environment and Natural Resources, University of Vermont. https://engagethechain.org/resources/measure-chain-tools-assessing-ghg-emissions-agricultural-supply-chains.

Robertson, A. D., Paustian, K., Ogle, S., Wallenstein, M. D., Lugato, E., & Cotrufo, M. F. (2019). Unifying soil organic matter formation and persistence frameworks: The MEMS model. *Biogeosciences, 16*(6), 1225–1248.

Robinson, T., Wint, G., Conchedda, G., Van Boeckel, T., Ercoli, V., Palamara, E., et al. (2014). Mapping the global distribution of livestock. *PLoS ONE, 9*(5), e96084.

Rotz, C. A., Corson, M. S., Chianese, D. S., & Coiner, C. U. (2009). *The integrated farm system model: Reference manual*. Park, PA: University USDA-ARS Pasture Systems and Watershed Management Research Unit. Available at: www.ars.usda.gov/SP2UserFiles/Place/19020000/ifsmreference.pdf. (Accessed 20 August 2021).

Sá, J. C.d. M., Lal, R., Cerri, C. C., Lorenz, K., Hungria, M., & de Faccio Carvalho, P. C. (2017). Low-carbon agriculture in South America to mitigate global climate change and advance food security. *Environment International, 98*, 102–112.

Saby, N. P. A., Bellamy, P. H., Morvan, X., Arrouays, D., Jones, R. J. A., Verheijen, F. G. A., et al. (2008). Will European soil-monitoring networks be able to detect changes in topsoil organic carbon content? *Global Change Biology, 14*(10), 2432–2442.

Sanderman, J., Hengl, T., & Fiske, G. J. (2017). Soil carbon debt of 12,000 years of human land use. *Proceedings of the National Academy of Sciences, 114*(36), 9575–9580.

Sandström, V., Valin, H., Krisztin, T., Havlík, P., Herrero, M., & Kastner, T. (2018). The role of trade in the greenhouse gas footprints of EU diets. *Global Food Security, 19*, 48–55.

Schader, C., Muller, A., El-Hage Scialabba, N., Hecht, J., Isensee, A., Erb, K.-H., et al. (2015). Impacts of feeding less human food and arable crops to livestock on global food system sustainability. *Journal of the Royal Society Interface, 12*, 20150891.

Scholten, H., Verdouw, C., Beulens, A., & van der Vorst, J. (2016). Defining and analyzing traceability systems in food supply chains. In M. Espiñeira, & F. J. Santaclara (Eds.), *Advances in food traceability techniques and technologies: Improving quality throughout the food chain* (pp. 9–33). Elsevier.

Schuh, A. E., Lauvaux, T., West, T. O., Denning, A. S., Davis, K. J., et al. (2013). Evaluating atmospheric CO_2 inversions at multiple scales over a highly inventoried agricultural landscape. *Global Change Biology, 19*, 1424–1439.

See, L., Comber, A., Salk, C., Fritz, S., van der Velde, M., Perger, C., et al. (2013). Comparing the quality of crowdsourced data contributed by expert and non-experts. *PLoS ONE, 8*(7), e69958.

400 Section | C Case Studies

Shiga, Y. P., Tadić, J. M., Qiu, X., Yadav, V., Andrews, A. E., Berry, J. A., et al. (2018). Atmospheric CO_2 observations reveal strong correlation between regional net biospheric carbon uptake and solar-induced chlorophyll fluorescence. *Geophysical Research Letters, 45*, 1122–1132.

Sleutel, S., De Neve, S., & Hofman, G. (2003). Estimates of carbon stock changes in Belgian cropland. *Soil Use and Management, 19*(2), 166–171.

Smil, V. (1999). Crop residues: Agriculture's largest harvest: Crop residues incorporate more than half of the world's agricultural phytomass. *BioScience, 49*(4), 299–308.

Smith, P. (2018). Managing the global land resource. *Proceedings of the Royal Society B: Biological Sciences.* Available at: https://royalsocietypublishing.org/doi/full/10.1098/rspb.2017.2798.

Smith, P., Bustamante, M., Ahammad, H., Clark, H., Dong, H., Elsiddig, E. A., et al. (2014). Agriculture, forestry and other land use (AFOLU). In O. Edenhofer, R. Pichs-Madruga, Y. Sokona, E. Farahani, S. Kadner, K. Seyboth, & J. C. Minx (Eds.), *Climate change 2014: Mitigation of climate change. Contribution of working group III to the fifth assessment report of the intergovernmental panel on climate change.* Cambridge, United Kingdom: Cambridge University Press.

Smith, P., Davies, C. A., Ogle, S., Zanchi, G., Bellarby, J., Bird, N., et al. (2012). Towards an integrated global framework to assess the impacts of land use and management change on soil carbon: Current capability and future vision. *Global Change Biology, 18*, 2089–2101.

Smith, W. N., Desjardins, R. L., & Pattey, E. (2000). The net flux of carbon from agricultural soils in Canada 1970-2010. *Global Change Biology, 6*, 557–568.

Smith, P., Haberl, H., Popp, A., Erb, K.-H., Lauk, C., Harper, R., et al. (2013). How much land-based greenhouse gas mitigation can be achieved without compromising food security and environmental goals? *Global Change Biology, 19*(8), 2285–2302.

Smith, P., Martino, D., Cai, Z., Gwary, D., Janzen, H., Kumar, P., et al. (2008). Greenhouse gas mitigation in agriculture. *Philosophical Transactions of the Royal Society B: Biological Sciences, 363*(1492), 789–813.

Smith, P., Soussana, J.-F., Angers, D., Schipper, L., Chenu, C., Rasse, D. P., et al. (2019). How to measure, report and verify soil carbon change to realize the potential of soil carbon sequestration for atmospheric greenhouse gas removal. *Global Change Biology.* https://doi.org/10.1111/gcb.14815.

Sommer, S., & Petersen, S. (2004). Algorithms for calculating methane and nitrous oxide emissions from manure management. *Nutrient Cycling in Agroecosystems, 69*, 143–154. https://doi.org/10.1023/B:FRES.0000029678.25083.

Storm, I. M. L. D., Hellwing, A. L. F., Nielsen, N. I., & Madsen, J. (2012). Methods for measuring and estimating methane emission from ruminants. *Animals, 2*(2), 160–183.

Thompson, R. L., et al. (2014). TransCom N_2O model inter-comparison part 2: Atmospheric inversion estimates of N_2O emissions. *Atmospheric Chemistry and Physics, 14*, 6177–6194.

Tilman, D., Cassman, K. G., Matson, P. A., Naylor, R., & Polasky, S. (2002). Agricultural sustainability and intensive production practices. *Nature, 418*(6898), 671–677.

Tilman, D., et al. (2011). Global food demand and the sustainable intensification of agriculture. *PNAS, 108*(50), 20260–20264.

Tubiello, F. (2018). Greenhouse gas emissions due to agriculture. In P. Ferranti, E. Berry, & A. Jock (Eds.), *Encyclopedia of food security and sustainability* Elsevier.

Tubiello, F. N. (2019). Greenhouse gas emissions due to agriculture. In P. Ferranti, E. M. Berry, & J. R. Anderson (Eds.), *Encyclopedia of food security and sustainability* (pp. 196–205). Oxford: Elsevier.

Tubiello, F. N., Salvatore, M., Ferrara, A. F., Rossi, S., Biancalani, R., Condor Golec, R. D., et al. (2015). The contribution of agriculture, forestry and other land use activities to global warming, 1990-2012: Not as high as in the past. *Global Change Biology*, *21*, 2655–2660.

Tubiello, F., Salvatore, M., Rossi, S., Ferrara, A. F., Fitton, N., & Smith, P. (2013). The FAOSTAT database of greenhouse gas emissions from agriculture. *Environmental Research Letters*, *8*, 015009.

Van Amburgh, M. E., Collao-Saenz, E. A., Higgs, R. J., Ross, D. A., Recktenwald, E. B., Raffrenato, E., et al. (2015). The Cornell Net Carbohydrate and Protein System: Updates to the model and evaluation of version 6.5. *Journal of Dairy Science*, *98*, 6361–6380. https://doi.org/10.3168/jds.2015-9378.

Van Oost, K., Quine, T. A., Govers, G., De Gryze, S., Six, J., Harden, J. W., et al. (2007). The impact of agricultural soil erosion on the global carbon cycle. *Science*, *318*(5850), 626–629.

van Wesemael, B., Paustian, K., Andrén, O., Cerri, C., Dodd, M., Etchevers, J., et al. (2011). How can soil monitoring networks be used to improve predictions of organic carbon pool dynamics and CO_2 fluxes in agricultural soils? *Plant and Soil*, *338*(1), 247–259.

van Wesemael, B., Paustian, K., Meersmans, J., Goidts, E., Barancikova, G., & Easter, M. (2010). Agricultural management explains historic changes in regional soil carbon stocks. *PNAS*, *107* (33), 14926–14930.

Vanden Bygaart, A. J., Gregorich, E. G., Angers, D. A., & Stoklas, U. F. (2004). Uncertainty analysis of soil organic carbon stock change in Canadian cropland from 1991 to 2001. *Global Change Biology*, *10*, 983–994.

Vermeulen, S. J., Campbell, B. M., & Ingram, J. S. I. (2012). Climate change and food systems. *Annual Review of Environment and Resources*, *37*, 195–222.

Viglizzo, E. F., Ricard, M. F., Taboada, M. A., & Vázquez-Amábile, G. (2019). Reassessing the role of grazing lands in carbon-balance estimations: Meta-analysis and review. *Science of the Total Environment*, *661*, 531–542.

Vuichard, N., Ciais, P., Belelli, L., Smith, P., & Valentini, R. (2008). Carbon sequestration due to the abandonment of agriculture in the former USSR since 1990. *Global Biogeochemical Cycles*, *22* (4), GB4018.

Weiss, F., & Leip, A. (2012). Greenhouse gas emissions from the EU livestock sector: A life cycle assessment carried out with the CAPRI model. *Agriculture, Ecosystems and Environment*, *149*, 124–134.

West, T. O., Brandt, C. C., Wilson, B. S., Hellwinckel, C. M., Tyler, D. D., Marland, G., et al. (2008). Estimating regional changes in soil carbon with high spatial resolution. *Soil Science Society of America Journal*, *72*(2), 285–294.

West, T. O., Brown, M. E., Duren, R. M., Ogle, S. M., & Moss, R. H. (2013). Definition, capabilities and components of a terrestrial carbon monitoring system. *Carbon Management*, *4*(4), 413–422.

Wolf, J., West, T. O., Le Page, Y. L., Kyle, G. P., Zhang, X., Collatz, G. J., et al. (2015). Biogenic carbon fluxes from global agricultural production and consumption. *Global Biogeochemical Cycles*, *29*, 1617–1639.

Wolz, K. J., Lovell, S. T., Branham, B. E., Eddy, W. C., Keeley, K., Revord, R. S., et al. (2018). Frontiers in alley cropping: Transformative solutions for temperate agriculture. *Global Change Biology*, *24*(3), 883–894.

WRI. (2018). *Corporate value chain (scope 3) accounting and reporting standard: Supplement to the GHG protocol corporate accounting and reporting standard*. Washington, DC: World Resources Institute. https://ghgprotocol.org/sites/default/files/standards/Corporate-Value-Chain-Accounting-Reporing-Standard_041613_2.pdf.

402 Section | C Case Studies

Xiao, J., Zhuang, Q., Baldocchi, D. D., Law, B. E., Richardson, A. D., Chen, J., et al. (2008). Estimation of net ecosystem carbon exchange for the conterminous United States by combining MODIS and Ameri Flux data. *Agricultural and Forest Meteorology, 148*(11), 1827–1847.

Yan, X., Akiyama, H., Yagi, K., & Akimoto, H. (2009). Global estimations of the inventory and mitigation potential of methane emissions from rice cultivation conducted using the 2006 Intergovernmental Panel on Climate Change Guidelines. *Global Biogeochemical Cycles, 23*, GB2002.

Yue, Y., Ni, J., Ciais, P., Piao, S., Wang, T., Huang, M., et al. (2016). Lateral transport of soil carbon and land–atmosphere CO_2 flux induced by water erosion in China. *Proceedings of the National Academy of Sciences, 113*(24), 6617–6622.

Zhao, M., & Running, S. W. (2010). Drought-induced reduction in global terrestrial net primary production from 2000 through 2009. *Science, 329*(5994), 940–943.

Zheng, B., Campbell, J. B., Serbin, G., & Galbraith, J. M. (2014). Remote sensing of crop residue and tillage practices: Present capabilities and future prospects. *Soil and Tillage Research, 138*, 26–34.

Zimmermann, A., Benda, J., Webber, H., & Jafari, Y. (2018). *Trade, food security and climate change: Conceptual of linkages and policy implications.* Rome: Food and Agriculture Organization of United Nations. 48 pp. Licence: CC BY-NC-SA 3.0 IGO.

Chapter 12

Greenhouse gas balances in coastal ecosystems: Current challenges in "blue carbon" estimation and significance to national greenhouse gas inventories

Lisamarie Windham-Myers[a], James R. Holmquist[b], Kevin D. Kroeger[c], and Tiffany G. Troxler[d]

[a]*US Geological Survey Water Mission Area, Menlo Park, CA, United States,* [b]*Smithsonian Environmental Research Center, Edgewater, MD, United States,* [c]*US Geological Survey Woods Hole Coastal & Marine Science Center, Woods Hole, MA, United States,* [d]*Department of Earth and Environment/Institute of Environment, Florida International University, Miami, FL, United States*

1 Background

Coastal wetlands have received attention for their climate mitigation potential over the past decade (Nellemann et al., 2009), but have only recently been incorporated into inventory methodologies for estimation of national scale greenhouse gas emissions and removals (e.g., Crooks et al., 2018). High rates of atmospheric carbon uptake coupled with carbon-rich soil formation processes lead to some of the highest observed annual rates of long-term carbon sequestration, from centuries to millennia (McLeod et al., 2011). Tidally influenced wetlands are distinct in time and space, due to historical processes of ecosystem development and degradation (Redfield, 1965). They currently represent less than 2% of North American wetland acreage, but occupy a distinct terrestrial-aquatic linkage that is responsible for significant carbon storage (Windham-Myers et al., 2018). Geomorphically driven carbon burial has led to deep stocks of carbon-rich soils and rates of sequestration that are commensurate with relative sea level rise (Morris et al., 2016; Rogers et al., 2019). Future rates of carbon sequestration from the atmosphere and storage within coastal ecosystems are largely a function of both immediate and long-term feed-backs of carbon accumulation, deposition, and erosion along coastlines (e.g., Hopkinson, Cai, & Hu, 2012). As other land-use sectors (agriculture, forestry and other land use, AFOLU) have improved in regional estimation with new technologies, observation syntheses, and model intercomparison, it is likely that regional-scale coastal wetland carbon estimation will also see advances in the next decade (e.g., Fargione et al., 2018).

2 What limits traditional AFOLU estimation approaches in coastal ecosystems?

The spatial and temporal dynamics of coastal ecosystems are inherently com-plex, requiring significant assumptions in greenhouse gas estimation. The three pillars of greenhouse gas (GHG) inventory reporting in land-use sectors—maps, models, and measurements—are all notably limited in coastal settings. At the terrestrial-aquatic interface, wetlands, in general, are frequently narrow, linear, and patchy features comprising a landscape that is variable and complex in time and space. They are difficult to map with traditional remotely sensed methods into land-use categories suitable for modeling purposes, as the critical drivers are often subsurface. Especially difficult in coastal settings is detecting tidal influence, whereby subtle changes in flooding duration or porewater salinity can dramatically shift the magnitude and direction of carbon fluxes (e.g., Wilson et al., 2018). Effective detection of state changes in coastal settings are limited to remotely sensed thresholds such as extent of open water and veg-etation phenology (Couvillion & Beck, 2013). Coupled physical and biogeo-chemical models—whether process-based or statistically determined—are lacking in predictive power primarily due to the inherent complexity of the landscape and the limited measurements and metadata available to test model sensitivity and thresholds. As such, uncertainty in upscaling national GHG fluxes is significant for coastal wetlands and this uncertainty continues to grow in response to active coastal modification—both human and naturally induced (National Academies of Sciences, Engineering, and Medicine (NASEM), 2018) (Fig. 1).

Coastal wetlands are defined herein as inundated, vegetated ecosystems with hydrology and biogeochemistry influenced by sea levels, at timescales of tides to millennia (Windham-Myers, Crooks, & Troxler, 2019). Because their soil building capacity, and thus carbon storage pool, is so strongly driven by geomorphic processes, they are poorly represented by traditional terrestrial or oceanic models (Ward et al., 2020). Further, due to their complexity, the rel-atively limited data that do exist yield limited statistical power, as the coastal domain may be disaggregated into multiple critical categories such as salinity class, vegetation class, relative elevation class, etc. Among regions of the US coastline, independent carbon monitoring efforts have led to relatively high data density in some regions (Northern Gulf of Mexico) and very poor coverage in

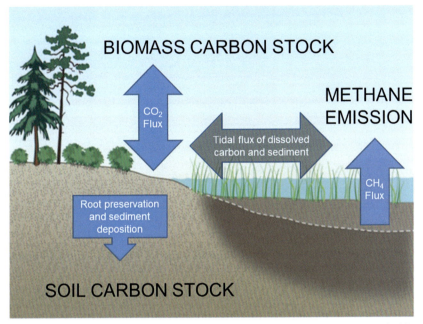

FIG. 1 Conceptual diagram of dominant atmospheric and hydrologic carbon exchanges in tidal wetlands that are related to UNFCCC NGGI accounting practice (soil stock change, biomass stock change, and methane fluxes). *(Source: Lisamarie Windham-Myers.)*

others (Pacific Northwest, Holmquist et al., 2018). Coastal wetland categories cover a range of salinities and are commonly referred to as mangroves (woody subtropical intertidal vegetation), marshes (emergent intertidal vegetation), and seagrass or submerged aquatic vegetation (SAV, at intertidal or subtidal elevations). These are widely distributed along coastlines, being predominantly present where coastal energy is limited, and thus poised to generate deep globally significant soil pools as the ecosystems grow vertically and adapt to rising tides (Chmura, Anisfeld, Cahoon, & Lynch, 2003; Duarte, Middleburg, & Caraco, 2005) (Fig. 2).

While coastal "blue carbon" ecosystems are geomorphically predisposed to carbon storage in soils, the potential is high for coastal wetland management to influence their GHG budgets (e.g., Pendleton, Donato, Murray, et al., 2012). Active daily fluxes —such as photosynthesis, respiration, lateral exchanges, methane production—can be altered in magnitude and direction by direct and indirect human influences. Climate drivers can be important in maintaining the notably high uptake of carbon into biomass (e.g., Osland et al., 2016), but the larger fluxes of concern to annual GHG estimation relate to net stock change in soil, net methane emissions, and net emissions from lateral fluxes of eroding soils (Fargione et al., 2018; Holmquist et al., 2018). For example, forest biomass carbon stocks are globally significant in mangrove forests, but the belowground

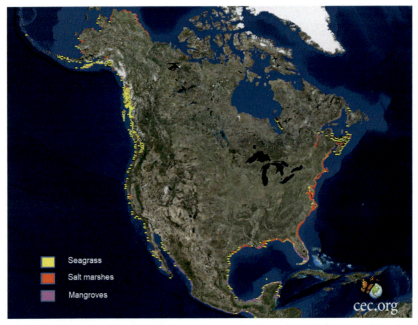

FIG. 2 North American distribution of coastal "blue carbon" ecosystems. *(From North American Blue Carbon, 2021—Commission for Environmental Cooperation (cec.org). www.cec.org/north-american-environmental-atlas/north-american-blue-carbon-2021/.)*

soil carbon pool is commonly an order of magnitude greater (Donato et al., 2011).

Remotely sensed products allow development of tools to track aboveground biomass dynamics. Simard et al. (2019) illustrate the use of structural and optical satellite data in assessing mangrove biomass patterns. Byrd et al. (2018) use multiscale optical (1–30 m) data to quantify patterns of aboveground biomass in emergent marshes within the continental United States. Feagin et al. (2020) have developed a spatial tool to map photosynthetic processes, specifically GPP, scaled to MODIS data (250×250 m) using relationships derived from eddy covariance data from a range of coastal wetland types, fresh to saline and woody to emergent. Productivity metrics are useful for model building but less important than assessing and mapping critical belowground characteristics and models of carbon fate. For example, the range of soil, methane, and lateral flux observations available within the United States does not cover with sufficient data density important conditions such as the current and past states of a wetland's hydrology (impounded, drained, ditched), vegetation structure (mangrove, tidal freshwater forests, and marshes), physical setting (erosional and depositional dynamics), or age (Holmquist, Windham-Myers, Bernal, et al., 2018). Whereas wetlands can be classified fairly easily by latitude and vegetation class (woody vs. nonwoody), these structural characteristics are less

Greenhouse gas accounting in coastal ecosystems **Chapter | 12** **407**

than dominant drivers of the significant soil-based fluxes, which are often more hydrologically driven. This inability to assess belowground drivers leads to significant speculation based on climate-driven observations, which dominate C cycling in terrestrial ecosystem models.

Given the fundamental shifts in environmental conditions during carbon exchanges among terrestrial and aquatic environments, a wide range of carbon pools, fluxes, and speciation is not represented well in model calibration, validation, or outputs (e.g., Kirwan & Blum, 2011). Lateral, vertical, and storage fluxes are difficult to constrain at site scales (e.g., Bogard et al., 2020) and impossible at larger scales from both top-down observations or Earth System modeling perspectives (e.g., Ward et al., 2020). In particular, we lack confidence in many scenarios of change as they are dependent on process-based understanding—such as the fate of carbon from eroded wetland soils (Enwright, Griffith, & Osland, 2016; Haywood, Hayes, White, & Cook, 2020). Despite these limitations, the emergence of NGGI guidance from the IPCC Wetlands Supplement supports a simple assessment (Tier 1) of GHG fluxes associated with land management decisions. To support this assessment, a large scientific and policy team was assembled to determine nationally relevant CONUS emission factors and support the US EPA (2017) to get coastal wetlands into the US NGGI for 2017. A complementary effort focused on feedback to the NGSTP on limitations in the Tier 1 approach and where additional information could be used to improve estimation of GHG emissions and removals. Thus, we focus here on a CONUS national scale inventory reporting effort based on annual scale estimation with bottom-up modeling. The most important finding of this effort is the need for inventory efforts to clarify the assumptions made based on the data available. Our strongest contribution was identifying and quantifying how the maps, modeling, and data point toward the greatest uncertainties to address in continued coastal GHG inventory efforts.

3 IPCC guidelines for national-scale estimation of coastal wetland carbon

As part of the comprehensive GHG estimation methodologies, the IPCC developed guidance for reporting of certain wetland types in the 2006 IPCC Guidelines for National GHG Inventories, Volume 4, Agriculture Forestry and Other Land Use (2006 IPCC Guidelines). The six land-use categories of the 2006 IPCC guidance include Forest Land, Cropland, Grassland, Wetlands, Settlements, and Other Lands (e.g., bare soil, rock, ice, etc.). The 2006 IPCC Guidelines use these categories for the purposes of estimating anthropogenic emissions and removals for managed lands. Emissions and removals are not reported for unmanaged lands, but the areas for those lands are tracked over time. The same six categories are used in the agreed UNFCCC Common Reporting Format (CRF) for submission of developed country (Annex I)

408 Section | C Case Studies

national GHG inventories. For each of the six land-use categories, emissions and removals from the following pools are estimated:

1. Living biomass (separate above- and belowground pools);
2. Dead organic matter (deadwood and litter);
3. Soil organic carbon (mineral- and organic-rich soils).

In addition, wood products such as timber used in construction or furniture, as well as wood products put in landfills, referred to as harvested wood products (HWP), are reported. A general assumption of AFOLU approaches with regard to estimation of CO2 emissions and removals is that the sum of C gains and losses is equivalent not only to net stock changes but also to total emissions and removals. Thus, the inventory compiler is not restricted to the use of C stock data or CO_2 flux data, as long as the general equations used (stock-difference or gain loss) are complete, accurate, and consistent, avoiding over- and underestimates and double-counting.

Notably, however, the coverage of the 2006 IPCC Guidelines on wetlands was restricted to peatlands drained and managed for peat extraction, conversion to flooded lands, and limited guidance for drained organic soils (IPCC, 2014). Significant land-use change in wetlands, and increasing data availability with which to estimate GHG emissions and removals associated with it, brought an invitation from the UNFCCC to the IPCC to "undertake further methodological work on wetlands, focusing on the rewetting and restoration of peatland." Agreeing to address the former gaps in methodological guidance, IPCC released the 2013 Supplement to the 2006 IPCC Guidelines for National Greenhouse Gas Inventories: Wetlands (herein "Wetlands Supplement"; IPCC, 2014). The Wetlands Supplement "extends the content of the 2006 IPCC Guidelines" by filling gaps in coverage and providing updated information reflecting scientific advances, including updating emission factors. It covers inland organic soils and wetlands on mineral soils, coastal wetlands including mangrove forests, tidal marshes, and seagrass meadows, and constructed wetlands for wastewater treatment.

In the case of coastal wetlands, a synthesis provided by Valiela, Bowen, and York (2001) reported that the most important human activities contributing to the loss of mangrove were shrimp culture (38%), forestry uses (26%), fish culture (14%), diversion of freshwater (11%), land reclamation (5%), and another 5% collectively to herbicides, agriculture, salt ponds, and coastal development. Significant land-use change combined with the significant carbon stock contained in soils and biomass of coastal wetlands subject to those land-use changes can produce CO_2 emissions of large magnitude (Kauffman et al., 2020). Alternatively, management that rewets or restores tidal action to coastal wetlands can result in active sequestration and long-term sinks of CO_2, with increasing areas of intact coastal wetlands contributing to a portfolio of anthropogenically derived atmospheric C reductions. The guidance provides estimation methodologies covering management activities with the largest global influence on

Greenhouse gas accounting in coastal ecosystems **Chapter | 12 409**

GHG emissions and removals on managed coastal wetlands (IPCC, 2014). Annex I countries, which include the United States, are encouraged to use the Wetlands Supplement in preparing their annual inventories under the Convention from 2015 (Decision 24/CP.19 paragraph 4).

Regardless of whether a land-use change occurs or not, it is "good practice" [Good Practice Guidance for Land Use, Land-use Change and Forestry (GPG-LULUCF)] to quantify and report significant emissions and removals resulting from management activities on coastal wetlands in line with their country-specific definition (Wetlands Supplement, Chapter 4, p. 4–6). Chapter 4 of the Wetlands Supplement provides guidance on estimating emissions and removals of greenhouse gases (CO_2, CH_4, and N_2O) associated with specific activities on managed coastal wetlands (Wetlands Supplement, Overview Chapter, p. O-8). Emissions and removals may be quantified as a rate or a stock change. For example, agricultural lands with organic soils that were formerly coastal wetlands will continuously emit CO_2 once drained (and CH_4 if extensive ditching accompanies the drainage; see Chapter 2, IPCC, 2014) and would be reported under the management activity of drained, wetland soil. Previously, terrestrial land-use conversions rarely considered the persistent nature of CO_2 emissions associated with drained lands that were formerly wetlands. As long as organic soils persist, CO_2 emissions continue, particularly when drainage depths are lowered to maintain agricultural production, allowing those emissions to continue at significant rates (Chapter 2, IPCC, 2014). Conversion to settlements is reported to cause CO_2 emissions when development practices remove anaerobic wetland soils, on average to a 1-m depth, and remove or otherwise expose those soils to aerobic conditions. In this case, emissions are quantified at the time of the conversion activity (i.e., instantaneous within the year that settlements in coastal wetlands were constructed). A similar assumption is applied for other coastal development activities like construction of aquaculture and salt production ponds. Regardless of activity, the methodological guidance for coastal wetlands provided that the area upon which the activity occurred in a year or changed to that activity within a year was the basis for applying the fundamental equation (GHG flux = emission factor * activity data).

The IPCC source/sink categories represent carbon dioxide, methane, and nitrous oxide. Current guidance is represented below (2013 IPCC Wetlands Supplement).

A. Carbon Dioxide (CO_2): Four distinct CO_2 sources/sinks associated with specific management activities that can be reported using the guidelines in the IPCC Wetland Supplement Forest Land, Cropland, Grassland, Wetlands, Settlements, and Other Land categories (CO_2) include:

    (1) Forest management in mangroves. Removal of wood occurs to different extents throughout the tropics where mangrove forests are harvested for fuelwood, charcoal, and construction (Ellison & Farnsworth, 1996; Walters et al., 2008). Natural disturbances are another form of biomass

410 Section | C Case Studies

carbon stock loss. There may also be conversion to forest land where mangrove replanting can take place on rewetted, or already saturated, soils. The guidance provides updates for data used to estimate C stock change specifically in mangrove living biomass and dead wood pools, relevant to the IPCC CO_2 source/sink categories that may fall under any land-use category, especially pertaining to mangrove forests, including: aboveground biomass, aboveground biomass growth, ratio of below-ground to aboveground biomass, C fraction of aboveground biomass, wood density, and litter and dead wood C stocks

(2) Extraction in mangroves, tidal marshes, and seagrass meadows (including excavation generally, and construction for aquaculture and salt production specifically). Extraction collectively refers to: Excavation of saturated soils above the local groundwater level, lead-ing to unsaturated soils and removal of biomass and dead organic mat-ter. Activities that lead to the excavation of soil often lead to loss of coastal wetlands. The excavated or dredged soil is also commonly used to help develop the coastal infrastructure where there is a need to raise the elevation of land in low-lying areas and/or contribute to new land areas for settlement. Aquaculture and salt production are also common activities in the coastal zone and similarly require excavation of soil and removal of biomass and dead organic matter to facilitate their construction. Global default C stock values that can be applied for this management activity include soil C stocks for mangroves, tidal marshes (in both mineral- and organic-rich soils), and seagrass meadows in mineral soils. All of these default factors lead to signif-icantly higher emissions than a loss of mangrove biomass C stocks from the same area. Excavation can result in significant emissions, for example, applying a point-change analysis for activity data derived from MODIS imagery, 21,000 ha were converted from mangrove for-est to aquaculture in Kalimantan, Indonesia (Rahman, Dragoni, Didan, Barreto-Munoz, & Hutabarat, 2013). Applying methodological guid-ance for conversion of mangrove forest to aquaculture, approximately 47 million metric tons (MMT) of CO_2 were emitted from Kalimantan aquaculture development during 2000 to 2010, and approximately 13MMT in 2002 alone. As an example of the magnitude of these emis-sions, it is equivalent to 70% of all GHG emissions in the SE FL region in 2009, roughly 6 million people. Construction is only the first phase in aquaculture and salt production. The second phase, termed "use," is when fish ponds, cages, or pens are stocked and fish produc-tion occurs, and salt production ponds in mangroves and tidal marshes represent the "use" of these facilities.

(3) Soil drainage in mangroves and tidal marshes (CO_2). Mangroves and tidal marshes have been diked and drained to create pastures,

Greenhouse gas accounting in coastal ecosystems Chapter | 12 **411**

croplands, and settlements since before the 11th century (Gedan, Silliman, & Bertness, 2009). The practice continues today on many coastlines. On some diked coasts, groundwater of reclaimed former wetlands is pumped out to maintain the water table at the required level below a dry soil surface, while on other coasts drainage is achieved through a system of ditches and tidal gates. Due to the substantial carbon reservoirs of coastal wetlands, drainage can lead to large CO_2 emissions. Tier 1 (global default) for soil drainage (7.9 ± 3 Mg CO_2/ha) is aggregated for organic and mineral soils and applied a single year loss.

(4) Rewetting, revegetation, and creation in mangroves, tidal marshes, and seagrass meadows. Rewetting is a prerequisite for vegetation reestablishment and/or creation of conditions conducive to revegetation. Revegetation can occur by natural recolonization, direct seeding, and purposeful planting. This activity is also used to describe the management activities designed to reestablish vegetation on undrained soils in seagrass meadows. Also included in this activity are mangroves and tidal marshes that have been created, typically by raising soil elevation or removing the upper layer of upland soil or dredge spoil, and grading the site until the appropriate tidal elevation is reached to facilitate reestablishment of the original vegetation. Alternatively, created wetlands with mangroves can be found where high riverine sediment loads lead to rapid sediment accumulation, so that previously subaqueous soils can be elevated above tidal influence. This naturally created land can be reseeded or purposefully vegetated. Once the natural vegetation is established, soil carbon accumulation is initiated at rates commensurate to those found in natural settings (Craft et al., 2003; Craft, Broome, & Campbell, 2002; Osland et al., 2012).

B. CH_4 Emissions from Rewetting of Mangroves and Tidal Marshes (CH_4)

Rewetting of drained soils, through reconnection of hydrology, shifts microbial decomposition from aerobic to anaerobic conditions, increasing the potential for CH_4 emissions (Harris, Milbrandt, Everham, & Bovard, 2010). A strong inverse relationship between CH_4 emissions and salinity of mangrove and tidal marsh soils exists (Poffenbarger, Needelman, & Megonigal, 2011; Purvaja & Ramesh, 2001). The global default CH_4 emission factors are 193.7 ± 100 kg CH_4 ha^{-1} yr^{-1} for coastal wetlands with salinity less than 18 psu (assumed to be tidal freshwater and brackish marsh and mangrove wetlands) and 1 kg CH_4 ha^{-1} yr^{-1} for coastal wetlands with salinity that exceeded 18 psu (assumed to be tidal saline water marsh and mangrove).

C. N_2O Emissions from Aquaculture (N_2O)

The most significant activity contributing to N_2O emissions from managed coastal wetlands is aquaculture. One-third of the global anthropogenic

N$_2$O emissions are from aquatic ecosystems, and nearly 6% of the anthropogenic N$_2$O-N emission is anticipated to result from aquaculture by 2030 at its current annual rate of growth (Hu, Lee, Chandran, Kim, & Khanal, 2012). In seagrass meadows, this direct N$_2$O source arises from N added to fish cages (offshore installations). N$_2$O is emitted from aquaculture systems primarily as a byproduct of the conversion of ammonia (contained in fish urea) to nitrate through nitrification and nitrate to N$_2$ gas through denitrification. The N$_2$O emissions are related to fish production (Hu et al., 2012). The global default emission factor is based on a synthesis developed by Hu et al. (2012) as kg N$_2$O-N per kg fish produced (0.00169 ± 0.001). This EF would be applied when the aquaculture was "in use" and would be additional to any CO$_2$ emission related to excavation associated with the "construction" of an aquaculture facility.

The Wetlands Supplement, like the 2006 IPCC Guidelines, generally provides guidance, with a series of decision trees to help guide the inventory compiler to produce national inventory estimates that are transparent, consistent, credible, complete, and accurate (TCCCA). Estimation methods are similar at three levels of detail, from Tier 1 (the default method) to Tier 3 (the most detailed method; IPCC, 2014). The provision of different tiers enables inventory compilers to use methods consistent with their resources and to focus their efforts on those categories of emissions and removals that contribute most significantly to national emission totals and trends. It is also worth noting that the IPCC released a 2019 Refinement on the guidance in the 2006 IPCC Guidelines and the 2013 Supplement, with some minor improvements for coastal wetlands (Lovelock et al., 2019).

4 Improving application of the IPCC NGGI guidelines in the United States

Due to spatial configuration, and needs for attribution in GHG accounting, the only current way to do regional accounting is "bottom-up" (statistical model-based estimates from measured observations attributed to mapped characteristics). For the NASA 2014–18 Carbon Monitoring System project, we focused on reducing uncertainties with each part of the bottom-up component estimates to achieve national assessments at the Tier 2 or Tier 3 level:

- Soil C stock
- Soil C net flux (annualized); accretion vs. erosion
- Plant Biomass stock
- Plant C net fluxes (annualized)
- Methane net fluxes (annualized)

Because N$_2$O data were in short supply, these data were not analyzed for the US NGGI, thus requiring use of the Tier 1 (global default) value. Distributions of all

5 of these values are reported by Holmquist, Windham-Myers, Bernal, et al. (2018), with an emphasis on the means and uncertainties associated with each. We focused on six "sentinel" sites and federal partners with the conterminous United States (CONUS) to represent broad characteristic coastlines at locations with a notable abundance of carbon stock and GHG flux data:

- Puget Sound, Washington, USA
- San Francisco Bay, California, USA
- Barataria Bay, Louisiana, USA
- Everglades, Florida, USA
- Chesapeake Bay, Maryland, USA
- Cape Cod, Massachusetts, USA

The general methodological approach provided in the IPCC Coastal Wetlands chapter was an "activity-based" approach. The intention of this was to reduce any potential additional burden on countries so that, at a minimum, GHG emissions and removals for managed coastal wetlands could be provided for the management activities that were expected to result in the largest sources and sinks. Also important was that countries may have more available data on management activities rather than comprehensive coverage of all managed and unmanaged coastal wetlands to use areas of land use and land-use change for the entire country in a way that follows "good practice."

Starting a new subsectoral GHG contribution also presents several opportunities for identifying additional sources and sinks of emissions to those outlined in the guidelines, which can potentially contribute significantly to meeting a country's nationally determined contribution (NDC). As an example, impoundment of fresh water within former saline intertidal wetlands has been found to be a prominent activity, comprising $\sim$0.5 million hectares within the contiguous United States and emitting as much as 12 MMt CO2e yr^{-1} of anthropogenic methane (Fargione et al., 2018; Kroeger, Crooks, Moseman-Valtierra, & Tang, 2017). This activity was absent from the 2013 Wetland Supplement, though the 2019 Refinement fills this gap by providing guidance for impounded waters in general.

Several challenges are also presented, related to identifying new sources of data and ensuring consistency with the existing inventory, applying additional, sector-specific quality assurance, including: (1) the difficulties in distinguishing managed from unmanaged coastal wetlands (as is common to Forest and Grassland categories of LULUCF), (2) the significant area of coastal wetlands under management within and outside of the land base, (3) ensuring consistency with the existing, land (terrestrial)-based inventory, and (3) using limited resources most efficiently. For the United States, seagrass (subtidal or intertidal benthic vegetated sites) was not reported on for the NGGI, due to their inability to be incorporated into the "managed land" framework. Therefore, only coastal wetlands—above the mean sea level—were considered terrestrial and thus capable of being included as "managed lands."

414 Section | C Case Studies

(1) Mapping: The importance of defining boundaries

Use of previously produced wetland boundaries was evaluated for representativeness and bias. Wetland maps exist through multiple data sources such as the USFWS National Wetlands Inventory, which provides a detailed delineated characterization of wetland types, including tidal status, which are updated at a rate of 2% per year. Because NWI is effectively a static map, another data source, NOAA's Climate Change Analysis Program was assessed given its salinity and vegetation classes updated roughly every 5 years at the scale of Landsat pixels (30 m × 30 m). After intercomparison across sources within the six sentinel sites, both sources were found to be notably underrepresentative of tidal wetland conditions (Holmquist, Windham-Myers, Bliss, et al., 2018). After the need for a physically defendable map boundary for tidal potential was established, a probabilistic tidal wetland boundary was produced in order to characterize wetlands as being tidal or nontidal. Presented as a fundamental data need in the IPCC Wetlands Supplement, GHG estimates reported for coastal lands are highly uncertain without a tidal wetland boundary, as processes of carbon accumulation in tidal and nontidal sites are extremely divergent (Fig. 3).

Further disaggregation of tidal wetland type was a key component of the ability to characterize data for upscaling. Mapping of wetland types, within those classified as "tidal," includes separation of salinity classes, vegetation types, and relative elevation. In particular, wetland condition/status was important to assessing differences in GHG flux rates. Notably, reduced tidal action (impounded or drained sites) or increased tidal action (restored or degrading sites) is significant in assessing the stability of wetland stock and flux rates and any deviation from the equilibrium (e.g., Kroeger et al., 2017). Current wetland delineation capabilities are for 2 salinity classes as coded by parts per thousand (ppt) in salinity units, with freshwater having zero ppt and marine waters having 35 ppt. CCAP classifies wetlands as having a mean salinity of either "palustrine" at 0–5 ppt ("fresh-oligohaline" in the Cowardin classification) or "estuarine" at 5–35 ppt ("mesohaline-saline" in the Cowardin classification). This bimodal classification was not found to improve classification of soil carbon stocks (Holmquist, Windham-Myers, Bliss, et al., 2018), or aboveground biomass stocks (Byrd et al., 2018). When the available data for methane emissions were split bimodally on these salinity criteria, ranges did vary but the wide range in each category suggested finer resolutions for salinity mapping were needed (Holmquist, Windham-Myers, Bernal, et al., 2018). Similarly, for the limited data available on soil stock change in wetlands remaining wetlands (e.g., accretion rates), the available mapped bimodal salinity classes were not definitively useful and thus not utilized for classifying soil carbon accretion rates (Holmquist, Windham-Myers, Bernal, et al., 2018). For both CCAP and NWI, vegetation types—emergent marsh (EM) or woody (forested or shrub-scrub, FO/SS)—were well mapped within both Palustrine and Estuarine salinity

Greenhouse gas accounting in coastal ecosystems Chapter | 12 **415**

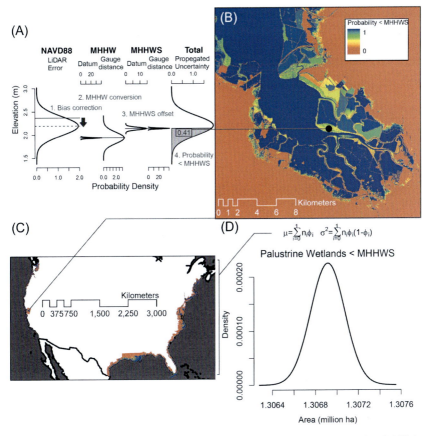

FIG. 3 A new approach to mapping the tidal wetland boundary with physical and probabilistic parameters. Tidal Mean Higher High Water Spring Tide, with 99% probability of being the tidal wetland boundary in 6 sentinel sites. Red indicates <1% probability of hydrology being tidal, and thus all other shades represent increasing probabilities of hydrology being tidal. *(Tidal boundary accessible at the Oak Ridge National Laboratories Data Archive. https://doi.org/10.3334/ORNLDAAC/1650.)*

classes but similarly did not improve classification of soil carbon stocks and fluxes (soil carbon accretion or methane), whereas biomass pools were markedly different (Simard, Fatoyinbo, & Pinto, 2010). Finally, neither CCAP nor NWI has an elevation categorization, but the probabilistic tidal boundary developed for the NASA CMS project enabled physical classification of wetlands below the mean higher high water for spring tides (MHHWS) and thus elevations to be flooded by tides at least once to twice per month. A revision of this analysis categorizes marsh elevation as above or below mean high water (MHW), allowing categorization into "low marsh" (flooded twice daily) and "high marsh" (flooded once or less daily, Holmquist & Windham-Myers,

2022). The influence of elevation, so apparent in vegetation distributions, may play a role in soil and emission factors but is currently being studied. With emerging global lidar datasets, this physical classification has global potential wherever tide gauges are available to validate the tidal range.

(1) Measurements and models:

A key question we tested at a national scale was whether the disaggregation of wetland types improved the certainty of each specific stock and flux. Specifically, to provide feedback to IPCC technical feasibility, we tested the three mappable characteristics for CONUS: salinity (at $<5>$ ppt), vegetation type (woody or herbaceous), and region (four climate zones: Subtropical, Warm Temperate, Cold Temperate, Mediterranean). Datasets were compiled for all 5 specific stock and flux datasets. Datasets were scrutinized and harmonized for avoiding bias and maximizing data utility.

Regarding soil carbon fluxes, results were more similar among wetlands than between these categories (Fig. 4). We found that none of these 3 characteristics improved the estimate of soil C stock (to 1 m). Similarly, net soil accretion (continued soil C accumulation in wetlands remaining wetlands) was not predictable from these three mappable characteristics. Vegetation biomass was the only metric that differed with vegetation class—whereby emergent marsh (nonwoody) wetlands do not accumulate aboveground biomass on a yearly basis, as do woody wetlands, such as mangrove ecosystems (Byrd et al., 2018). Methane fluxes were expected to respond to salinity variation at <18 ppt$>$ (e.g., Poffenbarger et al., 2011; Windham-Myers et al., 2018), but the mapped boundary at 5 ppt did not allow the salinity relationship to be utilized—the range instead was highly variable both in the Estuarine and Palustrine components of the tidal marsh.

Soil carbon stocks (to 1 m depth) had the greatest data density for CONUS but still were not sufficiently characterized to benefit from disaggregation. Deterministic boundaries require sufficient data density for statistical power. Thus, emission factors for other inventory carbon fluxes were assessed for whether disaggregation would allow for improved GHG estimation. The limited amount of data on annual methane emissions, as well as lateral flux of carbon and fate, led to very low predictability across the mapped classes. Further, the inability to model rates of methane flux or erosional emissions led to an inability to use more extant data—e.g., NPP—to reduce uncertainty (Fig. 5).

Some modeling approaches were used to assess the sensitivity of the estimates to modelable datasets. Among these, the Marsh Equilibrium Model (MEM, Morris, Sundareshwar, Nietch, Kjerfve, & Cahoon, 2002) was run in a virtual marsh scenario to bracket the range of potential carbon sequestration rates and their sensitivity to peak biomass and to sediment supply. The overwhelming driver was relative sea level rise—a geomorphic driver—and subsequent work sought to determine relative elevation (Z^*) in order to characterize the geomorphic setting (Morris & Callaway, 2019). Subsequent work also seeks to constrain the lateral flux estimate by national scale assessment of GPP

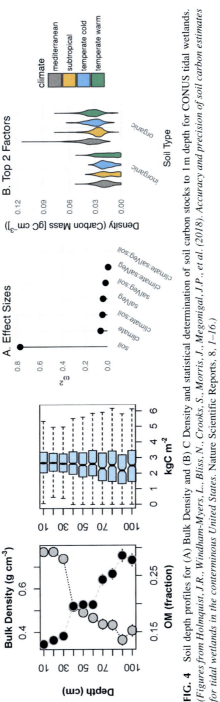

FIG. 4 Soil depth profiles for (A) Bulk Density and (B) C Density and statistical determination of soil carbon stocks to 1 m depth for CONUS tidal wetlands. *(Figures from Holmquist, J.R., Windham-Myers, L., Bliss, N., Crooks, S., Morris, J., Megonigal, J.P., et al. (2018). Accuracy and precision of soil carbon estimates for tidal wetlands in the conterminous United States. Nature Scientific Reports, 8, 1–16.)*

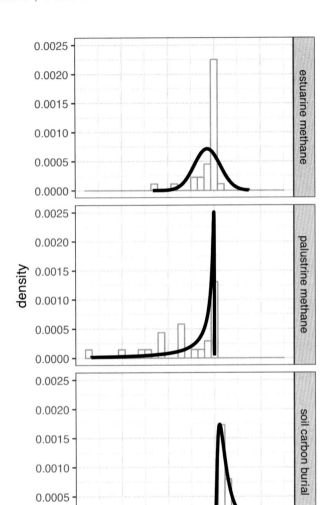

FIG. 5 Distributions of key GHG fluxes from Holmquist et al. (2018b), Net methane flux from (A) estuarine and (B) palustrine wetlands, in CO_2 equivalents, as well as (C) net soil carbon burial.

(Feagin et al., 2020) and tidal framework (Holmquist & Windham-Myers, 2022) (Fig. 6).

Review of the NASA CMS products was incorporated into the National Academy of Science Carbon Dioxide Removal study (NASEM, 2018). Improvements in spatial scaling would help us to refine estimates of coastal blue carbon approaches as a part of a portfolio of technological approaches for

Greenhouse gas accounting in coastal ecosystems Chapter | 12 **419**

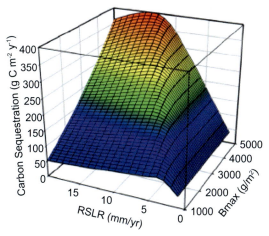

FIG. 6 Modeled sensitivity of carbon sequestration rates to changes in sea level rise, biomass and sediment supply. *(From Morris, J.T., & Callaway, J.C. (2019). Physical and biological regulation of carbon sequestration in tidal marshes. In L. Windham-Myers, S. Crooks, & T.G. Troxler (Eds.), A blue carbon primer: The state of coastal wetland carbon science, practice and policy. CRC Press.)*

achieving negative emissions and stabilization of atmospheric greenhouse gases to avoid the most catastrophic impacts of global climate change. Currently, CONUS-scale coastal blue carbon is estimated to potentially contribute 5.4 Gt CO_2 of carbon removal and reliable sequestration by 2100. This potential relies on addressing knowledge gaps on responses to drivers, such as SLR, temperature, and nutrient loading, and constraining uncertainties in coastal blue carbon to better predict and manage future trajectories so that acceleration of new opportunities for CO_2 removal could be addressed. Coastal blue carbon approaches are those management activities that include restoration of former wetlands, use and creation of nature-based features in coastal resilience projects, managed migration as sea levels rise, augmentation of engineered projects with carbon-rich materials, and management to prevent potential future losses and enhance gains in carbon capacity (NASEM, 2018). Coastal blue carbon is considered a valuable near-term, low-cost approach to achieving negative emissions relative to other negative emission technologies given the other benefits including coastal adaptation and other ecosystem services in addition to carbon mitigation (Barbier et al., 2011; Vegh et al., 2019).

5 Implications for the scale of GHG estimation

Scale dependence is not new to inventory reporting guidance but is paramount in bottom-up approaches of dynamic sectors such as coastal zones. Global scale estimates of GHG stocks and fluxes point to active exchanges, including storage, in coastal settings, but spatial representation is difficult because flux direction and magnitudes vary with drivers operating at scales of hours to millennia.

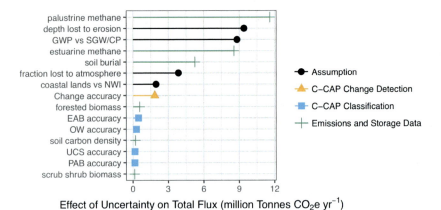

FIG. 7 Primary drivers of uncertainty in coastal wetland ("blue carbon") inventories ranked by dominance. *(Fig. 3 from Holmquist, J.R., Windham-Myers, L., Bernal, B., Byrd, K.B., Crooks, S., Gonneea, M.E., et al. (2018). Uncertainty in United States coastal wetland greenhouse gas inventorying.* Environmental Research Letters, 13(11), 115005.)

The spatial variability in Net Ecosystem Carbon Budgets (NECB) along the coast appears to be strongly geomorphic, where centimeters of elevation define the relationship of land to ocean and thus influence vertical, lateral, and storage fluxes. The capabilities for accuracy in GHG estimation are thus set by the inherent constraints and properties of spatial maps and temporal datasets available to the analyses. For example, the EPA GHG assessment for coastal lands was performed using IPCC guidance and required assumptions on emission factors, all at annual scales. The Holmquist, Windham-Myers, Bernal, et al. (2018) uncertainty analysis evaluated the influence of IPCC assumptions on relative uncertainty and results by emission factor. The guidance has now been incorporated into a Research Coordination Network to address key uncertainties (https://serc.si.edu/coastalCarbon) (Fig. 7).

When comparing GHG estimation constraints at different scales—North America vs. CONUS vs State, vs Site scales—it becomes apparent that mapping, modeling, and measurements play different but interconnected roles. Site-specific carbon budgets have only been attempted at well monitored sites such as Shark River Slough (Florida, Troxler et al., 2015), Plum Island Long-Term Ecological Research Site (Massachusetts, Forbrich, Giblin, & Hopkinson, 2018) and Rush Ranch (California, Bogard et al., 2020), and Waquoit Bay National Estuarine Research Reserve (Massachusetts, Abdul-Aziz et al., 2018; Gonneea et al., 2019; Wang, Kroeger, Ganju, Gonneea, & Chu, 2016). For all sites, the difficulty in closing an annual or decadal NECB emerges from the lack of continuous lateral flux measurements (how much organic carbon leaks from the system through hydrologic exchanges).

At state or regional scales, however, confidence exists for soil monitoring and likely effects of covered IPCC actions such as tidal reintroduction (rewetting) and preservation. While not predictable within a narrow uncertainty window, the certainty affixed to specific fluxes is greatest for soil C stock changes as wetlands remaining wetlands (under globally predictable sea level rise) are necessarily sequestering carbon in vertical accumulation (Callaway, 2019). With accretion estimates, sequestration can be modeled geomorphologically (Z dimension), and with landcover change, gain/erosion can be monitored with remote sensing (X,Y dimensions). This confidence is enabled by the relatively low uncertainty in soil carbon stocks (0.027 ± 0.01 SD) and low uncertainty in biomass stocks (RMSE $<20\%$). The primary uncertainties are for methane emissions in sites of low salinity (Poffenbarger et al., 2011; Windham-Myers et al., 2018) and for the fate of soil carbon in eroding wetlands (Crooks et al., 2018).

6 Implications for carbon cycle science on coastlines

Work to reduce the uncertainties in tidal wetland GHG inventory reporting has led to greater data sharing as well as data collection on carbon fluxes and associated metadata. The improvements are encouraged by the need to address this dynamic and leaky edge of terrestrial-aquatic exchanges (Ward et al., 2020) as well as synthesis efforts that make use of data across timescales and spatial scales. Advances within the past 2 years have included attention to millennial-scale relative elevation as a driver of soil carbon storage (Rogers et al., 2019), evidence of similarity in photosynthetic sensitivities across disparate coastal wetlands (woody and emergent; Feagin et al., 2020), and classification of mangrove soil carbon stocks by geomorphic setting (Rovai et al., 2018) and by depth (Kauffman et al., 2020). As coastal scientists move forward in 2020, greater data sharing is expected to improve statistical modeling and bring coastal wetland accounting into the same space as more integrated land-cover sectors such as forestry and agriculture. To achieve this, coastal wetland classifications will need more stringent metadata support on such factors as latitude and longitude (>4 significant digits decimal degrees), soil properties (texture and depth), tidal connectivity (presence of barriers), relative elevation in the tidal frame (dimensionless Z^*), and disturbance history.

7 Final thoughts

Coastal ecosystem science is increasingly vital to addressing global policy issues. The IPCC Task Force on National Greenhouse Gas Inventories (TFI) is supported by the best available science to develop methodological guidance for use by countries to quantify and report national-scale net GHG emissions associated with anthropogenic activities, maintain an "Emission Factor Database" http://www.ipcc-nggip.iges.or.jp/EFDB/) that provides updated and country-specific data for improved data quality and reduced uncertainties in estimates, produce available data, but can be significantly improved to reduce uncertainties in ecosystem carbon budgets and minimize cost, and provides new

422 Section | C Case Studies

methodologies and expanded data syntheses are needed—i.e., coupled ecosystem models, aquatic C transport, attribution of changes in nearshore water quality (boundary issues). Methodologies, data synthesis, and emission estimates associated with other current and emerging anthropogenic impacts are needed to address data and methodological gaps—for example, estimates of organic and inorganic C export, attribution of nutrient enrichment to sources of pollution, improved estimates of seagrass meadow area and areas of conversion, and differentiating upland from wetland nutrient sources are all important areas of improvement that can improve inventory estimates. As the intersection between science and policy strengthens, the scientific issues of incorporating better coverage of land-use GHG emissions and removals will improve, providing policy makers with more robust tools and information to achieve their goals of reducing the potential for catastrophic impacts of global GHG emissions.

Acknowledgments

We appreciate funding from the NASA Carbon Monitoring System program (#NNH14AY67) and insight from our NASA co-PIs Steve Crooks, Marc Simard, Kristin Byrd, Brian Bergamaschi, John Callaway, Judy Drexler, Rusty Feagin, Pat Megonigal, Jim Morris, Lisa Schile-Beers, John Takekawa, Don Weller, Isa Woo, and Meagan Gonneea. L. Windham-Myers's contributions were supported by the USGS Water Mission Area and USGS Land Carbon Program. Any use of trade, firm, or product names is for descriptive purposes only and does not imply endorsement by the US Government. J. Holmquist's contributions were supported through the Coastal Carbon Research Coordination Network (CCRCN; NSF DEB-1655622). K. Kroeger was supported by the USGS Coastal and Marine Hazards and Resources, and the USGS Land Carbon Programs. This is publication number 28 of the Sea Level Solutions Center in the Institute of Environment at Florida International University.

References

Abdul-Aziz, O. I., Ishtiaq, K. S., Tang, J., Moseman-Valtierra, S., Kroeger, K. D., Gonneea, M. E., et al. (2018). Environmental controls, emergent scaling, and predictions of greenhouse gas (GHG) fluxes in coastal salt marshes. *Journal of Geophysical Research – Biogeosciences, 123*(7), 2234–2256.

Barbier, E. B., Hacker, S. D., Kennedy, C., Koch, E. W., Stier, A. C., & Silliman, B. R. (2011). The value of 47 estuarine and coastal ecosystem services. *Ecological Monographs, 81*(2), 169–193. https://doi.org/10.1890/10-1510.1.

Bogard, M. J., Bergamaschi, B., Butman, D., Anderson, F., Knox, S., & Windham-Myers, L. (2020). Hydrologic export is a major component of coastal wetland carbon budgets. *Global Biogeochemical Cycles, 34*, e2019GB006430.

Byrd, K. B., Ballanti, L., Thomas, N., Nguyen, D., Holmquist, J. R., Simard, M., & Windham-Myers, L. (2018). A remote sensing-based model of tidal marsh aboveground carbon stocks for the conterminous United States. *ISPRS Journal of Photogrammetry and Remote Sensing, 139*, 255–271.

Callaway, J. C. (2019). Accretion: Measurement and interpretation of wetland sediments. In *A blue carbon primer* (pp. 81–92). CRC Press.

Greenhouse gas accounting in coastal ecosystems **Chapter | 12** **423**

Chmura, G. L., Anisfeld, S. C., Cahoon, D. R., & Lynch, J. C. (2003). Global carbon sequestration in tidal, saline wetland soils. *Global Biogeochemical Cycles, 17*, 1111.

Couvillion, B. R., & Beck, H. (2013). Marsh collapse thresholds for coastal Louisiana estimated using elevation and vegetation index data. *Journal of Coastal Research, 63*(suppl 1), 58–67.

Craft, C., Broome, S., & Campbell, C. (2002). Fifteen years of vegetation and soil developent after brackish water marsh creation. *Restoration Ecology, 10*, 248–258.

Craft, C., Megonigal, P., Broome, S., Stevenson, J., Freese, R., Cornell, J., et al. (2003). The pace of ecosystem development of constructed *Spartina alterniflora* marshes. *Ecological Applications, 13*(5), 1417–1432.

Crooks, S., Sutton-Grier, A. E., Troxler, T. G., Herold, N., Bernal, B., Schile-Beers, L., et al. (2018). Coastal wetland management as a contribution to the US National Greenhouse Gas Inventory. *Nature Climate Change, 8*(12), 1109–1112.

Donato, D. C., Kauffman, J. B., Murdiyarso, D., Kurnianto, S., Stidham, M., et al. (2011). Mangroves among the most carbon-rich forests in the tropics. *Nature Geoscience, 4*, 293–297.

Duarte, C. M., Middleburg, J., & Caraco, N. (2005). Major role of marine vegetation on the ocean carbon cycle. *Biogeosciences, 2*, 1–8.

Ellison, A. M., & Farnsworth, E. J. (1996). Anthropogenic disturbance of Caribbean mangrove ecosystems: Past impacts, present trends, and future predictions. *Biotropica, 28*(4), 549–565.

Enwright, N. M., Griffith, K. T., & Osland, M. J. (2016). Barriers to and opportunities for landward migration of coastal wetlands with sea-level rise. *Frontiers in Ecology and the Environment, 14*(6), 307–316.

Fargione, J. E., Bassett, S., Boucher, T., Bridgham, S. D., Conant, R. T., Cook-Patton, S. C., et al. (2018). Natural climate solutions for the United States. *Science Advances, 4*(11), eaat1869.

Feagin, R. A., Forbrich, I., Huff, T. P., Barr, J. G., Ruiz-Plancarte, J., Fuentes, J. D., … Kroeger, K. D. (2020). Tidal wetland gross primary production across the continental United States, 2000–2019. *Global Biogeochemical Cycles, 34*(2), e2019GB006349.

Forbrich, I., Giblin, A. E., & Hopkinson, C. S. (2018). Constraining marsh carbon budgets using long-term C burial and contemporary atmospheric CO_2 fluxes. *Journal of Geophysical Research – Biogeosciences, 123*(3), 867–878.

Gedan, K. B., Silliman, B. R., & Bertness, M. D. (2009). Centuries of human-driven change in salt marsh ecosystems. *Annual Review of Marine Science, 1*, 117–141.

Gonneea, M. E., Maio, C. V., Kroeger, K. D., Hawkes, A. D., Mora, J., Sullivan, R., et al. (2019). Salt marsh ecosystem restructuring enhances elevation resilience and carbon storage during accelerating relative sea-level rise. *Estuarine, Coastal and Shelf Science, 217*, 56–68.

Harris, R. J., Milbrandt, E. C., Everham, E. M., & Bovard, B. D. (2010). The effects of reduced tidal flushing on mangrove structure and function across a disturbance gradient. *Estuaries and Coasts, 33*(5), 1176–1185.

Haywood, B. J., Hayes, M. P., White, J. R., & Cook, R. L. (2020). Potential fate of wetland soil carbon in a deltaic coastal wetland subjected to high relative sea level rise. *Science of the Total Environment, 711*, 135185.

Holmquist, J. R., & Windham-Myers, L. (2022). A Conterminous USA-Scale Map of Relative Tidal Marsh Elevation. *Estuaries and Coasts*, pp. 1–19. https://doi.org/10.1007/s12237-021-01027-9.

Holmquist, J. R., Windham-Myers, L., Bernal, B., Byrd, K. B., Crooks, S., Gonneea, M. E., et al. (2018). Uncertainty in United States coastal wetland greenhouse gas inventorying. *Environmental Research Letters, 13*(11), 115005.

Holmquist, J. R., Windham-Myers, L., Bliss, N., Crooks, S., Morris, J., Megonigal, J. P., et al. (2018). Accuracy and precision of soil carbon estimates for tidal wetlands in the conterminous United States. *Nature Scientific Reports, 8*, 1–16.

424 Section | C Case Studies

Hopkinson, C. S., Cai, W. J., & Hu, X. (2012). Carbon sequestration in wetland dominated coastal systems—A global sink of rapidly diminishing magnitude. *Current Opinion in Environmental Sustainability*, *4*(2), 186–194.

Hu, Z., Lee, J. W., Chandran, K., Kim, S., & Khanal, S. K. (2012). Nitrous oxide (N2O) emission from aquaculture: A review. *Environmental Science & Technology*, *46*(12), 6470–6480.

IPCC. (2014). In T. Hiraishi, T. Krug, K. Tanabe, N. Srivastava, J. Baasansuren, M. Fukuda, & T. G. Troxler (Eds.), *2013 Supplement to the 2006 IPCC guidelines for national greenhouse gas inventories: Wetlands*. Switzerland: IPCC.

Kauffman, J. B., Adame, M. F., Arifanti, V. B., Schile-Beers, L. M., Bernardino, A. F., Bhomia, R. K., et al. (2020). Total ecosystem carbon stocks of mangroves across broad global environmental and physical gradients. *Ecological Monographs*, *90*, e01405.

Kirwan, M. L., & Blum, L. K. (2011). Enhanced decomposition offsets enhanced productivity and soil carbon accumulation in coastal wetlands responding to climate change. *Biogeosciences*, *8*(4), 987.

Kroeger, K. D., Crooks, S., Moseman-Valtierra, S., & Tang, J. (2017). Restoring tides to reduce methane emissions in impounded wetlands: A new and potent blue carbon climate change intervention. *Scientific Reports*, *7*(1), 1–12.

Lovelock, C., et al. (2019). *2019 Refinement to the 2006 IPCC Guidelines for National Greenhouse Gas Inventories. Volume 4: Agriculture, Forestry and Other Land Use*. https://www.ipcc-nggip.iges.or.jp/public/2019rf/pdf/4_Volume4/19R_V4_Ch07_Wetlands.pdf.

McLeod, E., Chmura, G. I., Bouillion, S., Salm, R., Bjork, M., et al. (2011). A blueprint for bluecarbon: Toward an improved understanding of the role of vegetated coastal habitats in sequestering CO2. *Frontiers in Ecology and the Environment*, *9*, 552–560.

Morris, J. T., Barber, D. C., Callaway, J. C., Chambers, R., Hagen, S. C., Hopkinson, C. S., ... Wigand, C. (2016). Contributions of organic and inorganic matter to sediment volume and accretion in tidal wetlands at steady state. *Earth's Future*, *4*(4), 110–121.

Morris, J. T., & Callaway, J. C. (2019). Physical and biological regulation of carbon sequestration in tidal marshes. In L. Windham-Myers, S. Crooks, & T. G. Troxler (Eds.), *A blue carbon primer: The state of coastal wetland carbon science, practice and policy* CRC Press.

Morris, J. T., Sundareshwar, P. V., Nietch, C. T., Kjerfve, B., & Cahoon, D. R. (2002). Responses of coastal wetlands to rising sea level. *Ecology*, *83*(10), 2869–2877.

National Academies of Sciences, Engineering, and Medicine (NASEM). (2018). *Negative emissions technologies and reliable sequestration: A research agenda*. Washington, DC: The National Academies Press. https://doi.org/10.17226/25259.

Nellemann, C., et al. (2009). *Blue carbon. A rapid response assessment 78*. United Nations Environment Programme, GRID-Arenal.

Osland, M. J., Enwright, N. M., Day, R. H., Gabler, C. A., Stagg, C. L., & Grace, J. B. (2016). Beyond just sea-level rise: Considering macroclimatic drivers within coastal wetland vulnerability assessments to climate change. *Global Change Biology*, *22*(1), 1–11.

Osland, M. J., Spivak, A. C., Nestlerode, J. A., Lessmann, J. M., Almario, A. E., Heitmuller, P. T., et al. (2012). Ecosystem development after mangrove wetland creation: Plant–soil change across a 20-year chronosequence. *Ecosystems*, *15*(5), 848–866.

Pendleton, L., Donato, D. C., Murray, B. C., et al. (2012). Estimating global 'blue carbon' emissions from conversion and degradation of vegetated coastal ecosystems. *PLoS One*, *7*, e43542.

Poffenbarger, H. J., Needelman, B. A., & Megonigal, J. P. (2011). Salinity influence on methane emissions from tidal marshes. *Wetlands*, *31*(5), 831–842.

Purvaja, R., & Ramesh, R. (2001). Natural and anthropogenic methane emission from coastal wetlands of South India. *Environmental Management*, *27*(4), 547–557.

Rahman, F., Dragoni, D., Didan, K., Barreto-Munoz, A., & Hutabarat, J. (2013). Detecting large scale conversion of mangroves to aquaculture with change point and mixed-pixel analyses of high-fidelity MODIS data. *Remote Sensing of Environment*, *130*, 96–107. https://doi.org/10.1016/j.rse.2012.11.014.

Redfield, A. C. (1965). Ontogeny of a salt marsh estuary. *Science*, *147*(3653), 50–55.

Rogers, K., Kelleway, J. J., Saintilan, N., Megonigal, J. P., Adams, J. B., Holmquist, J. R., et al. (2019). Wetland carbon storage controlled by millennial-scale variation in relative sea-level rise. *Nature*, *567*(7746), 91–95.

Rovai, A. S., Twilley, R. R., Castañeda-Moya, E., Riul, P., Cifuentes-Jara, M., Manrow-Villalobos, M., et al. (2018). Global controls on carbon storage in mangrove soils. *Nature Climate Change*, *8*(6), 534–538.

Simard, M., Fatoyinbo, L. E., & Pinto, N. (2010). Mangrove canopy 3D structure and ecosystem productivity using active remote sensing. *Remote Sensing of Coastal Environment*, 61–78.

Simard, M., Fatoyinbo, L., Smetanka, C., Rivera-Monroy, V. H., Castañeda-Moya, E., Thomas, N., & Van der Stocken, T. (2019). Mangrove canopy height globally related to precipitation, temperature and cyclone frequency. *Nature Geoscience*, *12*(1), 40–45.

Troxler, T. G., Barr, J. G., Fuentes, J. D., Engel, V., Anderson, G., Sanchez, C., et al. (2015). Component-specific dynamics of riverine mangrove CO2 efflux in the Florida coastal Everglades. *Agricultural and Forest Meteorology*, *213*, 273–282.

U.S. Environmental Protection Agency (EPA). (2017). *Inventory of U.S. greenhouse gas emissions and sinks 1990–2015*. U.S. Environmental Protection Agency. # EPA 430-P-17-001.

Valiela, I., Bowen, J. L., & York, J. K. (2001). Mangrove forests: One of the world's threatened major tropical environments: At least 35% of the area of mangrove forests has been lost in the past two decades, losses that exceed those for tropical rain forests and coral reefs, two other well-known threatened environments. *Bioscience*, *51*(10), 807–815.

Vegh, T., Pendleton, B., Murray, L., Troxler, T., Zhang, K., Guannel, G., et al. (2019). Ecosystem services and economic valuation: Co-benefits of coastal wetlands. In L. Windham-Myers, S. Crooks, & T. G. Troxler (Eds.), *A blue carbon primer: The state of coastal wetlands carbon science, practice and policy*. Boca Raton, FL: CRC Press. 480p.

Walters, B. B., Rönnbäck, P., Kovacs, J. M., Crona, B., Hussain, S. A., Badola, R., et al. (2008). Ethnobiology, socio-economics and management of mangrove forests: A review. *Aquatic Botany*, *89*(2), 220–236.

Wang, Z. A., Kroeger, K. D., Ganju, N. K., Gonneea, M. E., & Chu, S. N. (2016). Intertidal salt marshes as an important source of inorganic carbon to the coastal ocean. *Limnology and Oceanography*, *61*(5), 1916–1931.

Ward, N. D., Bond-Lamberty, B., Bailey, V., Butman, D., Canuel, E. A., Diefenderfer, H., et al. (2020). Representing the function and sensitivity of coastal interfaces in earth system models. *Nature Communications*, *11*, 1–14.

Wilson, B. J., Servais, S., Charles, S. P., Davis, S. E., Gaiser, E. E., Kominoski, J. S., et al. (2018). Declines in plant productivity drive carbon loss from brackish coastal wetland mesocosms exposed to saltwater intrusion. *Estuaries and Coasts*, *41*(8), 2147–2158.

Windham-Myers, L., Cai, W.-J., Alin, S., Andersson, A., Crosswell, J., Dunton, K. H., et al. (2018). Chapter 15: Tidal wetlands and estuaries. In N. Cavallaro, G. Shrestha, R. Birdsey, M. A. Mayes, R. Najjar, S. Reed, P. Romero-Lankao, & Z. Zhu (Eds.), *Second state of the carbon cycle report (SOCCR2): A sustained assessment report*. Washington, DC, USA: U.S. Global Change Research Program.

Windham-Myers, L., Crooks, S., & Troxler, T. G. (2019). *A blue carbon primer: The state of coastal wetland carbon science, practice and policy. 28 chapters, 93 contributing authors*. Boca Raton, Florida: CRC Press. 496 pp.

Chapter 13

Ocean systems

Peter Landschützer[a], Lydia Keppler[a,b,c], and Tatiana Ilyina[a]
[a]*Max Planck Institute for Meteorology, Hamburg, Germany, [b]International Max Planck Research School on Earth System Modelling (IMPRS-ESM), Hamburg, Germany, [c]Scripps Institution of Oceanography, University of California San Diego, La Jolla, CA, United States*

Key messages

- The ocean is a large and consistent net sink for carbon dioxide ($2.6 \pm 0.6 \, \mathrm{Pg\,C\,year^{-1}}$) and a smaller net source for methane (releases $13 \, \mathrm{Tg\,CH_4\,year^{-1}}$) and nitrous oxide ($3.4 \, \mathrm{Tg\,N\,year^{-1}}$).
- The ocean provides a large storage space for human emitted carbon and has stored $152 \pm 20 \, \mathrm{Pg\,C}$ from 1850 to 2007.
- The ocean carbon sink varies substantially on interannual to decadal timescales.

1 Summary

The ocean comprises $\sim71\%$ of the Earth's surface area and is in constant interaction with the atmosphere above and the land surface at the coastal interface, allowing a continuous exchange of greenhouse gases (GHGs) between the spheres. The ocean plays an important role in absorbing and storing carbon dioxide (CO_2) from fossil fuel combustion, land-use change, and cement production. Since the industrial revolution, the ocean has stored $\sim31\%$ of human emitted CO_2, adding to a total storage of anthropogenic CO_2 of $152 \pm 20 \, \mathrm{Pg\,C}$ ($\mathrm{Pg\,C} = \mathrm{Petagrams\ of\ carbon}$) from 1850 to 2007 (Gruber et al., 2019) and is currently removing about $2.6 \pm 0.6 \, \mathrm{Pg\,C}$ of excess CO_2 every year from the atmosphere (Friedlingstein et al., 2019). On longer timescales (i.e., centuries to millennia), the ocean carbon sink acts as a primary regulator of the Earth's climate. While the ocean carbon sink mitigates climate change, absorption of anthropogenic CO_2 leads to ocean acidification with potentially harmful effects for marine ecosystems. The ocean also contributes to the cycles of other greenhouse gases. Specifically, it is a weak source of methane. The contribution of the ocean to the net global methane budget, however, is substantially smaller than the oceanic uptake of CO_2. The ocean was a net source of methane (CH_4) to the atmosphere of $\sim13 \, \mathrm{Tg\,CH_4\,year^{-1}}$ ($\mathrm{Tg\,CH_4} = \mathrm{Teragrams\ of}$ CH_4) with a possible range of 9–$22 \, \mathrm{Tg\,CH_4\,year^{-1}}$ over the period

Balancing Greenhouse Gas Budgets. https://doi.org/10.1016/B978-0-12-814952-2.00004-6
Copyright © 2022 Elsevier Inc. All rights reserved.

428 Section | C Case Studies

2000–2017 (Saunois et al., 2020). Hence, the methane fluxes from the ocean to the atmosphere are an order of magnitude smaller than the anthropogenic emissions over the same period (Saunois et al., 2020). Likewise, the ocean comprises a natural source of nitrous oxide (N_2O) of ~3.4 Tg N year^{-1} between 2007 and 2016, although with a substantial possible range between 2.5 and 4.3 Tg N year^{-1} (Tian, Xu, Canadell, et al., 2020).

Complex interactions of different processes lead to a unique distribution of GHGs in the ocean. The surface ocean is further connected to the deep ocean through circulation and mixing, providing "storage space" for GHGs. This chapter outlines the processes governing the uptake or release of three dominant GHGs to the atmosphere. The following sections then continue to focus on the ocean carbon cycle, its preindustrial state, and its temporal evolution since preindustrial times. The chapter further discusses the contemporary ocean carbon sink and its variability in time and space and finally gives a future outlook.

2 The ocean as a sink/source of GHGs to the atmosphere

The exchange of a gas or the flux (F) at the atmosphere-ocean interface is dependent on two main factors: (I) the kinetic gas transfer velocity (k) and (II) the concentration difference of a gas in the surface water and the overlaying atmosphere ($C_o - C_a$) (Sarmiento & Gruber, 2006) and takes the form.

$$F = k\,(C_o - C_a) \tag{1}$$

Using Henry's law, the concentration difference term can be replaced with the partial pressure difference, taking into account the solubility of a gas in seawater. In this way, the air-sea flux can be estimated from measurable quantities. The gas transfer velocity is usually empirically determined as a function of the wind speed at 10 m height above the surface (Wanninkhof et al., 2013); however, various formulations have been suggested over time (Roobaert, Laruelle, Landschützer, & Regnier, 2018). The concentration difference or partial pressure difference term determines both the magnitude and the direction of the gas flux. If the concentration of a gas is larger in the ocean surface than its saturation concentration, the surface water is supersaturated and the ocean will release the gas to the atmosphere. If the concentration at the sea surface is smaller than its saturation concentration, the surface water is undersaturated and the ocean will take up the gas from the atmosphere (see, e.g., Sarmiento & Gruber, 2006).

Marine measurements of both CH_4 and N_2O are sparse, and hence large uncertainties exist regarding the air-sea exchange as well as the variability in time of both greenhouse gases (Ciais et al., 2013; Tian et al., 2020). However, efforts are underway to better constrain these fluxes from available measurements (Bange et al., 2009).

The ocean is in direct contact with the lithosphere and gases are introduced to the ocean, e.g., via river discharge or via sediments and the seafloor (Carpenter, Archer, & Beale, 2012; Saunois et al., 2020). In particular, the latter determines the marine CH_4 budget. Methane is introduced to the ocean via geological leaks at the seafloor, sediment production, thawing subsea permafrost, and destabilized hydrates. Furthermore, in coastal systems, water column production plays another important role in marine CH_4 production. CH_4 is then transported through the water column via diffusion or ebullition to the air-sea surface where it enters the atmosphere. There are, however, important processes at play throughout the water column that prohibit all of the methane in the ocean from reaching the atmosphere. Microbial processes and aerobic oxidation are the most important sinks of methane in the water column (Saunois et al., 2020). Furthermore, bubble dissolution and physical barriers such as the oceanic pycnocline prevent some methane from reaching the surface (Ciais et al., 2013). On the basis of a bottom-up approach, Saunois et al. (2020) estimate a marine CH_4 source of $13\,Tg\,CH_4\,year^{-1}$ with a possible range of 9–$22\,Tg\,CH_4\,year^{-1}$.

Nitrous oxide (N_2O) comprises another potent greenhouse gas (Ciais et al., 2013; Tian et al., 2020), whereas its in situ production in the ocean is largely unconstrained due to the complexity of the marine nitrogen cycle (Carpenter et al., 2012). Best estimates show that the ocean is a source of N_2O to the atmosphere as N_2O is produced in the ocean water column as a by-product of marine nitrification of ammonium to nitrate (Buitenhuis, Suntharalingam, & Le Quéré, 2018). Furthermore, N_2O is also an intermediate product of denitrification. N_2O production is mostly associated with processes in the open water column; however, hotspots for marine N_2O production exist in the low oxygenated waters of the Eastern Boundary upwelling systems (Battaglia & Joos, 2018). Overall, the global ocean comprises a small net source of N_2O to the atmosphere of $3.4\,Tg\,N\,year^{-1}$ with a possible range of 2.5–$4.3\,Tg\,N\,year^{-1}$ (Tian et al., 2020). Through atmospheric deposition into the ocean, Ciais et al. (2013) further report a small anthropogenic source of $0.2\,Tg\,N\,year^{-1}$ (range 0.1–$0.4\,Tg\,N\,year^{-1}$).

Quantitative assessments of the oceanic contribution to the N_2O and CH_4 budgets are still in their early stages due to gaps in observations and process understanding. In contrast, measurements of carbonate system parameters such as the partial pressure of CO_2 (pCO_2) at the sea surface are more numerous and fluxes as well as their variability in space and time are better constrained. Therefore, in the following, the chapter will continue to focus on CO_2. The two main pathways by which CO_2 enters the ocean are via the air-sea interface and from the land through river discharge. Unlike in the atmosphere, the present-day mean distribution of the pCO_2 in the ocean surface is strongly heterogeneous, ranging from a few µatm up to $500\,µatm$ in the equatorial upwelling region (Landschützer et al., 2014; Takahashi et al., 2009). Therefore, the regional pattern of the air-sea gas exchange is determined almost entirely by the oceanic processes. This pattern is controlled by the two main regulating group of mechanisms, referred to as pumps, i.e., the physical pump and the biological pumps (Heinze et al., 2015; Sarmiento & Gruber, 2006).

430 Section | C Case Studies

CO_2 entering the ocean via the air-sea interface dissolves in seawater to form carbonic acid and further reacts to bicarbonate and carbonate ion, summarized as dissolved inorganic carbon (DIC). The reactions can be summarized to:

$$CO_{2(aq)} + H_2O \rightleftharpoons H^+ + HCO_3^- \rightleftharpoons 2H^+ + CO_3^{2-} \tag{2}$$

Only a small fraction of gaseous CO_2 remains in the form of aqueous CO_2 ($CO_{2(aq)}$). The amount of CO_2 dissolving is largely temperature dependent (Sarmiento & Gruber, 2006; Weiss, 1974); hence, CO_2 is more soluble in cold waters, leaving the sea surface undersaturated with regard to the partial pressure of CO_2, providing a sink for atmospheric CO_2 in cold waters. DIC is distributed in the ocean by turbulent and mixing processes as a passive tracer (Heinze et al., 2015; Sarmiento & Gruber, 2006). In the North Atlantic, warm and salty surface water transported from the low latitudes cools and breaks the stratification. Close to the ice edges, when sea water freezes, salt is rejected, leading to the formation of cold and salty and hence dense water near the ice. This dense water sinks to the deep ocean. Through this process, DIC is brought to the ocean interior, increasing the deep water DIC concentration. Likewise, in ocean upwelling areas, DIC-rich water from deeper layers is brought back to the surface. At the surface, these waters warm and the reverse chemical reactions reintroduce DIC to the CO_2 pool, leaving the surface waters supersaturated, and CO_2 is released back to the atmosphere. Radiocarbon-based measurements reveal that deep waters in the Southern Ocean have not been in contact with the atmosphere on timescales substantially longer than the anthropogenic signature of CO_2 in the atmosphere (Sarmiento & Gruber, 2006). Therefore, waters upwelled in the Southern Ocean today still carry the natural radiocarbon signature. The solubility pump is sensitive not only to the temperature of seawater but also to changes in ocean mixing and the Meridional Overturning Circulation (Sarmiento & Gruber, 2006).

The second set of mechanisms affecting the distribution of surface DIC is governed by the biological pump. DIC at the sea surface is taken up by biota to form particulate matter (Heinze et al., 2015). In particular, phytoplankton is consuming CO_2 during their photosynthetic energy production forming organic carbon (Sarmiento & Gruber, 2006). The biomass produced by phytoplankton is then further utilized by organisms on higher trophic levels. CO_2, however, is not the only requirement for the production of particulate organic matter. Nutrient and light availability are essential components, closely tying the carbon cycle to the marine nutrient cycle. Approximately 25% of the particulate matter produced at the surface (Heinze et al., 2015) sinks to depth; however, only a tiny fraction reaches the seafloor. The majority of sinking particles is demineralized and eventually returns to the dissolved phase within the water column (Heinze et al., 2015). In addition to particulate organic matter, marine biota further produce dissolved organic carbon which is distinguished from particulate organic matter through the size of the particles. Marine organisms

Ocean systems Chapter | 13 **431**

further form calcium carbonate shells and skeletons, counteracting the organic carbon pump by releasing CO_2 during the shell-building process. The biological carbon pumps and their ability to remove dissolved carbon are tightly linked to ocean circulation (Sarmiento & Gruber, 2006). Upwelling regions, such as the equatorial Pacific, eastern boundary upwelling regions, or the Southern Ocean, supply surface waters with nutrients from deeper ocean layers. In combination with sufficient light availability and surface water stratification, plankton blooms occur, removing substantial amounts of dissolved organic carbon. Overall, the inorganic carbon pool is substantially larger than the organic carbon pool (Ciais et al., 2013).

3 Preindustrial (or natural) carbon budget based on inverse estimates

The contemporary ocean carbon sink consists of a natural flux component driven by the carbon discharge of rivers, an interhemispheric transport of carbon, that existed already in preindustrial times (Gruber et al., 2009). When taking marine measurements, the total contemporary (natural + anthropogenic) carbon is measured, which poses a challenge in estimating the anthropogenic contribution to present-day carbon budgets. Furthermore, both natural and anthropogenic carbon budgets further consist of a steady-state component as well as a nonsteady-state component. The ocean circulation and biological production create regional sources and sinks for CO_2 in the ocean, leading to substantial natural background air-sea CO_2 fluxes aside from the anthropogenic signal (Ciais et al., 2013).

The origin and distribution of preindustrial or natural fluxes can be estimated using ocean inverse approaches (Gloor et al., 2003; Mikaloff-Fletcher et al., 2007) combining observations of the dissolved inorganic carbon content and other tracers, e.g., from the Global Ocean Data Analysis Project (GLODAPv2) database (Lauvset et al., 2016; Olsen et al., 2016). Ocean inverse estimates are based on the definition of a tracer (e.g., C^*) that represents the sum of preindustrial and human-induced exchange of CO_2. This tracer is assumed to have no sources or sinks in the ocean interior and its distribution is based on ocean mixing and transport alone. Ocean inverse methods "reverse" this transport process and thereby imply the amount of carbon that has entered the ocean (Gloor et al., 2003; Gruber et al., 2009). Furthermore, ocean biogeochemical models inform us about the mechanisms determining the regional exchange fluxes between the ocean and the atmosphere. To estimate the preindustrial sink, preindustrial control simulations, i.e., when models are forced with a constant preindustrial atmospheric CO_2 molar fraction (287 ppm CO_2), help to illustrate regional sources and sinks of carbon dioxide in the ocean based on the physical and biogeochemical mechanisms controlling them in an unperturbed Earth system (e.g., Ilyina et al., 2013; Séférian et al., 2016), are commonly used.

In theory, the regional disequilibrium between the ocean and the atmosphere results in substantial annual background fluxes of 60 Pg C year^{-1} (Ciais et al., 2013); however, the overall net flux approximately amounts to zero in a steady-state climate and at steady atmospheric CO_2 conditions where the ocean only interacts with the atmosphere and biological activity does not substantially change in time. This is, however, not the case, since there is an unconstrained nonsteady-state component as well as an additional source of carbon entering the coastal ocean through rivers and eventually, at least in part, the open ocean. Estimates of this river transport to the coast at present range from 0.90 to 0.95 Pg C year^{-1} (Ciais et al., 2013; Regnier et al., 2013), whereas Regnier et al. (2013) propose that only ~0.75 Pg C year^{-1} enters the open ocean via the coast (see Fig. 1). Furthermore, there is an additional sink through burial in the ocean sediments of about 0.2 Pg C year^{-1} (Ciais et al., 2013; Regnier et al., 2013). Overall, estimates of the river induced natural outgassing of CO_2 from the open ocean to the atmosphere range from 0.23 Pg C year^{-1}, based on model results (Lacroix et al., 2019), through 0.45 Pg C year^{-1}

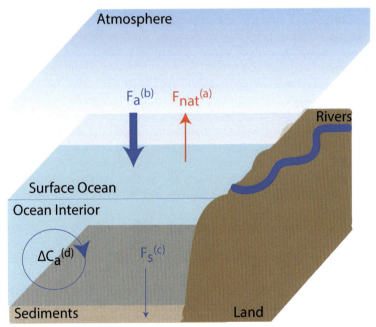

FIG. 1 Schematic of carbon fluxes in and out of the ocean and resulting ocean storage over the industrial period. (a) preindustrial outgassing of natural carbon induced by rivers ranges from 0.23 to 0.78 Pg C year^{-1} (Ciais et al., 2013; Jacobson et al., 2007; Lacroix, Ilyina, & Hartmann, 2019; Resplandy et al., 2018), (b) best estimate of the anthropogenic air-sea CO_2 flux of 2.6 ± 0.6 Pg C year^{-1} from Le Quéré et al. (2018), (c) sediment flux of 0.2 Pg C year^{-1} from Regnier et al. (2013) and Ciais et al. (2013), and (d) storage of anthropogenic CO_2 of 152 ± 20 Pg C from Gruber et al. (2019).

Ocean systems **Chapter | 13 433**

(Jacobson et al., 2007) to $0.70\,Pg\,C\,year^{-1}$ (Ciais et al., 2013), and $0.78\,Pg\,C\,year^{-1}$ (Resplandy et al., 2018); however, little is known about how variable these fluxes are in time.

Estimates of the natural net air-sea exchange from ocean inversions (Mikaloff-Fletcher et al., 2007) suggest regionally diverging patterns of substantial CO_2 uptake by the ocean in the midlatitudes in the Pacific Ocean, Indian Ocean, and Atlantic Ocean as well as the high latitude Atlantic Ocean ($\sim1.3\,Pg\,C\,year^{-1}$). As tropical waters move poleward, they cool and take up natural CO_2 (Mikaloff-Fletcher et al., 2007). In addition, Takahashi et al. (2009) attribute the strong uptake in the subtropical zones to strong winds between 40°S and 50°S, and low pCO_2 along the subtropical convergence zone. Here, cooled subtropical gyre waters with low pCO_2 meet the subpolar waters with biologically lowered pCO_2. In the high latitude North Atlantic, natural CO_2 is taken up because of the combination of cooling of warm subtropical waters moving north and the highly efficient biological carbon pump.

The tropics and the Southern Ocean release natural CO_2 from the ocean to the atmosphere (~0.9 and $\sim0.4\,Pg\,C\,year^{-1}$, respectively) (Gruber et al., 2009), i.e., where the ocean circulation brings up carbon-rich deep waters to the surface. Equatorial trade winds move water masses in the tropics toward the west where they divert further affected by the Earth's rotation (Sarmiento & Gruber, 2002). Because of the Coriolis force, water masses are diverted 90° to the right of the westward flow in the northern hemisphere (i.e., northward) and 90° to the left in the southern hemisphere (i.e., southward). This Ekman transport leads to diverging water masses, causing the upwelling of cold but carbon-rich waters from deeper layers to the surface. The upwelled and carbon-rich water warms at the surface, increasing the pCO_2 and leaving the surface waters supersaturated, which leads to CO_2 outgassing from the ocean.

In the Southern Ocean, the biological pump is relatively slow and inefficient due to low iron concentrations (Sarmiento & Gruber, 2002). The upwelling of carbon-rich deep waters dominated the preindustrial carbon budget in the Southern Ocean, making it a source of CO_2 to the atmosphere. The Southern Ocean is characterized by the strong eastward-flowing Antarctic Circumpolar Current, by zonal fronts that separate different water masses, and by upwelling and subduction areas. Hence, different processes are at play in the Southern Ocean. A lack of observations has historically hindered our understanding of the Southern Ocean biogeochemistry.

In the southernmost region, the Continental Zone just off the Antarctic coast, deep water formation occurs, creating the Antarctic Bottom Water. Just south of the Antarctic Polar Front, the eastward-flowing Antarctic Circumpolar Current induces intense surface divergence resulting in the Ekman transport of aged and carbon-rich deep water from the deep ocean, transporting natural carbon to the surface around 60°S. Further north, in the Polar Frontal Zone, the subduction and northward transport of Sub-Antarctic Mode Water provides a pathway for surface waters to the interior. In the northernmost region, the Sub-Tropical

434 Section | C Case Studies

Zone, Sub-Antarctic Surface Waters are transported northward and transform into Sub-Tropical Surface Waters (Gruber et al., 2009; Sabine et al., 2004; Talley et al., 2016).

Riverine carbon enters the ocean mainly in the northern hemisphere and the tropics (Gruber et al., 2009; Mikaloff-Fletcher et al., 2007), adding to the inter-hemispheric imbalance. The ocean general circulation models agree with the general pattern of tropical release of CO_2 and midlatitude uptake; however, they usually do not account for the preindustrial river source of carbon; hence, their preindustrial budget is balanced.

4 Anthropogenic perturbations and the contemporary global carbon sink

While the preindustrial ocean is estimated to be a source of carbon to the atmosphere, the present-day ocean accounts for the uptake of $\sim$25% of all human emitted emissions from fossil fuels, making it a net carbon sink (Friedlingstein et al., 2019). In simple terms, the present-day or contemporary ocean-atmosphere flux is represented by the sum of the preindustrial state and the anthropogenic fluxes, further assuming that sediment burial and river input have not changed substantially in time, i.e., the nonsteady-state components are negligibly small and can be expressed as:

$$F_{cont} = F_{anth} + F_{nat} \tag{3}$$

The continuous increase in the atmospheric CO_2 concentration since the industrialization (Dlugokencky & Tans, 2018) has changed the saturation concentration of surface ocean CO_2 that determines the exchange of CO_2 between the ocean and the atmosphere (see Eq. 1). Throughout the ocean, the anthropogenic component (F_{anth}, see Eq. 3) has increased and is responsible for the increase in the contemporary net uptake of CO_2.

Today, largely owing to the improvement of computer models, the increasing abundance of marine CO_2 system measurements, and international synthesis efforts (e.g., Bakker et al., 2016; Lauvset et al., 2016; Olsen et al., 2016; Takahashi, Sutherland, & Kozyr, 2018) (see Table 1), several independent constraints on the anthropogenic (F_{anth}) and contemporary carbon flux (F_{cont}) and their variability exist, unlike the variability in the natural component (F_{nat}) that is usually assumed constant in time for present-day carbon budgets (Friedlingstein et al., 2019; Gruber et al., 2009).

The Global Carbon Project estimates that the ocean currently takes up $2.6 \pm 0.6 \, \mathrm{Pg\,C\,year}^{-1}$ (Friedlingstein et al., 2019) of anthropogenic carbon from the atmosphere (see Fig. 1). This estimate is largely based on a collection of biogeochemistry model simulations forced with observationally derived fluxes of momentum, heat, and fresh water available from the climate reanalysis products (Friedlingstein et al., 2019). Models simulate the carbon cycle based on known processes governing the solubility, transport, and exchange of

TABLE 1 Overview of the key databases and synthesis efforts leading to novel constraints of the marine carbon sink.

Synthesis dataset	Key parameter	Observing programs	Reference
Surface Ocean CO_2 Atlas (SOCAT)	Surface ocean fugacity of CO_2 (fCO_2)	Repeat hydrography programs (e.g., WOCE, JGOFS); ship of opportunity program (VOS or SOOP)	Bakker et al. (2016)
Lamont Doherty Earth Observatory (LDEO) database	Surface ocean partial pressure of CO_2 (pCO_2)	Repeat hydrography program (e.g., WOCE, JGOFS); ship of opportunity program (VOS or SOOP)	Takahashi et al. (2018)
GLODAP	Dissolved inorganic carbon (DIC) in the ocean interior	Repeat Hydrography (e.g., WOCE, JGOFS, GO-ship)	Olsen et al. (2016)

CO_2 and biogeochemical processes; however, they are not directly constrained by CO_2 observations. The use of ocean biogeochemical models has a long history in global carbon budgets (Friedlingstein et al., 2019; Sarmiento et al., 2010) and until recently have been the prime constraint on the ocean carbon uptake and variability.

Traditionally, the contemporary air-sea exchange of CO_2 has been further constrained from atmospheric inverse approaches based on measurements of the atmospheric CO_2 content. Atmospheric inversions are powerful tools as they provide estimates for both the ocean and the land sink; however, air-sea flux estimates suffer from errors introduced due to the relative dominance of the land variability in atmospheric signals (Peylin et al., 2013).

Similarly, ocean inverse methods are used to estimate the ocean carbon sink (see e.g., Gruber et al., 2009; Mikaloff-Fletcher et al., 2006); however, historically, they have been limited in the time domain and resolution. Ocean interior estimates of the carbon inventory provide a strong constraint of the mean uptake of a certain reference year or a certain time period; hence, they are often used as a baseline to compare to fluxes from other methods (Wanninkhof et al., 2013) or to narrow down the spread between model ensembles (Quéré et al., 2018). Ocean inverse approaches combine physical modeling with observation-based model optimization (Gruber et al., 2019). Using in situ measurements of the dissolved carbon content of the ocean interior (e.g., Olsen et al., 2016; see Table 1), they rely on an observational constraint that is independent from the approaches using measurements of the surface ocean pCO_2. To overcome the time domain restriction, time-dependent inverse approaches have further been introduced, based on optimizing ocean transport models with hydrographic data (DeVries, Holzer, & Primeau, 2017).

Most recently, data-driven estimates based on measurements of the surface ocean pCO_2 emerged, which have the potential to constrain the mean and the variations in the oceanic uptake for CO_2. This is largely owing to the strong increase in the number of available measurements and the community effort to synthesize these measurements and combine them in large publicly available databases (see Table 1) such as the Lamont Doherty Earth Observatory (LDEO) database (Takahashi et al., 2018), or the Surface Ocean CO_2 Atlas database (SOCAT) (Bakker et al., 2016; Takahashi et al., 2018). Despite the vast abundance of available measurements, they are not uniformly distributed in time and space, providing a challenge to estimate the ocean carbon sink from direct measurements alone. In particular, in the southern hemisphere, observations remain sparse in time and space, largely owing to harsh sailing conditions in winter in the high latitude Southern Ocean, or limited commercial shipping routes that may serve as voluntary observing platforms. This strongly limits the global interpretation of shipboard-based measurements without any form of interpolation. Therefore, coinciding with the growing number of available measurements, various methods have emerged to interpolate and map shipboard pCO_2 data in space and time (Rödenbeck et al., 2015). These methods range from statistical techniques based on autocorrelation (Jones et al., 2015) to machine learning approaches such as neural networks (Landschützer et al., 2013).

Previous assessment studies have focused on the intercomparison between the fluxes derived from biogeochemical models, inverse and pCO_2-based estimates (e.g., Lenton et al., 2013; Schuster et al., 2013; Wanninkhof et al., 2013). These studies have found that the long-term mean uptake from 1990 through 2009 among different methods is broadly in agreement for different ocean regions, whereas Lenton et al. (2013) found that in the Southern Ocean no individual atmospheric inverse approach or ocean model matches the long-term mean sink derived from the climatological mean distribution calculated from surface ocean pCO_2 observations (Takahashi et al., 2009).

The most robust estimate for the oceanic uptake of anthropogenic carbon for the 2009–2018 period from the global carbon budget was $2.6 \pm 0.6 \, \mathrm{Pg \, C \, year^{-1}}$. Historically, this number is based on a set of ocean model simulations, verified by different methods used in the IPCC assessment (Ciais et al., 2013), namely trends in observed atmospheric O_2/N_2 concentrations (Manning & Keeling, 2006), an ocean inverse estimate (Mikaloff-Fletcher et al., 2006), and a method based on a penetration timescale for chlorofluorocarbons (McNeil et al., 2003). The latter 3 estimates suggest a mean ocean carbon uptake of $2.2 \pm 0.6 \, \mathrm{Pg \, C \, year^{-1}}$, in line with the majority of ocean models and other methodologies used by Wanninkhof et al. (2013). For the reference year 2000,

Wanninkhof et al. (2013) estimated an anthropogenic mean flux ranging from $1.9 \pm 0.3\,\mathrm{Pg\,C\,year^{-1}}$, based on the median and range of 6 ocean biogeochemical models, to $2.4 \pm 0.3\,\mathrm{Pg\,C\,year^{-1}}$ calculated from 10 ocean inversion estimates, including their median absolute deviation as uncertainty estimate (Wanninkhof et al., 2013).

Estimates of the oceanic mean uptake based on upscaled measurements of the surface ocean pCO_2 usually reveal lower fluxes than the methods discussed above, as the upscaled estimates constitute the uptake of contemporary carbon (i.e., F_{cont}). A study by Rödenbeck et al., 2015 illustrates that the contemporary air-sea CO_2 flux, estimated from 1992 to 2009 based on several methods upscaling surface ocean pCO_2 observations, only amounts to $1.75\,\mathrm{Pg\,C\,year^{-1}}$. Assuming that the preindustrial ocean is in disequilibrium with the atmosphere due to the carbon transported to the ocean from river systems and further assuming that this source of carbon to the ocean is steady in time, the mean carbon uptake from the observation-based estimates can be adjusted to the uptake of anthropogenic carbon by subtracting the natural river-derived carbon fluxes (which are of opposite sign to the uptake) from the contemporary uptake estimate. As these estimates range from $0.23\,\mathrm{Pg\,C\,year^{-1}}$ (Lacroix et al., 2019) through $0.78\,\mathrm{Pg\,C\,year^{-1}}$ (Resplandy et al., 2018) (see Fig. 1), the anthropogenic carbon uptake based on upscaled surface ocean pCO_2 observations results in 2.20–$2.53\,\mathrm{Pg\,C\,year^{-1}}$, which is broadly consistent with estimates from Wanninkhof et al. (2013) and Quéré et al. (2018) over slightly different time periods. Nevertheless, it has to be noted that pCO_2-based estimates differ in the estimated ocean area and whether they include marginal seas. Rödenbeck et al. (2015) do not explicitly calculate uncertainties; however, other studies suggest that including uncertainties from various sources, including the uncertainty from extrapolating measurements and uncertainties stemming from the air-sea gas transfer formulation, (Eq. 1) ranges from 30% (Wanninkhof et al., 2013) to over 37% (Landschützer et al., 2014) and up to 50% (Takahashi et al., 2009). Furthermore, errors derived from model subsampling suggest strong local discrepancies in decadal flux variations, particularly in the Southern Ocean (Gloege et al., 2021).

Besides the river correction, there are several other shortcomings that arise from the surface ocean observation-based estimates. Rödenbeck et al. (2015) show that some qualitative differences exist between air-sea CO_2 flux reconstructions based on surface ocean CO_2 measurements. Some methods reconstruct available surface observations better than others; however, fewer winter measurements exist in the high latitude Southern Ocean, limiting the opportunities to independently validate the results from various interpolation schemes. Furthermore, inconsistencies exist regarding the spatial extent of these estimates (Rödenbeck et al., 2015), with few covering the Arctic Ocean and with some extending further into the coastal ocean or seasonal sea-ice domain than others.

5 Regional marine carbon sink

Regionally, all large-scale ocean basins, i.e., the Pacific Ocean, the Atlantic Ocean, the Indian Ocean, and the Southern Ocean (see Fig. 2), contribute to the present-day ocean uptake of anthropogenic CO_2; however, substantial regional differences exist (see Fig. 3). The general air-sea flux pattern, however, has not changed since preindustrial times, with the tropics releasing CO_2 to the atmosphere and the midlatitudes and the high latitude Atlantic Ocean removing substantial amounts of CO_2 from the atmosphere (see Fig. 3). A notable exception is the Southern Ocean. Although the Southern Ocean is a source of natural carbon to the atmosphere, as discussed above, the uptake of rising anthropogenic CO_2 in the subduction zones overcompensated the outgassing of natural upwelled CO_2 further south, resulting in the Southern Ocean being a contemporary net marine carbon sink.

The Pacific Ocean (north of 44°S) is the largest ocean basin and also includes the region of most intense outgassing of CO_2. Intense carbon uptake in the present-day North Pacific (0.47 ± 0.13 Pg C year^{-1} from 1990 to 2009 (Ishii et al., 2014)), in particular along the Kuroshio current, is nearly canceled out by the intense trade wind driven upwelling of carbon-rich waters in the tropical Pacific (0.44 ± 0.14 Pg C year^{-1} from 1990 to 2009 (Ishii et al., 2014)) and the coastal Ekman transport along the California current system. The South Pacific is one of the least well-observed ocean basins, particularly toward the east of the basin, resulting in a larger spread between estimated fluxes from various methods. In particular, ocean inverse estimates suggest significantly larger ocean carbon uptake than a surface ocean pCO_2-based estimate and ocean biogeochemical models (Ishii et al., 2014). Nevertheless, there is general agreement that the mean flux illustrates a net oceanic sink for CO_2 in the South Pacific subtropics of 0.37 ± 0.08 Pg C year^{-1} (Ishii et al., 2014). In both the northern and southern subtropics as well as the tropical Pacific Ocean, an ocean

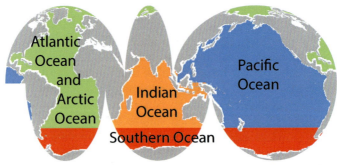

FIG. 2 Common coarse regional division of the Global Oceans following the RECCAP protocol into the Atlantic Ocean *(green)* including the Arctic Ocean, the Pacific Ocean *(blue)*, the Indian Ocean *(orange)*, and the Southern Ocean *(red)*. Note that in the literature several different boundaries for the Southern Ocean are applied.

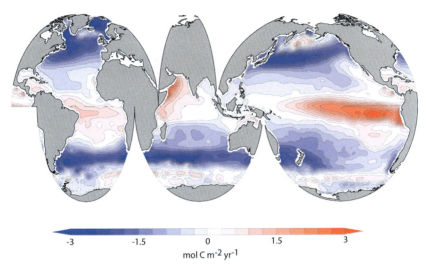

FIG. 3 Distribution of present-day source *(red)* and sink *(blue)* (F_{cont}) regions in the global ocean for CO_2 based on the extrapolation of measurements of the surface ocean partial pressure of CO_2 (Landschützer, Gruber, & Bakker, 2016).

inverse estimate suggests that the anthropogenic CO_2 component amounts to approximately half the magnitude of the natural CO_2 fluxes (Gruber et al., 2009); however, in all three Pacific Ocean subregions, the direction of the anthropogenic CO_2 flux component is directed toward more uptake in the subtropics and reduced outgassing in the tropics, making the Pacific Ocean a substantial present-day anthropogenic carbon sink.

The Atlantic Ocean (north of 44°S) including the Arctic Ocean also comprises a strong sink for atmospheric CO_2 of $0.61 \pm 0.06 \, \text{Pg C year}^{-1}$ (average from 1990 to 2009 (Schuster et al., 2013)) based on multiple estimates. Ocean biogeochemical models, ocean inverse as well as atmospheric inverse estimates, and surface pCO_2-based estimates all agree that both the northern and southern subtropics and the north subpolar Atlantic Ocean comprise a net CO_2 sink, whereas the tropical Atlantic Ocean is a small source of CO_2 to the atmosphere (Schuster et al., 2013). This agreement exists despite the sparse surface ocean pCO_2 measurement coverage in the South Atlantic Ocean. While agreement exists regarding the sign of the flux, there is less agreement regarding the strength of the Atlantic Ocean carbon sink. Atmospheric inverse estimates suggest a mean Atlantic Ocean carbon uptake (mean from 1990 to 2009) of $0.64 \pm 0.03 \, \text{Pg C year}^{-1}$, whereas ocean biogeochemical models suggest a substantially smaller sink of $0.37 \pm 0.03 \, \text{Pg C year}^{-1}$ over the same time period (Schuster et al., 2013). Unlike in the Pacific Ocean, based on an ocean inverse estimate, the anthropogenic flux component in the Atlantic Ocean is almost equal in magnitude compared to the natural CO_2 flux component (Gruber et al., 2009), with the subtropics and subpolar Atlantic Ocean regions

440 Section | C Case Studies

directed toward enhanced oceanic CO_2 uptake and in the tropics counteracting the natural CO_2 release.

The Indian Ocean (north of 44°S) can be roughly divided into a northern tropical (north of 18°S) part where CO_2 is released to the atmosphere (0.08 ± 0.04 Pg C year^{-1} average from 1990 to 2009 (Sarma et al., 2013)) and a southern subtropical region (south of 18°S) where a substantial amount of CO_2 is taken up from the atmosphere (0.37 ± 0.06 Pg C year^{-1} average from 1990 to 2009 (Sarma et al., 2013)), with encouraging agreement between inverse estimates and ocean biogeochemical model estimates. A climatological estimate based on surface ocean pCO_2 observations (Takahashi et al., 2009) shows the weakest CO_2 uptake of 0.24 ± 0.12 Pg C year^{-1} (reference year 2000); however, few measurements exist in the Indian Ocean in the 2000s (see e.g., Bakker et al., 2016). On the basis of an ocean inversion (Gruber et al., 2009), the anthropogenic CO_2 influx in the ocean is almost equal in magnitude compared to the natural fluxes. While this strengthens the subtropical carbon sink, it nearly balances the natural outgassing in the tropics (Gruber et al., 2009); however, owing to the additional influx of riverine carbon, the contemporary tropical Indian Ocean remains a source of CO_2 to the atmosphere (Gruber et al., 2009).

Estimates of the present-day Southern Ocean south of 44°S suggest this ocean basin to be a contemporary sink for carbon dioxide from the atmosphere. Ocean biogeochemical models and atmospheric inverse estimates (Lenton et al., 2013) suggest a Southern Ocean mean uptake (mean from 1990 to 2009) of 0.42 ± 0.07 Pg C year^{-1}. This estimate is larger than the one obtained based on a surface Ocean pCO_2-based climatology (Takahashi et al., 2009) of 0.27 ± 0.13 Pg C year^{-1} (reference year 2000), though within the combined uncertainty range, despite the sparse pCO_2 shipboard measurements. Present-day estimates regarding the strength of the Southern Ocean carbon sink further substantially depend on the regional definition. Recent estimates based on upscaled surface ocean pCO_2 observations suggest that south of 35°S the contemporary Southern Ocean carbon sink is as large as 1 Pg C year^{-1}, or approximately half of the annual oceanic CO_2 uptake from the atmosphere (Landschützer et al., 2016). Compared to preindustrial times, an ocean inverse estimate (Gruber et al., 2009) suggests that the Southern Ocean is the only ocean basin that has undergone a transformation from being a source of carbon to becoming a present-day net sink for CO_2 from the atmosphere (south of 44° S). The strong natural outgassing of CO_2 is more than compensated by the uptake of anthropogenic CO_2 (Gruber et al., 2009). In particular, the mode water formation zones between 45°S and 55°S effectively transport surface waters to the deeper layers (Gruber et al., 2019). Upwelled water from further south gets transported to the north where it is exposed to CO_2 in the atmosphere and high wind speed leading to an effective uptake of anthropogenic carbon by the ocean. Likewise, the conversion of subtropical water masses transported south into mode waters provides a sink for anthropogenic CO_2 (Gruber et al., 2019).

6 Storage of anthropogenic carbon

Deep water formation zones provide a pathway for anthropogenic CO_2 away from the surface to the interior ocean. Since the beginning of the industrial period, the ocean has stored a vast amount of CO_2 originating from fossil fuel emissions, land-use change, and cement production. At present, carbon inventories are estimated from ocean biogeochemical model simulations and several observation-based methods to calculate the amount of carbon stored as anthropogenic dissolved inorganic carbon in the deep ocean from various tracer fields (Khatiwala, Tanhua, et al., 2013). Traditional methods such as the Delta C* tracer method (Sabine et al., 2004) start from the present-day distribution of DIC in the ocean interior and back-calculate the anthropogenic accumulation of carbon in the ocean. Alternatively, the Transient Tracer Distribution method (Waugh et al., 2006) starts with a mathematical model linking the surface boundary conditions with interior tracers via ocean circulation, calibrated via traced observations (Khatiwala et al., 2013). A third method based on ocean interior measurements is the Greens Function approach (Khatiwala, Primeau, & Hall, 2009).

The ocean comprises a massive repository for carbon with a total stock size of about 37,200 Pg C (Keppler, Landschützer, Gruber, Lauvset, & Stemmler, 2020). The accumulated inventory of anthropogenic CO_2 in the ocean has been estimated to be $152 \pm 20 \, \mathrm{Pg\,C}$ from 1850 through 2007 (Gruber et al., 2019) based on the Delta C* method. A similar result of $155 \pm 30 \, \mathrm{Pg\,C}$ was obtained for the period 1750 through 2011 based on the Greens Function approach (Khatiwala et al., 2013), which is in agreement within $3 \, \mathrm{Pg\,C}$ with model estimates from the 5th Coupled Climate Model Intercomparison Project (CMIP5) when identical time periods (1751–1991) are compared (Bronselaer et al., 2017). Spatially, the largest column inventory, i.e., the largest concentration integrated over the water column, of anthropogenic carbon is linked to the deep-water formation zone in the high latitude North Atlantic (regionally $>100 \, \mathrm{mol\,m^{-2}}$) and the mode and intermediate water formation zones in the Southern Ocean (regionally $>60 \, \mathrm{mol\,m^{-2}}$) (Khatiwala et al., 2013). Due to the large area of the Southern Ocean, Frölicher et al. (2015) showed using the set of CMIP5 models, that the Southern Ocean south of 30°S accounts for $43 \pm 3\%$ of all anthropogenic CO_2 stored in the global ocean. Since 1800 through 1994, the global ocean has taken up $\sim48\%$ of all human emissions of CO_2 from fossil fuels, land-use change, and cement production, which highlights the important role of the global ocean to mitigate climate change (Sabine et al., 2004). Gruber et al. (2019) estimate based on measurements from the repeat hydrography program that the ocean has stored $\sim31\%$ of all human emitted CO_2 from 1994 through 2007. The storage rate of CO_2 has further been shown to vary substantially by region (see Fig. 4), likely linked to climate variability of the ocean circulation (Gruber et al., 2019).

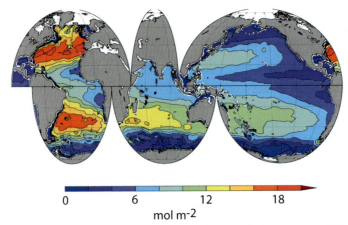

FIG. 4 Storage rate (column integral) of anthropogenic carbon in the ocean from 1994 to 2007 following Gruber et al. (2019) (output V101 in Gruber et al., 2019).

7 Variability of the ocean GHG uptake

Increasing atmospheric CO_2 concentrations lead to a change in the air-sea concentration difference, consequently shifting the system toward an overall more undersaturated ocean (see Eq. 1) increasing the flux of carbon through the air-sea interface. Because of the reaction of CO_2 with seawater, only a tiny fraction remains in the form of CO_2, whereas the majority adds to the DIC pool. The ratio of the relative change of pCO_2 to the relative change of DIC (Zeebe & Wolf-Gladrow, 2001) varies between 8 in low latitudes and 15 in the high latitudes, i.e., an 8%–15% change in pCO_2 only leads to a 1% change in DIC. The lower the so-called Revelle factor,[a] the more efficient the ocean in taking up anthropogenic CO_2 from the atmosphere. With the continuous uptake of CO_2, however, the surface ocean increase in DIC also increases the Revelle factor, and the ocean becomes more resistant to take up CO_2. Therefore, in the long run, the ocean is projected to lose its ability to effectively take up excess anthropogenic carbon.

In a simple atmosphere-ocean system without climatic oscillations, the ocean carbon sink at present is expected to steadily increase as the ocean tracks the increase in atmospheric CO_2 (Wanninkhof et al., 2013). This expected steady increase, however, is often masked by interannual to decadal variability (see Fig. 5), arising from changes in the ocean mixed layer dynamics, biological production, changes in ocean currents and transport, etc. (Wanninkhof et al., 2013). Fay and McKinley (2013) indeed show based on measurements of the surface ocean pCO_2 that short-term trends can vary substantially below or

---

a. The Revelle factor describes the ratio of the relative change of seawater pCO_2 to the relative change of dissolved inorganic carbon.

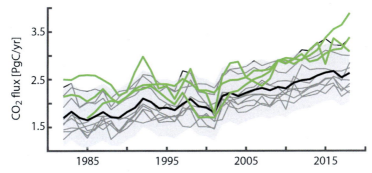

FIG. 5 Evolution of the global ocean carbon sink estimates from the Global Carbon Budget (GCB) (Friedlingstein et al., 2019). The *solid black line* represents the GCB best estimate with its uncertainty shading. *Blue lines* represent individual models (Aumont & Bopp, 2006; Berthet et al., 2019; Buitenhuis, Rivkin, Sailley, & Le Quéré, 2010; Doney et al., 2009; Hauck, Kohler, Wolf-Gladrow, & Volker, 2016; Law et al., 2017; Liu et al., 2019; Mauritsen et al., 2019; Schwinger et al., 2016) and *green lines* represent data-based estimates included in the GCB (Denvil-Sommer, Gehlen, Vrac, & Mejia, 2019; Landschützer et al., 2016; Rödenbeck et al., 2014). A steady-state river flux of $0.78\,PgC\,year^{-1}$ (Resplandy et al., 2018) has been added to the data-based estimates to adjust these to the anthropogenic CO_2 uptake.

above the fairly steady atmospheric growth rate; however, when timescales of 30 years and above are analyzed, oceanic pCO_2 trends closely follow the atmospheric growth rate. This is also consistent with the longest time series of measurements at fixed stations at Bermuda and Hawaii (Bates et al., 2014) where surface ocean pCO_2 increased at a rate of 1.69 ± 0.11 and $1.72 \pm 0.09\,\mu atm\,year^{-1}$ respectively, closely matching the atmospheric growth rate of 1.8 ppm over the 1990 through 2009 period.

While the present-day ocean carbon uptake and its long-term trend are well constrained, less agreement exists regarding its variability in time (Wanninkhof et al., 2013). In particular, estimates based on measurements of the surface ocean pCO_2 have challenged our traditional understanding of the marine carbon sink variability on decadal timescales. Previously, ocean carbon sink variability was largely attributed to the tropical Pacific Ocean and the El Niño Southern Oscillation (Le Quéré, Orr, Monfray, Aumont, & Madec, 2000), with repeating occurrences at timescales of 3–7 years. During positive phases, i.e., the El Niño phases, in which the equatorial trade winds weaken causing a deepening of the thermocline in the eastern tropical Pacific and reduced upwelling conditions there (Feely et al., 2006), the upward transport of dissolved carbon from the deep breaks down and the equatorial Pacific loses its outgassing signature. Likewise, during negative phases, i.e., the La Niña phases, stronger than average trade winds lead to a shallowing of the thermocline in the eastern part of the tropical Pacific and enhanced upwelling of carbon-rich waters from deeper layers enhancing the equatorial outgassing. The switch from El Niño to La Niña

444 Section | C Case Studies

conditions alone is responsible for a change in the equatorial CO_2 outgassing on the order of $0.5\,PgC\,year^{-1}$ (Feely et al., 2006).

Besides the Equatorial Pacific, substantial variability in the air-sea exchange of CO_2 has been reported in the North Atlantic Ocean, linked to variations in the North Atlantic Oscillation (NAO) (Schuster & Watson, 2007) and in the Atlantic Multidecadal Oscillation (Breeden & McKinley, 2016), the Pacific Ocean linked to the Pacific Decadal Oscillation (McKinley et al., 2006), and the Southern Ocean linked to the Southern Annular Mode (Le Quéré et al., 2007). In particular, the latter has raised a lot of concern, since Le Quéré et al. (2007) argued that because of the trend toward a positive index polarity of the Southern Annular Mode, the westerly wind belt strengthened and shifted southward, substantially altering the amount of natural carbon being upwelled to the surface. The resulting anomalous outgassing of natural carbon has led to a saturation of the Southern Ocean's ability to take up anthropogenic carbon. This saturation of the Southern Ocean carbon sink was supported by both model estimates and atmospheric inverse estimates (Le Quéré et al., 2007).

More recently, air-sea exchange estimates derived from surface ocean pCO$_2$ measurements revealed the dominance of variability on decadal timescales (DeVries et al., 2019; Hauck et al., 2020; Landschützer et al., 2016; McKinley et al., 2020; Rödenbeck et al., 2015). In particular, these estimates suggest that the ocean carbon uptake has substantially strengthened in the decade following the year 2000, increasing the mean CO_2 uptake by the ocean by more than $1\,PgC\,year^{-1}$ in 2010 relative to the year 2000 (Rödenbeck et al., 2015). While all ocean basins experienced an anomalous CO_2 sink increase in the early 2000s (Landschützer et al., 2016), the most remarkable increase was observed in the Southern Ocean. Roughly half of the global increase ($\sim 0.6\,PgC\,year^{-1}$ (Landschützer et al., 2015)) was attributed to the area south of 35°S. This is somewhat counterintuitive to previous findings, as the Southern Annular Mode continued its transition toward a more positive index polarity (Marshall, 2003). Instead, Gruber et al. (2019) link the evolution of the Southern Ocean carbon sink to the recent development of the large-scale weather pattern of the Southern Ocean, with a tendency toward a more zonally asymmetric pressure pattern causing a tilting of the westerly wind circulation. In particular, in the 2000s a strengthening of the Amundsen low in the Pacific sector of the Southern Ocean resulted in anomalous winds strengthening the westerly circulation, but likewise transporting fresh and cold water from the Antarctic continent toward lower latitudes. In contrast, anomalous high pressure in the Atlantic sector resulting in local wind strengths counteracting the general westerly circulation led to the transport of warm saline water from the low latitudes. Gruber et al. (2019) link these local variations with the signature of the zonal wavenumber 3 pattern (Raphael, 2004). This suggests that the Southern Ocean is more sensitive to local atmospheric variability than was previously recognized (Keppler & Landschützer, 2019).

While the magnitude of the decadal-scale variability in the uptake of CO_2 from the atmosphere as suggested by pCO_2-based estimates is unprecedented in ocean hindcast simulations, particularly in the Southern Ocean (Gruber, Landschützer, & Lovenduski, 2019), Quéré et al. (2018) further highlight that the variability, calculated from the standard deviation of the integrated air-sea CO_2 flux from 1985 through 2017 from the mean of 7 ocean model simulations of $\pm 0.29\,\mathrm{Pg\,C\,year}^{-1}$, is comparable to those calculated from two individual pCO_2-based methods that accurately match available observations (± 0.36 and $\pm 0.38\,\mathrm{Pg\,C\,year}^{-1}$ respectively—see also Fig. 5). Furthermore, the correlation between the model mean time series and the observation-based air-sea CO_2 flux time series of >0.75 suggests that processes driving the variability are well represented in the model simulations, whereas the actual magnitude of the variability is still under debate. Additionally, such model estimates are often based on a single ensemble simulation which may not entirely capture the chaotic nature of the climate system (Li & Ilyina, 2018; McKinley et al., 2016).

The ocean's ability to take up anthropogenic CO_2 from the atmosphere provides at first sight a valuable service to humanity as it abates man-made climate change. On the other hand, the additional uptake of CO_2 since the industrialization has already substantially affected the ocean carbon cycle and marine ecosystems. As CO_2 dissolves in seawater, the concentration of hydrogen ions $[H^+]$ increases, which in turn decreases the pH level of seawater ($pH = -\log_{10}([H^+])$) steadily acidifying ocean waters (Doney, Fabry, Feely, & Kleypas, 2009; Orr et al., 2005). The increasing uptake of CO_2 by the ocean will increase the amount of dissolved CO_2, while in turn CO_3^{2-} will decrease (Orr et al., 2005). This decrease of the carbonate ion concentration makes it more difficult for calcifying organisms to build their protective shells (Orr et al., 2005) with potential consequences for the entire food chain and the marine carbonate system (Ilyina & Zeebe, 2012; Ilyina, Zeebe, Maier-Reimer, & Heinze, 2009).

Another consequence of the marine uptake of CO_2 not further discussed here is variability on seasonal timescales. The ability of the ocean to take up CO_2 is dependent on temperature, dissolved carbon, biology, and the ocean's buffering capacity, which all vary by season. Both model- and observation-based studies (Hauck & Völker, 2015; Landschützer et al., 2018) have shown that the ocean uptake of CO_2 has significantly altered the seasonal cycle of CO_2 and will continue to do so in the future. As a consequence, the effects of ocean acidification will likely be visible earlier in time, as seasonal concentrations of dissolved CO_2 and pH will reach critical thresholds for calcifying organisms.

8 Future outlook

Traditionally, biogeochemical model-based assessments have been used to close the past and present marine carbon budgets (Quéré et al., 2018; Sarmiento et al., 2010) owing to the model internal consistency paired with their flexibility to close the marine carbon budget on various timescales and to their being continuously developed. The role of observations, largely represented through measurements of dissolved carbon in the ocean interior (Gruber et al., 2009), atmospheric inverse estimates (Peylin et al., 2013), through atmospheric O_2/N_2 trends (Manning & Keeling, 2006), and the distribution of CFCs (McNeil et al., 2003), helped to further test the long-term mean uptake for consistency (Quéré et al., 2018). More recently, estimates based on measurements of the surface ocean pCO_2 emerged as being applicable for reconstructing air-sea CO_2 fluxes at interannual to decadal timescales (Rödenbeck et al., 2015). Such observational estimates are crucial to constrain ocean carbon sink evolution in longer-term projections and near-term predictions (Li, Ilyina, Müller, & Landschützer, 2019) based on Earth System Models. One of the large remaining challenges for closing the marine carbon budget based on surface CO_2 observations, however, is heterogeneous sampling in space and time. Data paucity in oceanic regions such as the Southern Ocean is difficult to overcome with traditional approaches. The continuous deployment of autonomous measurement devices such as Argo floats equipped with biogeochemical sensors in these most remote ocean regions (Williams et al., 2017) will help in the future to better constrain local carbon budgets and decrease the uncertainty in today's estimates. Despite the opportunity to sample carbon system data in remote regions, biogeochemical floats further provide data all year round, which is particularly important in the Southern Ocean winter, where very few shipboard measurements exist. First results have already indicated that winter time sampling in the Southern Ocean is important to capture the full strength of the natural CO_2 outgassing signal (Bushinsky et al., 2019; Gray et al., 2018).

Furthermore, time-dependent ocean inverse estimates (DeVries et al., 2017) have been recently developed to provide the temporal evolution of the ocean carbon sink from ocean hydrographic observations. These data assimilating ocean circulation inverse models have further confirmed the strong decadal variations in the marine CO_2 uptake in the Southern Ocean (DeVries et al., 2017). Because of their process-based interpretation of the observations, they can further provide mechanistic understanding of the observed changes, making them a powerful tool to close future carbon budgets.

One remaining challenge that hampers today's marine CO_2 sink intercomparisons is the representation of natural background CO_2 fluxes and more specifically their variability in time. In order to make estimates of the anthropogenic carbon uptake (e.g., from model output and ocean interior estimates) comparable with the contemporary sink estimates (e.g., from surface ocean pCO_2 observations), a better representation of the carbon transport through the full aquatic continuum is required.

Future budgets will further benefit from model development, e.g., ocean model resolutions will increase and previously unresolved processes, such as the effect of eddies on the air-sea exchange of GHG, will become resolvable (Song et al., 2016). In addition, better representation of biology through

Ocean systems Chapter | 13 **447**

improved ecosystem model components or variable stoichiometric ratios of carbon to nitrate and phosphate has been proposed (Letscher, Moore, Teng, & Primeau, 2015) to better represent the complex biological carbon pump. Furthermore, as more awareness grows regarding the importance of closing the budgets for greenhouse gases other than CO_2 (e.g., Saunois et al., 2020), the modeling of their sources and sinks is expected to improve in the near future as well. Additionally, the growing availability of measurements and ongoing global synthesis efforts (e.g., Bakker et al., 2016; Lauvset et al., 2016; Olsen et al., 2016) will improve marine CO_2 budgets in the future.

While carbon budgets have a long tradition and are well constrained, other greenhouse gas budgets have only started to emerge (e.g., Saunois et al., 2016, 2020). However, with measurement and synthesis efforts underway to better constrain these fluxes from available measurements (e.g., Bange et al., 2009), the present uncertainties in marine CH_4 and N_2O budgets are expected to decrease in the future.

References

Aumont, O., & Bopp, L. (2006). Globalizing results from ocean in situ iron fertilization studies. *Global Biogeochemical Cycles*, *20*, GB2017.

Bakker, D. C. E., et al. (2016). A multi-decade record of high-quality fCO2 data in version 3 of the Surface Ocean CO_2 ATlas (SOCAT). *Earth System Science Data*, *8*, 383–413. https://doi.org/10.5194/essd-8-383-2016.

Bange, H. W., et al. (2009). MEMENTO: A proposal to develop a database of marine nitrous oxide and methane measurements. *Environmental Chemistry*, *6*, 195–197.

Bates, N., et al. (2014). A time-series view of changing ocean chemistry due to ocean uptake of anthropogenic CO2 and ocean acidification. *Oceanography*, *27*, 126–141.

Battaglia, G., & Joos, F. (2018). Marine N_2O emissions from mitrification and denitrification constrained by modern observations and projected in multimillennial global warming simulations. *Global Biogeochemical Cycles*, *32*, 92–121.

Berthet, S., Séférian, R., Bricaud, C., Chevallier, M., Voldoire, A., & Ethé, C. (2019). Evaluation of an online grid-coarsening algorithm in a global eddy-admitting ocean biogeochemical model. *Journal of Advances in Modeling Earth Systems*, *11*, 1759–1783. https://doi.org/10.1029/2019MS001644. submitted.

Breeden, M. L., & McKinley, G. A. (2016). Climate impacts on multidecadal pCO_2 variability in the North Atlantic. *Biogeosciences*, *13*, 3387–3396. https://doi.org/10.5194/bg-13-3387-2016.

Bronselaer, B., et al. (2017). Agreement of CMIP5 simulated and observed ocean anthropogenic CO2 uptake. *Geophysical Research Letters*, *44*, 12298–12305. https://doi.org/10.1002/2017GL074435.

Buitenhuis, E. T., Rivkin, R. B., Sailley, S., & Le Quéré, C. (2010). Biogeochemical fluxes through microzooplankton. *Global Biogeochemical Cycles*, *24*, GB4015. https://doi.org/10.1029/2009GB003601.

Buitenhuis, E. T., Suntharalingam, P., & Le Quéré, C. (2018). Constraints on global oceanic emissions of N_2O from observations and models. *Biogeosciences*, *15*, 2161–2175. https://doi.org/10.5194/bg-15-2161-2018.

Bushinsky, S. M., Landschützer, P., Rödenbeck, C., Gray, A. R., Baker, D., Mazloff, M. R., et al. (2019). Reassessing Southern Ocean air-sea CO_2 flux estimates with the addition of

448 Section | C Case Studies

biogeochemical float observations. *Global Biogeochemical Cycles, 33.* https://doi.org/10.1029/2019GB006176.

Carpenter, L., Archer, S. D., & Beale, R. (2012). Ocean-atmosphere trace gas exchange. *Chemical Society Reviews, 41,* 6473–6506. https://doi.org/10.1039/c2cs35121h.

Ciais, P., et al. (2013). Carbon and other biogeochemical cycles. In T. F. Stocker, D. Qin, G.-K. Plattner, M. Tignor, S. K. Allen, J. Boschung, & P. M. Midgley (Eds.), *Climate change 2013: The physical science basis. Contribution of working group I to the fifth assessment report of the intergovernmental panel on climate change.* Cambridge, United Kingdom and New York, NY: Cambridge University Press.

Denvil-Sommer, A., Gehlen, M., Vrac, M., & Mejia, C. (2019). LSCE-FFNN-v1: A two-step neural network model for the reconstruction of surface ocean pCO_2 over the global ocean. *Geoscientific Model Development, 12,* 2091–2105. https://doi.org/10.5194/gmd-12-2091-2019.

DeVries, T., Holzer, M., & Primeau, F. (2017). Recent increase in oceanic carbon uptake driven by weaker upper-ocean overturning. *Nature, 542,* 215–218.

DeVries, T., et al. (2019). Decadal trends in the ocean carbon sink. *Proceedings of the National Academy of Sciences of the United States of America.* https://doi.org/10.1073/pnas.1900371116.

Dlugokencky, E., & Tans, P. (2018). *Trends in atmospheric carbon dioxide.* National Oceanic & Atmospheric Administration, Earth System Research Laboratory (NOAA/ESRL). available at http://www.esrl.noaa.gov/gmd/ccgg/trends/global.html.

Doney, S., Fabry, V., Feely, R. A., & Kleypas, J. (2009). Ocean acidification: The other CO_2 problem. *Annual Review of Marine Science, 1,* 169–192.

Doney, S. C., Lima, I., Feely, R. A., Glover, D. M., Lindsay, K., Mahowald, N., et al. (2009). Mechanisms governing interannual variability in upper-ocean inorganic carbon system and air–sea CO_2 fluxes: Physical climate and atmospheric dust. *Deep Sea Research Part II, 56,* 640–655.

Fay, A. R., & McKinley, G. A. (2013). Global trends in surface ocean pCO_2 from in situ data. *Global Biogeochemical Cycles, 27,* 541–557. https://doi.org/10.1002/gbc.20051.

Feely, R. A., et al. (2006). Decadal variability of the air-sea CO_2 fluxes in the equatorial Pacific Ocean. *Journal of Geophysical Research, 111.* https://doi.org/10.1029/2005JC003129, C08S90.

Friedlingstein, P., Jones, M. W., O'Sullivan, M., Andrew, R. M., Hauck, J., Peters, G. P., et al. (2019). Global carbon budget 2019. *Earth System Science Data, 11,* 1783–1838. https://doi.org/10.5194/essd-11-1783-2019.

Frölicher, T. L., et al. (2015). Dominance of the Southern Ocean in anthropogenic carbon and heat uptake in CMIP5 models. *Journal of Climate, 28,* 862–886.

Gloege, L., et al. (2021). Quantifying errors in observationally-based estimates of ocean carbon sink variability. *Global Biogeochemical Cycles, 35.* https://doi.org/10.1029/2020GB006788, e2020GB006788.

Gloor, M., et al. (2003). A first estimate of present and preindustrial air-sea CO2 flux patterns based on ocean interior carbon measurements and models. *Geophysical Research Letters, 30,* 1010.

Gray, A., et al. (2018). Autonomous biogeochemical floats detect significant carbon dioxide outgassing in the high-latitude Southern Ocean. *Geophysical Research Letters, 45,* 9049–9057. https://doi.org/10.1029/2018GL078013.

Gruber, N., Clement, D., Carter, B. R., Feely, R. A., van Heuven, S., Hoppema, M., et al. (2019). *The oceanic sink for anthropogenic CO_2 from 1994 to 2007—The data (NCEI accession 0186034). Version 1.1.* National Oceanographic Data Center, NOAA. https://doi.org/10.25921/wdn2-pt10. Dataset.

Gruber, N., Landschützer, P., & Lovenduski, N. S. (2019). The variable Southern Ocean carbon sink. *Annual Review of Marine Science, 11*(1), 159–186. https://doi.org/10.1146/annurevmarine-121916-063407. 30212259.

Gruber, N., et al. (2009). Oceanic sources, sinks, and transport of atmospheric CO2. *Global Biogeochemical Cycles, 23*. https://doi.org/10.1029/2008GB003349, GB1005.

Gruber, N., et al. (2019). The oceanic sink for anthropogenic CO_2 from 1994 to 2007. *Science*, 1193–1199. https://doi.org/10.1126/science.aau5153.

Hauck, J., Kohler, P., Wolf-Gladrow, D., & Volker, C. (2016). Iron fertilisation and century-scale effects of open ocean dissolution of olivine in a simulated CO2 removal experiment. *Environmental Research Letters, 11*, 024007.

Hauck, J., & Völker, C. (2015). Rising atmospheric CO_2 leads to large impact of biology on Southern Ocean CO_2 uptake via changes of the Revelle factor. *Geophysical Research Letters, 42*, 1459–1464.

Hauck, J., et al. (2020). Consistency and challenges in the ocean carbon sink estimate for the global carbon budget. *Frontiers in Marine Science, 7*, 571720. https://doi.org/10.3389/fmars.2020.571720.

Heinze, C., et al. (2015). The ocean carbon sink—Impacts, vulnerabilities and challenges. *Earth System Dynamics, 6*, 327–358. https://doi.org/10.5194/esd-6-327-2015.

Ilyina, T., Six, K. D., Segschneider, J., Maier-Reimer, E., Li, H., & Núñez-Riboni, I. (2013). The global ocean biogeochemistry model HAMOCC: Model architecture and performance as component of the MPI-earth system model in different CMIP5 experimental realizations. *Journal of Advances in Modeling Earth Systems, 5*, 287–315. https://doi.org/10.1029/2012MS000178.

Ilyina, T., & Zeebe, R. E. (2012). Detection and projection of carbonate dissolution in the water column and deep-sea sediments due to ocean acidification. *Geophysical Research Letters, 39*. https://doi.org/10.1029/2012GL051272.

Ilyina, T., Zeebe, R. E., Maier-Reimer, E., & Heinze, C. (2009). Early detection of ocean acidification effects on marine calcification. *Global Biogeochemical Cycles, 23*. https://doi.org/10.1029/2008GB003278, GB1008.

Ishii, M., et al. (2014). Air-sea CO_2 flux in the Pacific Ocean for the period 1990–2009. *Biogeosciences, 11*, 709–734.

Jacobson, A. R., et al. (2007). A joint atmosphere-ocean inversion for surface fluxes of carbon dioxide: 2. Regional results. *Global Biogeochemical Cycles, 21*. https://doi.org/10.1029/2006GB002703, GB1020.

Jones, S. D., et al. (2015). A statistical gap-filling method to interpolate global monthly surface ocean carbon dioxide data. *Journal of Advances in Modeling Earth Systems, 7*, 1554–1575. https://doi.org/10.1002/2014MS000416.

Keppler, L., & Landschützer, P. (2019). Regional wind variability modulates the Southern Ocean carbon sink. *Scientific Reports, 9*, 7384. https://doi.org/10.1038/s41598-019-43826-y.

Keppler, L., Landschützer, P., Gruber, N., Lauvset, S., & Stemmler, I. (2020). Seasonal carbon dynamics in the near-global ocean. *Global Biogeochemical Cycles, 34*. https://doi.org/10.1029/2020GB006571. e2020GB006571.

Khatiwala, S., Primeau, F., & Hall, T. (2009). Reconstruction of the history of anthropogenic CO_2 concentrations in the ocean. *Nature, 462*, 346–349. https://doi.org/10.1038/nature08526.

Khatiwala, S., Tanhua, T., et al. (2013). Global ocean storage of anthropogenic carbon. *Biogeosciences, 10*, 2169–2191. https://doi.org/10.5194/bg-10-2169-2013.

Lacroix, F., Ilyina, T., & Hartmann, J. (2019). Oceanic CO_2 outgassing and biological production hotspots induced by pre-industrial river loads of nutrients and carbon in a global modelling approach. *Biogeosciences Discussions*. https://doi.org/10.5194/bg-2019-152.

450 Section | C Case Studies

Landschützer, P., Gruber, N., & Bakker, D. C. E. (2016). Decadal variations and trends of the global ocean carbon sink. *Global Biogeochemical Cycles, 30*, 1396–1417.

Landschützer, P., et al. (2013). A neural network-based estimate of the seasonal to inter-annual variability of the Atlantic Ocean carbon sink. *Biogeosciences, 10*, 7793–7815. https://doi.org/10.5194/bg-10-7793-2013.

Landschützer, P., et al. (2014). Recent variability of the global ocean carbon sink. *Global Biogeochemical Cycles, 28*, 927–949. https://doi.org/10.1002/2014GB004853.

Landschützer, P., et al. (2015). The reinvigoration of the Southern Ocean carbon sink. *Science, 349*, 1221–1224. https://doi.org/10.1126/science.aab2620.

Landschützer, P., et al. (2018). Strengthening seasonal marine CO_2 variations due to increasing atmospheric CO_2. *Nature Climate Change, 8*, 146–150.

Lauvset, S. K., et al. (2016). A new global interior ocean mapped climatology: The 1°X1° GLODAP version 2. *Earth System Science Data, 8*, 325–340.

Law, R. M., Ziehn, T., Matear, R. J., Lenton, A., Chamberlain, M. A., Stevens, L. E., et al. (2017). The carbon cycle in the Australian Community climate and earth system simulator (ACCESS-ESM1)—Part 1: Model description and pre-industrial simulation. *Geoscientific Model Development, 10*, 2567–2590. https://doi.org/10.5194/gmd-10-2567-2017.

Le Quéré, C., Orr, J. C., Monfray, P., Aumont, O., & Madec, G. (2000). Interannual variability of the oceanic sink of co2 from 1979 through 1997. *Global Biogeochemical Cycles, 14*(4), 1247–1265. https://doi.org/10.1029/1999GB900049.

Le Quéré, C., et al. (2007). Saturation of the Southern Ocean CO2 sink due to recent climate change. *Science, 316*, 1735–1738. https://doi.org/10.1126/science.1136188.

Lenton, A., et al. (2013). Sea-air CO_2 fluxes in the Southern Ocean for the period 1990–2009. *Biogeosciences, 10*, 4037–4054. https://doi.org/10.5194/bg-10-4037-2013.

Letscher, R. T., Moore, J. K., Teng, Y.-C., & Primeau, F. (2015). Variable C: N: P stoichiometry of dissolved organic matter cycling in the community earth system model. *Biogeosciences, 12*, 209–221. https://doi.org/10.5194/bg-12-209-2015.

Li, H., & Ilyina, T. (2018). Current and future decadal trends in the oceanic carbon uptake are dominated by internal variability. *Geophysical Research Letters, 45*(2), 916–925.

Li, H., Ilyina, T., Müller, W. A., & Landschützer, P. (2019). Predicting the variable ocean carbon sink. *Science Advances.* https://doi.org/10.1126/sciadv.aav6471.

Liu, X., Dunne, J. P., Stock, C. A., Harrison, M. J., Adcroft, A., & Resplandy, L. (2019). Simulating water residence time in the coastal ocean: A global perspective. *Geophysical Research Letters, 46*, 13910–13919. https://doi.org/10.1029/2019GL085097.

Manning, A. C., & Keeling, R. F. (2006). Global oceanic and land biotic carbon sinks from the Scripps atmospheric oxygen flask sampling network. *Tellus B, 58*, 95–116. https://doi.org/10.1111/j.1600-0889.2006.00175.x.

Marshall, G. J. (2003). Trends in the southern annular mode from observations and reanalyses. *Journal of Climate, 16*, 4134–4143.

Mauritsen, T., Bader, J., Becker, T., Behrens, J., Bittner, M., Brokopf, R., et al. (2019). Developments in the MPI-M earth system model version 1.2 (MPI-ESM1.2) and its response to increasing CO2. *Journal of Advances in Modeling Earth Systems, 11*, 998–1038. https://doi.org/10.1029/2018MS001400.

McKinley, G. A., Pilcher, D. J., Fay, A. R., Lindsay, K., Long, M. C., & Lovenduski, N. S. (2016). Timescales for detection of trends in the ocean carbon sink. *Nature, 530*, 469–472. https://doi.org/10.1038/nature16958.

McKinley, G. A., et al. (2006). North Pacific carbon cycle response to climate variability on seasonal to decadal timescales. *Journal of Geophysical Research, 111*. https://doi.org/10.1029/2005JC003173, C07S06.

Ocean systems **Chapter | 13 451**

McKinley, G. A., et al. (2020). External forcing explains recent decadal variability of the ocean carbon sink. *AGU Advances, 1*. https://doi.org/10.1029/2019AV000149, e2019AV000149.

McNeil, B., et al. (2003). Anthropogenic CO_2 uptake by the ocean based on the global chlorofluorocarbon data set. *Science, 299*, 235–239. https://doi.org/10.1126/science.1077429.

Mikaloff-Fletcher, S. E., et al. (2006). Inverse estimates of anthropogenic CO_2 uptake, transport, and storage by the ocean. *Global Biogeochemical Cycles, 20*, GB2002. https://doi.org/10.1029/2005GB002530.

Mikaloff-Fletcher, S. E., et al. (2007). Inverse estimates of the oceanic sources and sinks of natural CO_2 and the implied oceanic carbon transport. *Global Biogeochemical Cycles, 21*, GB1010.

Olsen, A., et al. (2016). The Global ocean data analysis project version 2 (GLODAPv2)—An internally consistent data product for the world ocean. *Earth System Science Data, 8*, 297–323.

Orr, J. C., et al. (2005). Anthropogenic Ocean acidification over the twenty-first century and its impact on calcifying organisms. *Nature, 437*, 681–686.

Peylin, P., et al. (2013). Global atmospheric carbon budget: Results from an ensemble of atmospheric CO_2 inversions. *Biogeosciences, 10*, 6699–6720. https://doi.org/10.5194/bg-10-6699-2013.

Quéré, L., et al. (2018). Global carbon budget 2018. *Earth System Science Data, 10*(4), 2141–2194. https://doi.org/10.5194/essd-10-2141-2018.

Raphael, M. N. (2004). A zonal wave 3 index for the Southern Hemisphere. *Geophysical Research Letters, 31*(23).

Regnier, P., Friedlingstein, P., Ciais, P., Mackenzie, F. T., Gruber, N., Janssens, I. A., et al. (2013). Anthropogenic perturbation of the carbon fluxes from land to ocean. *Nature Geoscience, 6*, 597–607. https://doi.org/10.1038/ngeo1830.

Resplandy, L., et al. (2018). Revision of global carbon fluxes based on a reassessment of oceanic and riverine carbon transport. *Nature Geoscience, 11*, 504–509. https://doi.org/10.1038/s41561-018-0151-3.

Rödenbeck, C., Bakker, D. C. E., Metzl, N., Olsen, A., Sabine, C., Cassar, N., et al. (2014). Interannual sea-air CO_2 flux variability from an observation-driven ocean mixed-layer scheme. *Biogeosciences, 11*(17), 4599–4613. https://doi.org/10.5194/bg-11-4599-2014.

Rödenbeck, C., et al. (2015). Data-based estimates of the ocean carbon sink variability—First results of the Surface Ocean pCO2 mapping intercomparison (SOCOM). *Biogeosciences, 12*, 7251–7278. https://doi.org/10.5194/bg-12-7251-2015.

Roobaert, A., Laruelle, G. G., Landschützer, P., & Regnier, P. (2018). Uncertainty in the global oceanic CO_2 uptake induced by wind forcing: Quantification and spatial analysis. *Biogeosciences, 15*, 1701–1720. https://doi.org/10.5194/bg-15-1701-2018.

Sabine, C. L., et al. (2004). The oceanic sink for anthropogenic CO_2. *Science, 305*(5682), 367–371. https://doi.org/10.1126/science.1097403.

Sarma, V. V. S. S., et al. (2013). Sea-air CO_2 fluxes in the Indian Ocean between 1990 and 2009. *Biogesciences, 10*(11), 7035–7052. https://doi.org/10.5194/bg-10-7035-2013.

Sarmiento, J. L., & Gruber, N. (2002). Anthropogenic carbon sinks. *Physics Today, 55*, 30–36.

Sarmiento, J., & Gruber, N. (2006). *Ocean biogeochemical dynamics*. 503 pp. Princeton, NJ: Princeton Univ. Press.

Sarmiento, J. M., et al. (2010). Trends and regional distributions of land and ocean carbon sinks. *Biogesciences, 7*, 2351–2367. https://doi.org/10.5194/bg-7-2351-2010.

Saunois, M., et al. (2016). The global methane budget 2000–2012. *Earth System Science Data, 8*, 697–751. https://doi.org/10.5194/essd-8-697-2016.

Saunois, M., et al. (2020). The global methane budget 2000–2017. *Earth System Science Data, 12*, 1561–1623. https://doi.org/10.5194/essd-12-1561-2020.

452 Section | C Case Studies

Schuster, U., & Watson, A. J. (2007). A variable and decreasing sink for atmospheric CO_2 in the North Atlantic. *Journal of Geophysical Research, 112.* https://doi.org/10.1029/2006JC003941, C11006.

Schuster, U., et al. (2013). Atlantic and Arctic sea-air CO_2 fluxes, 1990–2009. *Biogeosciences, 10,* 607–627. https://doi.org/10.5194/bg-10-607-2013.

Schwinger, J., Goris, N., Tjiputra, J. F., Kriest, I., Bentsen, M., Bethke, I., et al. (2016). Evaluation of NorESM-OC (versions 1 and 1.2), the ocean carbon-cycle stand-alone configuration of the Norwegian Earth System Model (NorESM1). *Geoscientific Model Development, 9,* 2589–2622.

Séférian, R., Gehlen, M., Bopp, L., Resplandy, L., Orr, J. C., Marti, O., et al. (2016). Inconsistent strategies to spin up models in CMIP5: Implications for ocean biogeochemical model performance assessment. *Geoscientific Model Development, 9,* 1827–1851. https://doi.org/10.5194/gmd-9-1827-2016.

Song, H., Marshall, J., Munro, D. R., Dutkiewicz, S., Sweeney, C., McGillicuddy, D. J., et al. (2016). Mesoscale modulation of air-sea CO_2 flux in Drake passage. *Journal of Geophysical Research, Oceans, 121,* 6635–6649. https://doi.org/10.1002/2016JC011714.

Takahashi, T., Sutherland, S. C., & Kozyr, A. (2018). *Global ocean surface water partial pressure of CO_2 database: Measurements performed during 1957–2017 (version 2018).* ORNL/CDIAC-160, NDP-088(v2018), Tech. Rep Oak Ridge, TN: Carbon Dioxide Information Analysis Center, Oak Ridge National Laboratory, U.S, Department of Energy. https://doi.org/10.3334/CDIAC/OTG.NDP088(V2013).

Takahashi, T., et al. (2009). Climatological mean and decadal change in surface ocean pCO_2, and net sea-air CO2 flux over the global oceans. *Deep-Sea Research Part II, 56,* 554–577.

Talley, L. D., et al. (2016). Changes in ocean heat, carbon content, and ventilation: A review of the first decade of GO-SHIP global repeat hydrography. *Annual Review of Marine Science, 8,* 185–215.

Tian, H., Xu, R., Canadell, J. G., et al. (2020). A comprehensive quantification of global nitrous oxide sources and sinks. *Nature, 586,* 248–256. https://doi.org/10.1038/s41586-020-2780-0.

Wanninkhof, R., et al. (2013). Global ocean carbon uptake: Magnitude, variability and trends. *Biogeosciences, 10,* 1983–2000. https://doi.org/10.5194/bg-10-1983-2013.

Waugh, D. W., et al. (2006). Anthropogenic CO_2 in the oceans estimated using transit time distributions. *Tellus B, 58,* 376–390.

Weiss, R. F. (1974). Carbon dioxide in water and seawater: The solubility of a non-ideal gas. *Marine Chemistry, 2,* 203–215.

Williams, N. L., et al. (2017). Calculating surface ocean pCO_2 from biogeochemical Argo floats equipped with pH: An uncertainty analysis. *Global Biogeochemical Cycles, 31,* 591–604. https://doi.org/10.1002/2016GB005541.

Zeebe, P. E., & Wolf-Gladrow, D. (2001). *CO_2 in seawater: Equilibrium, kinetics, isotopes.* Amsterdam: Elsevier.

Section D

Forward Looking

Chapter 14

Applications of top-down methods to anthropogenic GHG emission estimation

Shamil Maksyutov[a], Dominik Brunner[b], Alexander J. Turner[c], Daniel Zavala-Araiza[d,e], Rajesh Janardanan[a], Rostyslav Bun[f,g], Tomohiro Oda[h,i,j], and Prabir K. Patra[k]

[a]*National Institute for Environmental Studies, Tsukuba, Japan,* [b]*Empa, Swiss Federal Laboratories for Materials Science and Technology, Dübendorf, Switzerland,* [c]*Department of Atmospheric Sciences, University of Washington, Seattle, WA, United States,* [d]*Environmental Defense Fund, Amsterdam, The Netherlands,* [e]*Institute for Marine and Atmospheric Research Utrecht, Utrecht University, Utrecht, The Netherlands,* [f]*Department of Applied Mathematics, Lviv Polytechnic National University, Lviv, Ukraine,* [g]*Department of Transport, WSB University, Dąbrowa Górnicza, Poland,* [h]*The Earth From Space Institute, Universities Space Research Association, Columbia, MD, United States,* [i]*Department of Atmospheric and Oceanic Science, University of Maryland, College Park, MD, United States,* [j]*Graduate School of Engineering, Osaka University, Suita, Osaka, Japan,* [k]*Research Institute for Global Change, JAMSTEC, Yokohama, Japan*

Key messages

- The science of top-down emission estimation has experienced rapid technological developments over the past years and has been changing its main focus from global/continental scales to regional scales, taking advantage of the high volume of data collected from the dense regional observation networks and high-resolution atmospheric transport modeling capability.
- Validation of the bottom-up emission estimates in the range of scales from urban and province/state to national scale has become an important objective and application of top-down methods, especially effective for non-CO_2 greenhouse gases.
- Targeted atmospheric observations are applied for better quantification of the leaks and emission intensities at the facility level.
- New developments for atmospheric transport at increased spatial resolutions are introduced as demanded by the new spaceborne observations.
- Ways of reducing the dependence on prior fluxes are developed to address a need for more accurate emission attribution to the source sectors.

Balancing Greenhouse Gas Budgets. https://doi.org/10.1016/B978-0-12-814952-2.00006-X
Copyright © 2022 Elsevier Inc. All rights reserved.

1 Introduction

Due to renewed interest in emissions accounting and reduction inspired by the ratification of the Paris Agreement in 2015, the use of atmospheric observations and modeling to estimate emissions has grown. This interest has also led to the search for new methods of atmospheric observation and techniques to estimate emissions, including new inverse and transport modeling approaches, as well as the development of high-resolution emission inventories. The needs for top-down estimates range from emissions accounting and emissions reduction by facility operators to the validation of emission inventories by regional and national governments. Accordingly, the observing systems are developed depending on the target, such as the use of mobile observatories on ships, cars, or aircraft for facility-level emissions, or a combination of regional and global atmospheric observation networks for validating national scale emission inventories.

In this chapter, we discuss the development and application of new top-down methods for the purpose of climate mitigation by accounting and monitoring of greenhouse gas (GHG) emissions, e.g., (a) validation of the national greenhouse gas emission inventories, (b) detection of methane leak in the oil and gas industry based on aircraft, surface, and satellite observations, (c) observations and estimation of large point source GHG emissions from satellites, and (d) development of new transport modeling and gridded emission inventory techniques, suitable for processing the large amounts of high-resolution observations from both ground-based and spaceborne instruments.

2 Using inverse estimates of non-CO_2 GHG emissions in national reporting

In the context of national environmental policies and international treaties such as the Convention on Long-Range Transboundary Air Pollution (CLRTAP) or the United Nations Framework Convention on Climate Change (UNFCCC), developed countries are collecting detailed information on emissions of air pollutants, greenhouse gases, and ozone-depleting substances. These national emission inventories are generated "bottom-up" by combining emission factors for individual processes with statistical and economic data on the activity of those processes. An example is the emission factors for different types of vehicles and driving conditions (mass emitted per distance traveled) multiplied by the total distance traveled in the country by the different combinations of vehicles and driving conditions. Collecting this information in a comprehensive and complete manner is very demanding due to the vast number of processes to be considered and the heterogeneity of the underlying data. Countries are investing significant resources to compile this information year by year and to submit their inventories in the form of tables and reports under different international treaties. Under the Paris Agreement, signed in 2015, nearly all nations are committed to reducing greenhouse gas emissions according to Nationally Determined Contributions (NDCs). To build mutual trust and confidence, and to promote the effective implementation of the Paris Agreement, an Enhanced Transparency Framework was established, which requires all countries including those in the developing world to provide National Greenhouse Gas Inventory Reports on a regular basis. These reports should adhere to the principles of transparency, accuracy, completeness, consistency, and comparability (TACCC) already established under the UNFCCC, which will pose significant challenges for countries that have not yet established a corresponding reporting capacity.

The quality of the bottom-up reports critically depends on the accuracy and completeness of the input data and ultimately on the capacity of the individual countries to collect this information. The reporting is evaluated by external experts who check for compliance with the procedures defined, for example, by the IPCC Guidelines for National Greenhouse Gas Inventories (IPCC, 2006), but no evaluation against independent sources of information is made (with few exceptions).

Observations of atmospheric greenhouse gas concentrations can provide such an independent source of information. It is fully independent of the inventory collection process and has the advantage that no emissions can be overlooked or double-counted. However, the quantitative relationship between emissions over a given region and concentrations at observation sites downstream is nontrivial, as it is determined by atmospheric transport and, in the case of short-lived reactive species, by transport and chemistry. Therefore, to estimate emissions from atmospheric observations "top-down," the measurements need to be combined with models of atmospheric transport and chemistry in an inverse modeling framework.

The science of top-down emission estimation has seen rapid development over the past years and is increasingly shifting focus from global/continental to regional scales, taking advantage of the establishment of dense regional observation networks and high-resolution atmospheric transport models. Policy makers are increasingly interested in the methods to provide independent support for their officially reported emissions and to guide improvements in their inventories. The United Kingdom and Switzerland already successfully include top-down analyses in Annexes of their National Inventory Reports to UNFCCC, based on methods previously published in the scientific literature.

So far, these top-down estimates have been limited to non-CO_2 greenhouse gases for two reasons: First, relative uncertainties in emissions of non-CO_2 gases tend to be much larger compared to those in anthropogenic CO_2 (from burning fossil fuels). Methane (CH_4) and nitrous oxide (N_2O), for example, are released by a range of diffuse sources such as agricultural soils, landfills, manure management, or wastewater treatment plants, through microbiological processes that are sensitive to environmental conditions and are associated with large uncertainties. Similarly, emissions of halogenated greenhouse gases such

458   Section | D Forward Looking

as hydrofluorocarbons or SF_6 often come from poorly characterized leakages from individual installations or industries. Emissions of CO_2 from fossil fuel burning, in comparison, can be estimated rather precisely from fossil fuel import–export and production statistics. Second, top-down estimation of anthropogenic CO_2 is very challenging due to large variations in CO_2 caused by biospheric uptake and release (via photosynthesis and respiration), which are difficult to separate from anthropogenic signals.

As an example of how top-down emissions estimates can be used in national emissions reporting, Fig. 1 presents a figure from Annex 5 of the Swiss National Inventory Report of 2018 (FOEN, 2021), based on the methodology presented in Henne et al. (2016). The individual maps show a spatial representation of the difference between top-down estimates (posterior) and bottom-up reported emissions of CH_4 (prior) for the years 2013–16. The results suggest that the prior emissions were somewhat underestimated over the eastern parts (orange colors) and overestimated over the western parts of Switzerland. Overall, the top-down estimates of Swiss CH_4 emissions derived from observations at four measurement sites were in excellent agreement with the officially reported numbers. The top-down estimate of total Swiss CH_4 emissions for the period 2013–16 was $201 \pm 16\,kt\,yr^{-1}$ (1-σ confidence interval). The corresponding bottom-up values in the NIR decreased over the 4 years from 200 to $197\,kt\,yr^{-1}$ with a 1-σ uncertainty range of $\pm 18\,kt\,yr^{-1}$ for the reporting year 2016.

Such a good agreement cannot generally be expected due to significant uncertainties in both the bottom-up and top-down methods. Furthermore, in most countries, the observation network is not sufficiently dense to cover the whole country or it is even nonexistent. The examples for United Kingdom (Manning, O'Doherty, Jones, Simmonds, & Derwent, 2011) and Switzerland, however, clearly demonstrate the potential of top-down methods as an independent source of information to support inventory builders.

3   Methane emissions detection at facility and basin scale

Methane mitigation is a key component for effectively reducing the near-term rate of warming. Among the major global sources of anthropogenic CH_4 emissions, the oil and gas industry is an obvious target due to readily available emission control technologies as well as being, in comparison with other sources such as agriculture, more physically concentrated with fewer actors to consider, which helps facilitate the implementation of mitigation strategies.

Several countries [e.g., Mexico, Canada, as well as members of the Climate and Clean Air Coalition (CCAC)] and operators [e.g., members of the Oil and Gas Climate Initiative (OGCI)] have announced pledges to reduce methane emissions from the oil and gas industry. For the effectiveness of the goals and targets to be assessed, it is critical to understand the current magnitude of methane emission and emission patterns (e.g., spatial and temporal patterns,

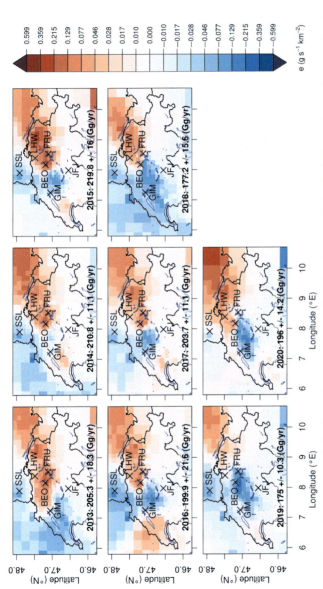

FIG. 1 Absolute difference between a posteriori and a priori mean annual emissions of methane over Switzerland. The top-down estimates are based on the assimilation of observations from four sites: Beromünster (BEO), Lägern Hochwacht (LHW), Jungfraujoch (JFJ), and Schauinsland (SSL). The numbers given in the plots refer to the total a posteriori Swiss emissions and their uncertainty (1σ level) for the given year. (*Adapted from FOEN: Switzerland's greenhouse gas inventory 1990–2019, Submission of April 2021 under the united nations framework convention on climate change and under the Kyoto protocol, Tech. Rep., Federal Office for the Environment (FOEN), 3003 Bern, Switzerland, 2021.*)

460 Section | D Forward Looking

emissions distributions) and to implement empirical monitoring and quantification schemes that can track the progress.

Atmospheric observations (i.e., top-down approaches) have been used to estimate regional methane emissions (i.e., basin-scale; width of 50–200 km) and this type of approach can provide valuable information in terms of (i) assessing the overall magnitude of regional emissions (ii) identifying emission hotspots, and (iii) characterizing spatial emission distributions.

In terms of estimating emissions from oil and gas infrastructure, top-down methods have mainly relied on airborne measurements and techniques based primarily on mass balance (Conley et al., 2017; Negron, Kort, Conley, & Smith, 2020; Johnson, Tyner, Conley, Schwietzke, & Zavala-Araiza, 2017; Karion et al., 2015, 2013; Peischl et al., 2015, 2016; Petron et al., 2014; Schwietzke et al., 2017; Smith et al., 2017; Yacovitch et al., 2018; Zavala-Araiza et al., 2021) or atmospheric transport models (Barkley et al., 2017).

In general terms, the mass balance approach is used to estimate the regional methane emissions flux (*Emissions*$_{CH4}$) based on the difference between downwind and background mole fraction of CH_4 (ΔX_{CH4}), integrating across the width of the plume ($-b$ to b), and from ground elevation (z_{ground}) to the top of the planetary boundary layer (PBL; z_{PBL}), multiplied by the perpendicular wind speed ($v cos\theta$):

$$Emissions_{CH_4} = \int_{-b}^{b} \Delta X_{CH_4}(vcos\theta)dx \int_{z_{ground}}^{z_{PBL}} n_{air}\, dz \qquad (1)$$

where n_{air} is the measured dry air molar density.

Uncertainties in the flux estimates are a combination of uncertainties in planetary boundary layer depth, wind speed, as well as uncertainty in the measurement of the CH_4 concentrations and selection of upwind boundary conditions. In addition, the mass balance approach is subject to the assumption of steady-state and homogeneous winds across the study region. By performing multiple flights, it is possible to significantly reduce the uncertainty in the regional estimates (Alvarez et al., 2018; Zavala-Araiza et al., 2015). During the Barnett Shale (Texas, US) regional study (Karion et al., 2015; Zavala-Araiza et al., 2015), eight flights and 17 downwind transects were performed, yielding a relative uncertainty of 17% on the total methane regional flux, versus the Uinta (Utah, US) study (Karion et al., 2015) with one fight, one downwind transect with a relative uncertainty of 54% on the total regional methane flux (Table 1).

Barkley et al. (2017) analyzed 10 flights over north-eastern Pennsylvania (US), six of the flights followed a box pattern surrounding the natural gas production region, while the remaining flights were used to characterize the spatial distribution of emissions via raster patterns. While only a subset of the box-pattern flights was suitable for the mass balance approach, all 10 flights were used for an inverse modeling approach. Although both approaches converged to

Applications of top-down methods **Chapter | 14** **461**

TABLE 1 Top-down estimates from aircraft-based studies in US oil and gas production regions (reference shown in brackets).

Study region	Days/flights/ downwind transects	Total flux (Mg CH$_4$/h)	Oil and gas emissions attribution approach
Bakken (Peischl et al., 2016)	3/3/5	28 ± 10	Subtraction of biogenic inventory
Barnett (Karion et al., 2015)	8/8/17	76 ± 13	Methane to ethane ratios
Fayetteville (Peischl et al., 2015)	1/1/2	39 ± 36	Subtraction of biogenic inventory, supported by ethane as a check
Fayetteville (Schwietzke et al., 2017)	2/2/4	31 ± 8	Subtraction of biogenic inventory, supported by ethane as a check
Haynesville (Peischl et al., 2015)	1/1/3	80 ± 54	Subtraction of biogenic inventory, supported by ethane as a check
NE Pennsylvania (Barkley et al., 2017)	4/4/7	20 ± 17	Subtraction of biogenic inventory, supported by ethane as a check
NE Pennsylvania (Peischl et al., 2015)	1/2	15 ± 12	Subtraction of biogenic inventory, supported by ethane as a check
San Juan (Smith et al., 2017)	5/5/5	62 ± 46	None
Uinta (Karion et al., 2013)	1/1/1	56 ± 30	Subtraction of biogenic inventory
Weld (Petron et al., 2014)	2/2/3	26 ± 14	Subtraction of biogenic inventory
West Arkoma (Peischl et al., 2015)	1/1/1	33 ± 30	Subtraction of biogenic inventory

Adapted from Table S2 in Alvarez, R., Zavala-Araiza, D., Lyon, D., Allen, D., Barkley, Z., Brandt, A., Davis, K., Herndon, S., Jacob, D., Karion, A., Kort, E., Lamb, B., Lauvaux, T., Maasakkers, J., Marchese, A., Omara, M., Pacala, S., Peischl, J., Robinson, A., Shepson, P., Sweeney, C., Townsend-Small, A., Wofsy, S., & Hamburg, S. (2018). Assessment of methane emissions from the US oil and gas supply chain, Science, 361, 186–188, 10.1126/science.aar7204; (uncertainties represent 95% confidence intervals).

462 Section | D Forward Looking

the same central estimate, the inverse modeling approach significantly reduced the uncertainty of the regional flux estimate (relative uncertainty on the total methane regional flux: 25% for the inverse modeling approach and 80% for mass balance approach).

Schwietzke et al. (2017) highlighted the relevance of raster pattern flights and characterizing emissions within production regions. Their study in the Fayetteville region (Arkansas, US) showed significant variability between the eastern and western sections, which would not have been evident with a more simple box-pattern approach.

For top-down approaches, one of the key challenges is attributing methane emissions to fossil sources (oil and gas) and biogenic sources (e.g., agriculture, waste sector). Early studies relied on assembling an inventory of biogenic sources and then subtracting it from the empirical estimate of total methane emissions. Recent studies have greatly improved the attribution through the use of isotopes (Cain et al., 2017) as well as hydrocarbon ratios (i.e., methane to ethane ratio; since ethane is only coemitted by fossil sources) (Smith et al., 2015).

Alvarez et al. (2018) found agreement between methane emission estimates based on atmospheric data (top-down approaches) and custom-built, spatially explicit inventories (bottom-up approaches) for nine oil and gas production basins in the United States. Reconciling top-down and bottom-up was accomplished by also improving the inventories, through incorporating a more complete count of facilities than past inventories, and more effectively accounting for the influence of large emission sources commonly known as superemitters (Brandt et al., 2014; Zavala-Araiza et al., 2015; Zavala-Araiza et al., 2017).

Future work should focus on assessing the temporal variability of emissions. Top-down estimates based on in situ airborne measurements provide a limited number of observations. While multiple flights can capture some of this variability, long-term atmospheric monitoring can provide additional information on the temporal distribution of emissions. Vaughn et al. (2018) reported that in the Fayetteville region, oil and gas emissions are systematically higher during the daytime—when top-down measurements take place, and lower at night. This is the consequence of a very particular type of operation during the production phase, which accounts for a higher fraction of production emissions than in any other basin. Thus this diurnal difference seems to be unique to this particular basin and there is no evidence that such variability occurs in other production basins (Alvarez et al., 2018).

Recent studies have also demonstrated the value of multiscale emission estimates (Alvarez et al., 2018; Vaughn et al., 2018; Yacovitch et al., 2018; Zavala-Araiza et al., 2015) where top-down approaches (e.g., in situ airborne measurements, tower networks, satellite remote sensing) and bottom-up approaches (e.g., ground-based, facility-wide measurements, on-site measurements) can work in concert to improve the characterization of methane emissions from oil and gas infrastructure and track mitigation efforts.

4 Large point source emission monitoring using satellite observations

Bottom-up emission inventories of greenhouse gases often are associated with large uncertainties owing to the probable inaccuracy and incompleteness in the information on source intensity, activity, and other statistical data. For this reason, large differences exist between different bottom-up estimates of emissions over various regions over the globe, especially over countries with emerging economies. A way to assess the inventories is by measuring the atmospheric concentration of these trace gases and relating them to emissions through atmospheric transport processes. Considering the practical limitations of the global coverage of surface observations, satellites have become increasingly important in this aspect (Duren & Miller, 2012). It has been demonstrated that the contribution of emissions from large sources to atmospheric mole-fractions can be detectable on scales of powerplants, urban areas, and oil and gas basins (Bovensmann et al., 2010; Kort, Frankenberg, Miller, & Oda, 2012; Sheng et al., 2018a, 2018b).

Janardanan, Maksyutov, Ito, Yoshida, and Matsunaga (2017) and Janardanan et al. (2016) proposed such a method to evaluate the regional greenhouse gas emission inventories using Greenhouse gases Observing SATellite (GOSAT) observations and a high-resolution atmospheric transport model. To evaluate anthropogenic emission inventories in various regions over the globe, they determined emission signatures from column-average observations of methane and carbon dioxide (XCH_4 and XCO_2) from the GOSAT satellite using high-resolution atmospheric transport model simulations. Using a high-resolution emission inventory [for CH_4 the Emission Database for Global Atmospheric Research (EDGAR) v4.2 FT2010 and for CO_2 the Open-source Data Inventory for Atmospheric Carbon dioxide (ODIAC)], the anomalies in the atmospheric mole fractions due to local emissions were simulated. These simulated values were used to separate GOSAT observations into background versus near-source observations. The local abundance in trace gas mole fractions due to anthropogenic emissions was estimated for both the simulations and the observations as the difference between the polluted observations and the surrounding cleaner background observations. Since this differencing is carried out for each month and ($10° \times 10°$ region) the spatial and temporal biases present in the satellite observations are reduced. These sets of simulated and observed anomalies were bin-averaged by categorizing the observations according to levels of model-simulated enhancements. Bin-averaged observation shows a good correlation with model simulation with a clear linear relationship, while the regression slope gives a regional factor that relates the simulated to the observed anomalies. Using this method, the satellite observation error, which is large compared to local abundance, is reduced by binning together many observations according to model-simulated enhancements over a large region.

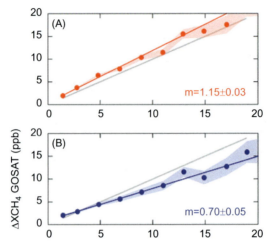

FIG. 2 Weighted regression between inventory-based (EDGAR, x-axis) and observed (GOSAT, y-axis) XCH$_4$ abundance averaged over 2 ppb bins for: (A) the Globe and (B) East Asia. The regression coefficient with the associated estimation error is given as "m." The shading represents the standard error in each bin. The colored lines and the gray lines show the regression model and the identity line respectively.

In their analysis, they found that the local enhancements of XCO$_2$ and XCH$_4$ calculated from GOSAT observations can be represented as a linear function of the inventory-based simulations globally and over large regions, having a sufficient number of observations, such as East Asia and North America. For methane over East Asia, the observed enhancements were 30% lower than suggested by the emission inventory used in the study, implying a potential overestimation of emissions in the inventory. On the contrary, in North America, the observations are approximately 28% higher than model predictions, suggesting an underestimation of emissions by the inventory (Fig. 2). In the case of CO$_2$ in East Asia, a difference of about 20% was found between the inventory-based and GOSAT observed anomalies, which is comparable to the reported uncertainties in the CO$_2$ emission inventories in that region. Janardanan et al. (2016, 2017) concluded that, as their results concur with several recent studies using other methodologies, their method underlines the fact that satellite observations provide an additional tool for bottom-up emission inventory verification.

With new satellite missions such as OCO-2 and TROPOMI, which have narrower observation footprints and considerably larger numbers of observations (by nearly 80 and 800 times more individual observations than GOSAT), the detection of local enhancements from human activities will be more reliable (Nassar et al., 2017) and a comparison to inventory estimates at finer spatial scales will be possible.

5 Precision and sampling requirements for future satellite observations

There is much interest in monitoring greenhouse gas emissions from space from both a climate policy and point source monitoring perspective. However, the observation requirements differ for climate policy and point source monitoring. Climate policy is typically aimed at characterizing annual mean emissions from entire countries and big regions or megacities, whereas point (or regional) source monitoring is aimed at characterizing the emissions from small, localized sources, which can be meaningful to businesses and regulators. Here we focus on the precision and sampling requirements for monitoring regional (e.g., basin-scale) or point source emissions.

We present a simple conceptual example to show the trade-offs between increased spatial coverage and improved instrument precision when estimating point or regional sources. This example can help guide the design of new satellite observations. This section is adapted from Section 4 in Jacob et al. (2016).

Our goal is to investigate the ability to observe the methane enhancement ΔX [ppb] from a source relative to the surrounding background. Single-observation instrument precisions σ [ppb] are taken from Table 1 in Jacob et al. (2016), and we make the optimistic assumption that the precision improves with the square root of the number of observations following the central limit theorem (Kulawik et al., 2016). To quantify a source, we set the required precision to $\Delta X/5$. Only a fraction, F of pixels is successfully retrieved because of clouds, unsuccessful spectral fits, or other factors. Thus, the time required to quantify the source is

$$ t = \frac{t_R}{FN} \left(\frac{5\sigma}{\Delta X} \right)^2 \tag{2} $$

A more general expression is presented as Eq. (12) in Jacob et al. (2016).

We will use the Barnett Shale in Texas as an example to illustrate this. The left panel of Fig. 3 shows a bottom-up estimate of methane emissions from the Barnett Shale that was compiled by the Environmental Defense Fund (Lyon et al., 2015). We will also use observing characteristics from SCIAMACHY to reduce the number of free parameters in Eq. (2). Specifically, we use $F = 0.09$ which is the fraction of successful SCIAMACHY retrievals and $t_R = 1$ day return time. The methane enhancement (ΔX) for the Barnett Shale is 8.5 ppb (or 0.47%).

The right panel in Fig. 3 shows how the time required to quantify methane emissions from the Barnett Shale (t) varies as the number of observations per overpass (N) and the instrument precision (σ) is varied. The black stars give estimates for observing characteristics of a few current and future satellites. From Fig. 3 we can see that SCIAMACHY would take about 1 year of observations to quantify the mean methane emissions from the Barnett Shale. GOSAT lies in a different region of the parameter space but results in a similar quantification

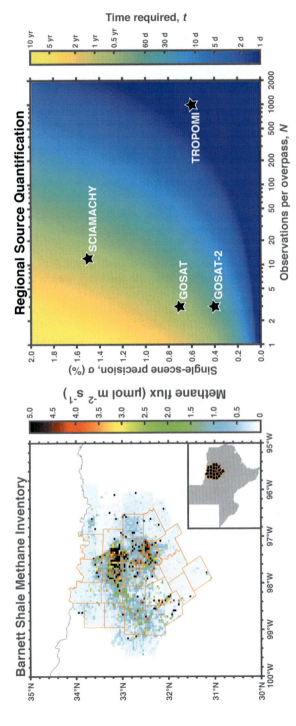

FIG. 3 Characterizing a regional source with satellite observations. (Left) Bottom-up inventory for the Barnett Shale in Texas, compiled by the Environmental Defense Fund (Lyon et al., 2015). (Right) The time required for detection and quantification of a regional source similar to the Barnett Shale using Eq. (1).

time. This is because SCIAMACHY had a larger number of observations per overpass, but they were quite noisy. In contrast, GOSAT had more precise observations but would only obtain a few observations over the Barnett on a given overpass. GOSAT-2 is expected to be more precise than GOSAT with a similar observing strategy; this will cut the quantification time down to 1–2 months. The observing strategy for TROPOMI results in a very dense set of observations with moderate precision. In theory, TROPOMI should be able to quantify methane emissions from the Barnett Shale in a single overpass.

The results from this analysis should be viewed as optimistic as they assume we (1) can accurately characterize the methane concentration upwind of the Barnett Shale, (2) simulate the mean flow, and (3) reduce the instrument error following the central limit theorem. In practice, these are nontrivial issues that may increase the quantification time. However, the trade-offs identified here should still be useful in guiding the precision and sampling requirements for new satellite observations. Other works such as Turner et al. (2018), Varon et al. (2018), and Sheng et al. (2018b) provide a more detailed treatment of uncertainties for a handful of focused case studies with satellite observations and models at spatial scales ranging from meters to kilometers.

6 Developing global high-resolution transport modeling capability for analysis of the satellite and ground-based observations of anthropogenic greenhouse gas emission

A comparison of CO_2 transport model simulations for continuous monitoring sites demonstrated that higher resolution transport models generally perform better in simulating the synoptic-scale variability of atmospheric CO_2 (Patra et al., 2008). Accordingly, most regional or national scale modeling efforts use either global Eulerian models with a higher resolution zoom region over country/region of interest, such as US (Peters et al., 2007; Turner et al., 2015), and Europe (van der Laan-Luijkx et al., 2017), or use regional Eulerian (Breon et al., 2015) or Lagrangian models (Manning et al., 2011; Henne et al., 2016; Rodenbeck, Gerbig, Trusilova, & Heimann, 2009) nested into a global model. Global-scale top-down estimates such as surface flux reanalyses (Chevallier et al., 2009; Rodenbeck, Houweling, Gloor, & Heimann, 2003) would also benefit from higher-resolution transport if that would be easily available. Extending the resolution of the global models is computationally demanding, and there is a gap between the resolution of the available global reanalysis data, such as $\sim 0.3°$ with GEOS-FP (Molod, Takacs, Suarez, & Bacmeister, 2015) and ERA-5 (Hersbach et al., 2018) and the transport model resolutions used currently in global scale inverse modeling, typically between $2°$ and $5°$ (e.g., Schuh et al., 2019).

Several groups have increased their model resolution by combining a global Eulerian model with a Lagrangian model for near-field transport simulations in global inverse modeling applications, such as the global model used by

468 Section | D Forward Looking

Rigby, Manning, and Prinn (2011), with 3 zoom regions at 0.38×0.56 degrees resolution, and the coupled model used by Zhuravlev, Ganshin, Maksyutov, Oshchepkov, and Khattatov (2013) and Shirai et al. (2017) with an effective horizontal resolution of $1°$ globally. The horizontal resolution of Lagrangian models can be increased to the kilometer scale at both global (Ganshin et al., 2012) and continental scale (He et al., 2018), thus providing a low-cost alternative to the global massively parallel high-resolution Eulerian transport models such as NICAM (Niwa et al., 2017) and GEOS-Chem HP (Eastham et al., 2018).

To illustrate the detail of the global high-resolution coupled model as developed by Ganshin et al. (2012), we present the model equations, model design, and some results in this section. The contribution of the surface fluxes to the concentration simulated by the Lagrangian model at the receptor (observation location) is usually considered proportional to the integral of the residence time of all particles at each grid cell multiplied by the flux corresponding to that grid (Lin et al., 2003; Seibert & Frank, 2004). Understanding the use of backward transport by Lagrangian models to simulate the concentration at an observation point at a given observation time can be facilitated by looking into a discussion of equivalence between backward and forward transport by Hourdin and Talagrand (2006). The concentration of CO_2 in the Lagrangian model at any receptor point (corresponding to an observation site) can be written as follows (Holzer & Hall, 2000; Lin et al., 2003):

$$C(x_r, t_r) = \int_{t_0}^{t_r} dt \left(\int_s ds\, I(x_r, t_r \,|\, x, t) \delta(x_s - x) S(x, t) + \int_V dV I(x_r, t_r \,|\, x, t_0) C(x, t_0) \right)$$

(3)

where $C(x_r, t_r)$ is the concentration at receptor point x_r at time t_r; $C(x, t_0)$ is the initial concentration field at time t_0, which is obtained from the background fields simulated by the Eulerian model; $I(x_r, t_r \,|\, x, t)-$ is the influence function or Green's function linking initial concentrations $C(x, t_0)$ and surface sources $S(x, t)$ to the simulated concentrations; and dV and ds are volume and surface elements. The first term of Eq. (3) denotes the concentration change at the receptor from sources/sinks on the surface during the time interval between initialization and observation. Delta function $\delta(x_s - x)$ is used to account for use of a surface source in the first term rather than for volumetric (x_s is surface topography). The second term refers to the contribution to the concentration at the receptor point by the advection of CO_2 from the background tracer field $C(x, t_0)$, which is provided by a Eulerian model. From a Lagrangian viewpoint, the influence function corresponds to the transition probability along the air mass trajectories $x_n(t)$, calculated by the Lagrangian model as follows:

$$I(x_r, t_r \,|\, x, t) = \frac{1}{N} \sum_{n=1}^{N} \delta(x_n(t) - x)$$

(4)

where N is the number of air parcels emitted in the backward direction from the receptor point, and δ is the delta function representing the presence or absence of parcel n at location x. In their coupled model, Ganshin et al. (2012) used FLEXPART (run in backward time mode) as the Lagrangian particle dispersion model and NIES-TM as the Eulerian global transport model (see their Introduction for details). The background CO_2 values on the model grid are obtained by NIES-TM. They used a 2-day length of backward transport in FLEXPART, following a notice by Gloor et al. (2001) that a period on the order of 1.5 days is representative of the time when the imprint of surface fluxes is significantly influencing the concentration at the observation point.

As Lagrangian models do not have numerical diffusion, the surface flux contribution to the simulated concentration can be sampled by particles at any spatial resolution and is only limited by the resolution of the available surface fluxes. The resolution of the wind data is important but for many flat terrain locations, i.e., not influenced by coastal or mountain slopes circulations, the winds are adequately represented by low resolution (0.5°–1.0°) reanalyses, and still provide capability of reproducing the fine temporal structure of the concentartionconcentration time series resembling the observations, as shown in Fig. 4. The high-resolution 30 arc sec surface fluxes prepared

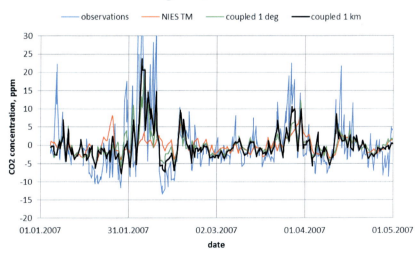

FIG. 4 Coupled 1-km resolution model simulation results for Egham, London (representative 4-month time series). *(From Ganshin, A., Oda, T., Saito, M., Maksyutov, S., Valsala, V., Andres, R. J., Fisher, R. E., Lowry, D., Lukyanov, A., Matsueda, H., Nisbet, E. G., Rigby, M., Sawa, Y., Toumi, R., Tsuboi, K., Varlagin, A., and Zhuravlev, R.: A global coupled Eulerian-Lagrangian model and 1 × 1 km CO2 surface flux dataset for high-resolution atmospheric CO2 transport simulations, Geoscientific Model Development, 5, 231–243, https://doi.org/10.5194/gmd-5-231-2012, 2012.)*

470 Section | D Forward Looking

by Ganshin et al. (2012) were a combination of the ODIAC inventory (Oda & Maksyutov, 2011) and the VISIT model mosaic simulation. A mosaic of vegetation net ecosystem exchange (NEE) was combined with 0.5° fluxes for each vegetation type with the VISIT model (Saito, Ito, & Maksyutov, 2014) and a vegetation typology map at 30 arc sec resolution.

Because each of the global high-resolution flux fields (biosphere and fossil fluxes) requires a sizable amount (several GB) of computer memory, it is inconvenient to operate the model with multiple layers of data at this resolution. To reduce the memory and disk storage requirements, several techniques were used:

(a) The high-resolution surface flux fields at a given point are calculated in the model using a combination of data fields at high and medium resolutions. In the case of fossil fuel emissions, high-resolution annual mean fluxes were multiplied by a medium-resolution ($1° \times 1°$) spatially varying factor that represents the seasonal cycle at a monthly time scale. This factor is derived from seasonally varying fossil-fuel-emissions data (Andres, Gregg, Losey, Marland, & Boden, 2011) at a resolution of $1° \times 1°$, normalized to the annual mean.

(b) A land-cover map at 1 km resolution was used to spatially redistribute the biospheric fluxes given at medium resolution and simulated separately for each of the 15 vegetation types. The flux data at each model time step are obtained by linearly interpolating between the monthly fields, except for biospheric fluxes, which are provided with a daily time step.

(c) For fossil fuel fluxes, sparse matrix storage was used to reduce the memory demand, because only $\sim 1\%$ of the elements in the matrix for anthropogenic emissions have nonzero values. To speed up the element search in the sparse matrix, 1-dimensional look-up tables were employed that contain indices for the first and last nonzero elements for each longitude. It is possible to use the same approach for biospheric fluxes; however, this has little effect on storage requirements because about 30% of the elements have nonzero values in this case.

Further development of global inverse modeling techniques using variational optimization of surface fluxes at high resolution requires the development of the adjoint of the coupled model. As introduced in Chapter 4 of this book, inverse models optimize the sum of misfits between the model simulations and observations. The transport model adjoint is needed in the variational optimization schemes for accurate estimation of the cost function gradient with respect to the time-varying two-dimensional field of surface fluxes. An important development toward the capability of global inversion and data assimilation based on high-resolution tracer transport was reported by Belikov et al. (2016) who developed an adjoint version of a coupled Eulerian–Lagrangian model, which is based on a combination of the global Eulerian model, National Institute for Environmental Studies Transport Model (NIES-TM) (Belikov et al., 2013) and the Lagrangian particle dispersion model, FLEXPART (Stohl et al., 2005). A popular method of adjoint code construction involves applying automatic differentiation tools to the source code of the program. Belikov et al. (2016) started with deriving an integrated model code that combines NIES-TM code, Lagrangian component, that utilized precompiled receptor sensitivity matrixes to simulate concentration at observation sites with given surface fluxes, and a coupler that combines simulations by the NIES-TM model and Lagrangian transport model. For the Eulerian part, the discrete adjoint was hand coded directly from the original NIES TM code. The adjoint model of the coupler and the Lagrangian component were derived using the automatic differentiation software TAF compiler (Giering, Kaminski, & Slawig, 2005). The major advantages of the coupled model include the capability of high resolution and fast computation for the high-resolution Lagrangian component. The Lagrangian component of the forward and adjoint models uses precompiled sensitivity matrixes, which are read from the disk at the beginning of the simulation. These matrices can in principle be calculated by another Lagrangian model, other than FLEXPART. The CPU time needed by the coupled model and its adjoint model is largely decided by the time needed to run the Eulerian component, thus variational optimization can be achieved at the almost same amount of CPU time as in the case of using the Eulerian model with its adjoint (Belikov et al., 2016). This approach was used to implement a high spatial resolution (0.1 degree) global methane flux inversion based on variational optimization (Wang et al., 2019), which demonstrates a path toward bringing the high-resolution performance of the regional scale models (Henne et al., 2016; Manning, O'Doherty, Jones, Simmonds, & Derwent, 2011) to global scale inversions. An example of utilizing inverse model estimates for an independent comparison with the national reports to the UNFCCC can be found in Janardanan et al. (2020). They reported country level top-down estimates of methane emissions using a high-resolution inverse model, utilizing surface, ship, aircraft, and satellite observations. Major emitting countries all showed the difference of top-down with bottom-up emissions within the model uncertainty estimates. For example, the posterior estimate for India and China was 24.2 ± 5.3 and 45.73 ± 8.6 Tg CH_4 year^{-1} while the national report was 20.1 and 54.3 Tg. These two countries have the largest difference between posterior and prior, yet they fall within the uncertainty limits. This study also demonstrates the model performance by independent check with aircraft observations over India, where the posterior fit showed better match with altitude-binned observations, especially in the boundary layer.

7 Developing high-resolution emission inventories for inverse modeling

Given the use of Bayesian inference, one way to improve the inverse carbon emission estimates is to improve the prior emission information, which are

472 Section | D Forward Looking

often given using spatially explicit emission inventories (EIs), such as gridded emission datasets. Spatially explicit EIs, which are a key product used in atmospheric modeling applications, are developed for research purposes (Gurney et al., 2012, 2020; Janssens-Maenhout et al., 2019; Gately & Hutyra, 2017; Oda, Maksyutov, & Andres, 2018; Bun, Matolych, Boychuk, Dmytriv, & Yaremchyshyn, 2010; Bun et al., 2018, and others). Spatially explicit EIs for fossil fuel CO_2 emissions need to be updated regularly (e.g., annual basis) in order to make carbon budget analyses using atmospheric observation (e.g., GCP). Downscaled emission datasets such as ODIAC (Open-source Data Inventory for Anthropogenic CO_2) have the advantage of updating emission fields quickly compared to multimodeling emission inventories as emission estimation and emission spatial modeling are done separately (e.g., Oda et al., 2019). In addition, the use of satellite remote sensing data for emission spatial modeling potentially allows us to incorporate timely emission pattern changes (e.g., Oda & Maksyutov, 2011; (Oda, Maksyutov, & Andres, 2018)), rather than fixed spatial emission distributions. ODIAC emission data product has been updated on annual basis (Oda & Maksyutov, 2015). In order to utilize the dense CO_2 data that become available from intensive ground-based observations and satellite programs, fine-scale EIs are required to adequately prescribe high-resolution atmospheric CO_2 simulations for forward and inverse calculations. However, as suggested by the recent studies, the errors and uncertainties are thought to be large especially at finer spatial scales (e.g., Gately & Hutyra, 2017; Gurney et al., 2019; Hogue, Marland, Andres, Marland, & Woodard, 2016; Oda et al., 2019). The errors and uncertainties associated with the spatial estimates are unique to the approaches employed (e.g., proxy data, see Chapter 3) and often difficult to assess. In addition, the downscaled emissions have a difficulty to help quantifying the emission changes driven by the local climate actions (Oda et al., 2019).

One way to improve the downscaled emissions is to compare and calibrate the downscaled models to a multiresolution model (Oda et al., 2019). GESAPU (geoinformation technologies, spatiotemporal approaches, and full carbon accounting for improving the accuracy of GHG inventories) is a multiresolution, high-definition spatially explicit emission inventory (Bun, Matolych, Boychuk, Dmytriv, & Yaremchyshyn, 2010; Bun et al., 2018; Charkovska et al., 2019). Being among other multiresolution modeling emissions such as Gurney et al. (2012), GESAPU is a national scale EI for Poland and Ukraine. The later emission inventory approach on a fine spatial scale is not initially based on a regular grid (see Chapter 3). Instead, the emission processes in all categories of human activity specified by IPCC (2006) Guidelines are analyzed at the scale of individual facilities (emission sources/sinks), classified into point-, line-, and area-type sources, depending on their emission intensity and physical size as compared to the territory under investigation. The emissions from such very diverse emission sources are combined to a grid only in the final stage to calculate total emissions, and the grid size can be chosen arbitrarily, depending on the need of the analysis. Examining downscaled emissions such as the global high-resolution (1 km) fossil fuel carbon dioxide (CO_2) gridded emission inventory ODIAC using an EI like GESAPU should allow us to characterize the errors and biases in downscaled subnational emissions (Oda et al., 2019).

Given the rich emission data granularity that GESAPU offers, the differences between GESAPU and ODIAC could be largely attributable to the disaggregation errors in ODIAC. Oda et al. (2019) identified the sources of errors in the ODIAC emissions due to its emission disaggregation and characterized the potential disaggregation biases seamlessly across different scales (national, subnational/regional, and urban policy-relevant scale). Fig. 5 shows the emission comparison at the province level over Poland. Three spatially explicit EIs (GESAPU, ODIAC, and EDGAR) are sampled using provincial maps. Given the spatial scale of this analysis (average area of Poland province is 140 km^2), this comparison is less impacted by the large differences due to the fine-scale spatial pattern, but showing the skill of downscaling method. With state-level estimates or subnational level emission estimates available, we can do the same analysis. The knowledge of the improved spatial distribution of emissions can be transferred to other species with the same emission sectoral category.

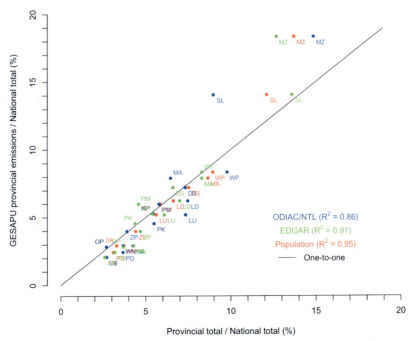

FIG. 5 A comparison of the % share of the provincial ODIAC emissions (hence, nightlight), population and GESAPU emissions in Poland.

474 Section | D Forward Looking

8 Summary

Under the Paris Agreement, each country is obliged to reduce its greenhouse gas emissions according to its Nationally Determined Contributions (NDCs). To build confidence and mutual trust in greenhouse gas emissions reporting by the parties, the Paris Agreement advises to use 2006 IPCC guidelines, and further updates and refinements to it, in the reporting process. The 2006 IPCC inventory guidelines (IPCC, 2006) mention the use of atmospheric observations as an additional measure for the quality assurance and quality control (QA/QC) of the reported emission inventories. Several countries, including Switzerland and UK, use atmospheric observations and inverse modeling of several non-CO_2 GHGs to support the QA/QC component of national emission inventory compilation. The national emission estimates derived from the inverse modeling of atmospheric transport are based on observations made at high-temporal resolution at several sites within a country, and some countries, such as India, are additionally using satellite observations (Ganesan et al., 2017). The methane emission estimates for Switzerland are based on measurements made at three tower sites and one mountain site and are close to the total of the inventory estimates, but still show discrepancies at subnational scales. In the Swiss example, a regional atmospheric transport model is used, and a high-resolution gridded version of the national emission inventory was developed to be used as a prior estimate for the inverse model.

Monitoring progress on emission reduction targets at regional and facility scales is aided by facility-level emission detection and quantification. Atmospheric observations can be employed for the quantification and detection of emissions from industrial facilities, where observations are made from mobile platforms close to emission sources, to reduce and separate interference from adjacent sources. In this approach, the emissions are estimated using a simple mass balance inversion, employing observations taken upwind and downwind of the emitting facility. An example of this type of emission detection was undertaken in the United States, where several projects on estimating emissions from oil and gas production, transmission, and distribution facilities were completed by the Environment Defense Fund and collaborating institutions. As a result of the studies, facility operators could obtain new data on gas leakage integrated on facility/plant scale, which helped initiating more detailed inventory of the leaks from production facilities at the component basis.

In addition to observations by mobile sources and stationary monitoring, a number of newly available satellite observations can be used for quantifying emissions at both facility level and national scale. To utilize the satellite observations effectively, new modeling approaches are being developed, including high-resolution transport modeling tools, high-resolution emission inventories, and inversion techniques suitable for using large volumes of relatively low precision satellite data. At the national scale, the emissions are estimated by either comparing the satellite-observed local concentration enhancements to those

Applications of top-down methods **Chapter | 14** **475**

simulated by a transport model implemented at the resolution of the satellite observation footprint or by applying a regional scale inverse model. In both cases, there is a need for a gridded emission dataset, such as EDGAR or ODIAC. An example of a new approach to provide national inventories as a gridded emission dataset is the GESAPU inventory developed for Poland. It uses high-resolution geospatial information available from open sources to disaggregate the national totals by sector and category into a spatially resolved high-resolution gridded emission dataset.

References

Alvarez, R., Zavala-Araiza, D., Lyon, D., Allen, D., Barkley, Z., Brandt, A., et al. (2018). Assessment of methane emissions from the US oil and gas supply chain. *Science, 361,* 186–188. https://doi.org/10.1126/science.aar7204.

Andres, R., Gregg, J., Losey, L., Marland, G., & Boden, T. (2011). Monthly, global emissions of carbon dioxide from fossil fuel consumption. *Tellus Series B: Chemical and Physical Meteorology, 63,* 309–327. https://doi.org/10.1111/j.1600-0889.2011.00530.x.

Barkley, Z., Lauvaux, T., Davis, K., Deng, A., Miles, N., Richardson, S., et al. (2017). Quantifying methane emissions from natural gas production in North-Eastern Pennsylvania. *Atmospheric Chemistry and Physics, 17,* 13941–13966. https://doi.org/10.5194/acp-17-13941-2017.

Belikov, D. A., Maksyutov, S., Sherlock, V., Aoki, S., Deutscher, N. M., Dohe, S., Griffith, D., et al. (2013). Simulations of column-averaged CO_2 and CH_4 using the NIES TM with a hybrid sigma-isentropic (sigma-theta) vertical coordinate. *Atmospheric Chemistry and Physics, 13,* 1713–1732.

Belikov, D. A., Maksyutov, S., Yaremchuk, A., Ganshin, A., Kaminski, T., Blessing, S., et al. (2016). Adjoint of the global Eulerian-Lagrangian coupled atmospheric transport model (A-GELCA v1.0): Development and validation. *Geoscientific Model Development, 9,* 749–764. https://doi.org/10.5194/gmd-9-749-2016.

Bovensmann, H., Buchwitz, M., Burrows, J., Reuter, M., Krings, T., Gerilowski, K., et al. (2010). A remote sensing technique for global monitoring of power plant CO2 emissions from space and related applications. *Atmospheric Measurement Techniques, 3,* 781–811. https://doi.org/10.5194/amt-3-781-2010.

Brandt, A., Heath, G., Kort, E., O'Sullivan, F., Petron, G., Jordaan, S., et al. (2014). Methane leaks from North American natural gas systems. *Science, 343,* 733–735. https://doi.org/10.1126/science.1247045.

Breon, F., Broquet, G., Puygrenier, V., Chevallier, F., Xueref-Remy, I., Ramonet, M., et al. (2015). An attempt at estimating Paris area CO2 emissions from atmospheric concentration measurements. *Atmospheric Chemistry and Physics, 15,* 1707–1724. https://doi.org/10.5194/acp-15-1707-2015.

Bun, R., Matolych, B., Boychuk, Kh., Dmytriv, K., & Yaremchyshyn, O. (2010). *Information technologies for creation of cadastre of greenhouse gas emissions of Lviv region* (pp. 1–272). Lviv: Ukropol.

Bun, R., Nahorski, Z., Horabik-Pyzel, J., Danylo, O., See, L., Charkovska, N., et al. (2018). Development of a high-resolution spatial inventory of greenhouse gas emissions for Poland from stationary and mobile sources. *Mitigation and Adaptation Strategies for Global Change.* https://doi.org/10.1007/s11027-018-9791-2.

476 Section | D Forward Looking

Cain, M., Warwick, N., Fisher, R., Lowry, D., Lanoiselle, M., Nisbet, E., et al. (2017). A cautionary tale: A study of a methane enhancement over the North Sea. *Journal of Geophysical Research-Atmospheres, 122*, 7630–7645. https://doi.org/10.1002/2017JD026626.

Charkovska, N., Halushchak, M., Bun, R., Nahorski, Z., Oda, T., Jonas, M., et al. (2019). A high-definition spatially explicit modelling approach for national greenhouse gas emissions from industrial processes: Reducing the errors and uncertainties in global emission modelling. *Mitigation and Adaptation Strategies for Global Change*. https://doi.org/10.1007/s11027-018-9836-6.

Chevallier, F., Maksyutov, S., Bousquet, P., Breon, F., Saito, R., Yoshida, Y., et al. (2009). On the accuracy of the CO2 surface fluxes to be estimated from the GOSAT observations. *Geophysical Research Letters, 36*. https://doi.org/10.1029/2009GL040108.

Conley, S., Faloona, I., Mehrotra, S., Suard, M., Lenschow, D., Sweeney, C., et al. (2017). Application of Gauss's theorem to quantify localized surface emissions from airborne measurements of wind and trace gases. *Atmospheric Measurement Techniques, 10*, 3345–3358. https://doi.org/10.5194/amt-10-3345-2017.

Duren, R. M., & Miller, C. E. (2012). Measuring the carbon emissions of megacities. *Nature Climate Change, 2*. https://doi.org/10.1038/nclimate1629.

Eastham, S., Long, M., Keller, C., Lundgren, E., Yantosca, R., Zhuang, J., et al. (2018). GEOS-Chem high performance (GCHP v11-02c): A next-generation implementation of the GEOS-Chem chemical transport model for massively parallel applications. *Geoscientific Model Development, 11*, 2941–2953. https://doi.org/10.5194/gmd-11-2941-2018.

FOEN. (2021). Switzerland's greenhouse gas inventory 1990–2019. In *Submission of April 2021 under the united nations framework convention on climate change and under the Kyoto protocol, Tech. Rep.* Bern, Switzerland: Federal Office for the Environment (FOEN).

Ganesan, A. L., Rigby, M., Lunt, M. F., Parker, R. J., Boesch, H., Goulding, N., et al. (2017). Atmospheric observations show accurate reporting and little growth in India's methane emissions. *Nature Communications, 8*, 836. https://doi.org/10.1038/s41467-017-00994-7.

Ganshin, A., Oda, T., Saito, M., Maksyutov, S., Valsala, V., Andres, R. J., et al. (2012). A global coupled Eulerian-Lagrangian model and 1×1 km CO_2 surface flux dataset for high-resolution atmospheric CO_2 transport simulations. *Geoscientific Model Development, 5*, 231–243. https://doi.org/10.5194/gmd-5-231-2012.

Gately, C. K., & Hutyra, L. R. (2017). Large uncertainties in urban-scale carbon emissions. *Journal of Geophysical Research-Atmospheres, 122*, 11242–11260. https://doi.org/10.1002/2017jd027359.

Giering, R., Kaminski, T., & Slawig, T. (2005). Generating efficient derivative code with TAF—Adjoint and tangent linear Euler flow around an airfoil. *Future Generation Computer Systems, 21*, 1345–1355. https://doi.org/10.1016/j.future.2004.11.003.

Gloor, M., Bakwin, P., Hurst, D., Lock, L., Draxler, R., & Tans, P. (2001). What is the concentration footprint of a tall tower? *Journal of Geophysical Research-Atmospheres, 106*, 17831–17840. https://doi.org/10.1029/2001JD900021.

Gurney, K. R., Liang, J., Patarasuk, R., Song, Y., Huang, J., & Roest, G. (2020). The Vulcan Version 3.0 high-resolution fossil fuel CO_2 emissions for the United States. *Journal of Geophysical Research: Atmospheres, 125*(19). https://doi.org/10.1029/2020JD032974, e2020JD032974.

Gurney, K. R., Liang, J., O'Keeffe, D., Patarasuk, R., Hutchins, M., Huang, J., et al. (2019). Comparison of global downscaled versus bottom-up fossil fuel CO2 emissions at the urban scale in four U.S. urban areas. *Journal of Geophysical Research-Atmospheres, 124*, 2823–2840. https://doi.org/10.1029/2018jd028859.

Gurney, K., Razlivanov, I., Song, Y., Zhou, Y., Benes, B., & Abdul-Massih, M. (2012). Quantification of fossil fuel CO2 emissions on the building/street scale for a large US City. *Environmental Science & Technology, 46*, 12194–12202. https://doi.org/10.1021/es3011282.

He, W., van der Velde, I. R., Andrews, A. E., Sweeney, C., Miller, J., Tans, P., et al. (2018). CTDAS-Lagrange v1.0: A high-resolution data assimilation system for regional carbon dioxide observations. *Geoscientific Model Development, 11*, 3515–3536. https://doi.org/10.5194/gmd-11-3515-2018.

Henne, S., Brunner, D., Oney, B., Leuenberger, M., Eugster, W., Bamberger, I., et al. (2016). Validation of the Swiss methane emission inventory by atmospheric observations and inverse modelling. *Atmospheric Chemistry and Physics, 16*, 3683–3710.

Hersbach, H., de Rosnay, P., Bell, B., Schepers, D., Simmons, A., Soci, C., et al. (2018). *ERA Report Series: 27. Operational global reanalysis: Progress, future directions and synergies with NWP.* ECMWF.

Hogue, S., Marland, E., Andres, R., Marland, G., & Woodard, D. (2016). Uncertainty in gridded CO2 emissions estimates. *Earth's Future, 4*, 225–239. https://doi.org/10.1002/2015EF000343.

Holzer, M., & Hall, T. (2000). Transit-time and tracer-age distributions in geophysical flows. *Journal of the Atmospheric Sciences, 57*, 3539–3558. https://doi.org/10.1175/1520-0469(2000) 057<3539:TTATAD>2.0.CO;2.

Hourdin, F., & Talagrand, O. (2006). Eulerian backtracking of atmospheric tracers. I: Adjoint derivation and parametrization of subgrid-scale transport. *Quarterly Journal of the Royal Meteorological Society, 132*, 567–583. https://doi.org/10.1256/qj.03.198.A.

IPCC. (2006). IPCC guidelines for national greenhouse gas inventories. In H. S. Eggleston, L. Buendia, K. Miwa, T. Ngara, & K. Tanabe (Eds.), *IPCC national greenhouse gas inventories programme, intergovernmental panel on climate change IPCC*. Japan: c/o Institute for Global Environmental Strategies.

Jacob, D. J., Turner, A. J., Maasakkers, J. D., Sheng, J., Sun, K., Liu, X., et al. (2016). Satellite observations of atmospheric methane and their value for quantifying methane emissions. *Atmospheric Chemistry and Physics, 16*, 14371–14396. https://doi.org/10.5194/acp-16-14371-2016.

Janardanan, R., Maksyutov, S., Ito, A., Yoshida, Y., & Matsunaga, T. (2017). Assessment of anthropogenic methane emissions over large regions based on GOSAT observations and high resolution transport modeling. *Remote Sensing, 9*. https://doi.org/10.3390/rs9090941.

Janardanan, R., Maksyutov, S., Oda, T., Saito, M., Kaiser, J., Ganshin, A., et al. (2016). Comparing GOSAT observations of localized CO2 enhancements by large emitters with inventory-based estimates. *Geophysical Research Letters, 43*, 3486–3493. https://doi.org/10.1002/2016GL067843.

Janardanan, R., Maksyutov, S., Tsuruta, A., Wang, F., Tiwari, Y. K., Valsala, V., et al. (2020). Country-scale analysis of methane emissions with a high-resolution inverse model using GOSAT and surface observations. *Remote Sensing, 12*(3), 375. https://doi.org/10.3390/rs12030375.

Janssens-Maenhout, G., Crippa, M., Guizzardi, D., Muntean, M., Schaaf, E., Dentener, F., et al. (2019). EDGAR v4.3.2 global atlas of the three major greenhouse gas emissions for the period 1970-2012. *Earth System Science Data Discussions, 2019*, 1–52. https://doi.org/10.5194/essd-2018-164.

Johnson, M., Tyner, D., Conley, S., Schwietzke, S., & Zavala-Araiza, D. (2017). Comparisons of airborne measurements and inventory estimates of methane emissions in the Alberta upstream oil and gas sector. *Environmental Science & Technology, 51*, 13008–13017. https://doi.org/10.1021/acs.est.7b03525.

478 Section | D Forward Looking

Karion, A., Sweeney, C., Kort, E., Shepson, P., Brewer, A., Cambaliza, M., et al. (2015). Aircraft-based estimate of total methane emissions from the Barnett Shale Region. *Environmental Science & Technology*, *49*, 8124–8131. https://doi.org/10.1021/acs.est.5b00217.

Karion, A., Sweeney, C., Petron, G., Frost, G., Hardesty, R., Kofler, J., et al. (2013). Methane emissions estimate from airborne measurements over a western United States natural gas field. *Geophysical Research Letters*, *40*, 4393–4397. https://doi.org/10.1002/grl.50811.

Kort, E., Frankenberg, C., Miller, C., & Oda, T. (2012). Space-based observations of megacity carbon dioxide. *Geophysical Research Letters*, *39*. https://doi.org/10.1029/2012GL052738.

Kulawik, S., Wunch, D., O'Dell, C., Frankenberg, C., Reuter, M., Oda, T., et al. (2016). Consistent evaluation of ACOS-GOSAT, BESD-SCIAMACHY, CarbonTracker, and MACC through comparisons to TCCON. *Atmospheric Measurement Techniques*, *9*, 683–709. https://doi.org/10.5194/amt-9-683-2016.

Lin, J., Gerbig, C., Wofsy, S., Andrews, A., Daube, B., Davis, K., et al. (2003). A near-field tool for simulating the upstream influence of atmospheric observations: The stochastic time-inverted Lagrangian transport (STILT) model. *Journal of Geophysical Research-Atmospheres*, *108*. https://doi.org/10.1029/2002JB001978|10.1029/2002JD003161.

Lyon, D., Zavala-Araiza, D., Alvarez, R., Harriss, R., Palacios, V., Lan, X., et al. (2015). Constructing a spatially resolved methane emission inventory for the Barnett Shale region. *Environmental Science & Technology*, *49*, 8147–8157. https://doi.org/10.1021/es506359c.

Manning, A., O'Doherty, S., Jones, A., Simmonds, P., & Derwent, R. (2011). Estimating UK methane and nitrous oxide emissions from 1990 to 2007 using an inversion modeling approach. *Journal of Geophysical Research-Atmospheres*, *116*, D02305. https://doi.org/10.1029/2010JD014763.

Molod, A., Takacs, L., Suarez, M., & Bacmeister, J. (2015). Development of the GEOS-5 atmospheric general circulation model: Evolution from MERRA to MERRA2. *Geoscientific Model Development*, *8*, 1339–1356. https://doi.org/10.5194/gmd-8-1339-2015.

Nassar, R., Hill, T., McLinden, C., Wunch, D., Jones, D., & Crisp, D. (2017). Quantifying CO2 emissions from individual power plants from space. *Geophysical Research Letters*, *44*, 10045–10053. https://doi.org/10.1002/2017GL074702.

Negron, G., Kort, E. A., Conley, S. A., & Smith, M. L. (2020). Airborne assessment of methane emissions from offshore platforms in the US Gulf of Mexico. *Environmental Science & Technology*, *54*(8), 5112–5120. https://doi.org/10.1021/acs.est.0c00179.

Niwa, Y., Fujii, Y., Sawa, Y., Iida, Y., Ito, A., Satoh, M., et al. (2017). A 4D-Var inversion system based on the icosahedral grid model (NICAM-TM 4D-Var v1.0)—Part 2: Optimization scheme and identical twin experiment of atmospheric CO2 inversion. *Geoscientific Model Development*, *10*, 2201–2219. https://doi.org/10.5194/gmd-10-2201-2017.

Oda, T., Bun, R., Kinakh, V., Topylko, P., Halushchak, M., Marland, G., et al. (2019). Errors and uncertainties in a gridded carbon dioxide emissions inventory. *Mitigation and Adaptation Strategies for Global Change*. https://doi.org/10.1007/s11027-019-09877-2.

Oda, T., & Maksyutov, S. (2011). A very high-resolution (1 km x 1 km) global fossil fuel CO_2 emission inventory derived using a point source database and satellite observations of nighttime lights. *Atmospheric Chemistry and Physics*, *11*, 543–556. https://doi.org/10.5194/acp-11-543-2011.

Oda, T., & Maksyutov, S. (2015). *Open-source data inventory for anthropogenic CO2 (ODIAC) emission dataset (ODIAC2016)*. Available Tsukuba, Japan: National Institute for Environmental Studies. http://db.cger.nies.go.jp/dataset/ODIAC/.

Oda, T., Maksyutov, S., & Andres, R. J. (2018). The open-source data inventory for anthropogenic carbon dioxide (CO_2), version 2016 (ODIAC2016): A global, monthly fossil-fuel CO_2 gridded emission data product for tracer transport simulations and surface flux inversions. *Earth System Science Data*, *10*(1), 87–107. https://doi.org/10.5194/essd-10-87-2018.

Patra, P. K., Law, R. M., Peters, W., Roedenbeck, C., Takigawa, M., Aulagnier, C., Baker, I., et al. (2008). TransCom model simulations of hourly atmospheric CO2: Analysis of synoptic-scale variations for the period 2002–2003. *Global Biogeochemical Cycles*, *22*, GB4013. https://doi.org/10.1029/2007GB003081.

Peischl, J., Karion, A., Sweeney, C., Kort, E., Smith, M., Brandt, A., et al. (2016). Quantifying atmospheric methane emissions from oil and natural gas production in the Bakken shale region of North Dakota. *Journal of Geophysical Research-Atmospheres*, *121*, 6101–6111. https://doi.org/10.1002/2015JD024631.

Peischl, J., Ryerson, T. B., Aikin, K. C., De Gouw, Gilman, J. B., ... Parrish, D. D. (2015). Quantifying atmospheric methane emissions from the Haynesville, Fayetteville, and northeastern Marcellus shale gas production regions. *Journal of Geophysical Research: Atmospheres*, *120* (5), 2119–2139. https://doi.org/10.1002/2014JD022697.

Peters, W., Jacobson, A. R., Sweeney, C., Andrews, A. E., Conway, T. J., Masarie, K., ... Tans, P. P. (2007). An atmospheric perspective on North American carbon dioxide exchange: CarbonTracker. *Proceedings of the National Academy of Sciences of the United States of America*, *104*(48), 18925–18930. https://doi.org/10.1073/pnas.0708986104.

Petron, G., Karion, A., Sweeney, C., Miller, B., Montzka, S., Frost, G., et al. (2014). A new look at methane and nonmethane hydrocarbon emissions from oil and natural gas operations in the Colorado Denver-Julesburg Basin. *Journal of Geophysical Research-Atmospheres*, *119*, 6836–6852. https://doi.org/10.1002/2013JD021272.

Rigby, M., Manning, A., & Prinn, R. (2011). Inversion of long-lived trace gas emissions using combined Eulerian and Lagrangian chemical transport models. *Atmospheric Chemistry and Physics*, *11*, 9887–9898. https://doi.org/10.5194/acp-11-9887-2011.

Rodenbeck, C., Gerbig, C., Trusilova, K., & Heimann, M. (2009). A two-step scheme for high-resolution regional atmospheric trace gas inversions based on independent models. *Atmospheric Chemistry and Physics*, *9*, 5331–5342.

Rodenbeck, C., Houweling, S., Gloor, M., & Heimann, M. (2003). Time-dependent atmospheric CO2 inversions based on interannually varying tracer transport. *Tellus Series B: Chemical and Physical Meteorology*, *55*, 488–497. https://doi.org/10.1034/j.1600-0889.2003.00033.x.

Saito, M., Ito, A., & Maksyutov, S. (2014). Optimization of a prognostic biosphere model for terrestrial biomass and atmospheric CO2 variability. *Geoscientific Model Development*, *7*, 1829–1840. https://doi.org/10.5194/gmd-7-1829-2014.

Schuh, A. E., Jacobson, A. R., Basu, S., Weir, B., Baker, D., Bowman, K., et al. (2019). Quantifying the impact of atmospheric transport uncertainty on CO2 surface flux estimates. *Global Biogeochemical Cycles*, *33*, 484–500. https://doi.org/10.1029/2018GB006086.

Schwietzke, S., Petron, G., Conley, S., Pickering, C., Mielke-Maday, I., Dlugokencky, E., et al. (2017). Improved mechanistic understanding of natural gas methane emissions from spatially resolved aircraft measurements. *Environmental Science & Technology*, *51*, 7286–7294. https://doi.org/10.1021/acs.est.7b01810.

Seibert, P., & Frank, A. (2004). Source-receptor matrix calculation with a Lagrangian particle dispersion model in backward mode. *Atmospheric Chemistry and Physics*, *4*, 51–63. https://doi.org/10.5194/acp-4-51-2004.

Sheng, J., Jacob, D., Turner, A., Maasakkers, J., Benmergui, J., Bloom, A., et al. (2018a). 2010–2016 Methane trends over Canada, the United States, and Mexico observed by the GOSAT satellite: Contributions from different source sectors. *Atmospheric Chemistry and Physics*, *18*, 12257–12267. https://doi.org/10.5194/acp-18-12257-2018.

480 Section | D Forward Looking

Sheng, J., Jacob, D., Turner, A., Maasakkers, J., Sulprizio, M., Bloom, A., et al. (2018b). High-resolution inversion of methane emissions in the southeast US using SEAC(4)RS aircraft observations of atmospheric methane: Anthropogenic and wetland sources. *Atmospheric Chemistry and Physics, 18*, 6483–6491. https://doi.org/10.5194/acp-18-6483-2018.

Shirai, T., Ishizawa, M., Zhuravlev, R., Ganshin, A., Belikov, D., Saito, M., et al. (2017). A decadal inversion of CO2 using the global Eulerian-Lagrangian coupled atmospheric model (GELCA): Sensitivity to the ground-based observation network. *Tellus Series B: Chemical and Physical Meteorology, 69*. https://doi.org/10.1080/16000889.2017.1291158.

Smith, M., Gvakharia, A., Kort, E., Sweeney, C., Conley, S., Faloona, I., et al. (2017). Airborne quantification of methane emissions over the four corners region. *Environmental Science & Technology, 51*, 5832–5837. https://doi.org/10.1021/acs.est.6b06107.

Smith, M., Kort, E., Karion, A., Sweeney, C., Herndon, S., & Yacovitch, T. (2015). Airborne ethane observations in the Barnett shale: Quantification of ethane flux and attribution of methane emissions. *Environmental Science & Technology, 49*, 8158–8166. https://doi.org/10.1021/acs.est.5b00219.

Stohl, A., Forster, C., Frank, A., Seibert, P., & Wotawa, G. (2005). Technical note: The Lagrangian particle dispersion model FLEXPART version 6.2. *Atmospheric Chemistry and Physics, 5*, 2461–2474.

Turner, A., Jacob, D., Benmergui, J., Brandman, J., White, L., & Randles, C. (2018). Assessing the capability of different satellite observing configurations to resolve the distribution of methane emissions at kilometer scales. *Atmospheric Chemistry and Physics, 18*, 8265–8278. https://doi.org/10.5194/acp-18-8265-2018.

Turner, A., Jacob, D., Wecht, K., Maasakkers, J., Lundgren, E., Andrews, A., et al. (2015). Estimating global and north American methane emissions with high spatial resolution using GOSAT satellite data. *Atmospheric Chemistry and Physics, 15*, 7049–7069. https://doi.org/10.5194/acp-15-7049-2015.

van der Laan-Luijkx, I. T., van der Velde, I. R., van der Veen, E., Tsuruta, A., Stanislawska, K., Babenhauserheide, A., et al. (2017). The CarbonTracker Data Assimilation Shell (CTDAS) v1.0: implementation and global carbon balance 2001–2015. *Geoscientific Model Development, 10*, 2785–2800. https://doi.org/10.5194/gmd-10-2785-2017.

Varon, D., Jacob, D., McKeever, J., Jervis, D., Durak, B., Xia, Y., et al. (2018). Quantifying methane point sources from fine-scale satellite observations of atmospheric methane plumes. *Atmospheric Measurement Techniques, 11*, 5673–5686. https://doi.org/10.5194/amt-11-5673-2018.

Vaughn, T., Bell, C., Pickering, C., Schwietzke, S., Heath, G., Petron, G., et al. (2018). Temporal variability largely explains top-down/bottom-up difference in methane emission estimates from a natural gas production region. *Proceedings of the National Academy of Sciences of the United States of America, 115*, 11712–11717. https://doi.org/10.1073/pnas.1805687115.

Wang, F., Maksyutov, S., Tsuruta, A., Janardanan, R., Ito, A., Sasakawa, M., Machida, T., et al. (2019). Methane emission estimates by the global high-resolution inverse model using national inventories. *Remote Sensing, 11*, 2489. https://doi.org/10.3390/rs11212489.

Yacovitch, T., Neininger, B., Herndon, S., van der Gon, H., Jonkers, S., Hulskotte, J., et al. (2018). Methane emissions in the Netherlands: The Groningen field. *Elementa-Science of the Anthropocene, 6*. https://doi.org/10.1525/elementa.308.

Zavala-Araiza, D., Alvarez, R., Lyon, D., Allen, D., Marchese, A., Zimmerle, D., et al. (2017). Super-emitters in natural gas infrastructure are caused by abnormal process conditions. *Nature Communications, 8*. https://doi.org/10.1038/ncomms14012.

Zavala-Araiza, D., Lyon, D., Alvarez, R., Davis, K., Harriss, R., Herndon, S., et al. (2015). Reconciling divergent estimates of oil and gas methane emissions. *Proceedings of the National*

Academy of Sciences of the United States of America, *112*, 15597–15602. https://doi.org/10.1073/pnas.1522126112.

Zavala-Araiza, D., Omara, M., Gautam, R., Smith, M. L., Pandey, S., Aben, I., & Hamburg, S. P. (2021). A tale of two regions: Methane emissions from oil and gas production in offshore/onshore Mexico. *Environmental Research Letters*, *16*(2). https://doi.org/10.1088/1748-9326/abceeb, 024019.

Zhuravlev, R. V., Ganshin, A. V., Maksyutov, S. S., Oshchepkov, S. L., & Khattatov, B. V. (2013). Estimation of global CO2 fluxes using ground-based and satellite (GOSAT) observation data with empirical orthogonal functions. *Atmospheric and Oceanic Optics*, *26*, 507–516. https://doi.org/10.1134/S1024856013060158.

Chapter 15

Earth system perspective

Lesley Ott[a] and Abhishek Chatterjee[a,b]
[a]*NASA Goddard Space Flight Center, Greenbelt, MD, United States,* [b]*Universities Space Research Association, Columbia, MD, United States*

1 Introduction and background: What is an earth system model?

Earth System models incorporate complex representations of the atmosphere, oceans, land, and ice to diagnose and predict environmental change across different time and space scales. These models have evolved over time from simple, theoretical systems of equations to complex software packages run on some of the world's largest supercomputers. Modern modeling infrastructures focus on a modular structure that supports a unified framework (Fig. 1) to be configured many ways to suit different needs. For example, the model could be configured to simulate atmospheric flow at very high spatial resolution to predict short-term changes in weather systems (i.e., days to weeks to months) with simplifications that assume changes in ocean circulation over centennial timescales to be neglected. For century-long climate prediction simulations, a more complex set of processes is required, and the model is run at much coarser resolutions. Despite decades of progress in computing and software engineering, the availability of computational resources still plays a large role in how model experiments are framed, what processes they can include, and what types of output and services they can provide to users.

Atmospheric general circulation models (AGCMs) describe the state of the atmosphere using equations that govern the conservation of mass, momentum, and thermal energy. These equations relate the behavior of temperature, pressure, wind velocity, and moisture and are solved iteratively over many grid boxes covering the Earth. In addition to the fundamental primitive equation that comprises the model's dynamical core, modern general circulation models (GCMs) contain parameterizations that seek to represent processes like thunderstorms, cloud formation, boundary layer turbulence, land-surface hydrology, and radiative transfer that are too complex to be explicitly represented at global scales but whose influence must be considered to realistically depict the evolution of weather. These

Balancing Greenhouse Gas Budgets. https://doi.org/10.1016/B978-0-12-814952-2.00014-9
Copyright © 2022 United States Government as represented by the Administrator of the National
Aeronautics and Space Administration. Published by Elsevier Inc. All Other Rights Reserved.

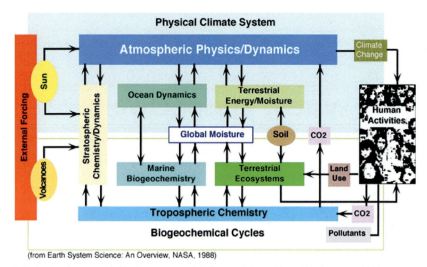

(from Earth System Science: An Overview, NASA, 1988.)

FIG. 1 Bretherton et al. Earth system model figure. *(Adapted from NASA report, 1988.)*

parameterizations also provide information on a much broader set of parameters (for example, soil moisture, surface and top-of-atmosphere radiation, cloud fraction, and optical thickness). Grid box sizes, referred to as the model's resolution, range from single to hundreds of kilometers depending on the application.

When combined with observations for initialization, AGCMs are commonly used to predict the evolution of weather systems 10–14 days in the future. Early numerical weather prediction experiments that demonstrated forecasts based on the direct use of observations were unrealistic because small errors in the observed initial state grow quickly over time. Additionally, weather observations are often discontinuous in space and time. Through the mathematical process of data assimilation, observations of the atmospheric state are merged with a modeled background state to create an optimal *"analysis"* that represents the best estimate of the current weather. After the analysis is completed, the AGCM is integrated forward in time to produce a forecast. Most weather prediction centers produce new analyses and forecasts multiple times per day, continually ingesting observations from a global network of conventional (e.g., weather balloons, surface stations, aircraft) and satellite observations to revise forecasts. The greatest skill occurs in the first few days of the forecast, when the impact of the initialization is greatest. The chaotic nature of the atmosphere means that precise predictions of events like cold fronts and rainstorms are not possible beyond 2 weeks (Lorenz, 1969; Zhang et al., 2019). Weather forecasts are typically produced at as high a spatial resolution as is computationally feasible, currently ~10-km globally, to provide detailed information on severe weather. Traditionally, weather forecasts have been produced using an imposed sea surface temperature boundary condition that does not vary during the forecast period, though this is evolving with some centers now producing forecasts using more complex ocean models (e.g., Mogensen, Balmaseda, & Weaver, 2012).

Some aspects of the state of the climate can be predicted months in advance when an AGCM is coupled with an ocean GCM (OGCM). While short-term weather forecasts derive skill from a realistic initial picture of the atmosphere, seasonal forecasts estimate the atmospheric response to relatively slow varying changes in boundary conditions (e.g., sea surface temperatures, soil moisture). For example, temperature forecasts over North America draw considerable skill from land surface initialization (Koster et al., 2010). Nearly a dozen organizations worldwide make seasonal predictions of climate using ensembles of coupled atmosphere-ocean GCM (AOGCM) simulations that are generated by perturbing initial conditions and/or model physical parameters (e.g., Kirtman & Min, 2009). The use of ensemble predictions provides a valuable metric for assessing forecast confidence. While numerical weather forecasts have been made since the 1950s, seasonal forecasts have only been made operationally since the 1990s. Though they are considered much less mature, seasonal forecasts are of great interest because of their potential to provide early warnings of a wide array of dangerous and costly hazards. Because of the added complexity and computational cost of simulating both the atmosphere and ocean, seasonal prediction models are typically run at coarser resolutions ranging from 50 to 100 km.

2 Carbon cycle modeling in the context of earth system models

Earth system models (ESMs) combine AOGCMs with modules that describe the evolution of the cryosphere, land biosphere, and atmospheric chemistry to predict changes over decadal and century timescales. They are best known for their use in coupled model intercomparison projects (CMIPs) that support the Intergovernmental Panel on Climate Change (IPCC), the United Nations body responsible for assessing the science of climate change. ESM experiments can be run in several ways. First, they can use as input specified atmospheric concentrations of greenhouse gases whose changes over time have been estimated by separate models that take into different scenarios of human activity. Second, they can be run with an interactive carbon cycle that influences the concentration of CO_2 over time, adding a higher degree of complexity to simulations. Models with interactive biogeochemistry have only been included in IPCC CMIPs since 2013 (Taylor, Stouffer, & Meehl, 2012). The most recent IPCC CMIP incorporates both types of simulations in its standard set of experiments and modeling groups

486 Section | D Forward Looking

are invited to participate in a wide range of voluntary additional experiments which seek to investigate key processes in greater detail (Eyring et al., 2016). Several of these experiments are directly related to greenhouse gas changes including one that seeks to quantify the response of the carbon cycle to climate change (Jones et al., 2016) and another that examines the impact of land-use and land-cover change (Lawrence et al., 2016). ESMs continue to evolve in the complexity of processes represented, although the wide range of spatial and temporal scales that need to be captured, even for carbon cycle processes (Fig. 2), makes this an extremely challenging enterprise. Because of the added complexity, length of simulations, and large data volumes generated, ESM simulations are typically run at resolutions of hundreds of kilometers, much coarser than the resolutions used for NWP or seasonal prediction.

High-quality information on current and historical regional greenhouse gas budgets plays a critical role in evaluating and improving ESMs. Results from the 2007 CMIP3 showed that large model differences in land carbon uptake results in differences of several hundred ppm of CO_2 by 2100, dramatically altering the rate of warming the planet experienced (Friedlingstein et al., 2006). Regional land flux differences between models, particularly in the tropics, were particularly large. Despite substantial model development and increases in the complexity of processes represented, results from the 2013 CMIP5 continued to show such large differences between models that it was not possible to predict whether the land would be a source or sink of carbon by 2100 (Friedlingstein et al., 2014). Evaluation of CMIP5 models showed that the difference between model predictions was related to errors in their estimates of present-day CO_2 concentrations, demonstrating the importance of rigorous model evaluation (Hoffman et al., 2014).

Though less mature, carbon cycle modules are being adapted to make predictions of carbon flux on shorter weather to seasonal timescales. The European Centre for Medium-Range Weather Forecasts (ECMWF) computes land-atmosphere within its Integrated Forecast System (IFS), using information from its weather forecast as input to the carbon module. These estimates are combined with near-real-time estimates of fire emissions that are estimated from satellite observations of fire radiative power (Kaiser et al., 2012) and inventory-based estimates of fossil fuel emissions and ocean carbon uptake to produce high-resolution forecasts of atmospheric carbon dioxide (Agusti-Panareda et al., 2014). Such forecasts are mainly used by the science community, who can use the spatial distributions of trace gases to plan field campaigns, to aid in evaluation of satellite retrievals of greenhouse gases (O'Dell et al., 2018) and as boundary conditions to regional models (e.g., Gourdji et al., 2012; Schuh et al., 2010). The major challenges for such forecasts are the bias in the underlying flux model, which can introduce substantial errors in atmospheric concentrations (Agusti-Panareda et al., 2016), and the long latency of fossil fuel and other emission datasets, which are often not available for a year or more (Weir et al., 2021).

Earth system perspective **Chapter | 15** **487**

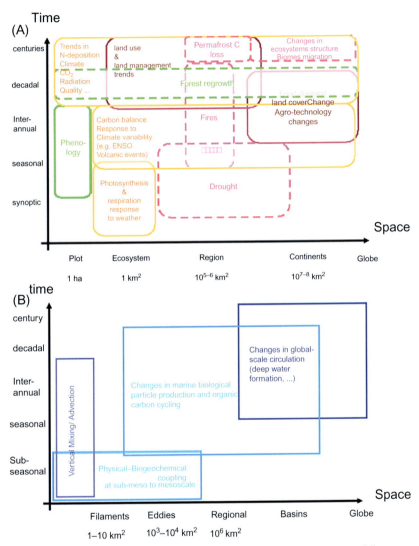

FIG. 2 (A) and (B) Ciais et al. (2014), Figure 1C–D—example showing the range of diverse terrestrial and ocean carbon cycle processes that occur on various time and space scales and need to be incorporated in Earth System models, as well as those that are constrained by observations in a DA context.

Seasonal forecasts of carbon flux represent a relatively new area of research with the potential to support a wider group of users with information 1–9 months in advance. Such predictions make use of either the current state of the climate through use of observationally based ocean climate indices or seasonal forecast meteorology fields from the AOGCMs described above. Forecasts of ocean chlorophyll produced by a model of ocean biology and biogeochemistry show skill in predicting anomalous patterns observed by satellites at lead times of 1–3 months (Rousseaux, Gregg, & Ott, 2021). Changes in tropical land primary production and fire risk that are associated with different phases of ENSO may also be predictable (Betts, Jones, Knight, Keeling, & Kennedy, 2016), gaining skill from the lagged impact of vegetation soil moisture initialization (Lee et al., 2020). Statistical predictions of fire frequency that make use of a combination of climate information and recent fire activity also demonstrate the potential to predict fire activity and indicate that in certain regions, the strength of the early part of the fire season can be a useful predictor of late season activity (Chen et al., 2020). More research is needed to assess forecast skill in the extratropics, where tropical ocean temperatures have less predictive power. A limiting factor of predictions using seasonal forecast meteorology is the substantial bias inherent in such products, though this can be minimized with bias correction techniques (Arsenault, Brissette, & Martel, 2018). Seasonal to interannual predictions of human carbon emissions are typically driven by the quality of predictions of economic indicators, such as GDP, which are most reliable over short timescales and in years without significant economic downturns.

Seasonal forecasts of greenhouse gas fluxes have the potential to bridge the gap in latency between observationally constrained estimates, which are typically delayed by several years, and the current time. This could aid in interpretation of observed changes in atmospheric greenhouse gas observations, helping to separate natural variability from human contributions. In addition, output from many flux models is accompanied by valuable information that could be used to inform a variety of stakeholders about diverse topics like fisheries, ecosystem health, and wildfire mitigation. Interannual to decadal scale predictions not only receive less attention and funding, but also show promise for prediction (e.g., Lovenduski, Bonan, Yeager, Lindsay, & Lombardozzi, 2019; Lovenduski, Yeager, Lindsay, & Long, 2019; Payne et al., 2017).

3 Data assimilation in earth system models

Data assimilation is *"an analysis technique in which the observed information is accumulated into the background state by taking advantage of consistency constraints with laws of time evolution and physical properties"* (Bouttier & Courtier, 1999). Although this definition is merely a few decades old, its roots can be traced back to the late eighteenth century when both Gauss and Legendre are credited with simultaneously discovering the core principles behind data assimilation (DA). The fundamental principle in data assimilation is to minimize the squared departure between an estimate, and observations and/or background information, very similar to the approach adopted in a least-squares framework. Today, DA is more generally cast in a probabilistic framework that formalizes the conjunction of the two states of information (e.g.,

Tarantola, 2005), and the least-squares method is derived as a special case of this much more general probabilistic framework.

3.1 Data assimilation related to numerical weather prediction

The popularity and recognition of DA has primarily stemmed from its application to weather forecasting. Sasaki's trilogy of papers in the Monthly Weather Review (Sasaki, 1970a, 1970b, 1970c) and the subsequent work of Lorenc (1986) and Ledimet and Talagrand (1986) demonstrated the inherent ability of DA techniques to bring disparate sources of information (i.e., models, various sources of data) to achieve the best analysis, with the analysis being 'better' than the individual pieces of information alone. The current applications of data assimilation have been to merge atmospheric weather observations with a modeled background state to create a realistic, globally continuous field for numerical weather prediction (e.g., Kalnay, 2003). Over time, such techniques have evolved substantially, helping to improve the skill in NWP models. Today, 5-day forecasts are as skillful as 3-day forecasts were in the early 1980s (Benjamin et al., 2018), in part because of their ability to ingest approximately more than 5 million observations every 6 h (Gelaro et al., 2017). Data assimilation methods have also been applied more broadly across the Earth system (Lahoz & Schneider, 2014). Assimilation of ocean observations provide a critical lower boundary condition for seasonal forecasts (e.g., Barnston et al., 1999; Stockdale, Anderson, Alves, & Balmaseda, 1998). As ocean data assimilation has evolved, the datasets assimilated have expanded from in situ temperature and salinity profiles to satellite-based estimates of altimetry, salinity, and sea surface temperature (e.g., Heimbach et al., 2019). The assimilation of ocean color data, which provides information on productivity, is also an active area of research.

3.2 Data assimilation related to land and ocean carbon cycle

Assimilation of data into land surface models can improve model deficiencies in representing the impact of rainfall on the near surface atmosphere. Additionally, such methods provide important information on root zone soil moisture and carbon flux that cannot be directly observed (e.g., dos Santos, Keppel-Aleks, De Roo, & Steiner, 2021). Efforts to directly assimilate vegetation observations such as leaf area index or solar induced fluorescence are increasing with a goal of either optimizing model parameters (Scholze, Kaminski, Rayner, Knorr, & Giering, 2007) or constraining flux components like gross primary production (MacBean et al., 2018). However, most current data assimilation systems are unable to take advantage of the increasing array of observations available by assimilating multiple land and vegetation observations simultaneously (Schimel et al., 2019).

490 Section | D Forward Looking

Like carbon cycle data assimilation for terrestrial models, assimilation of satellite data may also provide a constraint on properties related to ocean-atmosphere carbon flux over areas not directly observed by in situ measurements; however, such capabilities are still being developed. The assimilation of physical state variables such as sea surface temperature and salinity is relatively mature, supporting an array of forecasting and reanalysis efforts. However, most global models used to support near-real-time forecasting applications do not include complex representations of ocean biology or assimilate ocean color observations that provide information about productivity. Separate ocean biogeochemical models often rely on observationally derived surface forcing to estimate carbon flux and some include the ability to assimilate ocean color data (e.g., Gregg, 2008). Newer approaches that assimilate both physical state and ocean color data simultaneously are being developed (e.g., Carroll et al., 2020). Such methods are attractive because they have the greatest potential to simultaneously constrain surface conditions, upwelling, and ocean biology, which all influence ocean-atmosphere flux.

3.3 Data assimilation related to atmospheric carbon observations

Within the last two decades, increasing usage of data assimilation within carbon cycle research has been in two separate contexts. In an assimilation context, DA has been applied to generate consistent four-dimensional fields of atmospheric CO_2 concentrations (e.g., Chatterjee, Engelen, Kawa, Sweeney, & Michalak, 2013; Engelen, Serrar, & Chevallier, 2009) or to estimate parameters of biogeochemical models (e.g., Rayner et al., 2005). In this sense, the aim of the carbon DA system is to integrate atmospheric, terrestrial, and oceanic data together, along with underlying dynamical constraints, into a common analysis framework. Applications that make use of the assimilation framework are built on the premise of carbon cycle model development and/or the predictive properties of the carbon system. In addition, DA has gained more popularity in an inversion context (see Chapter 14) for inference of CO_2 sources and sinks using atmospheric CO_2 measurements (e.g., Chatterjee, Michalak, Anderson, Mueller, & Yadav, 2012; Chevallier, Breon, & Rayner, 2007; Crowell et al., 2019; Houweling et al., 2015). This application is based on an inverse modeling paradigm, in which the basic premise is that given a set of atmospheric CO_2 observations, and using a model of atmospheric transport, it is possible to infer information on the distribution of CO_2 fluxes at the surface of the Earth. Application to the CO_2 flux estimation problem simply intends to take advantage of the computational efficiency of a DA framework.

It is well accepted at this point that as carbon science becomes increasingly data rich, DA provides the optimal and most computationally efficient framework to process the high-density data for estimating CO_2 sources and sinks (Rayner, Michalak, & Chevallier, 2019). Simultaneously, rapid advancements are being made in the development of process-based models of the carbon cycle. Atmospheric CO_2 observations provide a valuable constraint on the total carbon budget, but not necessarily on the key signatures of anthropogenic and/or biogenic processes driving those concentration changes. One possible way of inferring information about the processes that influence the surface fluxes of CO_2 may be by examining or incorporating atmospheric measurements of other trace gases in the inversion process. Satellite observations of CO, a species emitted by incomplete combustion from fires, have been used to separate changes in CO_2 related to fires from other types of variability (e.g., Liu et al., 2017) while observations of NO_2, a short-lived pollutant emitted by fossil fuel combustion, have been used to estimate CO_2 emissions from power plants (e.g., Liu et al., 2020). The combination of CO, NO_2, and CO_2 data has also been used to identify the emissions fingerprint of different sources (Silva & Arelanno, 2017). While assimilation of heterogeneous observation types adds complexity and is currently in early stages, such studies demonstrate the potential for multispecies DA approaches for providing greater information about the processes governing the carbon cycle. Atmospheric assimilation of noncarbon trace gases has also advanced substantially, largely in support of air quality forecasting, helping to pave the way for multispecies approaches to carbon data assimilation.

Applications of such methods are also currently limited because observations of these gases are not made by the same satellite, which can lead to differences in the plume-scale enhancements observed by different satellites (Kuhlmann et al., 2019). Future satellites are being designed with this need in mind. NASA's GeoCarb, the world's first geostationary greenhouse gas satellite, is scheduled for launch in 2023 and will measure CO, CO_2, and CH_4 (Moore et al., 2018). ESA's multisatellite approach, called CO2M and scheduled for launch in 2026, will provide a much broader spatial coverage than current sensors and will measure CO_2 and CH_4 along with NO_2. Another challenge in multispecies data assimilation is developing appropriate error correlation statistics, which determine how information from one trace gas influences another.

Most noteworthy is that these data assimilation efforts are fragmented, occurring in different models, using differing data assimilation techniques, over divergent time and space scales. A major challenge over the coming years is bringing together these disparate modeling and data assimilation systems. Despite the complexity, such work is a major goal of groups worldwide because of the need for timely, high-quality information about greenhouse gas sources and sinks. Considerable effort is focused on merging land flux and atmospheric greenhouse gas data assimilation with the goal of better understanding the processes controlling land carbon uptake.

4 Future direction for carbon cycle science, earth system modeling, and DA applications

An immediate need will be to improve transport modeling capabilities in order to: (a) mitigate systematic biases that affect inversion results (e.g., Schuh et al.,

492 Section | D Forward Looking

2019) by correctly simulating key processes such as the timescales of stratosphere-troposphere exchange and planetary boundary layer dynamics, (b) allow better use of nighttime CO_2 measurements or measurements retrieved over complex terrains (e.g., Brooks et al., 2012), and (c) make better use of satellite data that provide observations over different time and space scales. Efforts have recently been made in improving certain aspects of the transport model, for example, the representation of vertical transport within the planetary boundary layer (e.g., Gerbig, Korner, & Lin, 2008), but the community still lacks pragmatic knowledge and accurate quantification of the real magnitude of the transport model uncertainties. Clearly large-scale efforts (i.e., sustained observations, funding, and theoretical advancements) are necessary to improve this critical piece within an atmospheric trace gas data assimilation framework.

Finally, within the next decade, the carbon cycle data assimilation applications will have to transition from an inversion to an assimilation framework. The traditional inversion framework does not support predictive capabilities associated with a true DA framework and lacks the ability to directly adjust the parameters of biogeochemical model components using atmospheric CO_2 observations. In an assimilation framework, however, it is quite feasible to identify which parameters of the biogeochemical model are less well understood and to subsequently design an observational network/system to constrain those parameters. A research framework adopting an integrated observation and modeling approach may provide the scientific basis for future atmospheric CO_2 mitigation strategies, but it will require a community-wide effort to advance current flux estimation systems and link them to prediction systems—the ultimate goal for the majority of Earth System modeling initiatives.

References

Agusti-Panareda, A., Massart, S., Chevallier, F., Balsamo, G., Boussetta, S., Dutra, E., et al. (2016). A biogenic CO_2 flux adjustment scheme for the mitigation of large-scale biases in global atmospheric CO_2 analyses and forecasts. *Atmospheric Chemistry and Physics, 16*(16), 10399–10418.

Agusti-Panareda, A., Massart, S., Chevallier, F., Boussetta, S., Balsamo, G., Beljaars, A., et al. (2014). Forecasting global atmospheric CO_2. *Atmospheric Chemistry and Physics, 14*(21), 11959–11983.

Arsenault, R., Brissette, F., & Martel, J. L. (2018). The hazards of split-sample validation in hydrological model calibration. *Journal of Hydrology, 566*, 346–362.

Barnston, A. G., Leetmaa, A., Kousky, V. E., Livezey, R. E., O'Lenic, E. A., Van den Dool, H., et al. (1999). NCEP forecasts of the El Nino of 1997-98 and its US impacts. *Bulletin of the American Meteorological Society, 80*(9), 1829–1852.

Benjamin, S. G., Brown, J. M., Brunet, G., Lynch, P., Saito, K., & Schlatter, T. W. (2018). 100 years of progress in forecasting and nwp applications. *Meteorological Monographs, 59*(1), 13.1–13.67. https://doi.org/10.1175/AMSMONOGRAPHS-D-18-0020.1.

Betts, R. A., Jones, C. D., Knight, J. R., Keeling, R. F., & Kennedy, J. J. (2016). El Nino and a record CO_2 rise. *Nature Climate Change, 6*(9), 806–810.

Brooks, B. G. J., Desai, A. R., Stephens, B. B., Bowling, D. R., Burns, S. P., Watt, A. S., et al. (2012). Assessing filtering of mountaintop CO_2 mole fractions for application to inverse models of biosphere-atmosphere carbon exchange. *Atmospheric Chemistry and Physics, 12*(4), 2099–2115.

Bouttier, F., & Courtier, P. (1999). *Data assimilation concepts and methods, ECMWF Metereological Training Course Lecture Series.* Available at https://www.ecmwf.int/en/elibrary/16928-data-assimilation-concepts-and-methods. (Accessed 24 June 2021).

Carroll, D., Menemenlis, D., Adkins, J. F., Bowman, K. W., Brix, H., Dutkiewicz, S., et al. (2020). The ECCO-darwin data-assimilative global ocean biogeochemistry model: Estimates of seasonal to multidecadal surface ocean pCO(2) and air-sea CO_2 flux. *Journal of Advances in Modeling Earth Systems, 12*(10), e2019MS001888.

Chatterjee, A., Engelen, R. J., Kawa, S. R., Sweeney, C., & Michalak, A. M. (2013). Background error covariance estimation for atmospheric CO_2 data assimilation. *Journal of Geophysical Research-Atmospheres, 118*(17), 10140–10154.

Chatterjee, A., Michalak, A. M., Anderson, J. L., Mueller, K. L., & Yadav, V. (2012). Toward reliable ensemble Kalman filter estimates of CO_2 fluxes. *Journal of Geophysical Research-Atmospheres, 117*. https://doi.org/10.1029/2012JD018176.

Chen, Y., Randerson, J. T., Coffield, S. R., Foufoula-Georgiou, E., Smyth, P., Graff, C. A., et al. (2020). Forecasting global fire emissions on subseasonal to seasonal (S2S) time scales. *Journal of Advances in Modeling Earth Systems, 12*(9), e2019MS001955.

Chevallier, F., Breon, F. M., & Rayner, P. J. (2007). Contribution of the orbiting carbon observatory to the estimation of CO_2 sources and sinks: Theoretical study in a variational data assimilation framework. *Journal of Geophysical Research-Atmospheres, 112*(D9). https://doi.org/10.1029/2006JD007375.

Ciais, P., Dolman, A. J., Bombelli, A., Duren, R., Peregon, A., Rayner, P. J., et al. (2014). Current systematic carbon-cycle observations and the need for implementing a policy-relevant carbon observing system. *Biogeosciences, 11*(13), 3547–3602.

Crowell, S., Baker, D., Schuh, A., Basu, S., Jacobson, A. R., Chevallier, F., et al. (2019). The 2015-2016 carbon cycle as seen from OCO-2 and the global in situ network. *Atmospheric Chemistry and Physics, 19*(15), 9797–9831.

dos Santos, T., Keppel-Aleks, G., De Roo, R., & Steiner, A. L. (2021). Can land surface models capture the observed soil moisture control of water and carbon fluxes in temperate-to-boreal forests? *Journal of Geophysical Research – Biogeosciences, 126*(4), e2020JG005999.

Engelen, R. J., Serrar, S., & Chevallier, F. (2009). Four-dimensional data assimilation of atmospheric CO_2 using AIRS observations. *Journal of Geophysical Research-Atmospheres, 114*. https://doi.org/10.1029/2008JD010739.

Eyring, V., Gleckler, P. J., Heinze, C., Stouffer, R. J., Taylor, K. E., Balaji, V., et al. (2016). Towards improved and more routine earth system model evaluation in CMIP. *Earth System Dynamics, 7*(4), 813–830.

Friedlingstein, P., Cox, P., Betts, R., Bopp, L., Von Bloh, W., Brovkin, V., et al. (2006). Climate-carbon cycle feedback analysis: Results from the (CMIP)-M-4 model intercomparison. *Journal of Climate, 19*(14), 3337–3353.

Friedlingstein, P., Mainshausen, M., Arora, V., Jones, C. D., Anav, A., Liddicoat, S. K., & Knutti, R. (2014). Uncertainties in CMIP5 climate projections due to carbon cycle feedbacks. *Journal of Climate, 27*(2), 511–526. https://doi.org/10.1175/JCLI-D-12-00579.1.

Gelaro, R., McCarty, W., Suarez, M. J., Todling, R., Molod, A., Takacs, L., et al. (2017). The modern-era retrospective analysis for research and applications, version 2 (MERRA-2). *Journal of Climate, 30*(14), 5419–5454.

494 Section | D Forward Looking

Gerbig, C., Korner, S., & Lin, J. C. (2008). Vertical mixing in atmospheric tracer transport models: Error characterization and propagation. *Atmospheric Chemistry and Physics*, *8*(3), 591–602.

Gregg, W. W. (2008). Assimilation of SeaWiFS ocean chlorophyll data into a three-dimensional global ocean model. *Journal of Marine Systems*, *69*(3–4), 205–225. https://doi.org/10.1016/j.jmarsys.2006.02.015.

Gourdji, S. M., Mueller, K. L., Yadav, V., Huntzinger, D. N., Andrews, A. E., Trudeau, M., et al. (2012). North American CO_2 exchange: Inter-comparison of modeled estimates with results from a fine-scale atmospheric inversion. *Biogeosciences*, *9*(1), 457–475.

Heimbach, P., Fukumori, I., Hills, C. N., Ponte, R. M., Stammer, D., Wunsch, C., et al. (2019). Putting it all together: Adding value to the global ocean and climate observing systems with complete self-consistent ocean state and parameter estimates. *Frontiers in Marine Science*, *6*, 55.

Hoffman, F. M., Randerson, J. T., Arora, V. K., Bao, Q., Cadule, P., Ji, D., et al. (2014). Causes and implications of persistent atmospheric carbon dioxide biases in earth system models. *Journal of Geophysical Research – Biogeosciences*, *119*(2), 141–162.

Houweling, S., Baker, D., Basu, S., Boesch, H., Butz, A., Chevallier, F., et al. (2015). An intercomparison of inverse models for estimating sources and sinks of CO2 using GOSAT measurements. *Journal of Geophysical Research-Atmospheres*, *120*(10), 5253–5266.

Jones, C. D., Arora, V., Friedlingstein, P., Bopp, L., Brovkin, V., Dunne, J., et al. (2016). C4MIP-the coupled climate-carbon cycle model intercomparison project: Experimental protocol for CMIP6. *Geoscientific Model Development*, *9*(8), 2853–2880.

Kaiser, J. W., Heil, A., Andreae, M. O., Benedetti, A., Chubarova, N., Jones, L., et al. (2012). Biomass burning emissions estimated with a global fire assimilation system based on observed fire radiative power. *Biogeosciences*, *9*(1), 527–554.

Kalnay, E. (2003). *Atmospheric modeling, data assimilation and predictability*. Cambridge: Cambridge University Press. 341 pp.

Kirtman, B. P., & Min, D. (2009). Multimodel ensemble ENSO prediction with CCSM and CFS. *Monthly Weather Review*, *137*(9), 2908–2930.

Koster, R. D., Mahanama, S. P. P., Yamada, T. J., Balsamo, G., Berg, A. A., Boisserie, M., et al. (2010). Contribution of land surface initialization to subseasonal forecast skill: First results from a multi-model experiment. *Geophysical Research Letters*, *37*. https://doi.org/10.1029/2009GL041677.

Kuhlmann, G., Broquet, G., Marshall, J., Clement, V., Loscher, A., Meijer, Y., et al. (2019). Detectability of CO2 emission plumes of cities and power plants with the Copernicus Anthropogenic CO_2 Monitoring (CO2M) mission. *Atmospheric Measurement Techniques*, *12*(12), 6695–6719.

Lahoz, W. A., & Schneider, P. (2014). Data assimilation: Making sense of Earth observation. *Frontiers in Environmental Science*, *2*. https://doi.org/10.3389/fenvs.2014.00016.

Lawrence, D. M., Hurtt, G. C., Arneth, A., Brovkin, V., Calvin, K. V., Jones, A. D., et al. (2016). The land use model intercomparison project (LUMIP) contribution to CMIP6: Rationale and experimental design. *Geoscientific Model Development*, *9*(9), 2973–2998.

Ledimet, F. X., & Talagrand, O. (1986). Variational algorithms for analysis and assimilation of meteorological observations – theoretical aspects. *Tellus A: Dynamic Meteorology and Oceanography*, *38*(2), 97–110.

Lee, E., Zeng, F. W., Koster, R. D., Ott, L. E., Mahanama, S., Weir, B., et al. (2020). Impact of a regional US drought on land and atmospheric carbon. *Journal of Geophysical Research – Biogeosciences*, *125*(8), e2019JG005599.

Liu, F., Duncan, B. N., Krotkov, N. A., Lamsal, L. N., Beirle, S., Griffin, D., et al. (2020). A methodology to constrain carbon dioxide emissions from coal-fired power plants using satellite observations of co-emitted nitrogen dioxide. *Atmospheric Chemistry and Physics*, *20*(1), 99–116.

Liu, J. J., Bowman, K. W., Schimel, D. S., Parazoo, N. C., Jiang, Z., Lee, M., et al. (2017). Contrasting carbon cycle responses of the tropical continents to the 2015-2016 El Nino. *Science*, *358*(6360), eaam5690.

Lorenc, A. C. (1986). Analysis-methods for numerical weather prediction. *Quarterly Journal of the Royal Meteorological Society*, *112*(474), 1177–1194.

Lorenz, E. N. (1969). '3 Approaches to atmospheric predictability. *Bulletin of the American Meteorological Society*, *50*(5), 345.

Lovenduski, N. S., Bonan, G. B., Yeager, S. G., Lindsay, K., & Lombardozzi, D. L. (2019). High predictability of terrestrial carbon fluxes from an initialized decadal prediction system. *Environmental Research Letters*, *14*(12), 124074.

Lovenduski, N. S., Yeager, S. G., Lindsay, K., & Long, M. C. (2019). Predicting near-term variability in ocean carbon uptake. *Earth System Dynamics*, *10*(1), 45–57.

MacBean, N., Maignan, F., Bacour, C., Lewis, P., Peylin, P., Guanter, L., et al. (2018). Strong constraint on modelled global carbon uptake using solar-induced chlorophyll fluorescence data. *Scientific Reports*, *8*, 1–12.

Mogensen, K., Balmaseda, M. A., & Weaver, A. (2012). The NEMOVAR ocean data assimilation system as implemented in the ECMWF ocean analysis for system 4. *ECMWF Technical Memorandum*, *668*, 1–59.

Moore, B., Crowell, S. M. R., Rayner, P. J., Kumer, J., O'Dell, C. W., O'Brien, D., et al. (2018). The potential of the geostationary carbon cycle observatory (GeoCarb) to provide multi-scale constraints on the carbon cycle in the Americas. *Frontiers in Environmental Science*, *6*, 109.

O'Dell, C. W., Eldering, A., Wennberg, P. O., Crisp, D., Gunson, M. R., Fisher, B., et al. (2018). Improved retrievals of carbon dioxide from orbiting carbon observatory-2 with the version 8 ACOS algorithm. *Atmospheric Measurement Techniques*, *11*(12), 6539–6576.

Payne, M. R., Hobday, A. J., MacKenzie, B. R., Tommasi, D., Dempsey, D. P., Fassler, S. M. M., et al. (2017). Lessons from the first generation of marine ecological forecast products. *Frontiers in Marine Science*, *4*, 289.

Rayner, P. J., Michalak, A. M., & Chevallier, F. (2019). Fundamentals of data assimilation applied to biogeochemistry. *Atmospheric Chemistry and Physics*, *19*(22), 13911–13932.

Rayner, P. J., Scholze, M., Knorr, W., Kaminski, T., Giering, R., & Widmann, H. (2005). Two decades of terrestrial carbon fluxes from a carbon cycle data assimilation system (CCDAS). *Global Biogeochemical Cycles*, *19*(2). https://doi.org/10.1029/2004GB002254.

Rousseaux, C. S., Gregg, W. W., & Ott, L. (2021). Assessing the skills of a seasonal forecast of chlorophyll in the global pelagic oceans. *Remote Sensing*, *13*(6), 1051.

Sasaki, Y. (1970a). Numerical variational analysis formulated under constraints as determined by long wave equations and a low-pass filter. *Monthly Weather Review*, *98*(12). https://doi.org/10.1175/1520-0493(1970)098<0884:nvafut>2.3.co;2.

Sasaki, Y. (1970b). Numerical variational analysis with weak constraint and application to surface analysis of severe storm gust. *Monthly Weather Review*, *98*(12). https://doi.org/10.1175/1520-0493(1970)098<0899:nvawwc>2.3.co;2.

Sasaki, Y. (1970c). Some basic formalisms in numerical variational analysis. *Monthly Weather Review*, *98*(12). https://doi.org/10.1175/1520-0493(1970)098<0875:sbfinv>2.3.co;2.

Schimel, D., Schneider, F. D., Bloom, A., Bowman, K., Cawse-Nicholson, K., Elder, C., et al. (2019). Flux towers in the sky: Global ecology from space. *New Phytologist*, *224*(2), 570–584.

Scholze, M., Kaminski, T., Rayner, P., Knorr, W., & Giering, R. (2007). Propagating uncertainty through prognostic carbon cycle data assimilation system simulations. *Journal of Geophysical Research-Atmospheres*, *112*(D17). https://doi.org/10.1029/2007JD008642.

496 Section | D Forward Looking

Schuh, A. E., Denning, A. S., Corbin, K. D., Baker, I. T., Uliasz, M., Parazoo, N., et al. (2010). A regional high-resolution carbon flux inversion of North America for 2004. *Biogeosciences, 7*(5), 1625–1644.

Schuh, A. E., Jacobson, A. R., Basu, S., Weir, B., Baker, D., Bowman, K., et al. (2019). Quantifying the impact of atmospheric transport uncertainty on CO2 surface flux estimates. *Global Biogeochemical Cycles, 33*(4), 484–500.

Silva, S. J., & Arelanno, A. F. (2017). Characterizing regional-scale combustion using satellite retrievals of CO, NO_2 and CO_2. *Remote Sensing, 9*(7). https://doi.org/10.3390/rs9070744.

Stockdale, T. N., Anderson, D. L. T., Alves, J. O. S., & Balmaseda, M. A. (1998). Global seasonal rainfall forecasts using a coupled ocean-atmosphere model. *Nature, 392*(6674), 370–373.

Tarantola, A. (2005). *Inverse problem theory and methods for model parameter estimation*. Society for Industrial and Applied Mathematics. 358 pp.

Taylor, K. E., Stouffer, R. J., & Meehl, G. A. (2012). An overview of CMIP5 and the experiment design. *Bulletin of the American Meteorological Society, 93*(4), 485–498.

Weir, B., Ott, L. E., Collatz, G. J., Kawa, S. R., Poulter, B., Chatterjee, A., … Pawson, S. (2021). Bias-correcting carbon fluxes derived from land-surface satellite data for retrospective and near-real-time assimilation systems. *Atmospheric Chemistry and Physics, 21*, 9609–9628. https://doi.org/10.5194/acp-21-9609-2021.

Zhang, F. Q., Sun, Y. Q., Magnusson, L., Buizza, R., Lin, S. J., Chen, J. H., et al. (2019). What is the predictability limit of midlatitude weather? *Journal of the Atmospheric Sciences, 76*(4), 1077–1091.

Index

Note: Page numbers followed by *f* indicate figures and *t* indicate tables.

A

Accounting tools, 219
Activity-based approach, 413
Activity data collection, 390–391
AGCMs. *See* Atmospheric general circulation models (AGCMs)
Aggregation errors, 137–139
Agricultural Life Cycle Inventory Generator (ALCIG) tool, 389
Agricultural systems
 bottom-up inventories, 383–384
 cropping systems (*see* Cropping systems)
 enteric emission inventory methods, 388*t*
 food security strategies, 377
 food systems, 377
 greenhouse gases emissions, 375–377, 376*f*, 378*t*
 inventories
  agricultural soil C stock changes, 384–387
  emissions from livestock systems, 387
  lateral transport, 387–389
  N_2O and CH_4 emissions, 384–387
  supply chains, 387–389
 IPCC guidelines, 377–378
 lateral transport, 383
 livestock systems (*see* Livestock systems)
 methane (CH_4), 377–378
 nitrous oxide (N_2O) emissions, 377–378
 regional greenhouse gases inventories, 390–392
 role, 377
 soil carbon inventory methods, 385*t*
 supply chains, 383
 top-down inversion methods, 389–390
 top-down methodologies, 383–384
Agriculture, 375, 407–408
AIMs. *See* Atmospheric inversion models (AIMs)
Airborne laser scanning (ALS), 209
Amazon fluxes, 278*t*
Amazon rainforest, 285–286
Animal feed, 375–377
Anthropogenic emissions, 4

Anthropogenic management, 220
Aquatic systems, 274–275
Arctic browning, 176–178
Arctic ecosystems
 arctic micrometeorological eddy covariance measurements, 165–166
 bottom-up methods, 172–173
 carbon cycling, 159
 chamber measurements, 163–165
 cold and remote, 161
 concentration measurements, 166–169
 freshwater/fluvial flux measurements, 169–170
 greenhouse gas budgets
  current status, 184–185
  estimation (*see* Arctic greenhouse gas)
  improvement, 185–186
  microbial activity, 162
  paleo-fire records, 163
  season carbon fluxes, 162–163
  warmer temperatures, 162
 land surface models, 173–176
 marine/oceanic fluxes, 171–172
 mean air temperature, 159–161
 organic carbon (C), 159–161
 permafrost carbon feedback, 159–161
 terrestrial ecosystem, 173–176
 top-down methods, 172–173
 uncertainty, 183–184
Arctic greenhouse gas
 arctic lakes, 180
 lake CH_4, 180
 lake CO_2, 180
 nitrous oxide, 182–183
 rivers and streams
  microbial processes, 181
  Ocean CH_4, 182
  Ocean CO_2, 182
 terrestrial CH_4, 177–178*t*, 178–180
 terrestrial CO_2, 176–178, 177–178*t*
Arctic greening, 176–178
The Arctic Ocean, 161
Arctic tundra, 161

497

498 Index

Artificial intelligence algorithms (AI), 75
The Atlantic Ocean, 439–440
Atmosphere, 483
Atmospheric deposition, 429
Atmospheric general circulation models
(AGCMs), 483–484
Atmospheric inversion models (AIMs), 217
Atmospheric inversions, 242–243, 318
Atmospheric observation-based method, 365
Atmospheric observations, 456
Atmospheric tracer transport model (ATM),
389
Autotrophic respiration, 71–72

B

Barnett Shale, 465–467
Bayesian inversion, 390
Bayes' theorem, 100, 103–110
Beromunster (BEO), 458
BigChain tool, 389
Bin-averaged observation, 463
Bioenergy production, 375
Biogeochemical process models, 212
Biomass, 72–73, 405–406
Bookkeeping approach, 66–70
Bookkeeping calculations, 245
Boreal forests
    anthropogenic source, 205–206
    aquatic system, carbon, 211–212
    biomes, 203
    carbon
        accounting, 212–216
        emissions, wildfire, 210–211
        in harvested wood products, 214–215
        synthesis, 218–221
    geographic scope, 204f
    greenhouse gases sources, 205
    hydrologic systems, 203–204
    impact, atmospheric GHG budget, 206–208
    managed vs. unmanaged forest lands,
        215–216
    National forest inventories, 213–214
    nations, 204–205
    pool and flux diagram, 207f
    regional-scale modeling, 217–218
    remote sensing role, 216
    sampling carbon fluxes, 210
    sampling carbon stocks, 208–209
    substantial stocks, 205
    warming, 206
Bottommoored funnels, 280
Bottom-up (BU) approaches

atmospheric CO_2 concentration, 59–60
bookkeeping methodology, 66–70
carbon monoxide (CO), 60–61
comparisons, 78
data-driven methodologies, 74–76
definitions, 63–64
discrepancies, 63–64
gain-loss methods, 76–77
global emission factors, 62
greenhouse gas (GHG) budgeting activities,
    60
hydrofluorocarbons (HFCs), 60–61
industrial emissions, 62
legacy effects, 61–62
objectives, 60–61
process-based methodology, 70–74
stock-change vs. flux-based accounting,
    64–66
top-down methods (TD), 61
uncertainties, 77–78
waste management practices, 59–60
Boundary layer turbulence, 483–484
Browning, 159–161
Bubble dissolution, 428–429

C

Carbonate system parameters, 429
Carbon balance, 274–275
Carbon budget assessment, 238
Carbon budget, temperate region
    adjustments, 253–255
    assessment
        atmospheric inversions, 257
        carbon stock changes, 255
        eddy flux observations, 255
        terrestrial biosphere models (TBMs),
            257
    atmospheric greenhouse gases, 237–238
    biogenic volatile organic compounds,
        252
    component fluxes, 238
    component flux estimation, 238–239
    components, 254t
    fire emissions, 251–252
    fluvial flux, 248–251
    land-use changes
        bookkeeping calculations, 245
        land-use datasets, 244
        process-based calculations, 245
    net carbon flux estimations, 238–239
        atmospheric inversions, 242–243
        carbon stock changes, 239–240

Index **499**

eddy covariance flux measurements, 240–241

terrestrial biosphere models, 241–242

policy decision-making

future perspective, 263

past decades, 262–263

policy-driven carbon budgets, 263–264

from river and lake, 251

sinks and sources, 238

uncertainties

components, 261–262

in observational methods, 258–261

Carbon-climate feedbacks, 313

Carbon cycle processes, 485–486

Carbon dioxide (CO_2), 3–4, 33, 59–60, 159, 275, 427–428

Carbon dioxide equivalents (CO_2eq), 61–62

Carbon Dioxide Information and Analysis Center (CDIAC), 34, 48–49

Carbon flux forecasts, 487–488

Carbon monoxide (CO), 60–61

Carbon pump, 430–431

Carbon stock, 239–240

C cycle, 313

CDIAC. *See* Carbon Dioxide Information and Analysis Center (CDIAC)

Cellular technology, 391

Central limit theorem, 465

Circumboreal biome, 218–219

Climate and Clean Air Coalition (CCAC), 458–460

Climate policy, 465

Cloud formation, 483–484

CLRTAP. *See* Convention on Long-range Transboundary Air Pollution (CLRTAP)

CMIPs. *See* Coupled model intercomparison projects (CMIPs)

Coastal ecosystems

carbon cycle science, implications, 421

Coastal ecosystems, AFOLU estimation, 404–407

GHG estimation, implications, 419–421

IPCC NGGI guidelines, improving application, 412–419

National scale estimation, IPCC guidelines, 407–412

CO_2 emission

activity data, 37–39

bp, 48

Carbon Dioxide Information and Analysis Center (CDIAC), 48–49

emission factors, 39

Emissions Database for Global Atmospheric Research (EDGAR), 49–50

energy statistics, 34–37

GESAPU modeling framework, 40*f*, 42

Global Carbon Project (GCP), 50

Global carbon project gridded fossil emissions dataset (GCP-GridFED), 50–51

International Energy Agency (IEA), 47–48

spatial and temporal emission disaggregation, 39–42

uncertainty

in AD, 45–46

climate change, 43

in EF, 46

fossil fuels, 43–45

Monte Carlo simulations, 45

sources, 47

spatial and temporal modeling, 46–47

vehicle miles traveled (VMT), 42

Community-wide trans-boundary infrastructure supply-chain carbon footprinting (CIF), 338–340, 348–358

Consumption-based footprinting (CBF), 338–340, 358–361

Convention on Long-range Transboundary Air Pollution (CLRTAP), 456–457

Cool Farm Tool (CFT), 389

Coupled model intercomparison projects (CMIPs), 485–486

Cropping systems

carbon, 379

net primary production (NPP), 379

nitrous oxide, 381

plant residues, 379–380

soil organic carbon (SOC), 380

Crowdsourcing, 391

Cryosphere, 485–486

Cultural diversity, 271–272

D

Data assimilation

atmospheric carbon observations, 490–491

defined, 488–489

fundamental principle, 488–489

future direction, 491–492

land carbon cycle, 489–490

numerical weather prediction, 489

ocean carbon cycle, 489–490

Deep learning (DL) algorithms, 75

Deforestation, 218–219, 288–289, 291

500 Index

Dissolved inorganic carbon (DIC), 279
Dissolved organic carbon (DOC), 279
Dynamic global vegetation models (DGVMs), 63–64, 70–71, 241–242

E

Earth system models (ESMs), 322
  atmospheric general circulation models (AGCMs), 483–485
  carbon cycle modeling, 485–488
  data assimilation (*see* Data assimilation)
  defined, 483
  future direction, 491–492
  general circulation models (GCMs), 483–484
  ocean general circulation models (OGCMs), 485
  optimal analysis, 484–485
Ecosystem inventories, 238
Ecosystem metabolism, 277
Ecosystem respiration, 237–238
Eddy covariance (EC), 74–75, 279
  flux measurements, 240–241
  flux technique, 210
EDGAR. *See* The Emissions Database for Global Atmospheric Research (EDGAR)
Emission Factor Database, 421–422
Emission factors (EFs), 39
Emissions Database for Global Atmospheric Research (EDGAR), 34, 49–50
Ensemble Kalman filter, 110–111
Enzyme-kinetic approach, 71–72
Error covariance matrices, 115–117
The European Centre for Medium-Range Weather Forecasts (ECMWF), 486
European Commission (EC), 49–50

F

Fertilizer, 375–377
Fine-scale process, 212–213
Flux-based accounting, 64–65
Forest degradation program, 288–289
Forest inventory, 239–240
Forest management, 218–219
Freshwater ecosystems, 281*t*
Fuel loads, 210–211

G

Gain-loss method, 214
General circulation models (GCMs), 483–484
Geographic information system (GIS), 389

GHG. *See* Greenhouse gas (GHG)
Global Carbon Project (GCP), 50
Global carbon project gridded fossil emissions dataset (GCP-GridFED), 50–51
Global Ecosystem Dynamics Investigation (GEDI) instrument, 209
Global Livestock Environmental Assessment Model—Interactive (GLEAM-I) tool, 389
Grassland management, 64
Greenhouse gas (GHG) emissions, 272
  emissions inventories, 33–34
  fossil fuel-based energy systems, 33
  global biogeochemical cycling, 33
  global climate system, 31–33
  multiple spatial issues, 33
  nationally determined contributions (NDCs), 33
  sources of emission estimates, 32*t*, 34
Greenhouse gases (GHGs), 3–4, 375, 404, 427–428, 485–486
  budgeting, 59–61
  carbon budget, 18–20
  cumulative carbon budget, 18, 19*f*
  data assimilation techniques, 14–15
  extraction-based emissions, 21
  global and regional carbon, 18
  global baseline measurements, 16
  global biogeochemical cycles, 5*f*
  global stocktake, 12–13
  ground-based networks, 14–15
  Kyoto protocol, 4–9
  net-zero emissions policy goals, 12–13
  optical imagery, 14
  Paris agreement, 4–9
  remote sensing, 14
  resolution sampling, 16
  satellite measurements, 15–16
  sources and sinks constrain, 9–12
  space-based monitoring, 14–15
  territorial-based emissions accounting, 21
Greenhouse gases Observing SATellite (GOSAT) observations, 463
Grid box sizes, 483–484

H

Halogenated gases, 3–4
Harvested wood products (HWP), 214–215, 408
Henry's law, 428
Hybridized observation-reanalysis products, 73
Hydrocarbon ratios, 462
Hydrofluorocarbons (HFCs), 60–61, 65–66

Index **501**

I

Ice, 483
The Indian Ocean, 440
In situ airborne measurements, 462
Intergovernmental Panel on Climate Change
(IPCC), 407
International Energy Agency (IEA), 34, 47–48
The International Panel on Climate Change
(IPCC), 214–215
International Recommendations for Energy
Statistics (IRES), 37

J

Joint Research Centre (JRC), 49–50
Jungfraujoch (JFJ), 458

K

Kalman filters, 105–106

L

Lagern-Hochwacht (LHW), 458
Lagrangian models, 467
Lamont Doherty Earth Observatory (LDEO)
database, 436
Land, 483
Land biosphere, 485–486
Land-surface hydrology, 483–484
Landuse change (LUC), 294
Land-use datasets, 244
Land use, land-use change, and forestry
(LULUCF), 363–366
Lithosphere, 428–429
Livestock grazing, 375
Livestock systems
global anthropogenic greenhouse gas
emissions, 382
grazing lands, 381
livestock enteric fermentation, 381
nutrients, 382
production of methane, 382
undigested organic C, 382
Loss of additional sink capacity (LASC),
63–64, 67–68

M

Machine learning (ML) algorithms, 75
Managed forest, 219–220
Managed *vs.* unmanaged forest, 215–216
Marine carbon sink, 435t
Marine ecosystems, 3–4

Marine nitrogen cycle, 429
Market globalization, 285
Markov chain Monte Carlo, 108–110
Mass balance approach, 460–462
Methane (CH_4), 3–4, 59–60, 159, 275, 405–406,
427–428, 457–458
Microbial processes, 275
Micrometeorological eddy covariance,
276–277
Micrometeorological flux measurements, 238
Microwave sensors, 238
Mitigation, 59–60, 95, 206, 263–264, 337–340,
347–348, 356, 358, 363, 367–368, 488,
492
Modern modeling infrastructures, 483
Monte Carlo simulations, 45
Multitemporal remote sensing, 216

N

National forest inventory (NFI) programs, 213
National Greenhouse Gas Inventory (NGGI),
407
National Inventory Reporting (NIR), 64
Nationally determined contribution (NDC),
413, 456–457
Nature-based solution, 339, 366
Net Ecosystem Carbon Budgets (NECB),
419–420
Net ecosystem exchange (NEE), 469–470
Net-zero emissions, 219–220
Neural networks (NN) algorithms, 75
Nitrogen-based urea fertilizer, 59–60
Nitrogen trifluoride (NF_3), 65–66
Nitrous oxide (N_2O), 3–4, 59–60, 275, 429,
457–458
Northern Hemisphere, 237–238

O

Observation error covariance, 114–115
Ocean general circulation models (OGCMs),
485
Ocean inverse methods, 435–436
Oceans, 427–428, 483
Ocean system
anthropogenic carbon storage, 441
anthropogenic perturbations, 434–437
atmospheric CO_2 concentrations, 442
biogeochemical model-based assessments,
445–446
carbonate ion concentration, 445
carbon budgets, 447

502 Index

Ocean system *(Continued)*
 climatic oscillations, 442–443
 contemporary global carbon sink, 434–437
 decadal-scale variability, 445
 ocean acidification, 445
 ocean carbon sink variability, 443–444
 preindustrial/natural fluxes, 431–434
 regional marine carbon sink, 438–440
 sink/source of greenhouse gases, 428–431
 Southern Ocean carbon sink, 444
 time-dependent ocean inverse estimation,
  446
OGCMs. *See* Ocean general circulation models
  (OGCMs)
Oil and Gas Climate Initiative (OGCI), 458–460
Optimization methods
 ensemble methods
  ensemble Kalman filter, 110–111
  Markov chain Monte Carlo, 108–110
 generic inversion framework, 103*f*, 104
 gradient methods, 106–110
 Kalman filters, 105–106

P

Particulate inorganic carbon (PIC), 279
Particulate organic carbon (POC), 279
Perfluorocarbons (PFCs), 65–66
Permafrost carbon, 159, 160*f*
 feedback, 159–161
Pesticides, 375–377
Photosynthesis, 71–72, 237–238
Policy relevance, 366
Preindustrial/natural fluxes, 431
Prior error covariance, 112–114
Process-based approaches, 70–74
Process-based models, 238
Proxy data, 45–46
Purely territorial carbon accounting (PTA)
 approach, 338

Q

Quality assurance, 474
Quality control, 474

R

Radiative transfer, 483–484
Radiocarbon-based measurements, 430
Reanalysis products, 73
Remote sensing, 14, 219, 273, 319–322, 320*f*
Revelle factor, 442
Riverine carbon, 434

S

Schauinsland (SSL), 458
Seasonal forecasts, 485, 487–488
Semiarid ecosystems
 ecology, 312–313
 future perspectives, 326
 greenhouse gas budget
  atmospheric inversion monitoring, 318
  components, 314–316
  land surface modeling, 322–323
  remote sensing, 319–322
  in situ based methodologies, 316–318
  soil erosion, 323–326
 seasonal/interannual variability, 311–312
 threats, 313–314
 transitional zones, 311
Shell-building process, 430–431
Simple mass balance inversion, 474
Software engineering, 483
Soil erosion, 323–326
Stock-change, 64–65
Sulfur hexafluoride (SF_6), 65–66
Sun-induced chlorophyll fluorescence (SIF),
  321–322
Supercomputers, 483
Superemitters, 462

T

Terrestrial biosphere models (TBMs), 217,
  241–242
Thunderstorms, 483–484
Top-down approaches
 application
  CH_4 fluxes, 130–132
  fossil fuel emissions, 126–130
  GHG fluxes, 132–137
  land biosphere, 120–126
 atmospheric modeling
  atmospheric transport, 97–98
  surface fluxes *vs.* atmospheric mixing
   ratios, 100–117
 atmospheric transport models, 88
 boundary conditions, 117–120
 flux estimates from inversions, 139–140
 greenhouse gases measurements, 88
  ground-based measurements, 91–95
  satellite measurement, 95–99
 inverse modeling, 88–89
 inversion concepts
  Bayes' theorem, 100, 103–110
  inverse problem, 100
  nomenclature, 101*t*

optimization methods (*see* Optimization methods)
posterior uncertainty, 111–115
steps, 100
prior flux
error covariance matrices, 115–117
observation error covariance, 114–115
prior error covariance, 112–114
sources of error
aggregation errors, 137–139
transport errors, 135–137
Top-down emission estimation methods
application, 456
atmospheric observations, 457
development, 456
ground-based observations, 467–471
high-resolution emission inventories, 471–473
high-resolution transport modeling, 467–471
inverse estimation, greenhouse gas emissions, 456–457
mean annual emissions, 459f
methane leak detection, 458–462
non-CO_2 greenhouse gases, 457–458
policy makers, 457
satellite observations
large point source emission monitoring, 463–464
sampling requirements, 465–467
Swiss CH_4 emissions, 458
Top-down methods (TD), 61
Total community-wide carbon footprinting (TCF), 338–340, 362–363
Transparency, accuracy, completeness, consistency, and comparability (TACCC), 456–457
Transport errors, 135–137
Transport modeling approaches, 456
Tropical countries, 291t
Tropical ecosystems
bottom-up methods, 283–285
carbon cycling, 272–273
carbon storage, 272–273
future opportunities, 296–298
general description, 271–272
greenhouse gas budget
chamber measurements, 275–276
components, 273–275
concentration measurements, 277–279

estimation by sector, 289–292
forest deforestation, 288–289
forest degradation, 288–289
freshwater/fluvial flux, 279–283
micrometeorological eddy covariance measurements, 276–277
land surface processes, 285–288
reducing uncertainty, 293–296
terrestrial ecosystem, 285–288
top-down methods, 283–285
uncertainty, 293–296
Tundra, 161, 163

U

Uncertainty, 219
United Nations Framework Convention on Climate Change (UNFCCC), 214–215, 456–457
Unmanaged forest, 219–220
Urban carbon accounting
cities to initiatives, 366–370
climate policies, 337–338
community-wide trans-boundary infrastructure supply-chain carbon footprinting (CIF), 339–340, 348–358
consumption- based footprinting (CBF), 339–340, 358–361
global nongovernmental organizations, 337
greenhouse gas (GHG) emissions, 337
land use, land-use change, and forestry (LULUCF), 363–366
protocols, 338–339
purely territorial carbon accounting (PTA) approach, 338–348
total community-wide carbon footprinting (TCF), 339–340, 362–363
urban green spaces, 339
US Department of Energy/Energy Information Administration (US DOE/EIA), 34
US Forest Inventory and Analysis (FIA) program, 213–214

V

Vegetation optical depth (VOD), 240

W

Water-air equilibration approach, 279–280
Wetlands Supplement, 408

Printed in the United States
by Baker & Taylor Publisher Services